Analytical Subsurface Environment

Rafid Al Khoury · Cor Kasbergen

Analytical Subsurface Environment

Heat Transfer, Groundwater Flow and Contaminant Transport

Springer

Rafid Al Khoury
Faculty of Civil Engineering and Geosciences
Delft University of Technology
Delft, Zuid-Holland
The Netherlands

Cor Kasbergen
Faculty of Civil Engineering and Geosciences
Delft University of Technology
Delft, Zuid-Holland
The Netherlands

ISBN 978-3-031-93909-9 ISBN 978-3-031-93910-5 (eBook)
https://doi.org/10.1007/978-3-031-93910-5

This Springer imprint is published by the registered company Springer Nature Switzerland AG
The registered company address is: Gewerbestrasse 11, 6330 Cham, Switzerland

To Our Families

Preface

The study of heat transfer, groundwater flow, and contaminant transport in the subsurface environment is central to advancing our understanding of numerous environmental and engineering challenges. Accurate modelling of these processes is essential for designing effective environmental remediation strategies, properly exploiting and managing natural resources, and ensuring the safety and sustainability of engineering projects.

In this textbook, we offer a comprehensive collection of analytical solutions to engineering and applied physics problems, with special emphasis on heat transfer, groundwater flow, and contaminant transport in shallow subsurface environments. Our objective is to present solutions to even complex physical problems in a clear, step-by-step manner, ensuring accessibility without requiring advanced mathematical expertise. With this aim, we strive to resolve many of the inherent ambiguities found in existing analytical models.

Existing analytical models are usually complex, and many students and researchers find them difficult to grasp. The complexity often arises from their intricate mathematical formulations and abstract presentations. While these models provide valuable theoretical insights, their limitations become evident when applied to more complex, real-world problems. As a result, there is a growing demand for more adaptable and practical approaches that account for the diverse geometry and boundary conditions encountered in engineering and environmental applications. This textbook is designed to bridge that gap by presenting refined analytical methods and tools, enhancing both the understanding and applicability of these solutions in modern contexts.

To accomplish this, we have selected the Fourier method as the foundation of our primary approach. The Fourier method's versatility makes it especially suited for tackling the wide variety of problems encountered in engineering and environmental science. Its ability to transform complex differential equations into more manageable forms allows for the accurate modelling of phenomena like heat transfer, fluid flow, and contaminant transport. Additionally, its inherent capability in addressing measured time-dependent boundary conditions makes it a crucial tool for solving real-world problems where traditional analytical methods fall short.

Through this approach, we aim not only to provide clear analytical solutions but also to promote a deeper appreciation for the mathematical structures that govern physical processes. By simplifying the derivation process and clarifying the underlying assumptions, we hope to make this powerful technique accessible to a broader audience of engineers and scientists. We are confident that the knowledge gained through the study of this book will inspire innovative solutions to some of the most pressing challenges in the field of engineering and applied sciences.

We trust that this textbook will become an essential resource for students, instructors, and researchers alike, motivating a deeper appreciation for the elegance and utility of analytical modelling. The content of this book is particularly well-suited for courses in civil, environmental, geothermal, geotechnical, and agricultural engineering, as well as in hydrology, geoscience, and geophysics. Its comprehensive approach and practical examples will enrich the learning experience and provide valuable insights on analytical modelling across these diverse fields. Moreover, practitioners working in a range of engineering and environmental disciplines will also find it invaluable for its practical insights and applications.

As with any significant undertaking, the completion of this textbook would not have been possible without the support of many individuals. We wish to express our sincere appreciation to the heads of our sections at Delft University of Technology: Prof. L. J. (Bert) Sluys, Head of the Computational Mechanics Chair, and Prof. S. M. J. G. (Sandra) Erkens, Head of the Pavement Engineering Chair, for their unwavering support. We also extend our thanks to the Springer publishing staff, namely Doris Bleier and Jayanthi Krishnamoorthi, for their efficient and patient handling of the book project. Additionally, we are deeply grateful to our family and friends for their endless encouragement and support. In particular, the first author would like to thank his wife, Sawsan, and two children, Zain and Maryam, for their continuous love and enthusiasm to see this book published. Also, many thanks go to his brother, Ray, for his valuable input on the artistic sketches of the book.

Delft, The Netherlands
January 2025

Rafid Al Khoury
Cor Kasbergen

Contents

II Heat Transfer

Introduction

R. Al Khoury and C. Kasbergen, *Analytical Subsurface Environment*,
https://doi.org/10.1007/978-3-031-93910-5_1

This chapter introduces the materials and modelling techniques discussed throughout this textbook. The chapter begins with an explanation of the subsurface environment as it pertains to the book's perspective. It then provides an overview of the three primary topics: heat transfer, groundwater flow, and contaminant transport in the subsurface environment. Following this, we discuss the modelling techniques commonly used for these topics. And subsequently, we briefly introduce the Fourier method, which will be utilized as the primary mathematical tool for solving the boundary value problems addressed in this book. Finally, we outline the book's theme and objectives and describe its structure.

1.1 Subsurface Environment

The subsurface environment encompasses layers of soil and rock, groundwater, and ecosystems situated beneath the surface of the earth. This vast environment spans from shallow soil layers to deep geological formations, forming a dynamic and interconnected system that is essential to various natural processes and human activities.

Shallow subsurface formations are essential sources of freshwater, ecosystem, agriculture, and industry. Over-extraction and contamination of groundwater pose significant threats to its availability and quality. Protecting groundwater requires sustainable management practice, pollution prevention, and rigorous monitoring to ensure its long-term viability and safety for all uses. Thus, modelling the dynamics of groundwater flow is critical for managing and conserving this precious environment.

Subsurface formations also host enormous renewable energy sources that can contribute in producing sustainable environment and reducing greenhouse gas emissions. Therefore, studying geothermal heat transfer in subsurface formations is essential for assessing energy reserves, optimizing extraction techniques, and transitioning towards sustainable energy production.

Subsurface studies play a crucial role in identifying and mitigating environmental risks associated with contamination. Understanding the transport and fate of pollutants in groundwater and soil helps in developing strategies to remediate contaminated sites and protect ecosystems and human health. These studies are fundamental for designing effective pollution control measures, monitoring environmental quality, and ensuring the safe use and disposal of hazardous materials.

Subsurface environment processes are intricately linked to groundwater flow, which acts as a primary driver in shaping the behaviour and distribution of heat and contaminants. Groundwater flow influences the rate and direction of heat transfer, playing a crucial role in thermal energy transport within the subsurface. Similarly, it governs the movement of contaminants, influencing their advection, dispersion, adsorption, and eventual fate. The interplay between these processes is dynamic and complex, as groundwater flow often amplifying or mitigating the impacts of heat and contaminant transport in the subsurface environment. Understanding this relationship is essential for accurately modelling and predicting subsurface phenomena.

Overall, comprehensive knowledge of the dynamics of the subsurface environment processes is indispensable for sustainable resource management and environmental protection, which represents the primary goal of this book.

1.2 Heat Transfer in Subsurface Environment

Studying heat transfer in the subsurface environment is essential for understanding subsurface temperature distributions, which is crucial for various engineering applications and in predicting the effects of over-utilization of underground resources. This knowledge is vital for assessing environmental impacts, optimizing extraction techniques, and developing sustainable energy solutions.

Heat transfer in the subsurface environment primarily occurs through *conduction* and *convection* processes. The geothermal gradient in the earth crust, implied by the increase of temperature with depth, drives heat through conduction over geological timescales. This gradient is influenced by factors such as radioactive decay, tectonic activity, and the thermal properties of subsurface materials. On the local scale, temperature gradients in shallow subsurface formations may arise from various influences like solar radiation, geothermal heat flux, and the injection or extraction of groundwater and energy.

In porous medium domains, conduction is the primary mechanism of heat transfer in the solid phase, often referred to as the *solid matrix* in engineering. Heat is transmitted from warmer regions to cooler regions through atomic and molecular vibrations. The ability of a material to conduct heat, denoted as *thermal conductivity*, plays a crucial role in determining the rate of heat transfer through conduction. Materials with higher thermal conductivity, such as dense rocks, conduct heat more efficiently compared to materials with lower thermal conductivity, such as air-filled porous media. Additionally, *heat capacity* and *mass density* are important factors significantly influencing heat storage and conduction. Materials with higher heat capacity can store more thermal energy, impacting the rate at which temperature changes occur. Similarly, higher mass density can enhance heat storage and transfer due to the increased amount of material available to absorb and conduct heat.

On the other hand, heat convection involves the transfer of heat through the movement of fluids carrying thermal energy. In subsurface environments, fluids naturally occupy the pore spaces and fractures within soil and rock formations. The effectiveness of heat convection is influenced by the fluid's properties, such as thermal conductivity, specific heat capacity, density, and viscosity. They determine how much thermal energy the fluid can carry and how effectively it transports energy through the porous medium. Convection can occur as either *forced convection*, where external forces like groundwater flow drive fluid movement, or *natural convection*, where buoyancy forces due to density differences within the fluid create spontaneous flow patterns. This book focuses solely on studying forced convection. This process can significantly enhance heat transfer rates compared to conduction alone, especially in highly permeable materials where fluid flow is substantial.

Therefore, modelling both conduction and convection processes in porous media is essential for accurately understanding heat transfer in subsurface environments. This knowledge is crucial for efficiently designing and operating geothermal energy systems, ensuring the sustainable use of subsurface resources.

1.3 Groundwater Flow in Subsurface Environment

Groundwater flow in the subsurface environment is a natural process that plays a crucial role in hydrological cycles, ecosystem dynamics, and human activities. Understanding the mechanisms and characteristics of groundwater flow is essential for managing water resources, assessing contamination risks, and designing sustainable engineering solutions.

Groundwater flow is driven by a gradient in the *hydraulic heads* or *hydraulic potentials* between regions. Groundwater flows from areas of higher hydraulic head, such as recharge zones with high groundwater elevation, to areas of lower hydraulic head, such as discharge zones with low groundwater elevation. This movement follows the path of steepest *hydraulic gradient*.

Groundwater flow takes place within the pore spaces of the solid matrix, typically in layers known as *aquifers*. Aquifers are geologic formations that have sufficient *porosity* and *permeability* to store and transmit groundwater. Porosity refers to the volume of void space within the geological formation. It determines the amount of water a formation can store, with higher porosity indicating greater water-holding capacity. The permeability, also referred to as *hydraulic conductivity*, is a measure of the material's ability to transmit water under a hydraulic gradient. It quantifies the rate at which groundwater can flow through the aquifer, influenced by factors such as grain size and sorting. Aquifers with high hydraulic conductivity facilitate efficient groundwater movement, allowing water to flow more readily through the subsurface.

Other properties of aquifers, such as *storativity* and *transmissivity*, influence groundwater flow characteristics and storage capacity. Storativity, also known as *specific storage*, represents the ability of an aquifer to store water within its pore spaces across a unit area of the aquifer under a unit reduction in hydraulic head. It accounts for the compressibility of the aquifer material and the volume of water that can be released or absorbed as pressure changes. Aquifers with higher storativity can store larger volumes of water and exhibit greater resilience to fluctuations in groundwater levels, providing a buffer against short-term variations in recharge and withdrawal rates.

Transmissivity, on the other hand, is a measure of the ability of an aquifer to transmit water across a unit width under a unit hydraulic gradient. It combines the hydraulic conductivity of the aquifer with its thickness perpendicular to flow, representing the overall capacity of the aquifer to transmit groundwater. Aquifers with higher transmissivity can sustain higher rates of groundwater flow and are often targeted for groundwater extraction and well development due to their ability to yield significant volumes of water.

1.4 Contaminant Transport in Subsurface Environment

Contaminant transport in the subsurface environment is a dynamic process influenced by various factors, including the properties of the contaminant, the characteristics of the subsurface environment, and the interactions between the contaminant and the surrounding environment. The chemical composition of contaminants influences their behaviour in the subsurface, including *solubility* and *reactivity*. Soluble contaminants can dissolve in groundwater and migrate with the flow, while reactive contaminants may undergo chemical transformations or sorption onto soil particles.

The porosity and permeability of subsurface formations govern the movement of contaminants with the groundwater through the soil matrix. Highly permeable formations facilitate rapid contaminant migration, while low-permeability soil layers may act as barriers or attenuators of contaminant movement. Contaminant transport is driven by hydraulic gradients in the region and via migrating from areas of higher concentration to lower concentration. Changes in hydraulic gradients due to pumping, recharge, or topographic features can influence the path and rate of contaminant movement.

There are several mechanisms governing the transport of contaminants, including *advection*, *dispersion*, and *sorption*. Advection refers to the movement of contaminants with the flowing groundwater. Contaminants can be transported over long distances through advection, following the direction of groundwater flow and hydraulic gradients. Dispersion involves the spreading of contaminants due to physical processes such as *diffusion*. It causes the spreading of contaminant plumes, leading to dilution and attenuation of concentrations over time and distance. Sorption, on the other hand, is the process by which contaminants adhere to soil particles or organic matter in the subsurface. Sorption reduces contaminant mobility and can lead to long-term retention or slow release of contaminants back into the groundwater.

Contaminant transport studies help in assessing the potential risks posed by contamination in the ground, designing remediation strategies, and evaluating the effectiveness of groundwater management measures.

1.5 Modelling Subsurface Processes

Understanding subsurface processes is essential for various fields, including geosciences, hydrogeology, geoengineering, agriculture, and environmental engineering. The intricate interplay of heat transfer, groundwater flow, and contaminant transport beneath the earth's surface drives critical phenomena shaping our environment and the well-being of the ecosystem. Modelling these phenomena and processes provides valuable insights for addressing environmental challenges, optimizing resource management, and designing effective exploitation and remediation strategies.

Heat transfer models provide insight into the thermal gradients in the ground subsurface, helping with the efficient utilization of geothermal energy systems, regulating soil temperature, and protecting the ecosystem from hazardous thermal effects. Groundwater flow models elucidate the movement of water through porous media, crucial for understanding aquifer behaviour, predicting water availability, and managing groundwater resources. Contaminant transport models are indispensable for assessing the spread of pollutants, identifying potential sources of contamination, and devising strategies for mitigation and remediation.

In the field of subsurface process modelling, two mathematical methods are commonly used: *numerical* and *analytical*. Numerical modelling involves partitioning the domain into elements or grid points and solving the governing equations through approximate discretization methods, such as the finite element or finite difference approach. While powerful and versatile, numerical models often require advanced computational skills and resources and may lack clarity regarding the underlying physical processes. On the other hand, analytical modelling procedures employ mathematical techniques to derive closed-form solutions to governing equations, providing insight into the fundamental behaviour of subsurface processes. Analytical models offer simplicity, efficiency, and interpretability, making them particularly valuable for theoretical investigations, sensitivity analyses, and rapid scenario evaluations. Additionally, they are well-suited for everyday engineering practice, requiring neither specialized expertise nor access to supercomputers.

This book is dedicated to introducing the principles and methodologies of analytical modelling techniques for subsurface processes. By emphasizing analytical approaches, we aim to provide the reader with a deeper understanding of the underlying physics governing heat transfer, groundwater flow, and contaminant transport in subsurface environments. Through a systematic procedure for formulating analytical models, this book equips students, researchers, and practitioners with the tools necessary to analyse and interpret subsurface phenomena and processes effectively. Moreover, through step-by-step illustrations of the mathematical procedures for solving the governing differential equations, we aim to promote the skills and proficiency in solving complex boundary value problems describing physical processes. By fostering skill development, our endeavour is not only meant to promote innovation and advancement in this field, but also to contribute to the broader objectives of sustainable resource management and environmental protection.

1.6 The Fourier Method

Analytical modelling of subsurface processes offers a diverse toolkit of methods, particularly suited for linear systems, each with its distinctive strengths and applications. Over the years, many analytical methods have been devised to address the complexity of the transient field problems by solving them in their *strong-form* time domain. While addressing partial differential equations in the time domain holds physical elegance and conceptual simplicity, the resulting solutions are typically limited to basic step– or continuous–functions as those depicted in ◘ Fig. 1.1. This limitation makes the time

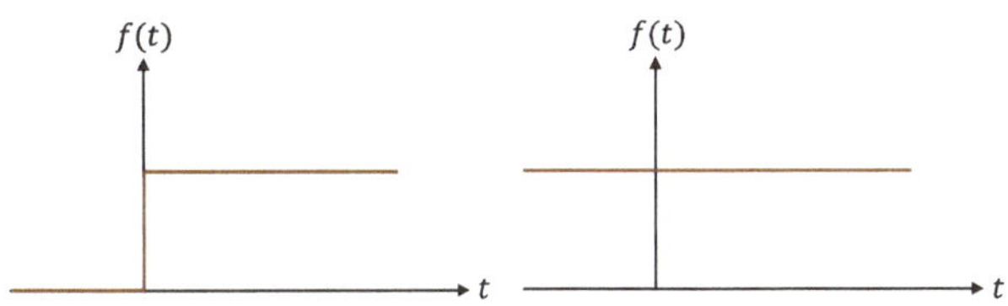

◘ **Fig. 1.1** Step-function (left), and continuous function (right)

domain formulations suitable merely for academic exercises. Moreover, solutions in the time domain frequently involve intricate integrals of mathematical expressions that can be challenging to translate into practical computer codes. In view of these limitations, the *transform methods* emerge as a an important substitute to the time domain solution methods.

The *Laplace transform* constitutes one of the most employed methods in addressing initial and boundary value problems related to subsurface processes. Basically, by transforming the governing partial differential equations into a simpler form of differential equations expressed in the Laplace domain, complex problems can be tackled and solved. This method is particularly suitable for *initial value problems*, offering elegant solutions to various physical processes, where the initial condition plays a major role in the transient behaviour of the system at later stages. However, for problems involving transient boundary conditions, this method is limited in describing the complex transient behaviour of the system. Its efficacy is most pronounced for boundary conditions featuring relatively simple time variations, notably those characterized by continuous or step-functions, as depicted in ◘ Fig. 1.1. Moreover, while the Laplace transform is straightforward in the forward transform, its inverse transform from the Laplace domain to the original time domain is, in most physical cases, mathematically cumbersome.

The Fourier method emerges as a robust alternative to the Laplace transform, effectively addressing its limitations in solving *boundary value problems*. Similar to the Laplace transform, the forward Fourier transform reduces the governing partial differential equations into a simpler form of differential equations, rendering them more amenable to solution. Then, through the inverse Fourier transform, the solution in the original time domain can be reconstructed. The distinct advantages lie in its versatility. On one hand, it can easily handle boundary conditions with arbitrary time variations, as shown schematically in ◘ Fig. 1.2. This is crucial as time signals in practice are rarely uniform enough to be represented by a step or continuous functions. And, on the other hand, the inverse Fourier transform is rather straightforward. This

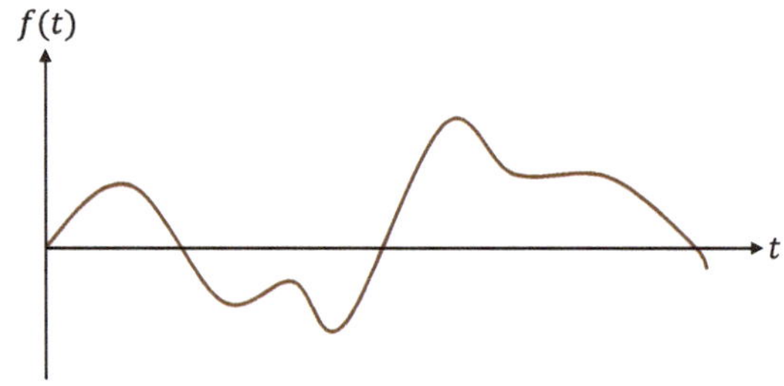

◘ **Fig. 1.2** Arbitrary function

is especially manifested if use is made of the *discrete Fourier transform* (DFT), notably the *fast Fourier transform* (FFT). The main disadvantage, though, is that it fails to address the initial conditions of the problem. Yet, in many practical cases the initial conditions can be added algebraically to the final solutions.

This book is devoted to the use of the Fourier method for solving boundary value problems in applied physics and engineering, in particular those related to simulating subsurface environmental processes. This method is a powerful technique that extends its versatility from signal processing of the input data to solving the governing partial differential equations (PDEs).

In signal processing (referred to as *Fourier analysis*), this method involves transforming signals from the time domain to the frequency domain, articulated in terms of a *frequency spectrum* expressing the range of frequency components of the signal and their corresponding magnitudes. This transformation is typically performed through techniques such as the *Fourier transform* or its derivatives, including the discrete Fourier transform and the fast Fourier transform. These techniques break down complex signals into simpler sinusoidal components, facilitating examination and interpretation by engineers.

In the context of solving PDEs, the Fourier method works by synthesizing complicated functions into a series of elementary functions using the *trigonometric Fourier series*, or other closely related types of representations. The application of the Fourier method commonly arises when addressing boundary value problems associated with linear homogeneous partial differential equations. This method provides a structured representation that helps in solving intricate mathematical challenges inherent in various scientific and engineering contexts. The use of the Fourier method for solving partial differential equations in association with signal processing of input data is commonly referred to as *spectral analysis*.

Spectral analysis offers a mathematical approach that transforms partial differential equations into equations in the frequency domain. This transformation often leads to a more manageable ordinary differential equations (ODEs) that are easier to solve. The resulting differential equations can in most cases be handled using the *separation of variables* method in association with *eigenfunction expansion* and the *superposition principle*.

Employing spectral analysis involves the transformation of the input signals and the governing equations from the time domain to the frequency domain. The transformation is efficiently done via the fast Fourier transform (FFT) algorithm, which has four main advantages: (1) it can handle measured data with arbitrary variations in time, (2) the summation of the Fourier series terms is done in an exceptionally efficient manner, (3) the behaviour of the system is analysed at each frequency independently, and (4) the inverse transform from the frequency domain back to the time domain is rather straightforward, done via the *inverse FFT* (IFFT) algorithm. The outcome is a solution amenable to implementation in computer codes, overcoming the challenges of solving intricate integrals normally associated with the strong-form time domain method and the inverse Laplace transform.

1.7 Theme and Objective

The goal of this textbook is to provide clear and systematic mathematical solution procedures for addressing complex boundary value problems. It serves as an in-depth guide for formulating and solving boundary value problems in areas such as heat transfer, groundwater flow, and contaminant transport in subsurface environments. Emphasizing analytical solutions using the Fourier method, this textbook equips readers with the necessary knowledge and techniques to model physical processes across a range of disciplines, including geosciences, hydrology, geoengineering, agriculture, and environmental engineering.

The Fourier method provides a powerful mathematical approach for decomposing complex systems into simpler components, facilitating the derivation of analytical solutions and their implementation in computer codes. Through step-by-step explanations and worked examples, the book guides readers through the process of solving complex boundary value problems encountered in subsurface environmental modelling. Emphasis is placed on developing analytical skills and problem-solving techniques, enabling readers to address practical challenges such as groundwater contamination, thermal pollution, and geothermal resource assessment, in a relatively straightforward manner.

In addition to the mathematical procedure for solving partial differential equations, this textbook incorporates descriptions drawn from an engineering perspective. By bridging the gap between theory and practice, readers gain a deeper understanding of the relevance and applicability of analytical solutions in addressing environmental issues and engineering challenges in subsurface environments.

The book is enriched with educational features including end-of-chapter problems, designed to reinforce learning and enhance conceptual understanding. Additionally, a general computer programming algorithm is introduced in the Appendix to supplement analytical solutions and facilitate numerical simulations, allowing the reader to explore complex scenarios and validate theoretical predictions. The algorithm is designed to be adapted to any programming language, including Fortran, MATLAB, C++, and Python.

In addition to serving as a vital resource for students, researchers, and practitioners aiming to master the analytical skills required for modelling subsurface environmental processes, this textbook also acts as a guide for formulating and developing analytical models across various other fields of physics and scientific applications. By offering analytical solutions through the Fourier method and focusing on skills development for complex boundary value problems, it empowers readers to address intricate physical scenarios encountered in engineering and scientific contexts.

While this book emphasizes analytical modelling and solution methodology, it also provides essential theoretical and engineering context for the topics discussed. However, its primary focus is on presenting key concepts and formulations without delving deeply into the theoretical foundations or complex engineering applications. To support readers seeking a deeper understanding of both the theoretical foundations and practical implementations, we include a list of *References* and/or *Further Readings* at the end of each chapter.

1.8 Scope

The book is designed to appeal to a wide range of readers, covering topics that are interesting and relevant to many fields. It explores boundary value problems at the undergraduate, graduate, and research levels. The content is particularly useful for students and researchers in engineering and environmental science. The mathematical methods and techniques provided are also applicable to studies in applied mathematics, physics, chemistry, biology, and agriculture, making this textbook helpful for anyone working in these areas or in interdisciplinary fields.

1.9 Book Organization

The book is primarily divided into three main parts: heat transfer, groundwater flow, and contaminant transport. The chapters are structured to assist readers at all levels, from beginners learning to solve boundary value problems to advanced researchers exploring complex challenges in these areas. Each chapter begins with basic concepts and gradually moves toward more advanced and intricate scenarios, making the learning process smooth and effective. This progressive approach is reflected in the ordering of the chapters, ensuring a logical and accessible flow throughout the book.

Preliminaries

Boundary Value Problems

R. Al Khoury and C. Kasbergen, *Analytical Subsurface Environment*,
https://doi.org/10.1007/978-3-031-93910-5_2

Boundary value problems (BVPs) are a class of mathematical formulations that involve finding solution to differential equations subjected to certain prescribed conditions at the boundary of the domains. These problems arise in various disciplines, such as physics, engineering, environment, and many other fields, where the behavior of a system under certain boundary conditions needs to be studied and understood. The classification of BVPs depends on the nature of the specified boundary conditions, with common types including the Dirichlet BVP, the Neumann BVP and the Mixed BVP.

The aim of this chapter is to provide an overview of the type of equations that we will be dealing with throughout this book. We will address the basic concepts of boundary value problems and discuss common types of boundary conditions. Special attention is given to the Sturm–Liouville problems and the Bessel functions, with a focus on their use for solving boundary value problems. We also address the concept of eigenvalue problems and eigenfunction expansion for solving differential equations in terms of orthogonal function expansion. Prior to this exploration, we will review essential concepts related to differential equations, encompassing both ordinary and partial differential equations, along with studying systems of differential equations. The concept of the method of separation of variables for solving partial differential equations will also be highlighted.

2.1 Differential Equations

Differential equations (DEs) are important mathematical expressions, typically employed for modeling natural phenomena, including growth, decay, diffusion and motion of physical quantities. Primarily, they are mathematical equations that involve rates of change. They describe a change of a physical quantity (such as temperature u, hydraulic head h, or contaminant species c) with respect to one or more variables (such as spatial, x, y, z, and/or time t). Differential equations can be classified as *ordinary differential equations* (ODEs), which involve a single independent variable, or *partial differential equations* (PDEs), which involve multiple independent variables.

The order of a DE is that of the highest order derivatives in the equation. For example,

$$\alpha \frac{\partial^2 u}{\partial x^2} - \frac{\partial u}{\partial t} = 0 \tag{2.1}$$

is a second order differential equation. As u in this equation is a function of more than one independent variable (x and t) this equation is thus a partial differential equation (PDE). Should u be dependent on x only or t only, the equation is said to be an ordinary differential equation (ODE).

In literature, differential equations are written in different forms. For example, these three forms are exchangeable:

$$
\begin{aligned}
a\frac{d^2u}{dx^2} + b\frac{du}{dx} + cu = 0 \Leftrightarrow \quad & au_{xx} + bu_x + cu = 0 \\
\Leftrightarrow \quad & au'' + bu' + cu = 0
\end{aligned}
$$

In this book, we will primarily use the first form of differential equation notation since it is commonly used in engineering. However, during the derivation process, we may also employ the last formulation for conciseness.

2.2 Linear Differential Equations with Constant Coefficients

Linear differential equations with constant coefficients provide a versatile framework for expressing numerous problems in physics and engineering. The term *linearity* here refers to differential equations whose coefficients remain independent of the variables. For example, in the context of heat flow, linearity is observed when the thermal conductivity remains unaffected by the change of temperature in the system. Similarly, in contaminant transport in porous media, linearity applies when the diffusion coefficient remains constant despite variations in the quantity of the contaminant species.

Consider this second order ordinary differential equation

$$Pu'' + Qu' + Ru = 0 \tag{2.2}$$

where its coefficients P, Q and R are real constants. The coefficients can be a function of the independent variable x, but not on u, otherwise the equation becomes *nonlinear*. This kind of differential equations possesses the elegant property that their solutions can be expressed by elementary functions such as exponential, trigonometric and polynomial functions. For instance, a trial solution of the above differential equation can be expressed as

$$u = ce^{\kappa x}$$

where c and κ are constants, which need to be determined. Upon substituting $u' = c\kappa e^{\kappa x}$ and $u'' = c\kappa^2 e^{\kappa x}$ into (2.2) gives

$$cP\kappa^2 e^{\kappa x} + cQ\kappa e^{\kappa x} + cRe^{\kappa x} = 0$$

Dividing by $ce^{\kappa x}$ yields

$$P\kappa^2 + Q\kappa + R = 0$$

This equation is called the *characteristic equation* (or the *auxiliary equation*) of the differential Eq. (2.2) with κ representing its roots, which, in this case, can be determined by the quadratic formula, leading to

2

$$\kappa_1 = \frac{-Q + \sqrt{Q^2 - 4PR}}{2P}, \qquad \kappa_2 = \frac{-Q - \sqrt{Q^2 - 4PR}}{2P}$$

Depending of the value of $Q^2 - 4PR$, , the general solution of (2.2) can be expressed in either of three cases:

Case 1: if $Q^2 - 4PR > 0$, we obtain distinct real roots $\kappa_1 \neq \kappa_2$, and the general solution becomes

$$u = c_1 e^{\kappa_1 x} + c_2 e^{\kappa_2 x}$$

Case 2: if $Q^2 - 4PR = 0$, we obtain repeated real roots $\kappa_1 = \kappa_2 = \kappa$, and the general solution becomes

$$u = c_1 e^{\kappa x} + c_2 x e^{\kappa x}$$

where we included x in the second term because otherwise it will be equal to the first term and we end up having only one constant, which is not possible for a second order differential equation.

Case 3: if $Q^2 - 4PR < 0$, we obtain complex conjugate roots $\kappa_1 = a + ib$, $\kappa_2 = a - ib$, and the solution can be expressed as

$$u = c_1 e^{\kappa_1 x} + c_2 e^{\kappa_2 x}$$

Note that, the solutions in Case 1 and Case 3 can also be expressed in these two forms:

$$u = c_1 e^{-i\kappa_1 x} + c_2 e^{-i\kappa_2 x} \quad \text{or} \quad u = c_1 e^{i\kappa_1 x} + c_2 e^{i\kappa_2 x}$$

In ◘ Table 2.1 we give general solutions of some typical ordinary differential equations that we will be encountering in this book.

Similar to the linear ordinary differential equations, a linear partial differential equation of order two of a dependent variable u and two independent variables x and y might have the form

$$P\frac{\partial^2 u}{\partial x^2} + Q\frac{\partial^2 u}{\partial y^2} + R\frac{\partial u}{\partial x} + S\frac{\partial u}{\partial y} = G \tag{2.3}$$

where P, Q, R, S, G are coefficients which might depend on x and/or y, but not on u, since otherwise the equation becomes nonlinear. If $G = 0$ in (2.3), the equation is denoted as *homogeneous*, but if $G \neq 0$ it becomes *nonhomogeneous*. The general solution of a nonhomogeneous differential equation is given by the sum of two functions, such that

$$u = u_c(x, y) + u_p(x, y)$$

where u_c is called the *complementary function* (also referred to as complementary equation) that denotes the solution of the homogeneous equation, and u_p is the *particular solution* of the nonhomogeneous equation. In some literature, this solution is written as

Table 2.1 Typical ordinary differential equations and their general solutions

Linear equations	General solutions
1. $u' + \alpha u = 0$	$u = ce^{-\alpha x}$
2. $u' - \alpha u = 0$	$u = ce^{\alpha x}$
3. $u'' + \alpha^2 u = 0; \quad \alpha > 0$	$u = c_1 \cos(\alpha x) + c_2 \sin(\alpha x)$
4. $u'' - \alpha^2 u = 0; \quad \alpha > 0$	$u = \begin{cases} c_1 e^{-\alpha x} + c_2 e^{\alpha x} \\ c_1 \cosh(\alpha x) + c_2 \sinh(\alpha x) \end{cases}$
5. $au'' + bu' + \alpha u = 0$	$u = c_1 e^{\kappa_1 x} + c_2 e^{\kappa_2 x}$, κ are real roots
6. $au'' + bu' + \alpha u = 0$	$u = c_1 e^{\kappa_1 x} + c_2 e^{\kappa_2 x}$, κ are complex roots
7. $au'' + bu' + \alpha u = 0$	$u = c_1 e^{\kappa x} + c_2 x e^{\kappa x}$, κ are repeated roots
8. $x^2 u'' + xu' + (x^2 - \nu^2) u = 0$	$u = c_1 J_\nu(x) + c_2 Y_\nu(x)$, Bessel functions
9. $x^2 u'' + xu' - (x^2 + \nu^2) u = 0$	$u = c_1 I_\nu(x) + c_2 K_\nu(x)$, Modified Bessel functions
10. $x^2 u'' + xu' - \alpha^2 u = 0, \quad \alpha \geq 0$	$u = \begin{cases} u = c_1 x^{-\alpha} + c_2 x^\alpha, & \alpha \neq 0 \\ u = c_1 + c_2 \ln x, & \alpha = 0 \end{cases}$, Cauchy-Euler

Note that the coefficients in 8, 9 and 10 above are functions of x that led the solutions to be expressed by what is called *special functions*, denoted as Bessel, modified Bessel and Cauchy-Euler functions. The equation in 8 is known as the Bessel equation, the equation in 9 as the modified Bessel equation, and the equation in 10 as the Cauchy-Euler equation

Table 2.2 Trial particular solutions to some standard nonhomogeneous terms

Nonhomogeneous term (G in (2.3))	Trial form of u_p
1. c (any constant)	A (a constant)
2. $5x + 2$	$Ax + B$
3. e^{5x}	Ae^{5x}
4. $\sin(5x)$	$A\cos(5x) + B\sin(5x)$

$$u = u_h(x, y) + u_p(x, y)$$

where the subscript h represents the solution to the homogenous part of the DE.

The general solutions of some typical homogeneous differential equations are given in Table 2.1. The particular solution, on the other hand, can be obtained by mainly two methods: the *method of undetermined coefficients*, and the *method of variation of parameters*. In this book, we mostly need the first method. In Table 2.2 we give some trial particular solutions, u_p, to some nonhomogeneous functions representing G in (2.3). It must be noted that the trial function must not be a duplicate of any function included in the complementary solution u_c. For more details on this topic, see, for example, Edwards and Penney (2008) and Zill and Wright (2011).

2

Hereafter, we give a motivating example describing the procedure for solving nonhomogeneous differential equations.

- **Example 2.1: Nonhomogeneous ODE**

Solve this nonhomogeneous differential equation:

$$y'' - 2y' - 3y = 4e^{2x} \tag{2.4}$$

- **Solution**

This equation is nonhomogeneous due to the presence of the right-hand side term. Thus the solution implies the sum of two solution functions

$$y = y_c + y_p$$

where y_c is the complementary function, and y_p is the particular solution. The solution can be carried out in three steps:

Step 1: We first solve the associated homogenous equation

$$y_c'' - 2y_c' - 3y_c = 0 \tag{2.5}$$

As the coefficients of this DE are constant, the solution of (2.5) can be expressed as

$$y_c = ce^{\kappa x} \tag{2.6}$$

where κ is a constant that needs to be determined. Inserting (2.6) and its derivatives into (2.5) gives

$$c\kappa^2 e^{\kappa x} - 2c\kappa e^{\kappa x} - 3ce^{\kappa x} = 0$$

Dividing by $ce^{\kappa x}$ yields

$$\kappa^2 - 2\kappa - 3 = 0 \qquad \text{or} \qquad (\kappa - 3)(\kappa + 1) = 0$$

which gives $\kappa = 3$ and $\kappa = -1$. Thus, the general solution of the homogeneous part is

$$y_c = c_1 e^{-x} + c_2 e^{3x}$$

Note that we could also use $y_c = ce^{i\kappa x}$ or $y_c = ce^{-i\kappa x}$ as a trial solution instead of (2.6) and would arrive at the right solution.

Step 2: We next solve the nonhomogeneous equation

$$y_p'' - 2y_{p'} - 3y_p = 4e^{2x} \tag{2.7}$$

Following ▫ Table 2.2, item 3, the term on the right-hand side of (2.7) suggests that the particular solution might take an exponential trial function, as

$$y_p = Ae^{2x}$$

Then, upon substituting $y_{p'} = 2Ae^{2x}$ and $y_p'' = 4Ae^{2x}$ into the differential Eq. (2.7), we obtain

$$A = -\frac{4}{3}$$

This implies

$$y_p = -\frac{4}{3}e^{2x}$$

Step 3: By combining the complementary and particular solutions from the above two steps, the general solution becomes

$$\begin{aligned} y &= y_c + y_p \\ &= c_1 e^{-x} + c_2 e^{3x} - \frac{4}{3}e^{2x} \end{aligned}$$

2.3 Systems of Linear Differential Equations

In the preceding example we have addressed solutions to differential equations that involve only one dependent variable. While this type of differential equations finds widespread applications, many scenarios necessitate dealing with two or more dependent variables. The system of linear second-order differential equations

$$\begin{aligned} m_1\ddot{x}_1 &= -k_1 x + k_2(x_2 - x_1) \\ m_2\ddot{x}_2 &= -k_2(x_2 - x_1) \end{aligned}$$

describing the motion of two coupled spring-mass systems of equations, represents an often-used model in mechanical engineering. In this system of equations, m represents the mass, k the stiffness of the spring, and x the displacement, where $\ddot{x} = d^2x/dt^2$ representing the acceleration.

Here, we present two illustrative examples to demonstrate the methodology employed in solving both homogeneous and nonhomogeneous coupled differential equations.

Example 2.2: Homogeneous system of differential equations

Solve this system of linear homogeneous first-order differential equations

$$\begin{aligned} \frac{dx}{dt} - 2qy = 0 \\ \frac{dy}{dt} - 2qx = 0 \end{aligned} \tag{2.8}$$

where q is some constant.

Solution

As the coefficients of these ODEs are constant, the trial solution can be expressed as

$$x = a\,e^{\kappa t}, \quad y = b\,e^{\kappa t}$$

where a and b, and κ are some constants that need to be determined. Inserting these trial solutions into (2.8) gives

2

$$\kappa a e^{\kappa t} - 2qb e^{\kappa t} = 0$$
$$\kappa b e^{\kappa t} - 2qa e^{\kappa t} = 0$$

Dividing by $e^{\kappa t}$ and putting the results in a matrix form, we obtain

$$\begin{pmatrix} \kappa & -2q \\ -2q & \kappa \end{pmatrix} \begin{pmatrix} a \\ b \end{pmatrix} = 0 \tag{2.9}$$

This equation can be satisfied if the determinant is equal to zero, yielding

$$\kappa^2 - 4q^2 = 0 \quad \Rightarrow \quad \kappa = \mp 2q \tag{2.10}$$

which is a second-order polynomial with its roots $\kappa_1 = -2q$ and $\kappa_2 = 2q$. These roots are denoted as the *characteristic values* or *characteristic roots*, or more commonly, the *eigenvalues*. In this case, κ_1 and κ_2 are the two eigenvalues of the system, with which the solution of (2.9) is not trivial. The corresponding functions $e^{\kappa_1 t}$ and $e^{\kappa_2 t}$ are called the *eigenfunctions* of the system, with each satisfies the differential equations in (2.8).

Since the system of the differential equations is linear, the *superposition principle* implies that if u_1 and u_2 are solutions of a linear homogeneous equation, then any arbitrary linear combination of them, $c_1 u_1 + c_2 u_2$, is also a solution of the equation. It follows that the general solution of the system of differential Eqs. (2.8) can be expressed as

$$\begin{aligned} x(t) &= a_1 e^{-2qt} + a_2 e^{2qt} \\ y(t) &= b_1 e^{-2qt} + b_2 e^{2qt} \end{aligned} \tag{2.11}$$

in which $a_1 \dots b_2$ are four arbitrary constants. However, since the system of equations is second order, only two constants are needed. To eliminate the extra two constants, and because the system is coupled, they can be related by determining the corresponding eigenvectors of the system.

One way to do this, is by use of (2.9). The first row in this matrix–vector multiplication entails

$$\kappa a - 2qb = 0 \quad \text{or} \quad b = \kappa a / 2q$$

As we have two eigenvalues, b must have an equivalence for each of them, such that

$$b_1 = \frac{\kappa_1}{2q} a_1 = -a_1 \quad \text{and} \quad b_2 = \frac{\kappa_2}{2q} a_2 = a_2$$

where we used the relations in (2.10). Following this, the general solution of the coupled differential Eqs. (2.8) becomes

$$x(t) = a_1 e^{-2qt} + a_2 e^{2qt}$$
$$y(t) = -a_1 e^{-2qt} + a_2 e^{2qt}$$

Note that if we have considered the second row in (2.9) instead of the first row, we would also arrive at the right solution.

- **Example 2.3: Nonhomogeneous system of differential equations**

Solve this system of linear nonhomogeneous differential equations

$$\begin{aligned} &\frac{d^2y}{dt^2}+\frac{dx}{dt}-9x=t^2 \\ &\frac{dy}{dt}+\frac{dx}{dt}+x=0 \end{aligned} \tag{2.12}$$

- **Solution**

Since the system is nonhomogeneous due to the presence of t^2 in one of the differential equations, the general solution is expressed by a complementary part and a particular part, such that

$$\begin{aligned} x &= x_c + x_p \\ y &= y_c + y_p \end{aligned} \tag{2.13}$$

As explained in Example 2.1, three steps are necessary to solve this nonhomogeneous system of differential equations.

Step 1: The first step is to obtain the complementary solution of the homogeneous equations

$$\begin{aligned} &\frac{d^2y_c}{dt^2}+\frac{dx_c}{dt}-9x_c=0 \\ &\frac{dy_c}{dt}+\frac{dx_c}{dt}+x_c=0 \end{aligned} \tag{2.14}$$

As the coefficients are constant, the trial solutions are expressed as

$$x_c = a_c e^{\kappa t}, \quad y_c = b_c e^{\kappa t} \tag{2.15}$$

Substituting (2.15) into (2.14) yields

$$\begin{aligned} &\kappa^2 b_c e^{\kappa t} + \kappa a_c e^{\kappa t} - 9a_c e^{\kappa t} = 0 \\ &\kappa b_c e^{\kappa t} + \kappa a_c e^{\kappa t} + a_c e^{\kappa t} = 0 \end{aligned}$$

Dividing by $e^{\kappa t}$ and putting the results in a matrix form, we obtain

$$\begin{pmatrix} \kappa-9 & \kappa^2 \\ \kappa+1 & \kappa \end{pmatrix}\begin{pmatrix} a_c \\ b_c \end{pmatrix}=0 \tag{2.16}$$

This set of equations has a nonzero solution if and only if the determinant of the matrix is zero. This implies

2

$$(\kappa - 9)\kappa - (\kappa + 1)\kappa^2 = 0$$

Solving this equation, gives two complex conjugate eigenvalues and one zero eigenvalues, resulting in

$$\kappa_1 = -3i, \quad \kappa_2 = 3i, \quad \kappa_3 = 0$$

It follows that the complementary solutions yield

$$x_c = a_{c1}e^{-3it} + a_{c2}e^{3it} + a_{c3}$$
$$y_c = b_{c1}e^{-3it} + b_{c2}e^{3it} + b_{c3}$$

Taking the derivative of these trial functions and substituting into (2.14) gives

$$(-9b_{c1}e^{-3it} - 9b_{c2}e^{3it}) + (-3i\,a_{c1}e^{-3it} + 3i\,a_{c2}e^{3it}) - 9\,(a_{c1}e^{-3it} + a_{c2}e^{3it} + a_{c3}) = 0$$
$$(-3i\,b_{c1}e^{-3it} + 3i\,b_{c2}e^{3it}) + (-3i\,a_{c1}e^{-3it} + 3i\,a_{c2}e^{3it}) + (a_{c1}e^{-3it} + a_{c2}e^{3it} + a_{c3}) = 0$$

Then, rearranging these two equations results in

$$(-9b_{c1} - 3i\,a_{c1} - 9a_{c1})\,e^{-3it} + (-9b_{c2} + 3i\,a_{c2} - 9a_{c2})\,e^{3it} - 9a_{c3} = 0$$
$$(-3i\,b_{c1} - 3i\,a_{c1} + a_{c1})\,e^{-3it} + (3i\,b_{c2} + 3i\,a_{c2} + a_{c2})\,e^{3it} + a_{c3} = 0$$

Treating these equations as identities implies that the coefficients of like powers must be equal. This leads to six equations with six unknowns

Coefficients of e^{-3it}	Coefficients of e^{3it}	Constants e^0
$-9b_{c1} - (3i + 9)\,a_{c1} = 0$	$-9b_{c2} + (3i - 9)\,a_{c2} = 0$	$-9a_{c3} = 0$
$-3i\,b_{c1} - (3i - 1)\,a_{c1} = 0$	$3i\,b_{c2} + (3i + 1)\,a_{c2} = 0$	$a_{c3} = 0$

Solving these equations gives

$$b_{c1} = -\,(\frac{i+3}{3})\,a_{c1}$$
$$b_{c2} = \frac{(i-3)}{3}a_{c2}$$
$$a_{c3} = 0$$

Thus, the complementary solutions become

$$x_c = a_{c1}e^{-3it} + a_{c2}e^{3it}$$
$$y_c = -\,(\frac{i+3}{3})\,a_{c1}e^{-3it} + \frac{(i-3)}{3}a_{c2}e^{3it} + b_{c3}$$

Step 2: The second step is to find the particular solution by solving the nonhomogeneous equations:

$$\begin{aligned} &\frac{d^2y_p}{dt^2} + \frac{dx_p}{dt} - 9x_p = t^2 \\ &\frac{dy_p}{dt} + \frac{dx_p}{dt} + x_p = 0 \end{aligned} \tag{2.17}$$

According to Table 2.2, using the method of undetermined coefficients, an educated guess entails the use of a polynomial of the form

$$x_p = a_{p1}t^2 + a_{p2}t + a_{p3}$$
$$y_p = b_{p1}t^3 + b_{p2}t^2 + b_{p3}t$$

where for y_p we used a third order polynomial because it involves a second order derivative.

The derivative of these trial functions are

$$x'_p = 2a_{p1}t + a_{p2}$$
$$y'_p = 3b_{p1}t^2 + 2b_{p2}t + b_{p3}$$
$$y''_p = 6b_{p1}t + 2b_{p2}$$

Substituting these derivatives into (2.17) and rearranging, yields

$$-9a_{p1}t^2 + \left(2a_{p1} - 9a_{p2} + 6b_{p1}\right)t + a_{p2} - 9a_{p3} + 2b_{p2} = t^2$$
$$\left(a_{p1} + 3b_{p1}\right)t^2 + \left(2a_{p1} + a_{p2} + 2b_{p2}\right)t + a_{p2} + a_{p3} + b_{p3} = 0$$

Treating these equations as identities implies that the *coefficients of like powers* of t must be equal. This leads to six equations with six unknowns.

Coefficients of t^2	**Coefficients of t**	**Constants t^0**
$-9a_{p1} = 1$	$2a_{p1} - 9a_{p2} + 6b_{p1} = 0$	$a_{p2} - 9a_{p3} + 2b_{p2} = 0$
$a_{p1} + 3b_{p1} = 0$	$2a_{p1} + a_{p2} + 2b_{p2} = 0$	$a_{p2} + a_{p3} + b_{p3} = 0$

Solving these equations simultaneously gives

$$a_{p1} = -\frac{1}{9}, \quad a_{p2} = 0, \quad a_{p3} = \frac{2}{81}$$
$$b_{p1} = \frac{1}{27}, \quad b_{p2} = \frac{1}{9}, \quad b_{p3} = -\frac{2}{81}$$

Thus

$$x_p = -\frac{1}{9}t^2 + \frac{2}{81}$$
$$y_p = \frac{1}{27}t^3 + \frac{1}{9}t^2 - \frac{2}{81}t$$

Step 3: The third step is to bring the complementary and particular solutions together to obtain the required general solution of the system of differential equations, i.e.

$$x(t) = a_{c1}e^{-3it} + a_{c2}e^{3it} - \frac{1}{9}t^2 + \frac{2}{81}$$
$$y(t) = -\frac{(i+3)}{3}a_{c1}e^{-3it} + \frac{(i-3)}{3}a_{c2}e^{3it} + b_{c3} + \frac{1}{27}t^3 + \frac{1}{9}t^2 - \frac{2}{81}t$$

2.4 Separation of Variables

The linear differential equations that we have addressed in the above examples are ordinary, and their dependent variables (x and y) are depending on one independent variable (t). This class of differential equations has many applications in physics and engineering. Yet, there is an enormous number of linear partial differential equations where their dependent variables depend on more than one independent variable. The PDE (2.1) is an example of this kind of equations. It typically represents time-dependent heat conduction in a one-dimensional domain. This kind of linear, homogeneous partial differential equations can elegantly and efficiently be solved using the *method of separation of variables.*

Separation of variables is a mathematical method for solving partial differential equations by breaking down equations consisting of multiple independent variables into a set of independent equations, each of which depends on just one independent variable. For instance, if we have a linear field $u(x, y)$ that depends on the two independent spatial variables x and y, the separation of variables allows expressing it as

$$u(x, y) = X(x)Y(y)$$

where $X(x)$ is a function of x only and $Y(y)$ is a function of y only. Using this method, we can transform a linear partial differential equation having two independent variables into two ordinary differential equations each depending on one independent variable. Similarly, if the equation involves more than two variables, we can repeat this process to reduce it to an equivalent number of ordinary differential equations. For example, if we have a field function that is dependent on three independent variables (x, y, z), using the separation of variables, we can break down the involved partial differential equation into ordinary differential equations in two steps, such that

$$u(x, y, z) = X(x)F(y, z)$$

then

$$F(y, z) = Y(y)Z(z)$$

The separation of variables method is ideal for solving linear homogenous partial differential equations. Hereafter, we give a motivating example describing the use of the method for solving partial differential equations.

Example 2.4: Separation of variables

Solve the following homogeneous PDE

$$\frac{1}{\alpha}\frac{\partial u}{\partial t} - \frac{\partial^2 u}{\partial x^2} = 0 \tag{2.18}$$

where α is some positive constant.

Solution

Using the separation of variables, we assume that $u(x, t)$ can be expressed in terms of a product of two functions: one which is a function of x only, and another which is a function of t only, such that

$$u(x, t) = X(x)T(t) \tag{2.19}$$

Substituting (2.19) into (2.18) gives

$$X''T = \frac{1}{\alpha}XT'$$

Dividing both sides by XT, we obtain

$$\frac{X''}{X} = \frac{T\prime}{\alpha T} \tag{2.20}$$

Since the left-hand side is dependent on X only and the right-hand side is dependent on T only, it can be deduced that both sides are independent of each other, but to be equal, they must have some constant value in common, often termed the *separation constant*. Denoting this constant as μ, we obtain

$$\frac{X''}{X} = \frac{T'}{\alpha T} = \mu$$

This formulation allows separating it into two ODEs of the form

$$X'' - \mu X = 0 \quad \text{and} \quad T' - \alpha\mu T = 0 \tag{2.21}$$

As μ is any arbitrary constant, it can be zero, negative or positive; that is $\mu = 0$, $\mu = \kappa^2 > 0$, and $\mu = -\kappa^2 < 0$, where $\kappa > 0$. Note that μ is expressed as κ^2 for convenience. Following these, the corresponding ODEs in (2.21) are in turn,

$$X'' = 0 \quad \text{and} \quad T' = 0 \tag{2.22}$$

$$X'' - \kappa^2 X = 0 \quad \text{and} \quad T' - \alpha\kappa^2 T = 0 \tag{2.23}$$

$$X'' + \kappa^2 X = 0 \quad \text{and} \quad T' + \alpha\kappa^2 T = 0 \tag{2.24}$$

2

The general solutions of these ODEs are as follows.

Case 1 ($\mu = 0$): The ODEs in (2.22) can be solved by direct integration, yielding

$$X = c_1 + c_2 x \quad \text{and} \quad T = c_3$$

Thus, (2.19) gives

$$u = XT = (c_1 + c_2 x)\, c_3 = a_1 + b_1 x \tag{2.25}$$

where $a_1 = c_1 c_3$ and $b_1 = c_2 c_3$.

Case 2 ($\mu = \kappa^2$): In this case, the general solutions of the ODEs in (2.23) yield

$$X = c_4 e^{-\kappa x} + c_5 e^{\kappa x} \quad \text{and} \quad T = c_6 e^{\alpha\kappa^2 t} \tag{2.26}$$

Thus, (2.19) gives

$$u = XT = \left(c_4 e^{-\kappa x} + c_5 e^{\kappa x}\right) c_6 e^{\alpha\kappa^2 t}$$

Or

$$u = \left(a_2 e^{-\kappa x} + b_2 e^{\kappa x}\right) e^{\alpha\kappa^2 t} \tag{2.27}$$

where $a_2 = c_4 c_6$ and $b_2 = c_5 c_6$.
Note that the solution of X in (2.26) can also be expressed as

$$X = c_4 \cosh(\kappa x) + c_5 \sinh(\kappa x)$$

This solution is preferable for relatively small domains, while the exponential form is preferable for large domains.

Case 3 ($\mu = -\kappa^2$): In this case, the general solutions of the ODEs in (2.24) yield

$$X = c_7 \cos(\kappa x) + c_8 \sin(\kappa x) \quad \text{and} \quad T = c_9 e^{-\alpha\kappa^2 t}$$

Thus, (2.19) gives

$$u = XT = [c_7 \cos(\kappa x) + c_8 \sin(\kappa x)]\, c_9 e^{-\alpha\kappa^2 t}$$

Or

$$u = [a_3 \cos(\kappa x) + b_3 \sin(\kappa x)]\, e^{-\alpha\kappa^2 t} \tag{2.28}$$

where $a_3 = c_7 c_9$ and $b_3 = c_8 c_9$.

After presenting the three cases, the question arises: which of the three solutions is correct? The answer is that all are valid general solutions. However, when the PDE is associated to a boundary value problem (BVP), the correct solution is the one that produces a nontrivial result while satisfying both the PDE and the related initial and boundary conditions.

2.5 Sturm–Liouville Problems

Sturm–Liouville problems are a specific class of *boundary value problems* (BVP) encountered in the study of second-order linear differential equations. These problems focus on finding eigenfunctions and eigenvalues that satisfy given differential equations along with specified boundary conditions (BC). In general, differential equations that describe the behavior of physical systems, subject to boundary conditions and/or initial conditions (IC), are categorized as boundary value problems.

A one-dimensional differential equation with prescribed boundary conditions, given, for example, by

$$u'' + \kappa^2 u = 0, \quad u(0) = 0 \quad \text{and} \quad u(L) = 0, \qquad 0 \le x \le L \tag{2.29}$$

is a typical boundary value problem. If κ is constant (or function of x), this differential equation is said to be linear and homogeneous, and as the associated boundary conditions equal to zero at the endpoints, they are also homogenous. Should one or both boundary conditions be not identically zero, the boundary conditions become nonhomogeneous, and accordingly the whole boundary value problem becomes nonhomogeneous.

The solution of a BVP is typically expressed by two related solution types: *general solution* and *particular solution*. The general solution is a solution of the differential equation itself, expressed in terms of arbitrary constants, equal in number to the order of the equation. For instance, the general solution of the differential equation in (2.29) is

$$u(x) = c_1 \cos(\kappa x) + c_2 \sin(\kappa x) \tag{2.30}$$

where c_1 and c_2 are arbitrary constants, and κ is another constant that needs to be determined.

The particular solution, on the other hand, is one obtained by satisfying the involved initial and boundary conditions. If we apply the first boundary condition in (2.29) into (2.30) we can readily find that $c_1 = 0$, reducing the solution to

$$u(x) = c_2 \sin(\kappa x)$$

Applying the second boundary condition implies

$$c_2 \sin(\kappa L) = 0$$

Then, in order not to have a trivial solution, we must let

$$\sin(\kappa L) = 0 \quad \Rightarrow \quad \kappa = n\pi/L \tag{2.31}$$

where n is an integer. Following this, the solution for every n becomes

$$u_n(x) = c_2 \sin(\kappa_n x) = c_2 \sin\left(\frac{n\pi}{L}x\right) \tag{2.32}$$

This kind of BVPs constitutes an important class of boundary value problems, called the *Sturm–Liouville problem*. It always possesses the trivial solution $u(x) = 0$, but for certain values of the eigenvalues κ, there exist nontrivial solutions.

A boundary value problem of the form

$$\begin{aligned} &R(x)u'' + R'(x)u' + [P(x) + \mu Q(x)]u = 0, \quad a \le x \le b \\ &\text{BC:} \quad A_1 u(a) + B_1 u'(a) = 0 \\ &\qquad\quad A_2 u(b) + B_2 u'(b) = 0 \end{aligned} \tag{2.33}$$

is a generalized Sturm–Liouville problem that includes numerous boundary value problems as special cases. If P, Q, R and R' are real-valued functions that are continuous on the interval $[a, b]$, and $R(x) > 0$ and $P(x) > 0$ for every x in the interval, the boundary value problem is classified as a *regular Sturm–Liouville problem*. This kind of problems is fundamental in various fields of applied physics and engineering, such as heat transfer and diffusion. Solutions of such problems are essential for understanding the behavior of physical systems under given initial and boundary conditions.

Each value of μ in (2.33) where a nontrivial solution exists represents an *eigenvalue*, and the corresponding nontrivial solution represents an *eigenfunction*. The eigenfunction satisfies both the linear homogeneous differential equation in (2.33) and the associated boundary conditions. $Q(x)$ in (2.33) represents a weight function that multiplies the eigenvalues of the system.

The BVP (2.29) is a special case of a Sturm–Liouville problem, since comparing its governing equations with (2.33), we infer that $R(x) = Q(x) = A_1 = A_2 = 1$, $\mu = \kappa^2$, $a = 0$, $b = L$ and $P(x) = B_1 = B_2 = 0$. In this context, κ in (2.31) represents a set of eigenvalues of the system, and $\sin(n\pi x/L)$ in (2.32) are the associated eigenfunctions. Solving the Sturm–Liouville problem involves finding the eigenvalues and eigenfunctions that satisfy the boundary conditions.

2.6 Eigenfunction Expansion

One of the distinctive features of the Sturm–Liouville problems is that the eigenfunctions $\{\phi_m\}, m = 1, \ldots \infty$ form a complete set, such that any piecewise smooth function $f(x)$ can be expanded into an infinite series of normalized eigenfunctions, such that

$$f(x) = \sum_{m=1}^{\infty} c_m \phi_m(x) \tag{2.34}$$

where c_m are the *expansion coefficients* of the series. This expansion finds a resemblance in linear algebra where a vector can be expressed in terms of its components along different basis vectors. This technique is commonly denoted as the *ei-*

genfunction expansion (also known as the *generalized Fourier series*). One of the main tasks in this series presentation is the determination of the expansion coefficients.

Eigenfunctions associated with eigenvalues are not necessarily easy to deal with, unless they are *orthogonal*. In vector analysis, two vectors are said to be orthogonal if their dot (inner) product is zero, i.e. $\mathbf{u} \cdot \mathbf{v} \equiv (\mathbf{u}, \mathbf{v}) = 0$. Geometrically, this means that the two vectors are perpendicular to each other.

In mathematics, functions are considered a generalization of vectors and possess the properties of the vectors dot product. The inner product of two functions $f_1(x)$ and $f_2(x)$ on interval $[a, b]$ is

$$f_1 \cdot f_2 \equiv (f_1, f_2) = \int_a^b f_1(x) f_2(x)\, dx$$

And these two functions are said to be orthogonal on $[a, b]$ if

$$(f_1, f_2) = \int_a^b f_1(x) f_2(x)\, dx = 0$$

When dealing with eigenfunction expansion, we are basically interested in infinite sets of orthogonal functions. A set of eigenfunctions $\phi_n(x)$ and $\phi_m(x)$ are orthogonal on the interval $[a, b]$ with respect to a weight function $w(x)$ (equivalent to $Q(x)$ in (2.33)) provided that

$$\int_a^b w(x)\, \phi_n(x)\, \phi_m(x)\, dx = 0, \quad m \neq n \tag{2.35}$$

Utilizing this orthogonality property we can now determine the expansion coefficients in (2.34). Multiplying both sides of (2.34) by $\phi_n(x)$ and integrating with respect to a weight function $w(x)$ on the interval $[a, b]$, we obtain

$$\int_a^b w(x) f(x)\, \phi_n(x)\, dx = \sum_{m=1}^{n} c_m \int_a^b w(x)\, \phi_m(x)\, \phi_n(x)\, dx$$

According to (2.35), the integral on the right-hand side of this equation is zero, unless we have $m = n$, leading to

$$\int_a^b w(x) f(x)\, \phi_n(x)\, dx = c_n \int_a^b w(x) [\phi_n(x)]^2 dx$$

This yields

$$c_n = \frac{\int_a^b w(x) f(x)\, \phi_n(x)\, dx}{\int_a^b w(x) [\phi_n(x)]^2 dx} \tag{2.36}$$

In some literature, this expression is given by

$$c_n = \frac{(f, \phi_n)}{\|\phi_n\|^2} \tag{2.37}$$

where the denominator $\|\phi_n^2\|$ is called the *square norm*, defined as

$$\|\phi_n\|^2 \equiv \phi \cdot \phi \equiv (\phi, \phi) \equiv \int_a^b [\phi_n(x)]^2 dx$$

For example, the square norm of a sine function in the interval $[0, \pi]$ is

$$\|\phi_n\|^2 \equiv \int_0^\pi [\sin(nx)]^2 dx = \frac{\pi}{2}$$

Inserting this result in (2.36), and assuming $w(x) = 1$, yields

$$c_n = \frac{2}{\pi} \int_0^L f(x) \sin(nx)\, dx$$

which is a typical Fourier sine series coefficient.

2.7 Bessel Functions

Bessel functions are most widely encountered in addressing problems in applied physics and engineering, such as heat transfer and diffusive problems. They are particularly vital for describing problems exhibiting cylindrical symmetry.

Bessel functions arise as solutions to the *Bessel differential equation*, given by

$$x^2 y'' + xy' + (x^2 - \nu^2)y = 0 \tag{2.38}$$

The general solution of this differential equation is

$$y = c_1 J_\nu(x) + c_2 Y_\nu(x)$$

where $J_\nu(x)$ is called a *Bessel function of the first kind* of order ν, and $Y_\nu(x)$ is called a *Bessel function of the second kind* of order ν.

Within the same family of differential equations that bear a resemblance to (2.38), a differential equation expressed in the form

$$x^2 y'' + xy' - (x^2 + \nu^2)y = 0$$

is denoted the *modified Bessel equation* of order ν. The general solution of this differential equation is

$$y = c_1 I_\nu(x) + c_2 K_\nu(x)$$

where $I_\nu(x)$ is called a *modified Bessel function of the first kind* of order ν, and $K_\nu(x)$ is called a *modified Bessel function of the second kind* of order ν.

In applied physics and engineering, several forms of (2.38) are encountered, but can, in many circumstances, manipulated, such that their solutions are conveniently given by the Bessel functions. One of the forms is

$$x^2y'' + (2\mu + 1)xy' + (\kappa^2 x^{2r} + \lambda^2)y = 0 \quad (2.39)$$

where μ, κ, r, λ are constants. The general solution of this equation is given by

$$y = x^{-\mu}[c_1 J_{\alpha/r}(\frac{\kappa x^r}{r}) + c_2 Y_{\alpha/r}(\frac{\kappa x^r}{r})]$$

in which $\alpha = \sqrt{\mu^2 - \lambda^2}$ (see Spiegel 1971). Following this, it is possible to transform various forms of differential equations into the standard Bessel Eq. (2.38). For instance, a differential equation given by

$$x^2y'' + xy' + (\kappa^2 x^2 - \nu^2)y = 0 \quad (2.40)$$

is referred to as the *parametric Bessel equation* of order ν. To find the solution of this equation, we compare it with (2.39). If we let $r = 1, \mu = 0$, and $\lambda^2 = i^2\nu^2$, where i is the complex number, the general solution of this equation becomes

$$y = c_1 J_\nu(\kappa x) + c_2 Y_\nu(\kappa x)$$

Bessel functions play a major role in addressing boundary value problems of the Sturm–Liouville form, especially when expressed in cylindrical or polar coordinate systems. The general form of the Sturm–Liouville boundary value problem in the polar coordinate system is

$$R(r)\frac{d^2u}{dr^2} + R'(r)\frac{du}{dr} + \left[\lambda Q(r) + P(r)\right]u = 0, \quad 0 \le r \le R$$

$$\begin{aligned} \text{BC: } & A_1 u(0) + B_1 u'(0) = 0 \\ & A_2 u(R) + B_2 u'(R) = 0 \end{aligned} \quad (2.41)$$

The parametric Eq. (2.40) in the polar coordinate system is expressed as

$$r\frac{d^2u}{dr^2} + \frac{du}{dr} + (\kappa^2 r - \frac{\nu^2}{r})u = 0 \quad (2.42)$$

This parametric Bessel equation is a special case of the Sturm–Liouville problem (2.41), where $R(r) = r$, $P(r) = -\nu^2/r$, $Q(r) = r$ and $\lambda = \kappa^2$. Here, λ represents a set of the eigenvalues of the system and $Q(r) = r$ denotes the weight function. This equation forms what is called the *singular Sturm–Liouville problem*, since as $r \to 0$, $P(r) \to \infty$. Yet, as for the regular Sturm–Liouville problem, the singular Sturm–Liouville problem has an infinite set of eigenvalues and their associated eigenfunctions.

Thus, if we have a piecewise smooth function $f(r)$ on the interval $[0, R]$, it can be expanded to an infinite series of the form

$$f(r) = \sum_{n=1}^{\infty} c_n J_\nu(\kappa_n r) \tag{2.43}$$

denoted the *Bessel series of order* ν of $f(r)$, for which κ_n are its scaled positive roots of the Bessel function J_ν.

Similar to the trigonometric functions, the Bessel function, for a non-negative integer ν, form an orthogonal set on the interval $[0, R]$ with respect to a weight function $Q(r) = r$, such that

$$\int_0^R J_\nu(\kappa_n r)\, J_\nu(\kappa_m r)\, r\, dr = 0, \quad m \neq n$$

Employing this property, and upon multiplying both sides of (2.43) by $rJ_\nu(\kappa_n r)$, where r is a weight function, and then integrating from 0 to R, we can readily determine the expansion coefficients in (2.43), such that

$$c_n = \frac{\int_0^R r f(r)\, J_\nu(\kappa_n r)\, dr}{\int_0^R r J_\nu^2(\kappa_n r)\, dr} \tag{2.44}$$

where

$$\int_0^R r J_\nu^2(\kappa_n r)\, dr = \frac{R^2}{2} J_{\nu+1}^2(\kappa_n R)$$

Substituting this result into (2.44) gives

$$c_n = \frac{2}{R^2 J_{\nu+1}^2(\kappa_n R)} \int_0^R r f(r)\, J_\nu(\kappa_n r)\, dr$$

which represent the *Fourier–Bessel series coefficients* for a case where the roots of the Bessel function are determined from $J_\nu(\kappa_n r) = 0$.

2.8 Coupled Boundary Value Problems

The boundary value problem (2.29) is a form that describes heat flow in a one-dimensional homogeneous domain. Mathematically, this BVP is expressed by a single partial differential equation together with its corresponding boundary conditions. While this particular scenario finds widespread applications in engineering and applied physics, numerous applications exist where the physical system encompasses more than one component featuring distinct physical properties. The mathematical representation of such a system requires a set of coupled partial differential equations describing the behavior of the different components of the system, together with their interactions.

Here we give two motivating examples illustrating two techniques for solving coupled boundary value problems. In the first example we decouple the system using the *Diagonalization* method, and in the second we solve it via the eigenvalue method.

Example 2.5: Diagonalization method

Solve the following boundary value problem using the diagonalization method:

$$\alpha_1 \frac{d^2 u_1}{dx^2} + b_{12}(u_1 - u_2) = 0$$
$$\alpha_2 \frac{d^2 u_2}{dx^2} + b_{12}(u_2 - u_1) = 0 \tag{2.45}$$

$$\text{BC}: \quad u_1(0) = 2, \quad u_1(10) = 10$$
$$u_2(0) = 1, \quad u_2(10) = 5$$

This problem resembles a steady state heat conduction in a system consisting of two domains in thermal contact with each other. α_1 and α_2 are some thermal parameters and b_{12} is a parameter that controls the thermal interaction between the two domains. In this example we will not indulge in the physical meaning of these parameters, as our focus is on illustrating the mathematical procedure for solving coupled systems. Following this, we assume that

$$\alpha_1 = 1, \quad \alpha_2 = 2, \quad b_{12} = 10 \tag{2.46}$$

Solution

The diagonalization method is a mathematical technique utilized for solving coupled differential equations by means of decoupling. Solving a system of differential equations using diagonalization is valid provided that an $n \times n$ matrix having n linearly independent eigenvectors does exist. That is, systems with matrices having repeated eigenvalues cannot be solved by this method.

The system of differential equation in (2.45) can be written as

$$\mathbf{A}\mathbf{u}'' + \mathbf{B}\mathbf{u} = 0 \tag{2.47}$$

where

$$\mathbf{u} = \begin{pmatrix} u_1 \\ u_2 \end{pmatrix}, \quad \mathbf{A} = \begin{pmatrix} \alpha_1 & 0 \\ 0 & \alpha_2 \end{pmatrix}, \quad \mathbf{B} = \begin{pmatrix} b_{12} & -b_{12} \\ -b_{12} & b_{12} \end{pmatrix}$$

Since A can be inverted, (2.47) can be written as

$$\mathbf{u}'' + \mathbf{A}^{-1}\mathbf{B}\mathbf{u} = 0, \quad \text{or} \quad \mathbf{u}'' + \mathbf{M}\mathbf{u} = 0 \tag{2.48}$$

where

2

$$\mathbf{M} = \mathbf{A}^{-1}\mathbf{B} = \begin{pmatrix} \frac{1}{\alpha_1} & 0 \\ 0 & \frac{1}{\alpha_2} \end{pmatrix} \begin{pmatrix} b_{12} & -b_{12} \\ -b_{12} & b_{12} \end{pmatrix} = \begin{pmatrix} \frac{b_{12}}{\alpha_1} & -\frac{b_{12}}{\alpha_1} \\ -\frac{b_{12}}{\alpha_2} & \frac{b_{12}}{\alpha_1} \end{pmatrix}$$

This matrix is called the *coefficient matrix*. Substituting the values of the parameters in (2.46) into the coefficient matrix **M**, we obtain

$$\mathbf{M} = \begin{pmatrix} 10 & -10 \\ -5 & 5 \end{pmatrix}$$

According to the diagonalization method, the eigenvalues of the system are those obtained from the coefficient matrix. This implies solving for $\det(\mathbf{M} - \lambda\mathbf{I}) = 0$, with **I** the unit vector, i.e.

$$\det\begin{pmatrix} 10-\lambda & -10 \\ -5 & 5-\lambda \end{pmatrix} = 0$$

This gives $\lambda_1 = 0$ and $\lambda_2 = 15$. Since the eigenvalues are distinct, the eigenvectors are linearly independent, which makes the diagonalization method applicable to this problem.

The next step is to find the corresponding eigenvectors. The eigenvector associated with eigenvalue $\lambda_1 = 0$ can be determined via solving:

$$(\mathbf{M} - \lambda_1\mathbf{I}) \cdot \mathbf{v}_1 = 0$$

or

$$\begin{pmatrix} 10 & -10 \\ -5 & 5 \end{pmatrix} \begin{pmatrix} v_{1,1} \\ v_{1,2} \end{pmatrix} = 0$$

This gives

$$10v_{1,1} - 10v_{1,2} = 0, \quad \text{or} \quad -5v_{1,1} + 5v_{1,2} = 0$$

where in both cases we find that $v_{1,1} = v_{1,2}$, and thus the first vector can be written as

$$v_1 = m_1 \begin{pmatrix} 1 \\ 1 \end{pmatrix}$$

in which m_1 is an arbitrary constant that can have any value. Here, we choose $m_1 = 1$, and thus our first eigenvector is

$$v_1 = \begin{pmatrix} 1 \\ 1 \end{pmatrix}$$

Going through the same procedure for the eigenvector associated with the second eigenvalue $\lambda_2 = 15$, we have

$$(\mathbf{M} - \lambda_2 \mathbf{I}) \cdot \mathbf{v}_2 = 0$$

or

$$\begin{pmatrix} 10-15 & -10 \\ -5 & 5-15 \end{pmatrix} \begin{pmatrix} v_{2,1} \\ v_{2,2} \end{pmatrix} = \begin{pmatrix} -5 & -10 \\ -5 & -10 \end{pmatrix} \begin{pmatrix} v_{2,1} \\ v_{2,2} \end{pmatrix} = 0$$

This gives

$$-5v_{2,1} - 10v_{2,2} = 0, \quad \text{or} \quad v_{2,1} = -2v_{2,2}$$

If we assign $v_{2,2} = 1$, the second eigenvector becomes

$$v_2 = m_2 \begin{pmatrix} -2 \\ 1 \end{pmatrix}$$

Notice that the choice of $v_{2,2} = 1$ was arbitrary; only the ratio is important. We can let $m_2 = 1$, and the second eigenvector will then becomes

$$v_2 = \begin{pmatrix} -2 \\ 1 \end{pmatrix}$$

Now, we put the two eigenvectors together in a matrix form, such that

$$\mathbf{P} = \begin{pmatrix} 1 & -2 \\ 1 & 1 \end{pmatrix}$$

Having determined the eigenvalues and their corresponding eigenvectors, we can now proceed with the diagonalization method. By inspection of the second equation of (2.48) we can promptly realize that if we diagonalize matrix **M**, we can decouple this system of equations. The resulting matrix should contain the eigenvalues on its diagonal. To do this, we first let $\mathbf{u} = \mathbf{PU}$, then upon substituting into the system of differential equations in (2.48), we obtain

$$\mathbf{PU}'' + \mathbf{MPU} = 0, \quad \text{or} \quad \mathbf{U}'' + \mathbf{P}^{-1}\mathbf{MPU} = 0, \quad \text{or} \quad \mathbf{U}'' + \mathbf{DU} = 0 \tag{2.49}$$

where

$$\mathbf{D} = \mathbf{P}^{-1}\mathbf{MP} = \begin{pmatrix} 1/3 & 2/3 \\ -1/3 & 1/3 \end{pmatrix} \begin{pmatrix} 10 & -10 \\ -5 & 5 \end{pmatrix} \begin{pmatrix} 1 & -2 \\ 1 & 1 \end{pmatrix} = \begin{pmatrix} 0 & 0 \\ 0 & 15 \end{pmatrix}$$

which is indeed a matrix with its entries on the diagonal are the eigenvalues ($\lambda_1 = 0$, and $\lambda_2 = 15$), and the rest are zero. From the third equation in (2.49), the system of equations can now be expressed as

$$\mathbf{U}'' + \begin{pmatrix} 0 & 0 \\ 0 & 15 \end{pmatrix} \mathbf{U} = 0, \quad \text{or} \quad \begin{pmatrix} U_1'' \\ U_2'' \end{pmatrix} + \begin{pmatrix} 0 & 0 \\ 0 & 15 \end{pmatrix} \begin{pmatrix} U_1 \\ U_2 \end{pmatrix} = 0$$

Obviously this system of equations is uncoupled and can be written as

$$U_1'' = 0$$
$$U_2'' + 15U_2 = 0$$

The solutions of these ODEs are

$$\begin{aligned} U_1 &= c_1 + c_2 x \\ U_2 &= c_3 \sin(\sqrt{15}x) + c_4 \cos(\sqrt{15}x) \end{aligned} \tag{2.50}$$

Going back to the original variables

$$\mathbf{u} = \mathbf{PU} = \begin{pmatrix} 1 & -2 \\ 1 & 1 \end{pmatrix} \begin{pmatrix} U_1 \\ U_2 \end{pmatrix} = \begin{pmatrix} U_1 - 2U_2 \\ U_1 + U_2 \end{pmatrix} \tag{2.51}$$

Upon substituting (2.50) into the right-hand side of (2.51), we obtain the general solutions of u_1 and u_2, such that

$$\begin{aligned} u_1(x) &= c_1 + c_2 x - 2c_3 \sin(\sqrt{15}x) - 2c_4 \cos(\sqrt{15}x) \\ u_2(x) &= c_1 + c_2 x + c_3 \sin(\sqrt{15}x) + c_4 \cos(\sqrt{15}x) \end{aligned} \tag{2.52}$$

These solutions can also be written in terms of the eigenvectors, as

$$\mathbf{u} = c_1 \begin{pmatrix} 1 \\ 1 \end{pmatrix} + c_2 \begin{pmatrix} 1 \\ 1 \end{pmatrix} x + c_3 \begin{pmatrix} -2 \\ 1 \end{pmatrix} \sin(\sqrt{15}x) + c_4 \begin{pmatrix} -2 \\ 1 \end{pmatrix} \cos(\sqrt{15}x)$$

Applying the boundary conditions in (2.45) into the general solutions in (2.52) gives

$$u_1(0) = 2 = c_1 - 2c_4$$
$$u_2(0) = 1 = c_1 + c_4$$
$$u_1(10) = 10 = c_1 + 10\,c_2 - 2\,c_3 \sin(\sqrt{15}.10) - 2\,c_4 \cos(\sqrt{15}.10)$$
$$u_2(10) = 5 = c_1 + 10\,c_2 + c_3 \sin(\sqrt{15}.10) + c_4 \cos(\sqrt{15}.10)$$

Solving these equations simultaneously yields

$$c_1 = \frac{4}{3}, \quad c_2 = \frac{8}{15}, \quad c_3 = \frac{\cos(10\sqrt{15}) - 5}{3\sin(10\sqrt{15})}, \quad c_4 = -\frac{1}{3}$$

Inserting these constants into (2.52), we obtain the final solution of this BVP, such that

$$\begin{aligned} u_1(x) &= \frac{4}{3} + \frac{8}{15}x - 2\frac{\cos(10\sqrt{15}) - 5}{3\sin(10\sqrt{15})} \sin(\sqrt{15}x) + \frac{2}{3}\cos(\sqrt{15}x) \\ u_2(x) &= \frac{4}{3} + \frac{8}{15}x + \frac{\cos(10\sqrt{15}) - 5}{3\sin(10\sqrt{15})} \sin(\sqrt{15}x) - \frac{1}{3}\cos(\sqrt{15}x) \end{aligned} \tag{2.53}$$

Example 2.6: Eigenvalue method

In this example we solve the BVP of the previous example using the eigenvalue method for solving coupled linear differential equations. Recall the BVP:

$$\begin{aligned}
&\alpha_1 \frac{d^2u_1}{dx^2} + b_{12}(u_1 - u_2) = 0\\
&\alpha_2 \frac{d^2u_2}{dx^2} + b_{12}(u_2 - u_1) = 0\\
&\text{BC}: \quad u_1(0) = 2, \quad u_1(10) = 10\\
&\qquad\quad u_2(0) = 1, \quad u_2(10) = 5
\end{aligned} \tag{2.54}$$

$$\alpha_1 = 1, \quad \alpha_2 = 2, \quad b_{12} = 10$$

Solution

As the coefficients of these differential equations are constant, the trial solutions can be

$$u_1 = a\,e^{\kappa x}, \quad u_2 = c\,e^{\kappa x}$$

where κ is a constant that needs to be determined from solving the differential equations, and a and c are constants to be determined from the boundary conditions. Note that we could also use $u_1 = ae^{-\kappa x}$ or $u_1 = ae^{-i\kappa x}$, and the same for u_2.

Inserting these trial solutions into (2.54) gives

$$\begin{aligned}
&\alpha_1\kappa^2 a\,e^{\kappa x} + b_{12} a\,e^{\kappa x} - b_{12} c\,e^{\kappa x} = 0\\
&\alpha_2\kappa^2 c\,e^{\kappa x} + b_{12} c\,e^{\kappa x} - b_{12} a\,e^{\kappa x} = 0
\end{aligned}$$

Dividing by $e^{\kappa x}$ and putting the result in a matrix form, yields

$$\begin{pmatrix} \alpha_1\kappa^2 + b_{12} & -b_{12} \\ -b_{12} & \alpha_2\kappa^2 + b_{12} \end{pmatrix} \begin{pmatrix} a \\ c \end{pmatrix} = 0 \tag{2.55}$$

This system of equations has a nontrivial solution if and only if its determinant is zero, yielding

$$(\alpha_1\kappa^2 + b_{12})\,(\alpha_2\kappa^2 + b_{12}) - b_{12}^2 = 0 \tag{2.56}$$

Substituting the values of the coefficients in (2.54) into (2.56), gives

$$(\kappa^2 + 10)\,(2\kappa^2 + 10) - 100 = 0$$

Solving this fourth order polynomial, we obtain $\kappa = 0$, $\kappa = 0$, $\kappa = -i\sqrt{15}$ and $\kappa = i\sqrt{15}$, which represent the four eigenvalues of the system. As indicated in Table 2.1, as we have two *repeated eigenvalues* (two times $\kappa = 0$), we need to include x in one of the repeated eigenfunctions, such that: $axe^{\kappa x} = axe^{0.x} = ax$. It follows

$$
\begin{aligned}
u_1(x) &= a_1 + a_2 x + a_3 e^{-i\sqrt{15}x} + a_4 e^{i\sqrt{15}x} \\
u_2(x) &= c_1 + c_2 x + c_3 e^{-i\sqrt{15}x} + c_4 e^{i\sqrt{15}x}
\end{aligned}
\tag{2.57}
$$

As the system is coupled, the coefficients must be related. From (2.55) we know that

$$
\left(\alpha_1 \kappa^2 + b_{12}\right) a - b_{12} c = 0, \quad \text{or} \quad c = \frac{\kappa^2 + 10}{10} a
$$

Considering the eigenvalues $\kappa = 0$, $\kappa = 0$, $\kappa = -i\sqrt{15}$ and $\kappa = i\sqrt{15}$, it entails

$$
c_1 = a_1, \quad c_2 = a_2, \quad c_3 = -\frac{1}{2} a_3, \quad c_4 = -\frac{1}{2} a_4
$$

Hence, the general solutions become

$$
\begin{aligned}
u_1(x) &= a_1 + a_2 x + a_3 e^{-i\sqrt{15}x} + a_4 e^{i\sqrt{15}x} \\
u_2(x) &= a_1 + a_2 x - \frac{1}{2} a_3 e^{-i\sqrt{15}x} - \frac{1}{2} a_4 e^{i\sqrt{15}x}
\end{aligned}
\tag{2.58}
$$

Note that if we have considered the second row in (2.55), we would also get the right solution.

Applying the boundary conditions in (2.54) to the general solutions in (2.58) gives

$$
\begin{aligned}
u_1(0) &= 2 = a_1 + a_3 + a_4 \\
u_2(0) &= 1 = a_1 - \frac{1}{2} a_3 - \frac{1}{2} a_4 \\
u_1(10) &= 10 = a_1 + 10 a_2 + a_3 e^{-i\sqrt{15}.10} + a_4 e^{i\sqrt{15}.10} \\
u_2(10) &= 5 = a_1 + 10 a_2 - \frac{1}{2} a_3 e^{-i\sqrt{15}.10} - \frac{1}{2} a_4 e^{i\sqrt{15}.10}
\end{aligned}
$$

Solving these equations simultaneously, results to

$$
a_1 = \frac{4}{3}, \quad a_2 = \frac{8}{15}, \quad a_3 = -\frac{2\left(e^{i10\sqrt{15}} - 5\right)}{3\left(e^{-i10\sqrt{15}} - e^{i10\sqrt{15}}\right)}, \quad a_4 = \frac{2\left(e^{-i10\sqrt{15}} - 5\right)}{3\left(e^{-i10\sqrt{15}} - e^{i10\sqrt{15}}\right)}
$$

Inserting these constants into (2.58), we obtain the final solution for this BVP, as

$$
\begin{aligned}
u_1(x) &= \frac{4}{3} + \frac{8}{15} x - \frac{2}{3\left(e^{-i10\sqrt{15}} - e^{i10\sqrt{15}}\right)} \left[\left(e^{i10\sqrt{15}} - 5\right) e^{-i\sqrt{15}x} + \left(e^{-i10\sqrt{15}} - 5\right) e^{i\sqrt{15}x}\right] \\
u_2(x) &= \frac{4}{3} + \frac{8}{15} x + \frac{1}{3\left(e^{-i10\sqrt{15}} - e^{i10\sqrt{15}}\right)} \left[\left(e^{i10\sqrt{15}} - 5\right) e^{-i\sqrt{15}x} - \left(e^{-i10\sqrt{15}} - 5\right) e^{i\sqrt{15}x}\right]
\end{aligned}
\tag{2.59}
$$

The reader can readily verify that the two solutions in (2.53) and (2.59) give identical results.

Alternative method

Here, we solve this BVP using an alternative procedure to handle the eigenvalue method. We first write the system of differential equations in (2.54) in a matrix form, leading to

$$\begin{pmatrix} u_1'' \\ u_2'' \end{pmatrix} = \begin{pmatrix} -\frac{b_{12}}{\alpha_1} & \frac{b_{12}}{\alpha_1} \\ \frac{b_{12}}{\alpha_2} & -\frac{b_{12}}{\alpha_2} \end{pmatrix} \begin{pmatrix} u_1 \\ u_2 \end{pmatrix} \tag{2.60}$$

This equation can be written in a more concise way, yielding

$$\mathbf{u}'' = \mathbf{A}\mathbf{u} \tag{2.61}$$

The eigenvalue method states that for a linear homogeneous system of differential equations with constant coefficients, the trial solution is given by

$$\mathbf{u} = \mathbf{v}e^{\lambda x}$$

where $\mathbf{v}$ are the nonzero eigenvectors and λ are the eigenvalues of the system. Taking the second derivative of the trial solution and substituting into (2.61) yields

$$\lambda^2 \mathbf{v}e^{\lambda x} = \mathbf{A}\mathbf{v}e^{\lambda x}$$

Cancelling out the nonzero scalar factor $e^{\lambda x}$, we get

$$\mathbf{A}\mathbf{v} = \lambda^2 \mathbf{v} \tag{2.62}$$

In this equation, we know only the coefficient of matrix $\mathbf{A}$, and both $\mathbf{v}$ and λ are unknown and need to be determined. We write (2.62) in this from:

$$\left(\mathbf{A} - \lambda^2\right)\mathbf{v} = 0$$

The nontrivial solution of this equation can only be obtained by letting the determinate equals zero, that is

$$\det\left(\mathbf{A} - \lambda^2 \mathbf{I}\right) \equiv \left|\mathbf{A} - \lambda^2 \mathbf{I}\right| = 0 \tag{2.63}$$

where $\mathbf{I}$ is the unit matrix. Inserting the values of the coefficient matrix (see (2.60)) gives

$$\mathbf{A} = \begin{pmatrix} -10 & 10 \\ 5 & -5 \end{pmatrix}$$

Thus, (2.63) becomes

$$\begin{vmatrix} -10 - \lambda^2 & 10 \\ 5 & -5 - \lambda^2 \end{vmatrix} = 0$$

or

$$\left(-10 - \lambda^2\right)\left(-5 - \lambda^2\right) - 50 = 0$$

Solving this fourth order polynomial gives $\lambda_1 = \lambda_2 = 0$, $\lambda_3 = -i\sqrt{15}$, and $\lambda_4 = i\sqrt{15}$, which are identical to the eigenvalues that we obtained in the previous solution procedure. Obviously, we have two repeated roots and two complex conjugates.

For each eigenvalue there is a corresponding eigenvector:

Case 1: $\lambda_1 = 0$. The eigenvector associated with this eigenvalue can be determined by solving:

$$\left(\mathbf{A} - \lambda_1^2 \mathbf{I}\right) \cdot \mathbf{v}_1 = 0$$

or

$$\begin{pmatrix} -10 & 10 \\ 5 & -5 \end{pmatrix} \begin{pmatrix} \mathrm{v}_{1,1} \\ \mathrm{v}_{1,2} \end{pmatrix} = 0$$

This gives

$$-10\mathrm{v}_{1,1} + 10\mathrm{v}_{1,2} = 0, \quad \text{or} \quad 5\mathrm{v}_{1,1} - 5\mathrm{v}_{1,2} = 0$$

where in both cases we find that $\mathrm{v}_{1,1} = \mathrm{v}_{1,2}$, and thus the first vector can be written as

$$\mathrm{v}_1 = c_1 \begin{pmatrix} 1 \\ 1 \end{pmatrix}$$

in which c_1 is an arbitrary constant that can have any quantity.

Case 2: $\lambda_2 = 0$. Similar to Case 1, this gives

$$\mathrm{v}_2 = c_2 \begin{pmatrix} 1 \\ 1 \end{pmatrix}$$

Case 3: $\lambda_3 = -i\sqrt{15}$. The eigenvector associated with this eigenvalue can be determined by solving:

$$(\mathbf{A} - \lambda_3^2 \mathbf{I}) \cdot \mathbf{v}_3 = 0$$

or

$$\begin{pmatrix} 5 & 10 \\ 5 & 10 \end{pmatrix} \begin{pmatrix} \mathrm{v}_{3,1} \\ \mathrm{v}_{3,2} \end{pmatrix} = 0$$

This gives

$$\mathrm{v}_3 = c_3 \begin{pmatrix} 1 \\ -\frac{1}{2} \end{pmatrix}$$

Case 4: $\lambda_4 = i\sqrt{15}$. The eigenvector associated with this eigenvalue can be determined by solving:

$$\left(\mathbf{A}-\lambda_4^2\mathbf{I}\right)\cdot\mathbf{v}_4=0$$

or

$$\begin{pmatrix} 5 & 10 \\ 5 & 10 \end{pmatrix}\begin{pmatrix} v_{4,1} \\ v_{4,2} \end{pmatrix}=0$$

This gives

$$v_4=c_4\begin{pmatrix} 1 \\ -\frac{1}{2} \end{pmatrix}$$

Hence, the general solution of the system in (2.54) is

$$\mathbf{u}(x)=c_1\begin{pmatrix} 1 \\ 1 \end{pmatrix}+c_2\begin{pmatrix} 1 \\ 1 \end{pmatrix}x+c_3\begin{pmatrix} 1 \\ -\frac{1}{2} \end{pmatrix}e^{-i\sqrt{15}}+c_4\begin{pmatrix} 1 \\ -\frac{1}{2} \end{pmatrix}e^{i\sqrt{15}} \tag{2.64}$$

where we multiplied the second term by x because we have repeated eigenvalues. In a scaler form, this equation can be written as

$$u_1(x)=c_1+c_2x+c_3e^{-i\sqrt{15}x}+c_4e^{i\sqrt{15}x}$$
$$u_2(x)=c_1+c_2x-\frac{1}{2}c_3e^{-i\sqrt{15}x}-\frac{1}{2}c_4e^{i\sqrt{15}x}$$

which are identical to the general solutions (2.58). Upon applying the boundary conditions in (2.54) to these general solutions, we must obtain the solution of the BVP, as that given in (2.59).

2.9 Boundary Conditions

When dealing with modeling a physical system, the general solution of the pertinent differential equation does not reveal the behavior of the studied system. It contains arbitrary constants that can take any value, representing a family of solutions. The uniqueness of the solution unfolds when prescribed values are imposed at the boundaries of the domain. Thus, combining partial differential equations with relevant boundary conditions would constitute a boundary value problem describing the behavior of the physical system.

In this book, the boundary value problems will typically be expressed, for instance, for a 1D domain, as

$$\frac{1}{\alpha}\frac{\partial u}{\partial t}-\frac{\partial^2 u}{\partial x^2}=0,\quad 0\le x\le L$$

$$\begin{aligned} \text{IC}: &\quad u(x,0)=0 \\ \text{BC}: &\quad u(0,t)=U_1(t) \\ &\quad u(L,t)=U_2(t) \end{aligned}$$

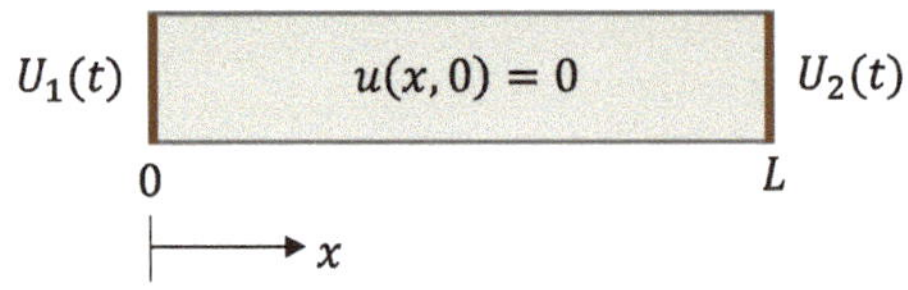

Fig. 2.1 A scheme of a 1D domain subjected to initial and boundary conditions

where IC stands for the initial condition and BC for the boundary conditions. This BVP might represent heat flow in a one-dimensional solid of length L, subjected to an initial temperature $u(x, 0) = 0$ and prescribed time-dependent temperatures $U_1(t)$ and $U_2(t)$ at $x = 0$ and $x = L$, respectively, see Fig. 2.1. The governing partial differential equation is homogeneous, but because the boundary conditions are not, this boundary value problem is classified as nonhomogeneous.

There are several types of boundary conditions associated with boundary value problems. The most relevant boundary conditions that we will be dealing with in this book will be outlined hereafter.

Dirichlet boundary condition: prescribed value

The Dirichlet boundary condition (also known as the *boundary condition of the first kind* or the *essential boundary condition*) implies prescribing the value of the dependent variable on some points or surface areas on the boundary of a domain.

In a heat flow problem, it implies prescribing a temperature on boundary points or areas, as that shown schematically in Fig. 2.1. It might be constant, time-dependent, or a mix of both, such that for a 1D domain spanning in the range $0 \leq x \leq L$, the Dirichlet boundary conditions might be

$$u\,(0, t) = U_1(t), \quad u\,(L, t) = U_2$$

where, in this case, $U_1(t)$ is a time-dependent temperature prescribed at endpoint $x = 0$, and U_2 is a constant temperature prescribed at endpoint $x = L$.

Similarly, the Dirichlet boundary condition for contaminant transport, in a semi-infinite domain, $x \geq 0$, can be

$$c\,(0, t) = C_0(t)$$

where c is the concentration of the contaminant species, and C_0 is some prescribed concentration at $x = 0$, which can be a function of time or constant.

For a groundwater flow problem, the Dirichlet boundary condition in a semi-infinite domain might be

$$h\,(0, t) = H_0(t)$$

where h is the hydraulic head and H_0 is some prescribed head at $x = 0$, which can be a function of time or constant.

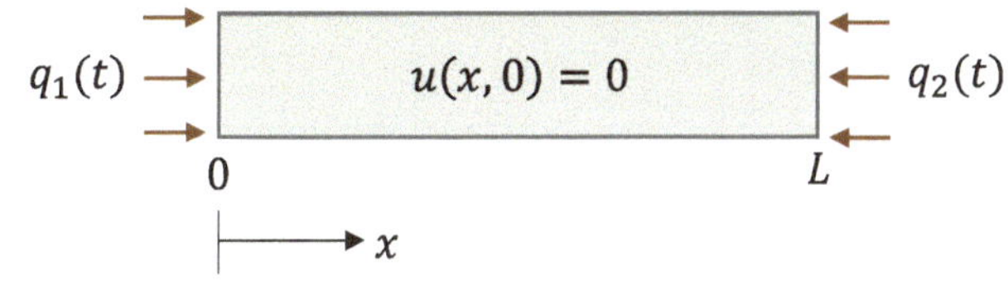

Fig. 2.2 Neumann boundary conditions

Neumann boundary condition: applied flux

The Neumann boundary condition (also known as the *boundary condition of the second kind* or the *natural boundary condition*) implies imposing the value of the spatial derivative of the dependent variable on some points or surface areas on the boundary of a domain.

In heat transfer, it implies specifying the value of heat flux per unit area per unit time normal to a surface area on the boundary. Using Fourier's law, the Neumann boundary condition on the end boundaries of a 1D domain, spanning in the range $0 \leq x \leq L$ (see Fig. 2.2), can be expressed as

$$-\lambda A_x \frac{\partial u}{\partial x}\bigg|_{x=0} = q_1(t), \qquad -\lambda A_x \frac{\partial u}{\partial x}\bigg|_{x=L} = q_2$$

where λ is the *thermal conductivity* and A_x is the cross-sectional area of the domain normal to the x-direction. For this case, $q_1(t)$ is a time-dependent heat flux, and q_2 is a constant heat flux.

If the surface area at $x = 0$ is effectively *insulated*, then $q = 0$, and the boundary condition on this location is expressed as

$$\frac{\partial u}{\partial x}\bigg|_{x=0} = 0$$

A similar boundary condition can be applied for contaminant transport and groundwater flow problems by replacing the heat flux, q, by a concentration flux J_c and a rechanrge flow rate q_w, respectively.

Robin boundary condition

The Robin boundary condition (also known as the *boundary condition of the third kind*) corresponds to a linear combination of the value of the dependent variable and its derivative on the boundary. This implies imposing a linear function combining a Dirichlet and a Neumann boundary condition on a specific point or area of the boundary.

A boundary condition of this kind is frequently used in contaminant transport problems. For a 1D domain $x \geq 0$, the Robin boundary condition can be expressed as

2

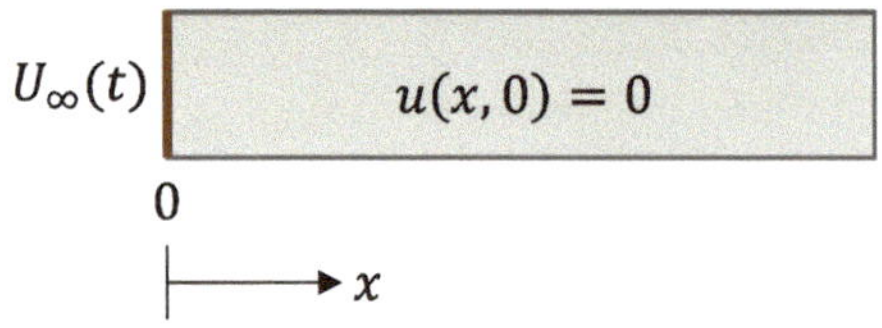

Fig. 2.3 Robin boundary condition

$$-D\frac{\partial c}{\partial x}\bigg|_{x=0} = v[C_\infty - c\,(0,t)]$$

where c is the concentration of the contaminant, D is the diffusion coefficient, v is the fluid velocity and C_∞ is some function describing the concentration on the surrounding environment. It can be a constant or a function of time.

A similar boundary condition is applied in groundwater flow problems, which can be expressed as

$$T\frac{\partial h}{\partial x}\bigg|_{x=0} = b[H_\infty - h\,(0,t)]$$

where T is the transmissivity coefficient, H_∞ is the hydraulic head of the feeding aquifer, and b is some leakage coefficient.

In heat flow problems, this boundary condition is associated with the *Newton's law of cooling*, which states that the heat flux across a boundary surface of a domain is proportional to the temperature difference between the body and its surrounding medium. In a 1D domain $x \geq 0$ (see Fig. 2.3), the Newton's law of cooling implies

$$-\lambda\frac{\partial u}{\partial x}\bigg|_{x=0} = \overline{h}[U_\infty - u\,(0,t)]$$

where U_∞ is the temperature of the surrounding medium, which can be air, water, or a solid domain; and $\overline{h}$ is some thermal interaction coefficient.

Mixed boundary condition

A mixed boundary condition implies specifying different boundary conditions at disjoint parts of the boundary of a domain. The most common application of this kind of boundary conditions is applying a Dirichlet boundary condition on some part of the boundary and a Neumann boundary condition on another (not connected) part of the boundary.

Cauchy boundary condition

A Cauchy boundary condition specifies both the value and the normal derivative of the dependent variable on a boundary of the domain. This corresponds to imposing both a Dirichlet and a Neumann boundary condition on a specific point or area of the boundary.

This type of boundary conditions is not common in heat flow, contaminant transport and groundwater flow problems.

Homogeneous boundary conditions

We note that the above given Dirichlet and Neumann boundary conditions are nonhomogeneous since they are not identically zero. Should these boundary conditions be

$$u(0,t) = 0, \quad \left.\frac{\partial u}{\partial x}\right|_{x=0} = 0$$

$$c(0,t) = 0, \quad \left.\frac{\partial c}{\partial x}\right|_{x=0} = 0$$

$$h(0,t) = 0, \quad \left.\frac{\partial h}{\partial x}\right|_{x=0} = 0$$

they are said to be homogeneous.

Similarly, the above given Robin boundary conditions are nonhomogeneous, but if $U_\infty = 0$, $C_\infty = 0$ or $H_\infty = 0$; i.e.

$$\lambda \left.\frac{\partial u}{\partial x}\right|_{x=0} = \bar{h}\, u(0,t)$$

$$D \left.\frac{\partial c}{\partial x}\right|_{x=0} = v\, c(0,t)$$

$$T \left.\frac{\partial h}{\partial x}\right|_{x=0} = b\, h(0,t)$$

they become homogeneous.

Problems

1 What is the difference between the general solution of a partial differential equation and the solution of a boundary value problem?
2 What makes the method of eigenfunction expansion attractive for solving boundary value problems?
3 If the nonhomogeneous term of a differential equation is $10\cos(5x)$, what is the suitable trial particular solution u_p for it? **Hint**: see Table 2.2.
4 Verify that the BVP $y''(x) + \lambda y(x) = 0$, $y'(0) = 0$, $y'(L) = 0$, $0 \le x \le L$ is a Sturm-Liouville system. And, find the eigenvalues and corresponding eigenfunction of the system.
5 Prove that the eigenfunctions in Problem 4 are orthogonal.
6 What Fourier series we get if the boundary conditions in Problem 4 were $y'(0) = 0, y(L) = 0$?

2

7 Show that the general solution of

$$xy'' + y\prime + \kappa y = 0 \text{ is } y = c_1 J_0\left(2\sqrt{\kappa x}\right) + c_2 Y_0\left(2\sqrt{\kappa x}\right)$$

Hint: multiply the differential equation by x and compare it with (2.39). See also Spiegel (1971).

8 Compare the diagonalization solution (2.53) with the eigenvalue solution (2.59) for $x = 0$, $x = 5$, $x = 8$ and $x = 10$.

9 The motion of two coupled spring-mass systems is given by a system of linear second-order differential equations, as

$$m_1 x_1^{''} = -k_1 x_1 + k_2(x_2 - x_1)$$

$$m_2 x_2^{''} = -k_2(x_2 - x_1)$$

where m_1 and m_2 represent the mass of the two systems and k_1 and k_2 are the stiffnesses of the corresponding springs. Solve this problem using the Diagonalization method.

10 Solve the system of differential equations of the previous problem using the eigenvalue method for solving linear homogeneous system of equations. Then, compare the two results by assuming that $m_1 = 1$, $m_2 = 2$, $k_1 = 4$ and $k_2 = 3$.

References

Henry EC, Penney DE (2008) Differential equations and boundary value problems. Computing and modeling, 4th ed. Pearson Education, Inc.

Spiegel MR (1971) Advanced mathematics for engineers and scientists. Schaum's outline series, McGraw Hill

Zill DG, Wright WS (2011) Advanced engineering mathematics, 4th ed. Jones and Bartlett Publishers

Further Reading

Xie, Wei-Chau. Differential Equations for Engineers. Cambridge University Press. 2010

Fourier Method

R. Al Khoury and C. Kasbergen, *Analytical Subsurface Environment*,
https://doi.org/10.1007/978-3-031-93910-5_3

The use of Fourier series for solving differential equations is often referred to as the Fourier method. The Fourier series is a technique for decomposing complex functions in terms of a series of sines and cosines, or other closely related types of elementary functions. The application of the Fourier method commonly arises when addressing boundary value problems associated with linear homogeneous partial differential equations. This method, together with the eigenfunction expansion, provides a structured procedure that helps in solving intricate mathematical challenges inherent in various scientific and engineering applications.

The Fourier transform is an extension of the Fourier series. It is basically an integral transform that extends its versatility from solving differential equations to signal analysis. It offers an approach to transform partial differential equations from the time (space) domain to equations expressed in the frequency (wavenumber) domain. This transformation often leads to a more manageable set of ordinary differential equations that are easier to solve. In signal processing (referred to as Fourier analysis, or spectral analysis), on the other hand, the Fourier transform approach involves transforming signals, whether analytical or digital, from the time domain to the frequency domain. This transformation is typically performed through techniques such as the continuous Fourier transform (CFT), discrete Fourier transform (DFT) and the fast Fourier transform (FFT). These techniques break down arbitrary signals into simpler sinusoidal components, facilitating examination and interpretation.

In solving the boundary value problems which we will be addressing in this book, we will make use of both the Fourier series and the Fourier transform. We refer to this coupling as the Fourier method. This chapter is devoted to introducing the Fourier method as a tool for solving time-dependent boundary value problems.

3.1 Motivation of Fourier Method

Prior to Fourier's work on "*The analytical theory of heat*" in 1822, the heat equation could only be solved for simple boundary conditions, mainly under sinusoidal heat source conditions. Fourier showed that the heat equation under any arbitrary boundary condition can be expressed in terms of a series of an infinite sum of trigonometric functions, now called by his name, the *Fourier series*. By this, he revolutionized the solution of boundary value problems involving linear partial differential equations. The representation of solutions by series has many attractive features, among which, they are relatively fast to converge, systematic and are remarkably suitable for computer implementation.

The set of functions that make up the terms in the Fourier series are typically determined by the eigenfunction expansion of series. As a motivation to the use of the Fourier method for solving boundary value problems with any arbitrary functions describing the boundary conditions, we consider a differential equation describing heat conduction in a one-dimensional domain expressed in the frequency domain, that is

$$\frac{d^2u}{dx^2} - \kappa^2(\omega)\,u = f(\omega)$$

where ω is the *radial frequency*. A straightforward solution of this nonhomogeneous equation can be obtained if $f(\omega)$ is a constant or a simple harmonic function. For example, the differential equation

$$\frac{d^2u}{dx^2} - \kappa^2(\omega)\,u = U_0 \sin(\omega x)$$

has the solution

$$u(x) = c_1 e^{\kappa x} + c_2 e^{-\kappa x} - \frac{U_0 \sin(\omega x)}{\kappa^2 + \omega^2}$$

Moreover, if the forcing function $f(\omega)$ is a linear combination of several simple harmonic functions, we can extend the above solution using the principle of superposition, to become

$$u(x) = \sum\nolimits_n c_1 e^{\kappa_n x} + c_2 e^{-\kappa_n x} - \frac{U_0 \sin(\omega_n x)}{\kappa_n^2 + \omega_n^2}$$

Yet, in reality, physical systems frequently incorporate forcing functions and boundary conditions that extend beyond simple harmonic functions. In particular, measured data typically appear as arbitrary functions of time. Figure 3.1 shows a schematic representation of measured temperature versus time. Clearly, this type of data cannot be directly treated as simple harmonics, and it is precisely for such data where the Fourier method demonstrates its effectiveness.

In this chapter, we will be exploring the Fourier's concept featuring that any arbitrary function of time can be depicted as an infinite series of trigonometric or exponential terms.

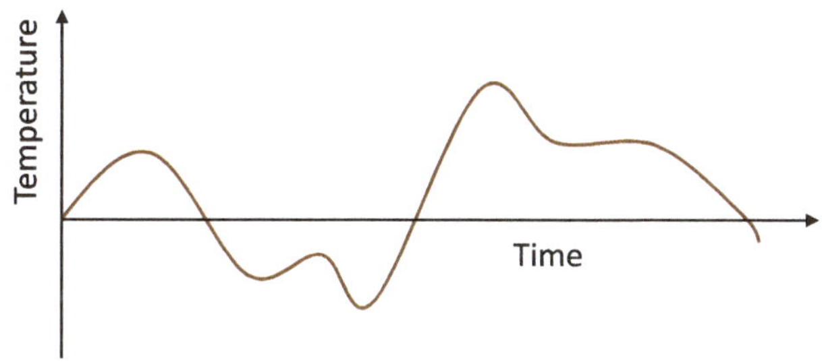

Fig. 3.1 Random data

This aspect of the Fourier method, demonstrated through both the Fourier series and the Fourier transform, holds significant potential. It equips us with the essential tools to tackle complex boundary value problems, especially those entailing arbitrary time-dependent boundary conditions.

3.2 Generalized Fourier Series

In ► Chap. 2 we established that when an orthogonal set of functions $\phi_m(x), m = 1, 2, \ldots$ exists, it is possible to represent a function $f(x)$ in the interval $[a, b]$ as a linear combination of these functions, forming an infinite series of the form

$$f(x) = \sum_{m=1}^{\infty} c_m \phi_m(x) \tag{3.1}$$

where c_m are the expansion coefficients. We also established that the determination of these coefficients can be achieved using the eigenfunction expansion of orthogonal functions. Multiplying both sides of (3.1) by $\phi_n(x)$ and integrating with respect to a weight function $w(x)$ on the interval $[a, b]$, we obtain

$$c_n = \frac{\int_a^b w(x) f(x)\, \phi_n(x)\, dx}{\int_a^b w(x) [\phi_n(x)]^2 dx} \tag{3.2}$$

in which we utilized the orthogonality property

$$\int_a^b w(x)\, \phi_n(x)\, \phi_m(x)\, dx = 0, \quad m \neq n$$

Hereafter, we will explore how such expansions are formed and how the expansion coefficients are obtained in applications. To illustrate this, we present two motivating examples underlining this procedure.

▪ **Example 3.1: 1D Dirichlet boundary value problem**

Consider the following linear one-dimensional Dirichlet boundary value problem:

$$\begin{aligned} &\frac{d^2u}{dx^2} + \mu u = 0, \quad 0 \le x \le L \\ &\text{BC:} \quad u(0) = 0 \\ &\qquad\;\;\; u(L) = 0 \end{aligned} \tag{3.3}$$

Solution

To solve this linear homogenous BVP, we seek a solution that is strictly nontrivial. We first need to consider the three possible forms of μ: zero, negative, and positive.

Case 1: $\mu = 0$

For $\mu = 0$, the ODE in (3.3) becomes $u'' = 0$ and the solution of this equation can be obtained by direct integration, giving

$$u(x) = c_1 x + c_2$$

in which c_1 and c_2 are the integration constants to be determined from the boundary conditions.

Applying the boundary conditions in (3.3) to this solution yields both constants equal to zero, such that

$$u(0) = 0 = c_2$$
$$u(L) = 0 = c_1 L \Rightarrow c_1 = 0$$

Thus, for $\mu = 0$, the only solution to this BVP is $u(x) = 0$, which is trivial.

Case 2: $\mu < 0$

For convenience we write $\mu = -\kappa^2$ with $\kappa > 0$. . Following this, the ODE in (3.3) becomes

$$u'' - \kappa^2 u = 0$$

The general solution of this ODE is (see Table 2.1, Chap. 2)

$$u(x) = c_1 e^{-\kappa x} + c_2 e^{\kappa x}$$

or

$$u(x) = c_1 \cosh(\kappa x) + c_2 \sinh(\kappa x)$$

where we note that for a finite domain the hyperbolic form is preferable.

Applying the boundary conditions in (3.3) to this solution yields

$$u(0) = 0 = c_1 \cosh(0) + c_2 \sinh(0) \Rightarrow c_1 = 0$$
$$u(L) = 0 = c_2 \sinh(\kappa L)$$

But as $\kappa > 0$, $\sinh \kappa L$ would never become zero, and thus c_2 must be zero. This would lead again to having only the trivial solution $u(x) = 0$.

Case 3: $\mu > 0$

As for the previous case, for convenience we write $\mu = \kappa^2$ with $\kappa > 0$. . The ODE in (3.3) of this case is thus

$$u'' + \kappa^2 u = 0$$

The general solution of this ordinary differential equation is (see Table 2.1, Chap. 2)

$$u(x) = a\cos(\kappa x) + b\sin(\kappa x)$$

in which a and b are constants to be determined from the boundary conditions.

Applying the two boundary conditions in (3.3), gives $a = 0$, and

$$b\sin(\kappa L) = 0$$

This equation entails that $b \neq 0$, otherwise it results to the trivial solution $u(x) = 0$. Thus $\sin(\kappa L)$ should be zero, implying

$$\kappa L = m\pi, \quad \text{or} \quad \kappa_n = m\pi/L, \qquad m = 1, 2, 3, \ldots$$

where $m = 0$ is excluded since this would make $u(x)$ identically zero, which is trivial.

Hence (3.3) possesses nontrivial solutions when its coefficient $\mu = \kappa^2$ is equal to $m^2\pi^2/L^2$, leading to

$$u_m(x) = b_m \sin(\frac{m\pi}{L}x), \quad m = 1, 2, \ldots \tag{3.4}$$

where the values

$$\mu_m = \frac{m^2\pi^2}{L^2}$$

for which the BVP in (3.3) has nontrivial solutions are denoted the *eigenvalues* (also called characteristic values) of the system, and the solutions in (3.4) are their corresponding *eigenfunctions* (or characteristic functions).

Following the superposition principle, if $u_1, u_2, \cdots, u_m$ are, individually, solutions of a given linear homogeneous differential equation, then $c_1u_1 + c_2u_2 + \cdots + c_mu_m$ is also a solution, such that the general solution can be expressed as

$$u(x) = \sum_{m=1}^{\infty} b_m \sin(\frac{m\pi}{L}x) \tag{3.5}$$

This function is typically the Fourier series of an odd function, denoted as the *Fourier sine series* with b_m are the *Fourier coefficients*.

To determine the b_m coefficients, we tempt, for sake of explanation, to re-write (3.5) in a more convenient form, where we assume that $L = \pi$ for the interval $(0, \pi)$, and $u(x) = f(x)$, leading to

$$f(x) = \sum_{m=1}^{\infty} b_m \sin(mx)$$

Multiplying both sides by $\sin(nx)$, where n is a positive integer, and integrating, we obtain

$$\int_0^{\pi} f(x) \sin(nx)\, dx = \sum_{m=1}^{\infty} b_m \int_0^{\pi} \sin(mx) \sin(nx)\, dx \tag{3.6}$$

Since the trigonometric functions are orthogonal, their integration on the right-hand side implies.

$$\int_0^{\pi} \sin(mx) \sin(nx)\, dx = \begin{cases} 0 & \text{for } n \neq m \\ \pi/2 & \text{for } n = m \end{cases}$$

Hence, (3.6) reduces to

$$\int_0^{\pi} f(x) \sin(nx)\, dx = \frac{\pi}{2} b_n$$

This gives

$$b_n = \frac{2}{\pi} \int_0^{\pi} f(x) \sin(nx)\, dx, \qquad n = 1, 2, \cdots \tag{3.7}$$

which typically represents the *Fourier sine series coefficients*. Following this, the Fourier series becomes

$$f(x) = \sum_{n=1}^{\infty} b_n \sin(nx) \tag{3.8}$$

where we replaced m with n.

Note that we could obtain these Fourier coefficients by directly applying (3.2). Comparing the differential equation in (3.3) to the standard formulation of the Sturm–Liouville equation (▶ Chap. 2) we can readily deduce that the weight function $Q(x) = w(x) = 1$. Hence

$$b_n = \frac{\int_0^{\pi} f(x) \sin(nx)\, dx}{\int_0^{\pi} [\sin(nx)]^2 dx} = \frac{2}{\pi} \int_0^{\pi} f(x) \sin(nx)\, dx$$

Example 3.2: 1D Neumann boundary value problem

Analogous to the previous example, let us consider the following linear one-dimensional Neumann boundary value problem:

$$\begin{aligned} &\frac{d^2u}{dx^2} + \mu u = 0, \quad 0 \leq x \leq L \\ &\text{BC:} \quad u'(0) = 0 \\ &\qquad\;\; u'(L) = 0 \end{aligned} \tag{3.9}$$

■ **Solution**

If we consider the three cases of μ given in the previous example, we find that Case 3 where $\mu = \kappa^2$ with $\kappa > 0$ would be the only one that possess nontrivial solutions. Thus, we can express the ODE in (3.9) as

$$u'' + \kappa^2 u = 0$$

with its solution given by

$$u(x) = a\cos(\kappa x) + b\sin(\kappa x) \tag{3.10}$$

Taking the derivative of (3.10) with respect to x, we obtain

$$u'(x) = -a\kappa\sin(\kappa x) + b\kappa\cos(\kappa x)$$

Applying the two boundary conditions in (3.9), gives $b = 0$, and

$$a\kappa\sin(\kappa L) = 0$$

Since a and κ must not be zero (otherwise it leads to the trivial solution $u(x) = 0$), $\sin(\kappa L)$ would consequently be zero, implying

$$\kappa L = m\pi, \quad \text{or} \quad \kappa_m = m\pi/L, \qquad m = 0, 1, 2, \ldots$$

Hence (3.10) possesses nontrivial solutions when its coefficient κ^2 is equal to $m^2\pi^2/L^2$, leading to

$$u_m(x) = a_m\cos(\frac{m\pi}{L}x), \quad m = 0, 1, 2, \ldots \tag{3.11}$$

where $\mu_m = \kappa_m^2 = m^2\pi^2/L^2$ are the eigenvalues of the system, and the solutions in (3.11) are their corresponding eigenfunctions. Note that, unlike the Dirichlet case, $\mu_0 = 0$ is an eigenvalue for this BVP, since $u(x) = 1$ is the corresponding eigenfunction. This eigenvalue is already incorporated in (3.11) as indicated by $m = 0$.

Following the superposition principle, the general solution of this boundary value problem is then

$$u(x) = \sum_{m=0}^{\infty} a_m\cos(\frac{m\pi}{L}x) \tag{3.12}$$

This function is the Fourier series of an even function, denoted as the *Fourier cosine series* with a_m its coefficients.

The a_m coefficients can be determined in the same way as we did for the previous case. Re-writing (3.12) in a more convenient form by letting $L = \pi$ and $u(x) = f(x)$, we obtain

$$f(x) = \sum_{m=0}^{\infty} a_m\cos(mx)$$

Multiplying both sides by $\cos(nx)$ and integrating, gives

$$\int_0^{\pi} f(x)\cos(nx)\,dx = \sum_{m=0}^{\infty} a_m \int_0^{\pi} \cos(mx)\cos(nx)\,dx \tag{3.13}$$

But since

$$\int_0^{\pi} \cos(mx)\cos(nx)\,dx = \begin{cases} 0 & \text{for } n \neq m \\ \pi/2 & \text{for } n = m \end{cases}$$

it gives, upon substituting into (3.13),

$$a_n = \frac{2}{\pi}\int_0^{\pi} f(x)\cos(nx)\,dx, \qquad n = 0,\ 1,\ 2,\ \cdots \tag{3.14}$$

which typically represents the Fourier cosine series coefficient. Accordingly, the Fourier series becomes

$$f(x) = \sum_{n=0}^{\infty} a_n \cos(nx) \tag{3.15}$$

3.3 Fourier Series

In the preceding examples, we perceived the representation of solutions of boundary value problems through sine and cosine Fourier series. The fundamental concept underlying this approach lies in the eigenfunction expansion, wherein the solution to differential equations is systematically decomposed into infinite series of orthogonal trigonometric functions. The coefficients of these terms can be easily obtained through a systematic mathematical procedure.

The subject of Fourier series is expansive, thoroughly covered in renowned mathematics textbooks such as those presented by Edwards and Penney (2008), Zill and Wright (2011), as well as Brown and Churchill (2012). In this context, we offer an overview of this intricate topic, tailored to addressing the specific challenges posed by the boundary value problems under consideration in this book.

3.3.1 Trigonometric Fourier Series

The Fourier method is widely applicable to diverse boundary value problems. In many cases, the solutions to these problems take the form of trigonometric Fourier series, which incorporate both sine and cosine terms.

If a function $f(x)$ has a period $2L$ in the range $(-L, L)$, its trigonometric Fourier series is expressed as

$$f(x) = \frac{a_0}{2} + \sum_{n=1}^{\infty} a_n \cos\left(\frac{n\pi x}{L}\right) + \sum_{n=1}^{\infty} b_n \sin\left(\frac{n\pi x}{L}\right), \qquad -L < x < L \tag{3.16}$$

with its Fourier coefficients (see (3.7) and (3.14))

$$
\begin{aligned}
a_0 &= \frac{1}{L}\int_{-L}^{L} f(x)\,dx, && n = 0 \\
a_n &= \frac{1}{L}\int_{-L}^{L} f(x)\cos\left(\frac{n\pi x}{L}\right)dx, && n = 1,2,3,\ldots \\
b_n &= \frac{1}{L}\int_{-L}^{L} f(x)\sin\left(\frac{n\pi x}{L}\right)dx, && n = 1,2,3,\ldots
\end{aligned}
\tag{3.17}
$$

Note that a_0 can be incorporated within the summation in (3.16) if we make $n = 0, 1, 2, \ldots$, such that

$$
f(x) = \sum_{n=0}^{\infty} a_n \cos\left(\frac{n\pi x}{L}\right) + \sum_{n=1}^{\infty} b_n \sin\left(\frac{n\pi x}{L}\right)
$$

but in this case it must be expressed as

$$
a_0 = \frac{1}{2L}\int_{-L}^{L} f(x)\,dx
$$

which is a mean value of $f(x)$ over the period $(-L, L)$.

If the interval of $f(x)$ is in the range $(-\pi, \pi)$, (3.16) is given by

$$
f(x) = \frac{a_0}{2} + \sum_{n=1}^{\infty} a_n \cos(nx) + \sum_{n=1}^{\infty} b_n \sin(nx), \qquad -\pi < x < \pi
$$

a form that is usually written in mathematical text books.

▪ Example 3.3: Deriving the Fourier coefficients

Derive the Fourier coefficients (3.17).

▪ Solution

These coefficients are obtained by first integrating both side of (3.16) from $-L$ to L to obtain a_0. Then, upon multiplying both sides of (3.16) by $\cos(m\pi x/L)$ and integrate the outcome, we obtain $a_n, n > 0$. Finally, if we multiply (3.16) by $\sin(m\pi x/L)$, and integrate, we obtain $b_n, n > 0$. This process can be achieved using these integral identities:

$$\begin{aligned}
&\int_{-L}^{L} \cos\left(\frac{n\pi x}{L}\right) dx = 0 \\
&\int_{-L}^{L} \sin\left(\frac{n\pi x}{L}\right) dx = 0 \\
&\int_{-L}^{L} \cos\left(\frac{n\pi x}{L}\right) \sin\left(\frac{m\pi x}{L}\right) dx = 0 \qquad (3.18) \\
&\int_{-L}^{L} \cos\left(\frac{n\pi x}{L}\right) \cos\left(\frac{m\pi x}{L}\right) dx = \begin{cases} 0 & m \neq n \\ L & m = n \end{cases} \\
&\int_{-L}^{L} \sin\left(\frac{n\pi x}{L}\right) \sin\left(\frac{m\pi x}{L}\right) dx = \begin{cases} 0 & m \neq n \\ L & m = n \end{cases}
\end{aligned}$$

Following this, we first integrate both sides of (3.16) from $-L$ to L, yielding

$$\int_{-L}^{L} f(x)\, dx = \frac{a_0}{2} \int_{-L}^{L} dx + \sum_{n=1}^{\infty} a_n \int_{-L}^{L} \cos\left(\frac{n\pi x}{L}\right) dx + \sum_{n=1}^{\infty} b_n \int_{-L}^{L} \sin\left(\frac{n\pi x}{L}\right) dx$$

Considering the first and second integrals in (3.18) gives

$$\int_{-L}^{L} f(x)\, dx = \frac{a_0}{2} \int_{-L}^{L} dx = \frac{a_0 x}{2}\bigg|_{-L}^{L} = La_0$$

Thus

$$a_0 = \frac{1}{L} \int_{-L}^{L} f(x)\, dx$$

Now, if we multiply (3.16) by $\cos(m\pi x/L)$ and integrate, we obtain

$$\begin{aligned}
\int_{-L}^{L} f(x) \cos\left(\frac{m\pi x}{L}\right) dx = {} & \frac{a_0}{2} \int_{-L}^{L} \cos\left(\frac{m\pi x}{L}\right) dx \\
& + \sum_{n=1}^{\infty} a_n \int_{-L}^{L} \cos\left(\frac{m\pi x}{L}\right) \cos\left(\frac{n\pi x}{L}\right) dx \\
& + \sum_{n=1}^{\infty} b_n \int_{-L}^{L} \cos\left(\frac{m\pi x}{L}\right) \sin\left(\frac{n\pi x}{L}\right) dx
\end{aligned}$$

Using the first and third integrals in (3.18), we obtain

$$\int_{-L}^{L} f(x) \cos\left(\frac{m\pi x}{L}\right) dx = \sum_{n=1}^{\infty} a_n \int_{-L}^{L} \cos\left(\frac{m\pi x}{L}\right) \cos\left(\frac{n\pi x}{L}\right) dx$$

Considering the fourth integral, we then get

$$\int_{-L}^{L} f(x) \cos\left(\frac{n\pi x}{L}\right) dx = a_n L$$

3

Thus

$$a_n = \frac{1}{L}\int_{-L}^{L} f(x)\cos\left(\frac{n\pi x}{L}\right) dx$$

Finally, if we multiply (3.16) by $\sin(m\pi x/L)$ and integrate, we obtain

$$\int_{-L}^{L} f(x)\sin\left(\frac{m\pi x}{L}\right) dx = \frac{a_0}{2}\int_{-L}^{L}\sin\left(\frac{m\pi x}{L}\right) dx + \\ + \sum_{n=1}^{\infty} a_n \int_{-L}^{L}\sin\left(\frac{m\pi x}{L}\right)\cos\left(\frac{n\pi x}{L}\right) dx \\ + \sum_{n=1}^{\infty} b_n \int_{-L}^{L}\sin\left(\frac{m\pi x}{L}\right)\sin\left(\frac{n\pi x}{L}\right) dx$$

Considering the second and third integrals in (3.18), we reach to

$$\int_{-L}^{L} f(x)\sin\left(\frac{m\pi x}{L}\right) dx = \sum_{n=1}^{\infty} b_n \int_{-L}^{L}\sin\left(\frac{m\pi x}{L}\right)\sin\left(\frac{n\pi x}{L}\right) dx$$

Making use of the fifth integral, we finally obtain

$$\int_{-L}^{L} f(x)\sin\left(\frac{n\pi x}{L}\right) dx = b_n L$$

Thus

$$b_n = \frac{1}{L}\int_{-L}^{L} f(x)\sin\left(\frac{n\pi x}{L}\right) dx$$

- **Example 3.4: Fourier series expansion**

Expand $f(x) = x^2, -L < x < L$ in a Fourier series.

- **Solution**

The Fourier series expansion is

$$f(x) = \frac{a_0}{2} + \sum_{n=1}^{\infty} a_n \cos\left(\frac{n\pi x}{L}\right) + \sum_{n=1}^{\infty} b_n \sin\left(\frac{n\pi x}{L}\right) \tag{3.19}$$

For $n = 0$, we have

$$a_0 = \frac{1}{L}\int_{-L}^{L} x^2 dx = \frac{2L^2}{3}$$

For $n \geq 1$, we have

$$a_n = \frac{1}{L}\int_{-L}^{L} x^2 \cos\left(\frac{n\pi x}{L}\right) dx, \quad n = 1, 2, \cdots$$

which, upon integration by parts, yields

$$a_n = \frac{4L^2(-1)^n}{n^2\pi^2}$$

and

$$b_n = \frac{1}{L}\int_{-L}^{L} x^2 \sin\left(\frac{n\pi x}{L}\right) dx$$
$$= 0$$

which is logic, since $f(x) = x^2$ is by definition symmetric about the y–axis, making a_n the only Fourier coefficients (see the discussion on the cosine and sine Fourier series below).

Substituting these coefficients into (3.19) we obtain

$$f(x) = \frac{L^2}{3} + \sum_{n=1}^{\infty} \frac{4L^2(-1)^n}{n^2\pi^2} \cos\left(\frac{n\pi x}{L}\right), \quad -L < x < L$$

Figure 3.2 shows the Fourier series expansion of $f(x) = x^2$ for $n = 2, 6$ and 20, together with its exact representation. The figure clearly shows that with increasing the number of summation, the solution becomes more accurate, and, in this case, approaching convergence by $n = 20$.

3.3.2 Complex Fourier Series

In the previous section, we have seen that any real function $f(x)$ can be expanded into a series of *real-valued* trigonometric functions. As we will be frequently encountering in the following chapters, this kind of Fourier expansion is important for solving boundary value problems. For signal anal-

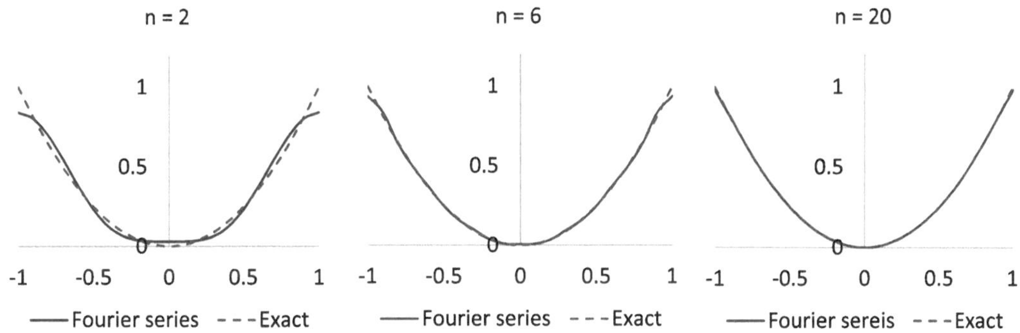

Fig. 3.2 Fourier expansion of $f(x) = x^2$ for different n

ysis, however, it is more convenient and efficient to represent the signal in an infinite series of *complex-valued* functions of a real variable, such as e^{inx}, where $i = \sqrt{-1}$.

Using Euler's formula:

$$e^{ix} = \cos(x) + i\sin(x) \quad \text{and} \quad e^{-ix} = \cos(x) - i\sin(x)$$

the trigonometric Fourier series in (3.16) for a signal periodic in $(-L, L)$ can be expressed in a complex form as

$$f(x) = \sum_{n=-\infty}^{\infty} c_n\, e^{in\pi x/L} \tag{3.20}$$

where the corresponding Fourier coefficients are given by

$$c_n = \frac{1}{2L}\int_{-L}^{L} f(x)\, e^{-in\pi x/L} dx, \quad n = 0, \pm 1, \pm 2, \cdots \tag{3.21}$$

This form of Fourier series is known as the *complex Fourier series*.

The complex Fourier coefficients c_n are related to the trigonometric coefficients a_n and b_n via

$$\begin{matrix} a_n = c_n + c_{-n} \\ b_n = i\,(c_n - c_{-n}) \end{matrix}, \quad n = 0, 1, 2, \cdots$$

We later will see that if we have a time-dependent signal, periodic in $(0, T)$, its complex Fourier series is given by

$$f(t) = \sum_{n=-\infty}^{\infty} c_n\, e^{2in\pi t/T}$$

with

$$c_n = \frac{1}{T}\int_{0}^{T} f(x)\, e^{-2in\pi t/T} dx, \quad n = 0, \pm 1, \pm 2, \cdots$$

3.3.3 Fourier Cosine and Sine Series

Functions in practice might be either even or odd, as depicted in ◘ Fig. 3.3. If a function $f(x)$ is even on the interval $(-L, L)$, i.e. $f(x) = f(-x)$ for all x in the interval, then the trigonometric Fourier series will contain only cosine terms. This special form of the Fourie series is referred to as the Fourier cosine series. This series is usually expressed as

$$f(x) = \frac{a_0}{2} + \sum_{n=1}^{\infty} a_n \cos\left(\frac{n\pi x}{L}\right) \tag{3.22}$$

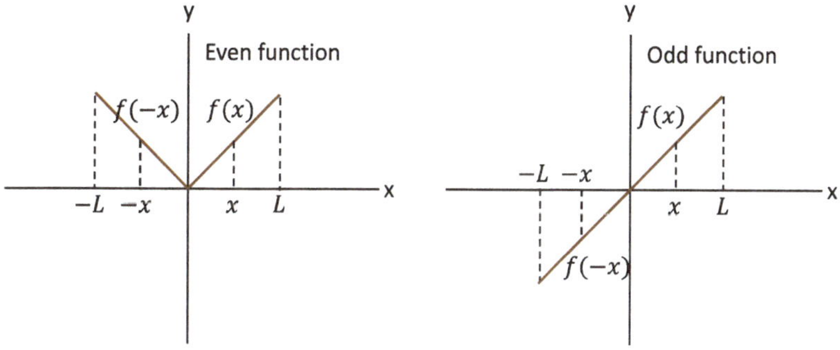

Fig. 3.3 An example of even and odd functions

where

$$
\begin{aligned}
a_0 &= \frac{2}{L}\int_0^L f(x)\,dx \\
a_n &= \frac{2}{L}\int_0^L f(x)\cos\left(\frac{n\pi x}{L}\right)dx, \quad n = 1, 2, \ldots
\end{aligned}
\tag{3.23}
$$

If, however, $f(x)$ is odd on the interval $(-L, L)$, i.e. $f(x) = -f(-x)$, the trigonometric Fourier series will contain only sine terms, denoted as the Fourier sine series. This series is expressed as

$$
f(x) = \sum_{n=1}^{\infty} b_n \sin\left(\frac{n\pi x}{L}\right) \tag{3.24}
$$

where

$$
b_n = \frac{2}{L}\int_0^L f(x)\sin\left(\frac{n\pi x}{L}\right)dx, \quad n = 1, 2, \ldots \tag{3.25}
$$

3.3.4 Half-Range Expansions

The trigonometric series of $f(x)$ discussed earlier is periodic in the range $(-L, L)$ with the origin as its midpoint, i.e. $-L < x < L$ (though, it can also range in the interval $0 < x < 2L$). However, in many cases, particularly in physical applications, $f(x)$ is limited to a finite interval $[0, L]$, as illustrated in Fig. 3.4a. The Fourier series in this case is referred to as the *half-range expansions*. Since the period of the trigonometric Fourier series is $2L$, we might assume that the half range function with period L is virtually spanning to $2L$, but we maintain our interest to be solely within period L. Yet, we lack evidence regarding how the function would look like

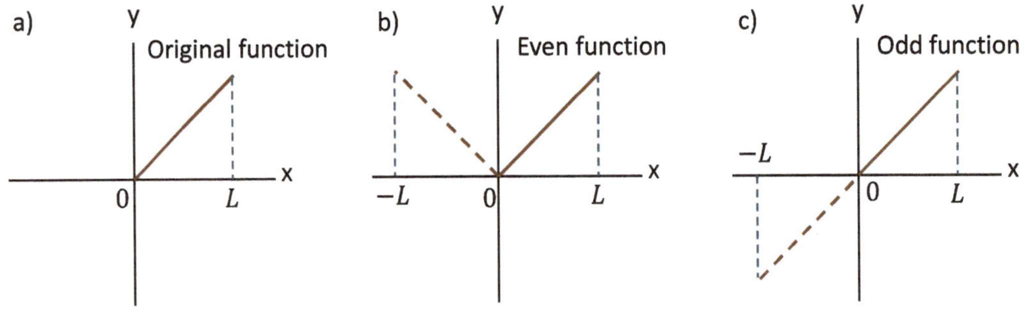

Fig. 3.4 Half range expansion. **a** original function, **b** virtual even function, **c** virtual odd function

on the virtual half. It might exhibit symmetry about the y–axis (Fig. 3.4b), or symmetry about the origin (Fig. 3.4c). In the former, the function becomes even, allowing the use of the Fourier cosine series, as expressed in (3.22)–(3.23). In the latter, the function becomes odd, and the Fourier sine series, given in (3.24)–(3.25), is applicable.

Noticing that the cosine and sine coefficients of a function periodic in the range $-L < x < L$ in (3.23) and (3.25) are defined on $0 < x < L$. Thus, for a half range function defined on $0 < x < L$, we simply set

$$
\begin{aligned}
a_0 &= \frac{2}{L}\int_0^L f(x)\,dx \\
a_n &= \frac{2}{L}\int_0^L f(x)\cos\left(\frac{n\pi x}{L}\right)dx, \quad n = 1, 2, 3, \ldots \\
b_n &= \frac{2}{L}\int_0^L f(x)\sin\left(\frac{n\pi x}{L}\right)dx, \quad n = 1, 2, 3, \ldots
\end{aligned}
$$

If, for example, the function in Fig. 3.4 is expressed as $f(x) = x$ in the interval $(0, L)$, then the coefficients of its expansion as an even function are

$$
\begin{aligned}
a_0 &= \frac{2}{L}\int_0^L x\,dx = L, \\
a_n &= \frac{2}{L}\int_0^L x\cos\left(\frac{n\pi x}{L}\right)dx = \frac{2L\,[\cos(n\pi) - 1]}{n^2\pi^2} = -\frac{4L}{n^2\pi^2}, \quad n = 1, 3, \cdots
\end{aligned}
$$

And the coefficients of its expansion as an odd function are

$$
b_n = \frac{2}{L}\int_0^L x\sin\left(\frac{n\pi x}{L}\right)dx = \frac{2L}{n\pi}, \quad n = 1, 2, \cdots
$$

3.3.5 Spatial Domain Versus Temporal Domain

In mathematics, the nature of a variable being a function of space or time holds equal significance. However, when dealing with practical applications in physics and engineering, the

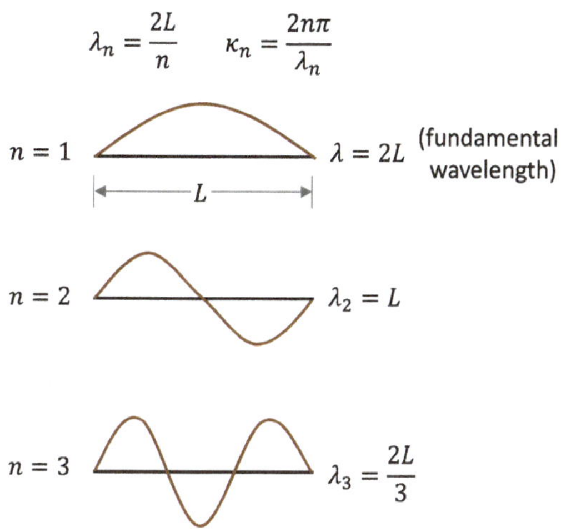

Fig. 3.5 Three eigenmodes of a sine function in the interval 0 to L

specific dependence of a variable becomes a crucial consideration.

In the preceding discussion, $f(x)$ is expanded in the spatial domain. In (3.4), for instance, the involved Fourier sine series exhibits a periodicity of L, and is defined in terms of the eigenvalues $\kappa_n = n\pi/L$. Figure 3.5 shows plots of the first three modes of the eigenfunctions with $n = 1, 2, 3$ in the interval $(0, L)$. The figure shows that these modes are analogous to what is physically occurring in standing waves in a string. As such, the eigenvalues κ_n in here denote the *wavenumbers*, which are functions of the *wavelengths* λ_n, such that

$$\kappa_n = \frac{n\pi}{L} = \frac{2n\pi}{\lambda_n}, \quad n = 1, 2, 3 \ldots$$

with

$$\begin{aligned} \kappa &= \frac{2\pi}{\lambda} && \text{(fundamental wavenumber)} \\ \lambda &= 2L && \text{(fundamental wavelenght)} \\ \lambda_n &= \frac{\lambda}{n} && \text{(wavelenght)} \end{aligned} \tag{3.26}$$

The wavenumber and the wavelength in the spatial domain have counterparts in the time domain; namely the *radial frequency* and the signal *period*, respectively. They are defined as

3

$$\begin{aligned}\omega_n &= n\omega = \frac{2n\pi}{T} \\ &= \frac{2n\pi}{N\Delta T}, \quad n = 1, 2, 3 \ldots N\end{aligned}$$

with

$$\omega = 2\pi f = \frac{2\pi}{T} \quad \text{(fundamental frequency)}$$

f : wave frequency

T : signal period

ΔT : sampling rate

(3.27)

Following these, Fourier series in space and time can be expressed as follows:

Spatial Fourier series defined over the interval $(0, \boldsymbol{\lambda} = 2\boldsymbol{L})$:

Trigonometric Fourier series:

$$\begin{aligned}u(x) &= \frac{a_0}{2} + \sum_{n=1}^{\infty} a_n \cos(\kappa_n x) + \sum_{n=1}^{\infty} b_n \sin(\kappa_n x) \\ &= \frac{a_0}{2} + \sum_{n=1}^{\infty} a_n \cos\left(\frac{2n\pi x}{\lambda}\right) + \sum_{n=1}^{\infty} b_n \sin\left(\frac{2n\pi x}{\lambda}\right)\end{aligned} \tag{3.28}$$

with

$$\begin{aligned}a_0 &= \frac{2}{\lambda}\int_0^{\lambda} u(x)\,dx \\ a_n &= \frac{2}{\lambda}\int_0^{\lambda} u(x)\cos\left(\frac{2n\pi x}{\lambda}\right)dx, \quad n = 1, 2, 3, \ldots \\ b_n &= \frac{2}{\lambda}\int_0^{\lambda} u(x)\sin\left(\frac{2n\pi x}{\lambda}\right)dx\end{aligned} \tag{3.29}$$

Complex Fourier series:

$$\begin{aligned}u(x) &= \sum_{n=-\infty}^{\infty} c_n\, e^{ik_n x} \\ &= \sum_{n=-\infty}^{\infty} c_n\, e^{2in\pi x/\lambda}\end{aligned} \tag{3.30}$$

with

$$\begin{aligned}c_n &= \frac{1}{\lambda}\int_0^{\lambda} u(x)\, e^{-ik_n x} dx \\ &= \frac{1}{\lambda}\int_0^{\lambda} u(x)\, e^{-2in\pi x/\lambda} dx, \quad n = 0, \pm 1, \pm 2, \cdots\end{aligned} \tag{3.31}$$

Temporal Fourier series defined over the interval $(0, T)$:

Trigonometric Fourier series:

$$\begin{aligned} u(t) &= \frac{a_0}{2} + \sum_{n=1}^{\infty} a_n \cos(\omega_n t) + \sum_{n=1}^{\infty} b_n \sin(\omega_n l) \\ &= \frac{a_0}{2} + \sum_{n=1}^{\infty} a_n \cos\left(\frac{2n\pi t}{T}\right) + \sum_{n=1}^{\infty} b_n \sin\left(\frac{2n\pi t}{T}\right) \end{aligned} \tag{3.32}$$

with

$$\begin{aligned} a_0 &= \frac{2}{T} \int_0^T u(t)\, dt \\ a_n &= \frac{2}{T} \int_0^T u(t) \cos\left(\frac{2n\pi t}{T}\right) dt, \quad n = 1, 2, 3, \ldots \\ b_n &= \frac{2}{T} \int_0^T u(t) \sin\left(\frac{2n\pi t}{T}\right) dt \end{aligned} \tag{3.33}$$

Following the definitions in (3.27), the first equation in (3.32) can also be written as

$$u(t) = \frac{a_0}{2} + \sum_{n=1}^{\infty} a_n \cos(n\omega\, t) + \sum_{n=1}^{\infty} b_n \sin(n\omega\, t)$$

where ω is the fundamental radial frequency.

Complex Fourier series:

$$\begin{aligned} u(t) &= \sum_{n=-\infty}^{\infty} c_n\, e^{i\omega_n t} \\ &= \sum_{n=-\infty}^{\infty} c_n\, e^{2in\pi t/T} \end{aligned} \tag{3.34}$$

with

$$\begin{aligned} c_n &= \frac{1}{T} \int_0^T u(t)\, e^{-i\omega_n t} dt \\ &= \frac{1}{T} \int_0^T u(t)\, e^{-2in\pi t/T} dt, \quad n = 0, \pm 1, \pm 2, \cdots \end{aligned} \tag{3.35}$$

Example 3.3: Fourier coefficients determination

Consider the following function:

$$f(x) = \begin{cases} 5 & 0 < x < 3 \\ -5 & -3 < x < 0 \end{cases}, \quad \text{Period} = 6$$

shown in Fig. 3.6. Using the trigonometric Fourier series:

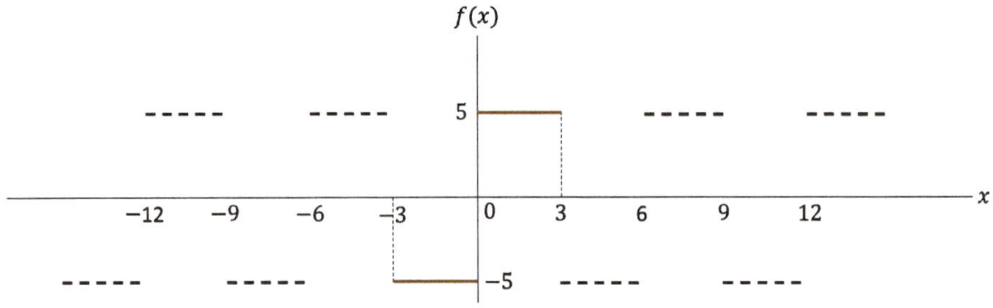

Fig. 3.6 Graph of this function

1 Determine the Fourier coefficients and write the corresponding Fourier series.
2 Define the function at the discontinuities $x = -3$, $x = 0$, and $x = 3$ such that the Fourier series will converge for $-3 \leq x \leq 3$.

▪ Solution

1 The Fourier coefficients are:

$$
\begin{aligned}
a_0 &= \frac{1}{L}\int_{-L}^{L} f(x)\, dx \\
&= \frac{1}{3}\left[\int_{-3}^{0} -5\, dx + \int_{0}^{3} 5\, dx\right] \\
&= 0 \\
a_n &= \frac{1}{L}\int_{-L}^{L} f(x)\cos\left(\frac{n\pi x}{L}\right) dx \\
&= \frac{1}{3}\left[\int_{-3}^{0} \left[-5\cos\left(\frac{n\pi x}{3}\right)\right] dx + \int_{0}^{3} \left[5\cos\left(\frac{n\pi x}{3}\right)\right] dx\right] \\
&= 0 \\
b_n &= \frac{1}{L}\int_{-L}^{L} f(x)\sin\left(\frac{n\pi x}{L}\right) dx \\
&= \frac{1}{3}\left[\int_{-3}^{0} \left[-5\sin\left(\frac{n\pi x}{3}\right)\right] dx + \int_{0}^{3} \left[5\sin\left(\frac{n\pi x}{3}\right)\right] dx\right] \\
&= \frac{10\left[1 - \cos(n\pi)\right]}{n\pi}
\end{aligned}
$$

The corresponding Fourier series is thus

$$
\begin{aligned}
f(x) &= \sum_n \frac{10\,[1-\cos(n\pi)\,]}{n\pi}\sin\left(\frac{\pi x}{3}\right)\\
&= \frac{20}{\pi}\left[\sin\left(\frac{\pi x}{3}\right)+\frac{1}{3}\sin(\pi x)+\frac{1}{5}\sin\left(\frac{5\pi x}{3}\right)+\frac{1}{7}\sin\left(\frac{7\pi x}{3}\right)+\cdots\right]
\end{aligned}
$$

2. According to the Dirichlet conditions of series convergence, if $f(x)$ is periodic in $(-L, L)$ and piecewise continuous in this period, together with its first derivative, then the series $f(x)$ and its coefficients converges to:

   1 $f(x)$ if x is a point of continuity, or.

   2 $\frac{f(x+0)+f(x-0)}{2}$ if x is a point of discontinuity.

This implies

$$
f(x) = \begin{cases} 0 & x=-3 \\ -5 & -3<x<0 \\ 0 & x=0 \\ 5 & 0<x<3 \\ 0 & x=3 \end{cases}
$$

Apparently, if x is a point of discontinuity, the Dirichlet conditions imply prescribing an average value for $f(x)$.

3.3.6 Frequency Spectrum

In the analysis of time-dependent signals, constructing their *frequency spectrum* is crucial because it reveals how different frequency components contribute to the overall signal. The spectrum shows the distribution of these frequencies and their corresponding magnitudes, helping identify dominant frequencies, noise, or patterns in the signal. This breakdown is important in fields like communications, audio processing, and signal analysis to interpret the underlying structure and behavior of complex signals. In particular, frequency spectrum of measured data is fundamental in solving boundary value problems involving boundary conditions with complex time-dependent signals.

If $f(t)$ is any arbitrary signal with period T, the relationship between the frequency and the complex Fourier coefficients c_n represents the frequency spectrum of $f(t)$. It is closely associated with the Fourier transform. Hereafter, we give a motivation example illustrating this relationship.

■ **Example 3.4: frequency spectrum**

Find the frequency spectrum of a periodic square pulse shown in ◘ Fig. 3.7. The pulse is given by

3

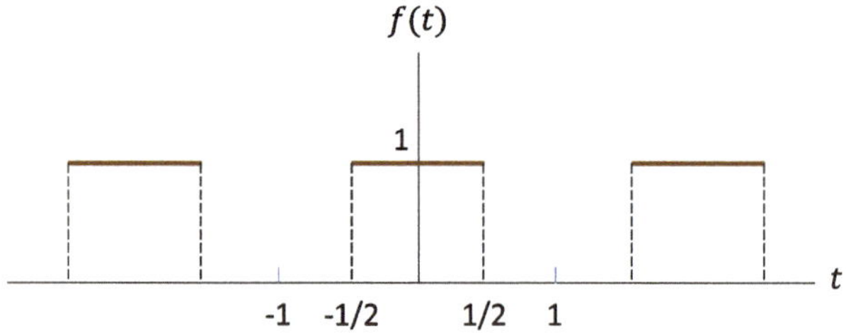

Fig. 3.7 Periodic square signal

$$f(t) = \begin{cases} 0 & -1 < t < -1/2 \\ 1 & -1/2 < t < 1/2 \\ 0 & 1/2 < t < 1 \end{cases}$$

- **Solution**

This square pulse has a period $T = 2$ and can be expanded using the complex Fourier series, such that

$$f(t) = \sum_{n=-\infty}^{\infty} c_n \, e^{2in\pi t/T} = \sum_{n=-\infty}^{\infty} c_n \, e^{i\omega_n t}$$

with its coefficients given by

$$\begin{aligned} c_n &= \frac{1}{T} \int_{-1}^{1} f(t) \, e^{-i\omega_n t} dt \\ &= \frac{1}{2} \int_{-1/2}^{1/2} 1 \cdot e^{-i\omega_n t} dt \\ &= -\frac{1}{2i\omega_n} (e^{-i\omega_n/2} - e^{i\omega_n/2}) \\ &= \frac{1}{\omega_n} \sin\left(\frac{\omega_n}{2}\right), \quad n = \pm 1, \pm 2, \cdots \end{aligned}$$

Note that we calculated the coefficients for $n \neq 0$, since for $n = 0$, this form will lead to infinity. Thus we determine the Fourier coefficient for $n = 0$, directly from

$$\begin{aligned} c_0 &= \frac{1}{2} \int_{-1/2}^{1/2} 1 \, dt \\ &= \frac{1}{2} \end{aligned}$$

which is the average magnitude.

Figure 3.8 shows the frequency spectrum of $f(t)$ of this pulse, where we plotted the magnitude of the Fourier coefficient $|c_n|$ versus $\omega_n = 2n\pi/T$.

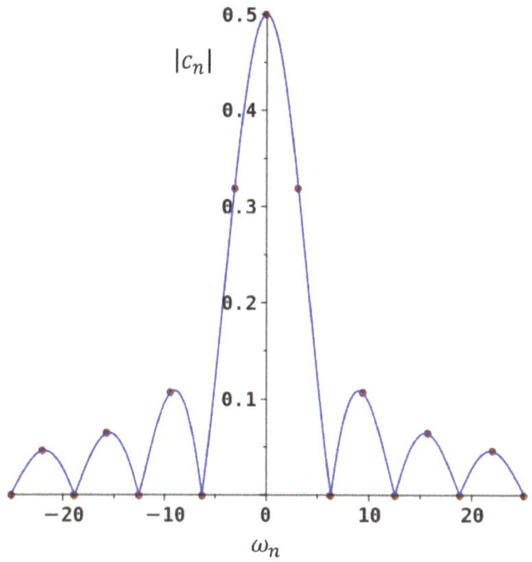

Fig. 3.8 Frequency spectrum of the square pulse

3.4 Fourier–Bessel Series

Similar to the Fourier series, where functions can be expanded in terms of an orthogonal set of trigonometric function, it is also possible to expand functions in terms of orthogonal Bessel functions.

In ▶ Sect. 2.7, it was shown that the parametric Bessel equation in the polar coordinate system is given by

$$r^2\frac{d^2u}{dr^2} + r\frac{du}{dr} + \left(\kappa^2 r^2 - \nu^2\right)u = 0 \tag{3.36}$$

And, its general solution is expressed by the Bessel functions, such that

$$u(r) = c_1 J_\nu(\kappa r) + c_2 Y_\nu(\kappa r)$$

where $J_\nu(\kappa r)$ is the *Bessel function of the first kind* of order ν, and $Y_\nu(\kappa r)$ is the *Bessel function of the second kind* of order ν. If the domain is bounded in the range $(0, R)$, the general solution becomes

$$u(r) = c\, J_\nu(\kappa r)$$

because $Y_\nu(\kappa r)$ is unbounded at $r = 0$. Thus, the orthogonality

$$\int_0^R J_\nu(\kappa_n r)\, J_\nu(\kappa_m r)\, r\, dr = 0, \quad m \neq n \tag{3.37}$$

holds provided that the eigenvalues κ^2 are defined by means of the homogeneous boundary condition at $r = R$, as typically

3

involved in the Sturm–Liouville boundary value problems. Accordingly, if we have a piecewise smooth function $f(r)$ on interval $(0, R)$, the generalized Fourier series can be expanded to an infinite series of the form

$$f(r) = \sum_{n=1}^{\infty} c_n J_\nu(\kappa_n r) \tag{3.38}$$

which is known as the *Fourier–Bessel series of order* ν of $f(r)$, for which κ_n are the scaled positive roots of the Bessel function J_ν.

Through the eigenfunction expansion (generalized Fourier series), we can readily determine the coefficients c_n, by first multiplying both sides of (3.38) by $J_\nu(\kappa_n r)$, and then, upon applying the orthogonality condition (3.37), it yields

$$c_n = \frac{\int_0^R r f(r)\, J_\nu(\kappa_n r)\, dr}{\int_0^R r J_\nu^2(\kappa_n r)\, dr} \tag{3.39}$$

The integration on the denominator is determined based on the type of boundary conditions and the resulting eigenfunctions. Recall the second boundary condition of the Sturm–Liouville BVP (see ▶ Chap. 2)

$$A_2 J_\nu(\kappa R) + B_2 \kappa J_\nu'(\kappa R) = 0 \tag{3.40}$$

The following three cases are most common:

Case 1:

If $A_2 = 1$ and $B_2 = 0$ in (3.40), the resulting eigenfunction is $J_\nu(\kappa_n r)$ and the eigenvalues are determined by solving for the roots of

$$J_\nu(\kappa_n R) = 0$$

Leading the denominator to become

$$\int_0^R r J_\nu^2(\kappa_n r)\, dr = \frac{R^2}{2} J_{\nu+1}^2(\kappa_n R)$$

Substituting this identity into (3.39), the Bessel coefficients of this case are thus given by

$$c_n = \frac{2}{R^2 J_{\nu+1}^2(\kappa_n R)} \int_0^R r f(r)\, J_\nu(\kappa_n r)\, dr$$

Case 2:

If $A_2 = h > 0$ and $B_2 = R$ in (3.40), the resulting eigenfunction is $h J_\nu(\kappa_n r) + \kappa_n R J_\nu\prime(\kappa_n r)$ and the eigenvalues are determined by solving for the roots of

$$h J_\nu(\kappa_n R) + \kappa_n R J_\nu\prime(\kappa_n R) = 0$$

leading to

$$\int_0^R r J_\nu^2(\kappa_n r)\, dr = \frac{(\kappa_n^2 R^2 - \nu^2 + h^2)}{2\kappa_n^2} J_\nu^2(\kappa_n R)$$

And the Fourier–Bessel coefficients become

$$c_n = \frac{2\kappa_n^2}{(\kappa_n^2 R^2 - \nu^2 + h^2)\, J_\nu^2(\kappa_n R)} \int_0^R r f(r) J_\nu(\kappa_n x)\, dr$$

Case 3:

If $A_2 = 0$, $B_2 = R$ in (3.40), and the Bessel order $\nu = 0$, the resulting eigenfunction is $J_0'(\kappa_n r)$ and the eigenvalues are determined by solving for the roots of

$$J_0'(\kappa_n R) = J_1(\kappa_n R) = 0$$

In this case, the series becomes

$$f(r) = c_0 + \sum_{n=1}^{\infty} c_n J_0(\kappa_n r)$$

with the Fourier–Bessel coefficients

$$c_0 = \frac{2}{R^2} \int_0^R r f(r)\, dr$$

$$c_n = \frac{2}{R^2 J_0^2(\kappa_n R)} \int_0^R r f(r)\, J_0(\kappa_n r)\, dr$$

The coefficients of these three cases are called the *Fourier–Bessel series coefficients*.

3.5 Fourier Integral

In the previous sections we discussed how Fourier series expansions were used to represent periodic functions, defined over finite intervals, such as $(-L, L)$ or $(0, T)$. In practice, however, functions, particularly measured signals, are arbitrarily distributed over time or space, and are commonly non-periodic. Figure 3.9 shows an example of a periodic rectangular function and its non-periodic counterpart. To make use of the orthogonal expansion of the Fourier series, the period of the non-periodic function is envisioned to approach infinity, i.e. $-\infty < 2L < \infty$. In this case, the summation of the series is replaced by integration over an infinite interval, and the Fourier series becomes a *Fourier integral*.

Conveniently, the Fourier integral possesses the same structure as the Fourier series. It allows expressing any sufficiently well-behaved function as an integral of trigonomet-

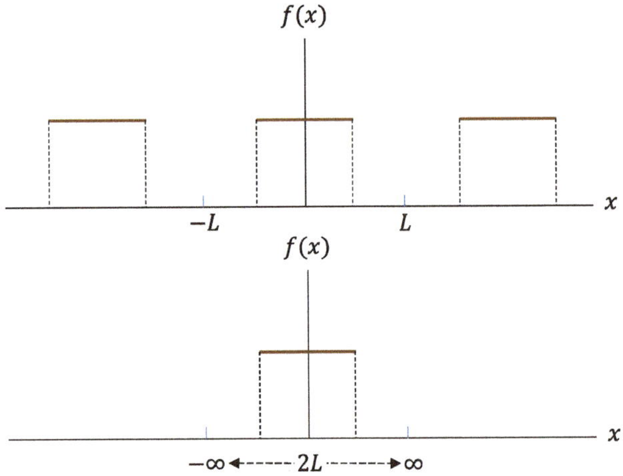

Fig. 3.9 Periodic and non-periodic functions

ric or exponential functions over the entire interval, such that, the Fourier integral of a function $f(x)$ defined on the interval $(-\infty, \infty)$ is given by

$$f(x) = \int_0^\infty [A(\alpha)\cos(\alpha x) + B(\alpha)\sin(\alpha x)]\, d\alpha$$

where

$$A(\alpha) = \frac{1}{\pi}\int_{-\infty}^{\infty} f(x)\cos(\alpha x)\, dx \tag{3.41}$$

$$B(\alpha) = \frac{1}{\pi}\int_{-\infty}^{\infty} f(x)\sin(\alpha x)\, dx$$

Similar to the Fourier series, the Fourier integral (3.41) also possesses an equivalent cosine, sine and complex form, such that:

Cosine Fourier integral

$$f(x) = \int_0^\infty A(\alpha)\cos(\alpha x)\, d\alpha$$

where (3.42)

$$A(\alpha) = \frac{2}{\pi}\int_0^\infty f(x)\cos(\alpha x)\, dx$$

Sine Fourier integral

$$f(x) = \int_0^\infty B(\alpha)\sin(\alpha x)\, d\alpha$$

where (3.43)

$$B(\alpha) = \frac{2}{\pi}\int_0^\infty f(x)\sin(\alpha x)\, dx$$

Complex (Exponential) Fourier integral

$$f(x) = \int_{-\infty}^{\infty} C(\alpha)\, e^{-i\alpha x}\, d\alpha$$

where (3.44)

$$C(\alpha) = \frac{1}{2\pi} \int_{-\infty}^{\infty} f(x)\, e^{i\alpha x}\, dx$$

Example 3.4: Fourier integral

Find the Fourier integral representation of

$$f(x) = \begin{cases} 1 & |x| < b \\ 0 & |x| > b \end{cases}$$

Solution

From (3.41) the Fourier coefficients are

$$A(\alpha) = \frac{1}{\pi} \int_{-b}^{b} \cos(\alpha x)\, dx$$

$$= \frac{1}{\pi} \frac{\sin(\alpha x)}{\alpha}\bigg|_{-b}^{b} = \frac{2}{\pi} \frac{\sin(\alpha b)}{\alpha}$$

$$B(\alpha) = \frac{1}{\pi} \int_{-b}^{b} \sin(\alpha x)\, dx$$

$$= -\frac{1}{\pi} \frac{\cos(\alpha x)}{\alpha}\bigg|_{-b}^{b} = 0$$

Thus

$$f(x) = \frac{2}{\pi} \int_{0}^{\infty} \frac{1}{\alpha} \sin(\alpha b)\, \cos(\alpha x)\, d\alpha \tag{3.45}$$

If $b = 2$, for example, we can find that for $-2 < x < 2$, the integral in (3.45) will produce $f(x) = 1$, and at $x = \pm 2$, this integral gives $f(x) = 1/2$, as expected on a discontinuity point.

Note that the integral in (3.45) is quite complex and involves special functions and conditions on x, such that the function behaves differently based on the value of x. In particular, the solution of this integral leads to the *Sine Integral function* (Si), and this can be discretized by piecewise evaluation of the Si function.

3.6 Fourier Transform

The Fourier cosine, sine and complex integrals (3.42)–(3.44) are basically a sort of integral transforms, often referred to as *Fourier transforms*. The Fourier integral is essentially a continuous version of the Fourier transform, known as the *continuous Fourier transform* (CFT). In the above examined cases, it facilitates the transformation of $f(x)$ from the spatial domain x to the α domain.

Commonly, the Fourier transform is utilized to transform time-dependent functions into functions expressed in the frequency domain. For example, the continuous Fourier transform of a non-periodic, time-dependent function $f(t)$ is expressed as

$$\begin{aligned} \hat{f}(\omega) &= \int_{-\infty}^{\infty} f(t)\, e^{-i\omega t}\, dt \qquad &\text{(Forward CFT)} \\ f(t) &= \frac{1}{2\pi} \int_{-\infty}^{\infty} \hat{f}(\omega)\, e^{i\omega t}\, d\omega \qquad &\text{(Inverse CFT)} \end{aligned} \tag{3.46}$$

where the first expression represents the *forward Fourier transform*, while the second is the *inverse Fourier transform*. These two expressions form what is known as *Fourier transform pairs*. In the forward transform, the function $f(t)$ in transformed from the time domain to the frequency domain, and in the inverse transform, the function $\hat{f}(\omega)$ is transformed back from the frequency domain to the time domain.

Comparing (3.46) with (3.44) we can easily observe the analogy between the two formulations, where the concept of Fourier transform pairs is introduced. In addition to the complex form, the Fourier transforms also have equivalent cosine and sine forms. However, our primary focus in this book will be on the complex form. As we will discuss later, the Fourier transform plays a fundamental role in both signal analysis and the solution of boundary value problems.

It is worth mentioning that variations in sign convention can be found in different sources and are sometimes a matter of convention or convenience in specific applications. The choice of sign convention is often related to the definition of the complex unit in the context of the Fourier transform. For instance, the Fourier transform pairs (3.46) are written in some literature as

$$\begin{aligned} \hat{f}(\omega) &= \int_{-\infty}^{\infty} f(t)\, e^{i\omega t}\, dt \\ f(t) &= \frac{1}{2\pi} \int_{-\infty}^{\infty} \hat{f}(\omega)\, e^{-i\omega t}\, d\omega \end{aligned}$$

And the same is applied to the scaling coefficient $1/2\pi$. Yet, what matters in here is that we must be consistent on what sort of convention we adopt.

3.6.1 Discrete Fourier Transform

The continuous transform is a powerful tool in analyzing functions, analytical in the whole time interval. However, in practice, analytic functions are rare, particularly in applied physics and engineering. Most measured time signals in engineering are digital, making it essential to have a transform technique that can effectively handle discrete data.

Consider a time function $F(t)$ with period T shown in Fig. 3.10. If the function values $F_0, F_1, \cdots, F_m, \cdots$ are measured at equally spaced points $t_0, t_1, \cdots, t_m, \cdots$ within its period T, the function is said to be discretized with discrete samples. Remarkably, the sine, cosine and exponential complex functions have the exceptional property that they are not only orthogonal over the interval of continuum functions, but also over a signal of discrete, equally spaced points. Thus, such a property allows us to expand functions given in discrete data to trigonometric or exponential series, giving rise to the *discrete Fourier transform* (DFT). In this, for a given data function $F(t)$ of N samples, the discrete form of the complex Fourier transform can be expressed as

$$\begin{aligned}
\hat{F}(\omega_n) &= \frac{1}{N}\sum_{m=0}^{N-1} F(t_m)\, e^{-i\omega_n t_m} \\
&= \frac{1}{N}\sum_{m=0}^{N-1} F(t_m)\, e^{-i\,2n\pi m/N}, \quad n = 0, 1, 2, \ldots, N-1 \qquad \text{(Forward DFT)} \\
F(t_m) &= \sum_{n=0}^{N-1} \hat{F}(\omega_n)\, e^{i\omega_n t_m} \\
&= \sum_{n=0}^{N-1} \hat{F}(\omega_n)\, e^{i\,2n\pi m/N}, \quad m = 0, 1, 2, \ldots, N-1 \qquad \text{(Inverse DFT)}
\end{aligned} \tag{3.47}$$

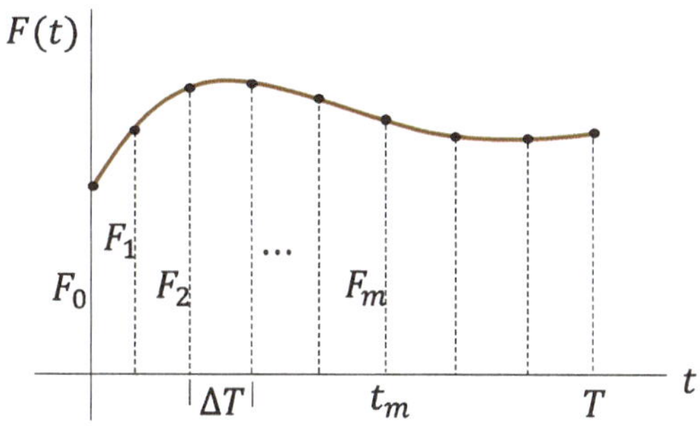

Fig. 3.10 A random time function $F(t)$

where

$F(t_m) = F_m$:	input signal value at time t_m (sec).
$t_m = m\Delta T$:	m th sampling instant (sec), $m = 0, 1, 2, \cdots, N-1$
ΔT:	sampling rate (sec).
$\hat{F}(\omega_n)$:	frequency spectrum at ω_n
$\omega = \frac{2\pi}{T} = \frac{2\pi}{N\Delta T}$:	fundamental frequency (rad/sec).
$\omega_n = n\omega = \frac{2n\pi}{N\Delta T}$:	n th frequency sample (rad/sec).
N:	number of frequency samples = number of time samples M
$T = N\Delta T$:	period of the signal.

Note that based on these definitions, the exponents in (3.47) are attained, such that

$$e^{-i\,\omega_n t_m} = e^{-i\,n\,\omega t_m} = e^{-i\,n\frac{2\pi}{N\Delta T}m\Delta T} = e^{\frac{-i\,2n\pi m}{N}}$$

In some literature these DFT pairs are written as (Doyle 2021):

$$\hat{F}(\omega_n) = \Delta T \sum_{m=0}^{N-1} F(t_m)\,e^{-i\omega_n t_m} = \Delta T \sum_{m=0}^{N-1} F(t_m)\,e^{-i\,2n\pi m/N} \quad \text{(Forward DFT)}$$

$$F(t_m) = \frac{1}{T}\sum_{n=0}^{N-1} \hat{F}(\omega_n)\,e^{i\omega_n t_m} = \frac{1}{N\Delta T}\sum_{n=0}^{N-1} \hat{F}(\omega_n)\,e^{i\,2n\pi m/N} \quad \text{(Inverse DFT)}$$

where both m and n range from 0 to $N-1$.

▪ Example 3.5: Discrete Fourier transform

Consider a digital data measured at equal time intervals with $N = 4$ samples. The magnitude of the measured data for each time step are given in this table.

m	0	1	2	3
f_m	1	2	2	1

Find the corresponding frequency domain $\hat{f}(\omega)$ using the DFT.

▪ Solution

According to (3.47), the forward transform is obtained by means of

$$\hat{f}(\omega_n) = \frac{1}{N}\sum_{m=0}^{N-1} f(t_m)\,e^{-i\,2n\pi m/N}, \quad n = 0, 1, 2, \ldots, N-1$$

This implies

$$\hat{f}(0) = \frac{1}{4}[1 \cdot e^{-i2 \cdot 0 \cdot \pi \cdot 0/4} + 2 \cdot e^{-i2 \cdot 0 \cdot \pi \cdot 1/4} + 2 \cdot e^{-i2 \cdot 0 \cdot \pi \cdot 2/4} + 1 \cdot e^{-i2 \cdot 0 \cdot \pi \cdot 3/4}] = \frac{3}{2}$$

$$\hat{f}(1) = \frac{1}{4}[1 \cdot e^{-i2 \cdot 1 \cdot \pi \cdot 0/4} + 2 \cdot e^{-i2 \cdot 1 \cdot \pi \cdot 1/4} + 2 \cdot e^{-i2 \cdot 1 \cdot \pi \cdot 2/4} + 1 \cdot e^{-i2 \cdot 1 \cdot \pi \cdot 3/4}] = -\frac{1}{4} - \frac{i}{4}$$

$$\hat{f}(2) = \frac{1}{4}[1 \cdot e^{-i2 \cdot 2 \cdot \pi \cdot 0/4} + 2 \cdot e^{-i2 \cdot 2 \cdot \pi \cdot 1/4} + 2 \cdot e^{-i2 \cdot 2 \cdot \pi \cdot 2/4} + 1 \cdot e^{-i2 \cdot 2 \cdot \pi \cdot 3/4}] = 0$$

$$\hat{f}(3) = \frac{1}{4}[1 \cdot e^{-i2 \cdot 3 \cdot \pi \cdot 0/4} + 2 \cdot e^{-i2 \cdot 3 \cdot \pi \cdot 1/4} + 2 \cdot e^{-i2 \cdot 3 \cdot \pi \cdot 2/4} + 1 \cdot e^{-i2 \cdot 3 \cdot \pi \cdot 3/4}] = -\frac{1}{4} + \frac{i}{4}$$

An important feature of this transform is that it is complex and the value of $\hat{f}(2)$ at $\frac{1}{2}N + 1$ is the *Nyquist frequency*, beyond which the transform generates conjugate values.

If we continue calculating the frequency values beyond $N - 1$, we obtain

$$\hat{f}(4) = \frac{1}{4}[1 \cdot e^{-i2 \cdot 4 \cdot \pi \cdot 0/4} + 2 \cdot e^{-i2 \cdot 4 \cdot \pi \cdot 1/4} + 2 \cdot e^{-i2 \cdot 4 \cdot \pi \cdot 2/4} + 1 \cdot e^{-i2 \cdot 4 \cdot \pi \cdot 3/4}] = \frac{3}{2}$$

$$\hat{f}(5) = \frac{1}{4}[1 \cdot e^{-i2 \cdot 5 \cdot \pi \cdot 0/4} + 2 \cdot e^{-i2 \cdot 5 \cdot \pi \cdot 1/4} + 2 \cdot e^{-i2 \cdot 5 \cdot \pi \cdot 2/4} + 1 \cdot e^{-i2 \cdot 5 \cdot \pi \cdot 3/4}] = -\frac{1}{4} - \frac{i}{4}$$

$$\hat{f}(6) = \frac{1}{4}[1 \cdot e^{-i2 \cdot 6 \cdot \pi \cdot 0/4} + 2 \cdot e^{-i2 \cdot 6 \cdot \pi \cdot 1/4} + 2 \cdot e^{-i2 \cdot 6 \cdot \pi \cdot 2/4} + 1 \cdot e^{-i2 \cdot 6 \cdot \pi \cdot 3/4}] = 0$$

which is obviously repeating itself.

Also, should we apply the inverse DFT in (3.47) on this outcome, we must obtain the original signal. The reader is encouraged to do this useful exercise as it gives an insight into how this simple, but powerful operation is achieved.

3.6.2 Fast Fourier Transform

The essence of the discrete Fourier transform is that it is suitable for utilization in computer software. Its main advantage is that it allows working with measured data in an equally feasible manner as working with analytical functions. The drawback of the DFT, however, is that the task of performing the series summation can be very large and time consuming. This has motivated scientists to find a well-organized way to conduct the series summation. Cooley and Tukey (1965) (see Brigham 1988) are credited for introducing the modern version of the *Fast Fourier Transform* (FFT) algorithm for efficiently performing the summation operation.

The FFT algorithm is designed to rearrange the terms of the discrete Fourier transform in an exceptionally effective manner, in such a way that this arrangement would drastically reduce the number of the required numerical operations. Since then, the fast Fourier transform has revolutionized the spectral analysis method, and other method involving large integers and polynomial multiplications. It has also facilitated the use of the Fourier transform for solving large systems of differential equations.

The Forward-inverse pair of the fast Fourier transform is analogous to that of the DFT, stated in (3.47). When N is a product of integers, the forward function $\hat{F}(t_m)$ of (3.47) is proved to be closely interdependent. This interdependence has been utilized intelligently to substantially reduce the amount of computation. To illustrate the computational efficiency of FFT, we need to consider the number of complex multiplication and additions required for a data with N samples. Using DFT, the direct calculation requires N^2 multiplications and $N(N-1)$ additions. However, when N is a power of 2, the FFT algorithm of Cooley and Tukey reduces the number of multiplications to $N/2\log_2 N$ and the number of additions to$N\log_2 N$. For example, with $N = 1024 = 2^{10}$ samples, the FFT achieves a computational reduction by a factor of over 130.

FFT is suitable for analyzing both measured and computed data signals. When dealing with a non-periodic measured data function $F(t)$, which is often the case, it is still possible to construct an FFT approximation. In this case, the function can be conceptualized to extend periodically beyond the interested interval. This is done by multiplying the signal function by a data window. In signal analysis, this technique is referred to as *windowing* where the measured signal is truncated, and only a small portion of the "imagined" infinite trace is visible through the window. Commonly used time windows include the *rectangle* window and the *Hanning* window.

In windowing, it is important to sample and truncate the signal properly. Otherwise, upon the inverse transform, the reconstructed signal might exhibit significant disturbances. Commonly encountered phenomena related to erroneous signal analysis are *aliasing* and *leakage*.

Aliasing

This phenomenon occurs when a discrete signal is truncated by a sampling time window with its bandwidth smaller than the highest frequency of the signal. The highest frequency that can be detected by DFT from a discrete and real time series is the *Nyquist frequency*, given by

$$f_{\mathrm{Nq}} = \frac{1}{2\,\Delta t} \tag{3.48}$$

in which Δt is the sample rate. For example, if the sample rate is 10 *s*, the Nyquist frequency is 0.05 Hz.

To perform an accurate signal analysis, the Nyquist frequency must be equal to or larger than the highest frequency $f_{\max}$ of the signal. From (3.48) we can readily deduce that the sample time interval must be within

$$\Delta t \le \frac{1}{2 f_{\max}} \tag{3.49}$$

which is called the *Nyquist condition*. Violation of this condition results in a distortion of the reconstructed time signal.

In practice, signals have finite durations and their frequency contents have no upper bound. Hence, some amount of aliasing always occurs when such functions are sampled. If the function is sampled at a high enough rate, determined by the bandwidth $[f_{\min}, f_{\max}]$ and the Nyquist condition (3.49), the original function can in theory be perfectly reconstructed from an infinite set of samples. Otherwise, higher frequency components merge into the frequency spectrum, inevitably leading to error in the spectral analysis of the system. Figure 3.11 shows a typical aliasing phenomenon in a sinusoidal function when using a sampling time interval larger than that specified by (3.49), and a Nyquist frequency (3.48) less than the maxi-

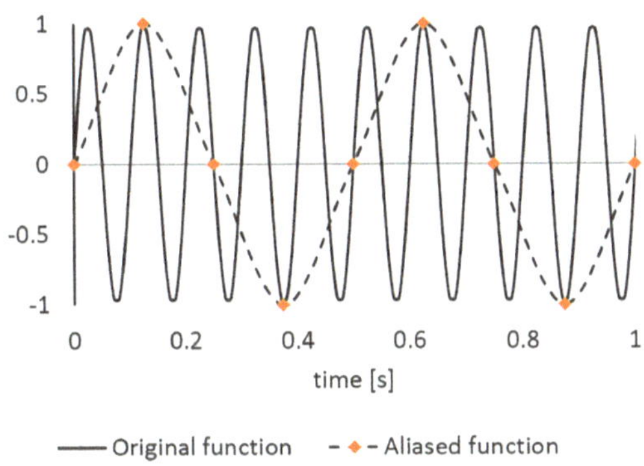

Fig. 3.11 Aliasing of a signal

3

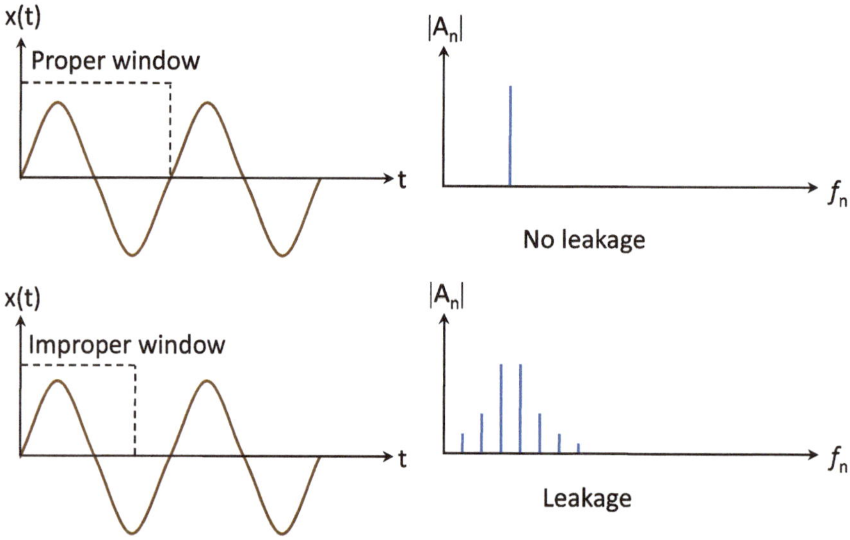

Fig. 3.12 Leakage of a signal

mum frequency, $f_{Nq} < f_{max}$. This figure illustrates that, due to aliasing, a high frequency signal appears like a low frequency signal in the reconstructed time signal.

Leakage

This phenomenon occurs when the sampling time window does not meet the periodicity of the signal. Thus, the time window must be an integer multiple of the period T of the signal, otherwise side lobs appear in the frequency domain.

The basic cause of this phenomenon is that, even if the measured signal is not periodic, the DFT treats it as a periodic function. So, if the time window matches the period of the signal exactly, the Fourier transform will be accurate. If this condition is not met, side lobes occur, and the transformed signal becomes inaccurate. Figure 3.12 shows schematically this effect.

The leakage can be minimized by changing the shape of the time window by, for example, multiplying the signal by a window function that tapers the amplitudes at the end of the signal. Leakage can also be minimized by using a triangular window, also known as *Bartlett*, or a *bell-shaped* time window. The latter has smooth discontinuities and thus produces little or no side lobes.

As the DFT represents a finite sample of an infinite signal on a finite period, a scheme for increasing the apparent period is necessary. A simple way of doing this is by *padding* the signal with zeros (Doyle 2021). This technique makes the signal flat at its end and prevents the occurrence of a wraparound error. Furthermore, padding in FFT makes the sample of the signal an integer power of 2, i.e. $N = 2^{\gamma}$.

3.7 Fourier Method

Beyond signal analysis, the Fourier method is a powerful mathematical technique for solving partial differential equations (PDEs) by expressing functions as a series of basic components. It works by transforming PDEs from the time (or spatial) domain to the frequency (or wavenumber) domain, making complex differential equation more feasible to solve. This transformation often reduces partial differential equations to ordinary differential equations, establishing the Fourier method as a versatile tool across various fields in physics and engineering.

The Fourier method is particularly effective in handling problems with periodic boundary conditions. Using the Fourier transform in (3.47), a function in space and time such as $u(x, t)$ can be decomposed into its sinusoidal components, such that

$$u(x,t) = \sum_{n} \hat{u}(x, \omega_n)\, e^{i\omega_n t} \tag{3.50}$$

where, here, the original function of space and time $u(x, t)$ has been transformed into a function of space and frequency $\hat{u}(x, \omega_n)$, dropping thus one of the physical dimensions: the time. By this, the problem for solving the partial differential equation has been transformed into finding the Fourier coefficients $\hat{u}(x, \omega_n)$, which is by far easier. This concept will be intensively adopted in this book for solving boundary value problems related to heat transfer, groundwater flow and contaminant transport.

For convenience, the transform in (3.50) is usually abbreviated as

$$u(x,t) = \sum_{n} \hat{u}_n\, e^{i\omega_n t}, \quad \text{or} \quad u(x,t) \Leftrightarrow \hat{u}(x, \omega_n), \quad \text{or} \quad u(x,t) \Leftrightarrow \hat{u}_n \tag{3.51}$$

where the symbol $\Leftrightarrow$ indicates "transformed into" and is valid in both directions, and the hat in $(\hat{u})$ denotes that the variable has been transformed.

Transient (time-dependent) boundary value problems, such as those describing heat transfer, are typically expressed in terms of space and time derivatives. The Fourier representation (also known as *spectral representation*) of the time derivative of (3.50) is

$$\frac{\partial u(x,t)}{\partial t} = \frac{\partial}{\partial t}\sum\nolimits_n \hat{u}(x,\omega_n)\, e^{i\omega_n t} = \sum\nolimits_n i\,\omega_n \hat{u}(x,\omega_n)\, e^{i\omega_n t} \tag{3.52}$$

This identity is usually abbreviated as

$$\frac{\partial u}{\partial t} \Rightarrow i\omega_n \hat{u}_n \quad \text{or} \quad i\omega\hat{u} \tag{3.53}$$

The spectral representation in (3.53) indicates that the time derivative of the state variable has been replaced by an algebraic expression, reducing thus the number of the derivatives in the governing differential equation.

Similarly, the Fourier representation of the spatial derivative of (3.50) is

$$\frac{\partial u(x,t)}{\partial x} = \frac{\partial}{\partial x}\sum\nolimits_n \hat{u}(x,\omega_n)\, e^{i\omega_n t} = \sum\nolimits_n \frac{\partial \hat{u}(x,\omega_n)}{\partial x} e^{i\omega_n t} \tag{3.54}$$

In shorthand notation, this identity can be expressed as

$$\frac{\partial u}{\partial x} \Rightarrow \frac{\partial \hat{u}_n}{\partial x}, \quad \text{or} \quad \frac{\partial \hat{u}}{\partial x} \tag{3.55}$$

where in this case there is no reduction in the order of the derivative because the transformation used in this example is specifically designed to convert the time domain into the frequency domain.

For higher order derivatives, the temporal and spatial derivatives are represented by

$$\begin{aligned} \frac{\partial^m u}{\partial t^m} &\Rightarrow i^m \omega_n^m \hat{u}_n, \quad \text{or} \quad i^m \omega^m \hat{u} \\ \frac{\partial^m u}{\partial x^m} &\Rightarrow \frac{\partial^m \hat{u}_n}{\partial x^m}, \quad \text{or} \quad \frac{\partial^m \hat{u}}{\partial x^m} \end{aligned} \tag{3.56}$$

If, however, our transform was targeting the spatial domain, we would get the spatial derivatives reduced to algebraic expressions in the *wavenumber* domain, and the time derivatives would exhibit no reduction, such that

$$\begin{aligned} \frac{\partial u}{\partial x} &\Rightarrow i\kappa_n \hat{u}_n, \quad \text{or} \quad i\kappa\hat{u} \\ \frac{\partial u}{\partial t} &\Rightarrow \frac{\partial \hat{u}_n}{\partial t}, \quad \text{or} \quad \frac{\partial \hat{u}}{\partial t} \end{aligned}$$

The boundary conditions are treated in the same manner. ◘ Table 3.1 shows some typical boundary conditions in heat transfer problems. Apparently, this transformation is straightforward, which adds to the advantages of the Fourier method.

The following example demonstrates the use of the Fourier method for solving a boundary value problem describing a one-dimensional heat flow in a semi-infinite bar.

Table 3.1 Typical boundary conditions in heat transfer problems

BC	Time domain	Frequency domain
Dirichlet	$u(0,t) = U(t)$	$\hat{u}(0,\omega) = \hat{U}(\omega)$
Neumann	$\lambda A \left.\frac{\partial u(x,t)}{\partial x}\right\vert_{x=0} = -q(t)$	$\lambda A \left.\frac{\partial \hat{u}(x,\omega)}{\partial x}\right\vert_{x=0} = -\hat{q}(\omega)$
Newton law of cooling	$\lambda \left.\frac{\partial u(x,t)}{\partial x}\right\vert_{x=0} = \overline{h}\left[u(0,t) - U_\infty(t)\right]$	$\lambda \left.\frac{\partial \hat{u}(x,\omega)}{\partial x}\right\vert_{x=0} = \overline{h}\left[\hat{u}(0,\omega) - \hat{U}_\infty(\omega)\right]$

Example 3.6: Fourier method for solving 1D heat conduction

The boundary value problem of heat conduction in a semi-infinite 1D bar subjected to a time-dependent temperature $U(t)$ at $x = 0$ is given by

$$\frac{1}{\alpha}\frac{\partial u(x,t)}{\partial t} = \frac{\partial^2 u(x,t)}{\partial x^2}, \quad 0 \le x < \infty, \quad t \ge 0$$

$$\text{IC}: \quad u(x,0) = 0$$
$$\text{BC}: \quad u(0,t) = U(t) \tag{3.57}$$

where α is a real and positive material constant. Solve this BVP using the Fourier method.

Solution

Using the discrete Fourier transform, the temperature in the time domain can be transformed to the frequency domain via

$$u(x,t) = \sum_n \hat{u}(x,\omega_n)\, e^{i\omega_n t}, \quad \text{or} \quad u(x,t) = \sum_n \hat{u}_n e^{i\omega_n t}, \quad \text{or} \quad u(x,t) \Rightarrow \hat{u}_n \tag{3.58}$$

In the same way, the boundary condition at $x = 0$ can be transformed via

$$U(t) = \sum_n \hat{U}(\omega_n)\, e^{i\omega_n t}, \quad \text{or} \quad U(t) = \sum_n \hat{U}_n e^{i\omega_n t}, \quad \text{or} \quad U(t) \Rightarrow \hat{U}_n \tag{3.59}$$

Substituting (3.58) into the heat equation in (3.57), yields

$$\frac{1}{\alpha}\frac{\partial}{\partial t}\left(\sum_n \hat{u}_n e^{i\omega_n t}\right) = \frac{\partial^2}{\partial x^2}\left(\sum_n \hat{u}_n e^{i\omega_n t}\right) \tag{3.60}$$

Conducting the time and spatial derivatives (see (3.52) and (3.56)), this equation can be expressed as

$$\frac{1}{\alpha}\sum_n i\omega_n \hat{u}_n e^{i\omega_n t} = \sum_n \frac{\partial^2 \hat{u}_n}{\partial x^2} e^{i\omega_n t} \tag{3.61}$$

This has to be true for all n, giving, after re-arranging

$$\frac{d^2 \hat{u}_n}{dx^2} - \frac{i\omega_n}{\alpha}\hat{u}_n = 0 \tag{3.62}$$

where it is obvious that the partial differential equation in (3.57) has been reduced to an ordinary differential equation with its time derivative term becoming algebraic.

As the coefficient of this equation is constant, the trial solution can be

$$\hat{u}_n = Ae^{\kappa x} \tag{3.63}$$

Substituting this expression into (3.62) gives the characteristic equation

$$\kappa^2 - \frac{i\omega_n}{\alpha} = 0 \tag{3.64}$$

This yields

$$\kappa = \pm\sqrt{\frac{i\omega_n}{\alpha}}$$

where here κ represents the *eigenvalues* of the system. In this case, we have two eigenvalues:

$$\kappa_1 = -\sqrt{\frac{i\omega_n}{\alpha}}, \quad \kappa_2 = \sqrt{\frac{i\omega_n}{\alpha}}$$

This relationship between eigenvalues and frequencies is commonly referred to as the *spectrum relation* of the system.

For every eigenvalue there is a corresponding eigenfunction, and the general solution of (3.62) can thus be expressed as

$$\hat{u}(x, \omega_n) = A_1 e^{\kappa_1 x} + A_2 e^{\kappa_2 x}$$

As the rod is semi-infinite, the second term on the right-hand side becomes unbounded at far distances, leading the general solution to be

$$\hat{u}(x, \omega_n) = A_1 e^{\kappa_1 x}, \quad \kappa_1 = -\sqrt{\frac{i\omega_n}{\alpha}}$$

Applying the boundary condition at $x = 0$, gives

$$\hat{u}(0, \omega_n) = \hat{U}_0 = A_1$$

Thus, the solution of the BVP (3.57) in the frequency domain is

$$\hat{u}(x, \omega_n) = \hat{U}_0 e^{\kappa_1 x}, \quad \text{or} \quad \hat{u}_n = \hat{U}_0 e^{\kappa_1 x} \tag{3.65}$$

Having solved the problem in the frequency domain, we are now ready to reconstruct the solution to the original time domain. Applying the inverse Fourier transform (see (3.50)), the solution of this BVP can then be expressed as

$$
\begin{aligned}
u(x,t) &= \sum_n \hat{u}_n\, e^{i\omega_n t} \\
&= \sum_n \hat{U}_0 e^{\kappa_1 x}\, e^{i\omega_n t} \\
&= \sum_n \hat{U}_0 e^{-\sqrt{\frac{i\omega_n}{\alpha}}x} e^{i\omega_n t}
\end{aligned}
\tag{3.66}
$$

This solution states that the heat in the bar is propagating in the $x > 0$ direction, away from the source, and as κ_1 is not linearly dependent on ω, the heat flow is dissipative.

In general, for problems with large number of eigenvalues, the solution can have the following spectral representation:

$$
u(x,t) = \sum_n \sum_m A_{mn} \hat{u}\,(\kappa_{mn} x)\, e^{i\omega_n t_m} \tag{3.67}
$$

where A_{mn} is the *amplitude spectrum*. This double summation might be computationally demanding, but the use of the FFT algorithm for summing over the frequencies ω_n facilitates the computation.

The use of the Fourier method together with the FFT for solving boundary value problems possesses several advantages, including:

(a) The behaviour of the time-dependent system is analysed at each frequency independently,
(b) The summation of the Fourier series is done in an exceptionally efficient manner,
(c) It is possible to handle measured data with any arbitrary variation in time, and
(d) Its inverse transform is straightforward, unlike other transform techniques, such as the Laplace transform.

Example 3.7: Heat flow in a semi-infinite domain

A semi-infinite solid bar is subjected to a rectangular temperature pulse, $U_0(t)$, applied at its boundary, $x = 0$ m, for nearly 200 h. The magnitude and time variation of the applied temperature are depicted in Fig. 3.13.

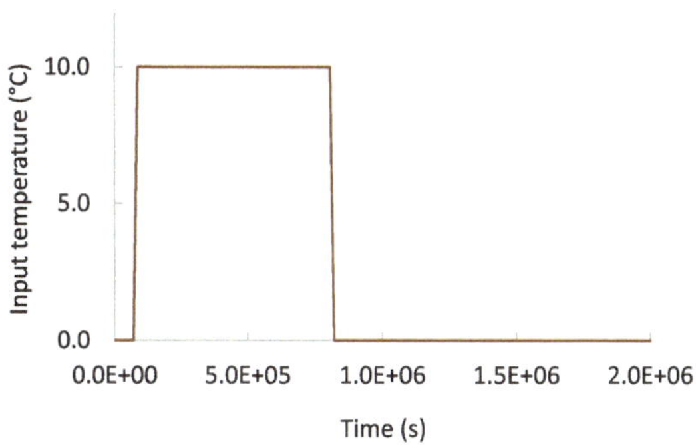

Fig. 3.13 Prescribed temperature at $x = 0$ m

3

Implement the solution derived in the previous example in a computer code, and find the temporal temperature distribution at $x = 0, 0.1, 0.5, 1.0$ and 2.0 m.

Solution

The temperature distribution along the bar at frequency ω_n can be obtained via

$$\hat{u}(x, \omega_n) = \hat{U}_0 e^{\kappa x}, \quad \kappa = -\sqrt{\frac{i\omega_n}{\alpha}} \tag{3.68}$$

see also (3.65). A detailed procedure for implementing the models discussed in this book is provided in the Appendix. Below, we provide an outline of a Fortran subroutine that describes the necessary steps and commands to implement the requested calculations.

Steps for the Fortran subroutine

Variable Declaration and Data Input

- Begin by declaring the necessary variables for the subroutine.
- Read in the input data for the calculations.

Perform Forward FFT

- Apply the forward Fast Fourier Transform (FFT) to the data to convert the applied temperature into the frequency domain. Almost all coding languages provide built-in subroutines for conducting forward and inverse FFT. Figure 3.14 shows the frequency range and the corresponding real and imaginary magnitudes of the applied temperature pulse.

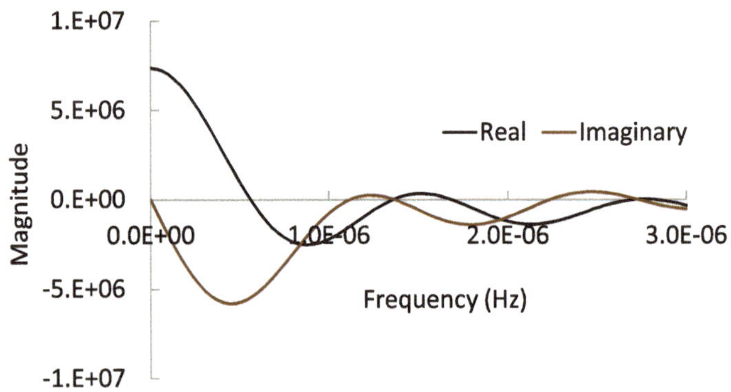

Fig. 3.14 Forward FFT of the prescribed temperature

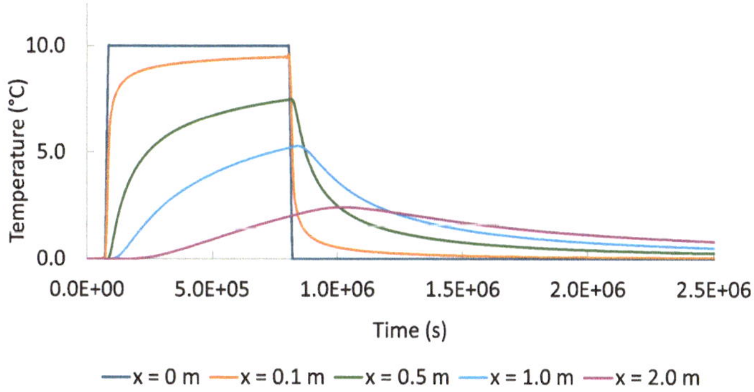

Fig. 3.15 Temporal temperature distribution along the bar

Temperature Calculation in the Frequency Domain

- Use (3.68) to calculate the temperature in the frequency domain.
- This calculation is performed for every frequency, as outlined in the "Calculation" block of the box below.

Store Results

- Save the results of the frequency-domain temperature calculation in a file.

Inverse FFT for Time-Domain Reconstruction

- The stored file is read by the inverse FFT subroutine to reconstruct the temperature back into the time domain.

Store Results and Close the Subroutine

- Save the results of the time-domain temperature in a file.

Plot the Results

- Figure 3.15 shows the calculated temperature distributions at the required output points along the bar. From this figure, it can be noted that the temperature clearly exhibits dissipation with distance.

3

```
      SUBROUTINE
C     Declare variables
      ⋮
C     Read data including FFT samples number and time step
      ⋮
C     Conduct forward FFT
      CALL FourierTransform (dt,NPT,fin,ffrq1)
C     Calculation
      DO N = 1,NPT/2                         ! NPT: number of FFT samples
         READ (ffrq1,*) FREQN,UREAL,UIMAG    ! READ the boundary temperature
         U_0 = DCMPLX(UREAL,UIMAG)           ! Û_0 in the frequency domain
         OMEGAN = 2.0d0*PI*FREQN             ! ω_n = 2πf_n , f_n: frequency
         KAPPA = CDSQRT(ICI*OMEGAN/ALPHA)    ! κ = √(iω_n/α)
         DO IOUTP=1,NOUTP                    ! NOUTP :number of output points
           U_HAT(N,IOUTP) = U_0*(CDEXP(-KAPPA*Z(IOUTP))
           ASUM(IOUTP) = ASUM(IOUTP)+U_HAT(N,IOUTP)
         ENDDO
      ENDDO
      ASUM(1:NOUTP) = -2.0d0*ASUM(1:NOUTP)! The magnitude of f_0

C     WRITE FFRQ2 FILES
      ⋮
C     Conduct inverse FFT and transform to time domain
      DO N = 1,NPT/2-1                       ! Mirrored conjugates after Nyquist frequency
(np2/2)
        READ (ffrq1,*) FREQN,UJUNK1,UJUNK2
        WRITE (ffrq2,*) FREQN,
&         DREAL(U_HAT(NPT/2-N,IOUTP)),-DIMAG(U_HAT(NPT/2-N,IOUTP))
      ENDDO
      CALL InvFastFourier (ioutput,ffrq2,ftmp)
C     Write the output files and close the files
      ⋮
      END SUBROUTINE
```

3.8 Key Observations and Conclusions

Upon reviewing the derivation of the solution to the boundary value problem (3.57), it becomes evident that the employed Fourier method is practical and does not require extensive mathematical skills. It is interesting to compare this solution with the conventional approach for solving heat conduction in a semi-infinite domain, commonly found in the literature. This traditional method states that, for a *continuous boundary condition* U_0, the temperature distribution in a semi-infinite domain is given by

$$u(x,t) = U_0\left[1 - \operatorname{erf}\left(\frac{x}{2\sqrt{\alpha t}}\right)\right] \tag{3.69}$$

where erf denotes the error function. In this context, it is important to highlight that the mathematical derivation leading to (3.69) is complex and demands a thorough understanding of Fourier integrals and their manipulations. To demonstrate the solution of this problem using Fourier integrals, we draw on a key part of the derivations given by Xie (2010) and Al-Khoury (2012) for solving heat conduction in a semi-infinite bar.

Recall the boundary value problem (3.57):

$$\begin{aligned} &\frac{1}{\alpha}\frac{\partial u(x,t)}{\partial t} = \frac{\partial^2 u(x,t)}{\partial x^2}, \quad 0 \le x < \infty, \quad t \ge 0 \\ &\text{IC}: \quad u(x,0) = 0 \\ &\text{BC}: \quad u(0,t) = U_0 \end{aligned} \tag{3.70}$$

where U_0 is continuous, steady state function. Since the domain is semi-infinite and our goal is to solve this problem without transforming it into the frequency domain, we must employ the Fourier integral. To achieve this, the boundary condition needs to be eliminated, effectively converting this boundary value problem into an initial value problem. This can be done by introducing a new variable, such that,

$$T(x,t) = u(x,t) - U_0 \tag{3.71}$$

This allows (3.70) to take the form

$$\begin{aligned} &\frac{1}{\alpha}\frac{\partial T(x,t)}{\partial t} = \frac{\partial^2 T(x,t)}{\partial x^2}, \quad 0 \le x < \infty, \quad t \ge 0 \\ &\text{IC}: \quad T(x,0) = -U_0 \\ &\text{BC}: \quad T(0,t) = 0 \end{aligned} \tag{3.72}$$

Using the separation of variables, the general solution of the partial differential equation in (3.72) leads to

3

$$T = \left[A\cos(\lambda x) + B\sin(\lambda x)\right]e^{-\alpha\lambda^2 t} \tag{3.73}$$

where λ is the separation constant, which represent the eigenvalues of the system; and A and B are arbitrary constants that can be functions of λ.

The boundary condition at $x = 0$ implies $A = 0$, and the solution simplifies to

$$T(x,t) = B(\lambda)\sin(\lambda x)\, e^{-\alpha\lambda^2 t} \tag{3.74}$$

Since U_0 is non-periodic, there are uncountably infinite eigenvalues in the system, leading the general solution to become

$$T(x,t) = \int_0^\infty B(\lambda)\sin(\lambda x)\, e^{-\alpha\lambda^2 t} d\lambda \tag{3.75}$$

At $t = 0$, the initial condition in (3.72) implies

$$\int_0^\infty B(\lambda)\sin(\lambda x)\, d\lambda = -U_0$$

This relationship is a typical Fourier sine integral with its coefficients given by

$$B(\lambda) = \frac{2(-U_0)}{\pi}\int_0^\infty \sin(\lambda\xi)\, d\xi \tag{3.76}$$

in which we used (3.43) and denoted the variable of integration by ξ for convenience. Substituting (3.76) into (3.75) yields

$$\begin{aligned} T(x,t) &= \int_0^\infty \left(\frac{-2U_0}{\pi}\int_0^\infty \sin(\lambda\xi)\, d\xi\right)\sin(\lambda x)\, e^{-\alpha\lambda^2 t} d\lambda \\ &= \frac{-2U_0}{\pi}\int_{\xi=0}^\infty \left(\int_{\lambda=0}^\infty \sin(\lambda\xi)\sin(\lambda x)e^{-\alpha\lambda^2 t} d\lambda\right) d\xi \end{aligned} \tag{3.77}$$

From trigonometry, we have

$$2\sin(\lambda x)\sin(\lambda\xi) = \cos[\lambda(x-\xi)] - \cos[\lambda(x+\xi)]$$

Applying this relationship to (3.77), and knowing that

$$\int_0^\infty \cos[\lambda(x-\xi)]\, e^{-\alpha\lambda^2 t} d\lambda = \sqrt{\frac{\pi}{4\alpha t}}\exp\left(\frac{-(x-\xi)^2}{4\alpha t}\right)$$

$$\int_0^\infty \cos[\lambda(x+\xi)]\, e^{-\alpha\lambda^2 t} d\lambda = \sqrt{\frac{\pi}{4\alpha t}}\exp\left(\frac{-(x+\xi)^2}{4\alpha t}\right)$$

we can readily obtain

$$T(x,t) = \frac{-U_0}{\sqrt{4\pi\alpha t}}\int_{\xi=0}^\infty \left(\exp\left(\frac{-(x-\xi)^2}{4\alpha t}\right) - \exp\left(\frac{-(x+\xi)^2}{4\alpha t}\right)\right) d\xi \tag{3.78}$$

This integral can be treated using the *error function* and the *complementary error function*, defined as

$$\mathrm{erf}(x) = \frac{2}{\sqrt{\pi}} \int_0^x e^{-t^2} dt, \quad \mathrm{erfc}\,(x) = 1 - \mathrm{erf}\,(x) = \frac{2}{\sqrt{\pi}} \int_x^\infty e^{-t^2} dt$$

To bring the integral in (3.78) to the error functions format, we introduce new variables, such as

$$a = -\frac{x - \xi}{\sqrt{4\alpha t}}, \quad \therefore \quad d\xi = \sqrt{4\alpha t}\, da \quad \text{for the first integral}$$
$$b = \frac{x + \xi}{\sqrt{4\alpha t}}, \quad \therefore \quad d\xi = \sqrt{4\alpha t}\, db \quad \text{for the second integral}$$

Following these, (3.78) can be expressed as

$$T(x,t) = \frac{-U_0}{\sqrt{\pi}} \left(\int_{-x/\sqrt{4\alpha t}}^{\infty} e^{-a^2} da - \int_{x/\sqrt{4\alpha t}}^{\infty} e^{-b^2} db \right)$$

Referring to the definition of the complementary error function given above, this equation can be written as

$$T(x,t) = \frac{-U_0}{2} \left(\mathrm{erfc}\,(-\frac{x}{\sqrt{4\alpha t}}) - \mathrm{erfc}\,(\frac{x}{\sqrt{4\alpha t}}) \right)$$

Having the relationship between the erf and the erfc stated above, and knowing that erf $(x) = -\mathrm{erf}(-x)$, it entails that

$$\begin{aligned} T(x,t) &= \frac{-U_0}{2} \left(-\mathrm{erf}\,(-\frac{x}{\sqrt{4\alpha t}}) + \mathrm{erf}\,(\frac{x}{\sqrt{4\alpha t}}) \right) \\ &= -U_0\, \mathrm{erf}\,(\frac{x}{\sqrt{4\alpha t}}) \end{aligned}$$

Substituting this outcome into (3.71), we finally reach to

$$\begin{aligned} u\,(x,t) &= U_0 + T(x,t) \\ &= U_0\,[1 - \mathrm{erf}\,(\frac{x}{\sqrt{4\alpha t}})] \end{aligned} \tag{3.79}$$

as given in (3.69).

Apparently, the solution of this relatively simple boundary value problem using Fourier integrals is, in fact, quite complex and requires considerable mathematical skills. Moreover, this solution is applicable to continuous boundary conditions. While it can be adapted to accommodate piecewise, rectangular functions through the superposition principle, it fails to provide solutions for arbitrary time-dependent functions. This limitation represents a significant drawback of the conventional method compared to spectral analysis, which effectively manages measured data for simulating real-world physical applications. Therefore, it is imperative to be apt in utilizing the Fourier method in conjunction with spectral analysis to effectively address boundary value problems with physical significance.

▪ Problems

1. Prove

$$\int_{-L}^{L} \sin\left(\frac{n\pi x}{L}\right) dx = \int_{-L}^{L} \cos\left(\frac{n\pi x}{L}\right) dx = 0, \quad n = 1, 2, \cdots$$

2. Prove

$$\int_{-L}^{L} \sin\left(\frac{n\pi x}{L}\right) \cos\left(\frac{m\pi x}{L}\right) dx = 0, \quad n, m = 1, 2, \cdots$$

3. Prove

$$\int_{-L}^{L} \cos\left(\frac{m\pi x}{L}\right) \cos\left(\frac{n\pi x}{L}\right) dx = \int_{-L}^{L} \sin\left(\frac{m\pi x}{L}\right) \sin\left(\frac{n\pi x}{L}\right) dx = \begin{cases} 0 & m \neq n \\ L & m = n \end{cases}$$

4. Find the eigenvalues and corresponding eigenfunctions of

   $u''(x) + \lambda u(x) = 0, \quad 0 < x < L$

   subjected to these boundary conditions:
   (a) $u(0) = 0, u\prime(L) = 0$
   (b) $u'(0) = 0, u(L) = 0$
   (c) $u(0) + u'(0) = 0, u'(L) = 0$

5. Consider the following function:

$$f(x) = \begin{cases} 5 & 0 < x < 3 \\ 0 & -3 < x < 0 \end{cases}, \quad \text{Period} = 6$$

   Using the trigonometric Fourier series:
   (a) Graph the above given function.
   (b) Determine the Fourier coefficients and write the corresponding Fourier series.
   (c) Define the function at the discontinuities $x = -3$, $x = 0$, and $x = 3$ such that the Fourier series will converge for $-3 \leq x \leq 3$.

6. Repeat Q. 5 above, but for this function:

   $f(x) = 5, \ -3 < x < 3$

7. Apply the inverse transform on the results obtained in Example 3.5 (Fourier transform section!!) and prove that you get back the original data.
   Hint: use (3.47).

8. Solve this BVP using the Fourier method:

$$\frac{1}{\alpha}\frac{\partial u(x,t)}{\partial t}=\frac{\partial^2 u(x,t)}{\partial x^2}$$

$$\text{IC}: \quad u(x,0)=0$$

$$\text{BC}: \quad \lambda \left.\frac{\partial u(x,t)}{\partial x}\right|_{x=0}=\bar{h}[u(0,t)-U_\infty(t)]$$

$$u(\infty,t)=0$$

References

Al-Khoury R (2012) Computational modeling of shallow geothermal systems. CRC Press

Brigham EO (1988) The fast Fourier transform and its applications. Prentice Hall

Brown JW, Churchill RV (2012) Fourier series and boundary value problems, 8th ed. McGraw-Hill, New York

Cooley JW, Tukey JW (1965) An algorithm for the machine calculation of complex Fourier series. Math Comput 19(90):297–301

Doyle FJ (2021) Wave propagation of structures, 3rd ed. Springer Nature, Zwitserlands

Edwards CH, Penney DE (2008) Differential equations and boundary value problems, computing and modeling, 4th ed. Pearson Education Inc., New Jersey

Xie W-C (2010) Differential equations for engineers. Cambridge University Press

Zill DG, Wright WS (2011) Advanced engineering mathematics, 4th ed. Jones and Bartlett Publisher, Canada

Heat Transfer

Heat Transfer

R. Al Khoury and C. Kasbergen, *Analytical Subsurface Environment*,
https://doi.org/10.1007/978-3-031-93910-5_4

Heat transfer plays a significant role in various physical systems, spanning from mechanical and electronic devices to geothermal energy in subsurface environment. In studying heat flow in a physical system, the amount of heat transferred to and from the system due to thermal interaction with other systems or with the environment is a quantity that is essential for engineering design and analysis. A quantitative description of the involved thermal processes requires knowledge of the concept of thermal equilibrium, thermal energy, and heat transfer. The laws of thermodynamics provide relationships between temperature, heat transfer, and energy, and thus it is vital that these laws are used as the basis for formulating mathematical models to describe heat flow in physical systems. This chapter highlights the physics of heat transfer, energy balance, and some relevant initial and boundary conditions.

4.1 Heat Transfer

The *zeroth law* of thermodynamics states that if two objects are separately in thermal equilibrium with a third object, then the two objects are in thermal equilibrium with each other. The basic physical quantity governing this law is *temperature*. As such, when two objects with different temperatures are put in thermal contact in an insulated environment, they would exchange heat (energy) between themselves until reaching thermal equilibrium, a point when both objects exhibit the same temperature.

The temperature is a scalar quantity, defined by its magnitude. The SI units of temperature are *Kelvin* (K) and *degree Celsius* (°C), which are related via

$$T(°\mathrm{C}) = T_a(\mathrm{K}) - 273.15 \tag{4.1}$$

where T_a is the Kelvin temperature, referred to also as the *absolute* temperature.

The *first law* of thermodynamics is a form of the law of conservation of energy, which states that an increase in energy in some form in a system must be encountered by a decrease of energy to some other form, no matter what type of energy is transformed. *Thermal energy* can be described as the sum of the *kinetic* and *potential* energy of the particles forming a body. Upon increasing the temperature, the particles move faster, increasing their kinetic energy. But, as the particles move fast, they depart from each other and hence increase the body's potential energy. This combination governs the total thermal energy in the body. ◘ Figure 4.1 shows schematically this phenomenon.

The driving force for heat transfer is the *temperature gradient*. If a body with temperature T is put in an environment

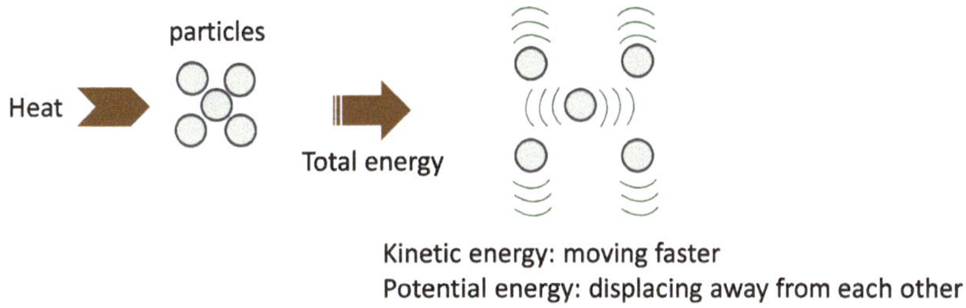

Fig. 4.1 A schematic representation of the concept of heat energy

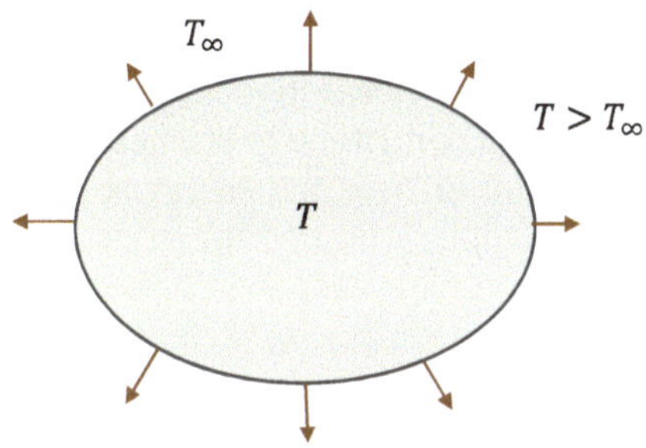

Fig. 4.2 A body with temperature T exists in an environment with temperature T_∞

with temperature T_∞ (see Fig. 4.2), heat will either flow out of the body or flow into it. The process of transferring thermal energy between bodies or systems in contact is referred to as *thermal energy transfer*. In engineering, the terms *heat flow* and *heat transfer* are utilized to describe thermal energy transfer, and the term *heat* refers to the thermal energy.

The magnitude of heat transfer is referred to as *heat flux* and is also known as *thermal flux*, *heat flux density*, or *heat flow rate*. It is defined as the amount of heat energy that flows per unit area per unit time. Unlike temperature, heat flux is a vector quantity that has both a magnitude and a direction. The rate of heat flow in a certain direction is proportional to the temperature gradient in that direction. The conservation of energy implies that only the heat flow across a boundary is important. That is, only the component of heat flux perpendicular to the boundary contributes to the heat flow process. This component is identified by the *unit outward normal vector* **n**. Figure 4.3 depicts a heat flux $\mathbf{q}_A$ making an angle θ with the normal to the surface at point A. It is only the component

$$q_A \cos(\theta) = q_A \cdot \mathbf{n} = q_A \cdot \frac{n}{|n|}$$

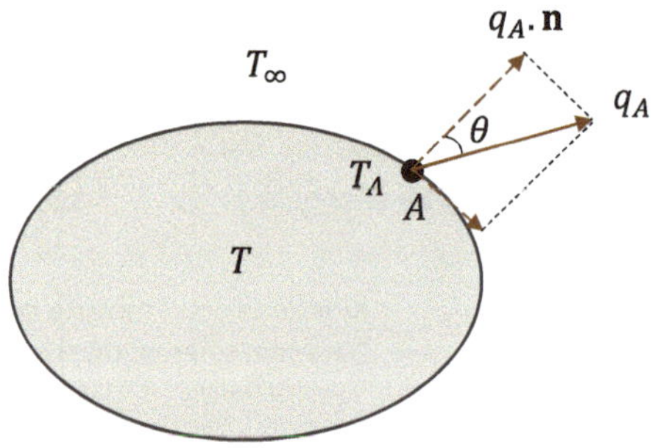

Fig. 4.3 Heat flux vector and its components

that contributes. The other component is parallel to the surface and thus does not contribute to the heat energy across the boundary. The SI unit of heat and energy is Joule (J) and that for heat flux is (Wm^{-2}).

4.2 Heat Transfer Mechanisms

Heat transfer in solids and fluids is characterized by three mechanisms, also referred to as modes: *conduction*, *convection,* and *radiation*. In most engineering applications, heat transfer is dominated by conduction, convection, or a combination of them. Heat radiation occurs in applications with highly elevated temperatures that are beyond the scope of this book.

4.2.1 Heat Conduction

Heat transfer by conduction occurs at the microscopic level, where atoms and molecules are energized by exposure to a heat source, causing them to oscillate. This oscillation facilitates the transfer of thermal energy to neighbouring atoms and molecules, which in turn, transfer some of their energy to adjacent particles.

At the macroscopic scale, heat is transferred from regions with high temperatures to regions with lower temperatures. Heat conduction at this scale is mathematically expressed by *Fourier's law*, which states that the rate of heat conduction in a medium is proportional to the temperature difference and the cross-sectional area normal to the direction of heat transfer, and inversely proportional to the distances between points in the domain with different temperatures, such that

$$q_\mathrm{d} \propto A \frac{\Delta T}{\Delta x}$$

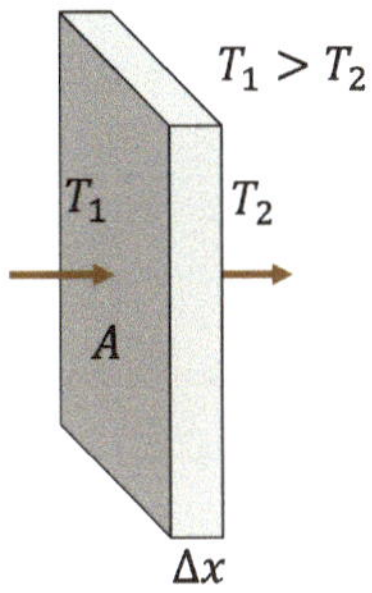

Fig. 4.4 A schematic one-dimensional heat conduction

where the subscript, d, stands here for heat conduction. The proportionality depends on the *thermal conductivity* of the substance, which is a measure of its capability to conduct heat. In a one-dimensional domain, such as the one depicted in Fig. 4.4, Fourier's law is given by

$$q_{\mathrm{d}}(x) = -\lambda A \frac{dT}{dx} \tag{4.2}$$

where λ ($\mathrm{Wm^{-1}K^{-1}}$) is the thermal conductivity of the material, and dT/dx is the temperature gradient between two regions along the x-axis. The minus sign in Fourier's law indicates that heat flows from the high temperature point to the low. Viewing it in the positive x-direction, $dT/dx \equiv (T_2 - T_1)/\Delta x$ is a negative quantity because $T_2 < T_1$, leading to a positive q_{d}.

In a three-dimensional domain, Fourier's law is expressed as

$$\mathbf{q}_d = -\nabla(\boldsymbol{\lambda} T) \cdot \mathbf{n} dA \tag{4.3}$$

where $\mathbf{n}$ is the normal directional vector at a point on the surface, see Fig. 4.5. In the Cartesian coordinate system, the heat conduction vector can be expressed in terms of its three components, such that

$$q_{\mathrm{d}}(x) = -\lambda_x A_x \frac{dT}{dx}, \quad q_{\mathrm{d}}(y) = -\lambda_y A_y \frac{dT}{dy}, \quad q_{\mathrm{d}}(z) = -\lambda_z A_z \frac{dT}{dz} \tag{4.4}$$

in which A_x, A_y, and A_z ($\mathrm{m^2}$) are the cross-sectional areas normal to the x, y, and z-directions, respectively, and λ_x, λ_y, and

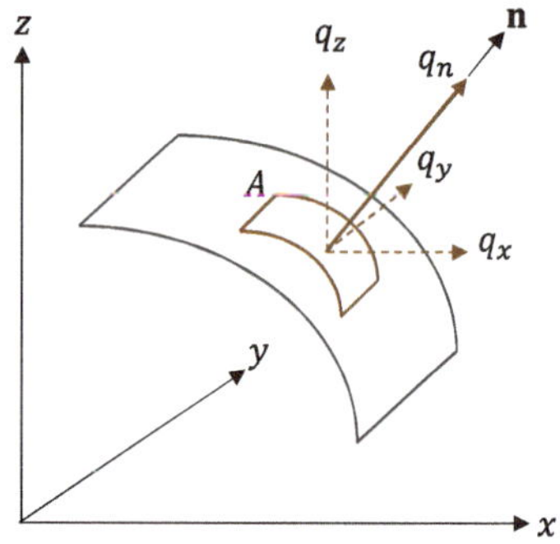

Fig. 4.5 Heat flux in a 3D domain

λ_z are the corresponding thermal conductivities. If the material is isotropic, the thermal conductivities in all directions are equal, giving $\lambda_x = \lambda_y = \lambda_z = \lambda$.

The concept of heat conduction has its equivalent in mass transport, where *diffusion* refers to the spread of particles in a fluid through random *Brownian* motion, moving from regions of higher concentration to regions of lower concentration. In many literatures, the terms diffusion and conduction are often used interchangeably, but in this book, conduction will specifically refer to heat flow, while diffusion will describe mass, species, or contaminant transport.

4.2.2 Heat Convection

Heat transfer by convection takes place at the macroscopic level, where a moving fluid carries thermal energy from one location to another. Convection is classified as either *natural* or *forced*, depending on the cause of the fluid movement. Natural convection arises from buoyancy forces generated by changes in the medium's mass density due to variations in temperature. Forced convection, on the other hand, is driven by some sort of external mechanical forces. This book focuses exclusively on forced heat convection. Figure 4.6 shows a schematic depiction of heat convection in a subsurface soil layer, induced by the flow of groundwater around a cylindrical heat source.

In a one-dimensional domain, heat convection can be expressed as

$$q_v(x) = \rho c v_x A_x T \tag{4.5}$$

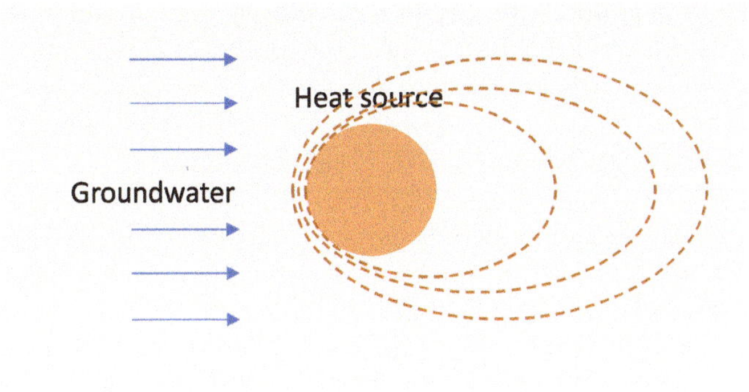

Fig. 4.6 A schematic heat convection in a 2D domain (top view)

in which v_x (ms^{-1}) is fluid velocity in the x-direction, ρ (kg m^{-3}) is the *mass density* and c (J kg^{-1}K^{-1}) is the *specific heat capacity*, with the subscript, v, standing here for heat convection. In a three-dimensional domain, the convection equation is written as

$$\mathbf{q}_v = \rho c \mathbf{v} A T. \tag{4.6}$$

One type of heat convection is expressed by *Newton's law of cooling*, a process occurring due to thermal interaction between a solid mass and a surrounding fluid. It is given by

$$q_\mathrm{v} = \bar{h}\,(T_\infty - T)\,dA \tag{4.7}$$

where $\bar{h}$ is a constant of proportionality, known as the *convective heat transfer coefficient*, and T_∞ (K) is the temperature of the surrounding environment. Figure 4.7 illustrates this concept.

As for the conduction-diffusion paradigm, the concept of heat convection has its equivalent in mass transport, where *advection* describes the flow of mass due to fluid motion. In this book, the term convection will be referred to heat transfer

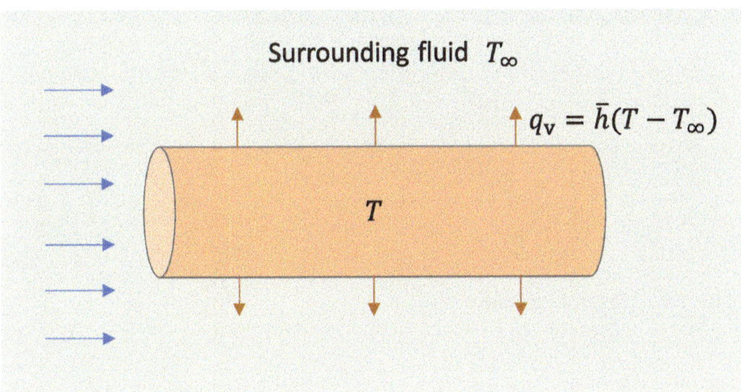

Fig. 4.7 A schematic representation of Newton's law of cooling

due to fluid flow with a controlled velocity, and advection will be used to describe the flow of mass, species, or contaminants.

4.3 Energy Balance and Heat Equation

Consider a medium with a mass m, mass density ρ, and specific heat c. A control volume of this medium with dimensions dx, dy, and dz is depicted in Fig. 4.8.

The conservation of energy states that the total energy of an isolated physical system remains constant. This implies that the *energy balance* of this control volume can be stated as.

$$\begin{pmatrix}\text{Rate of inflow heat}\\ q_{x1}\end{pmatrix}-\begin{pmatrix}\text{Rate of outflow heat}\\ q_{x2}\end{pmatrix}+\begin{pmatrix}\text{Rate of heat}\\ \text{generation } \dot{Q}\end{pmatrix}=\begin{pmatrix}\text{Rate of stored}\\ \text{energy } \dot{E}\end{pmatrix}$$

Mathematically, this can be expressed as

$$dq_{x1} - dq_{x2} + d\dot{Q} = d\dot{E}$$

where the dot $(\cdot)$ indicates the gradient in time $\partial/\partial t$. This equation essentially states that the change in total energy within the control volume over time is equal to the net rate of energy transfer into the control volume, minus the net rate of energy transfer out of the control volume, plus any energy generated within the control volume.

Net Heat Rate Entering/Leaving

The net heat flux due to heat entering and leaving the control volume in Fig. 4.8, in the x-direction, is

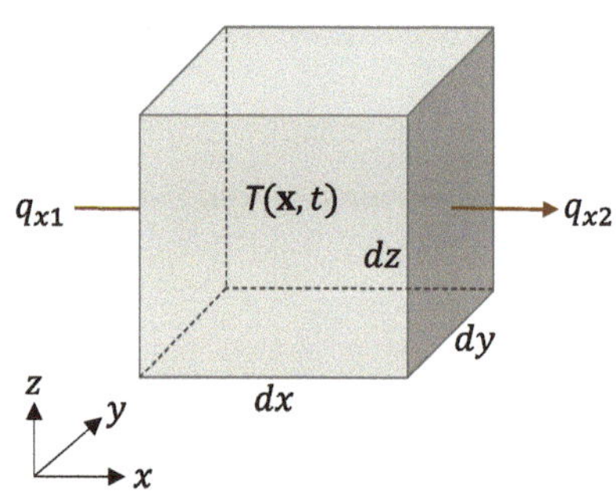

Fig. 4.8 Control volume

$$\begin{pmatrix} \textit{Rate of inflow heat} \\ q_{x1} \end{pmatrix} - \begin{pmatrix} \textit{Rate of outflow heat} \\ q_{x2} \end{pmatrix}$$

where

$$q_{x1} = q_x dy\, dz$$

in which q_x is the amount of heat flux entering the surface area $dydz$. Using Taylor series, the heat flux leaving the system can be expressed as

$$q_{x2} = \left(q_{x1} + \frac{\partial q_x}{\partial x} dx \right) dy\, dz.$$

Thus, the net heat flux (heat flow rate) in the control volume in the x-direction is

$$q_{x1} - q_{x2} = -\frac{\partial q_x}{\partial x} dxdy\, dz. \tag{4.8}$$

Similarly, the net heat flux crossing the other two directions are

$$\begin{aligned} q_{y1} - q_{y2} &= -\frac{\partial q_y}{\partial y} dxdy\, dz \\ q_{z1} - q_{z2} &= -\frac{\partial q_z}{\partial z} dxdy\, dz. \end{aligned} \tag{4.9}$$

Summing (4.8) and (4.9), the net heat flow rate in the control volume is hence

$$-\left(\frac{\partial q_x}{\partial x} + \frac{\partial q_y}{\partial y} + \frac{\partial q_z}{\partial z} \right) dV \quad \text{or} \quad -\nabla \cdot \mathbf{q}\, dV \tag{4.10}$$

where $dV = dxdydz$.

Accumulated Thermal Energy

The accumulated thermal energy per unit time is given by

$$\dot{E} = \dot{m}c\, \Delta T$$

where $\dot{E}$ (J) is the *rate of heat energy*, $\dot{m}$ (kg s^{-1}) is the *mass flow rate* and c (J kg^{-1}K^{-1}) is the specific heat capacity. In differential form, this term can be expressed as

$$\rho c \frac{\partial T}{\partial t} dV \tag{4.11}$$

where here we assume that ρc is constant, in which ρ (kg m^{-3}) is the mass density.

4

Net Production

The net production represents heat generated or lost per unit time, expressed as

$$\dot{Q}\,dV. \tag{4.12}$$

Production rate can be a source or a sink, but can also arise from thermal interactions with the surrounding environment, such as that arising due to Newton's law of cooling (4.7).

4.4 Heat Conduction–Convection Equation

Collecting the above three conservation modes, the macroscopic heat balance equation can be put as

$$\underbrace{\rho c \frac{\partial T}{\partial t} dV}_{\text{storage}} = \underbrace{-\nabla \cdot \mathbf{q}\, dV}_{\text{net heat flux}} + \underbrace{\dot{Q}\, dV}_{\text{production rate}} \tag{4.13}$$

where the heat flux gradient might arise due to heat conductions and convection, such that

$$\nabla \mathbf{q} = \nabla \mathbf{q}_{\mathrm{d}} + \nabla \mathbf{q}_{\mathrm{v}}. \tag{4.14}$$

Using Fourier's law (4.3) for heat conduction and (4.6) for heat convection, the heat balance equation can then be described as

$$\underbrace{\rho c \frac{\partial T}{\partial t}}_{\text{accumulation}} = \underbrace{\nabla \cdot (\boldsymbol{\lambda} \cdot \nabla T)}_{\text{heat conduction}} - \underbrace{\nabla \cdot (\rho c \mathbf{v} T)}_{\text{heat convection}} + \underbrace{\dot{Q}}_{\text{production rate}}. \tag{4.15}$$

For an isotropic domain, with $\boldsymbol{\lambda} = \lambda$, the conservation equation is usually given by

$$\underbrace{\frac{1}{\alpha}\frac{\partial T}{\partial t}}_{\text{accumulation}} = \underbrace{\nabla \cdot \nabla T}_{\text{heat conduction}} - \underbrace{\frac{\mathbf{v}}{\alpha} \nabla \cdot T}_{\text{heat convection}} + \underbrace{\frac{\dot{Q}}{\lambda}}_{\text{production rate}} \tag{4.16}$$

where $\alpha = \lambda / \rho c$ ($\mathrm{m^2 s^{-1}}$) is the *thermal diffusivity*.

In (4.16), $\nabla . \nabla = \nabla^2$ is the *Laplacian*, which, in the Cartesian coordinate system, is expressed as

$$\nabla^2 = \frac{\partial^2}{\partial x^2} + \frac{\partial^2}{\partial y^2} + \frac{\partial^2}{\partial z^2}$$

and in the cylindrical coordinate system, as

$$\nabla^2 = \frac{\partial^2}{\partial r^2} + \frac{1}{r}\frac{\partial}{\partial r} + \frac{1}{r^2}\frac{\partial^2}{\partial \theta^2} + \frac{\partial^2}{\partial z^2}, \quad \text{or} \quad \nabla^2 = \frac{1}{r}\frac{\partial}{\partial r}\left(r\frac{\partial}{\partial r}\right) + \frac{1}{r^2}\frac{\partial^2}{\partial \theta^2} + \frac{\partial^2}{\partial z^2}.$$

For an axial symmetric geometry, the Laplacian is given by

$$\nabla^2 = \frac{\partial^2}{\partial r^2} + \frac{1}{r}\frac{\partial}{\partial r} + \frac{\partial^2}{\partial z^2}, \quad \text{or} \quad \nabla^2 = \frac{1}{r}\frac{\partial}{\partial r}\left(r\frac{\partial}{\partial r}\right) + \frac{\partial^2}{\partial z^2}$$

and for an axial symmetric, planar 2D domain, it is given by

$$\nabla^2 = \frac{\partial^2}{\partial r^2} + \frac{1}{r}\frac{\partial}{\partial r}, \quad \text{or} \quad \nabla^2 = \frac{1}{r}\frac{\partial}{\partial r}\left(r\frac{\partial}{\partial r}\right).$$

4.5 Initial and Boundary Conditions

In solving a boundary value problem, specifying the initial and boundary conditions is essential for determining the temperature distribution in the physical system. Below we outline the initial condition and several typical boundary conditions for heat flow scenarios relevant to the boundary value problems addressed in this book.

Initial Condition

The initial condition of a transient problem specifies the state of a system at the beginning of the time. In heat transfer, it entails specifying some initially known temperature or a function of temperature with space, usually at $t = 0$. Typically, it is specified as

$$T(\mathbf{x}, 0) = T_0(\mathbf{x})$$

where $T_0(\mathbf{x})$ can be a constant value or a function of $\mathbf{x}$.

However, using spectral analysis, and as it implies transforming time into frequency, the initial condition does not play a role in the solution. To circumvent this "shortcoming" of the spectral analysis in dealing with initial conditions, we formulate the governing equations in terms of the change of temperature $u \equiv \Delta T$, such that

$$u(\mathbf{x}, t) = T(\mathbf{x}, t) - T_0(\mathbf{x}, 0).$$

Thus, in here, the initial condition can readily be given by

$$u(\mathbf{x}, 0) = 0.$$

Yet, as the problems that we are dealing with are linear, the initial conditions can be added algebraically to the calculated values in the post-processing.

Boundary Conditions

The boundary conditions imply prescribing temperatures and/or functions of temperatures at different parts of the domain's boundary. The relevant boundary conditions for heat flow problems are the Dirichlet, Neumann, and Robin.

Dirichlet Boundary Condition: Prescribed Temperature

The Dirichlet boundary condition (also known as the boundary condition of the *first kind* or the *essential* boundary condition) implies prescribing a known temperature $U(\mathbf{x}, t)$ on some points or surface areas on the boundary of a domain, typically expressed as

$$u(x_i, t) = U_i(t)$$

where $U_i(t)$ is some prescribed temperature at $x = x_i$. Depending on the problem, this temperature might be constant or time-dependent.

For a 1D domain spanning in the range $0 \leq x \leq L$ (see Fig. 4.9) the Dirichlet boundary conditions might be

$$u(0, t) = U_1(t), \quad u(L, t) = U_2$$

where, in this case, $U_1(t)$ is a time-dependent temperature at $x = 0$, and U_2 is a constant temperature at $x = L$.

Neumann Boundary Condition: Prescribed Heat Flux

The Neumann boundary condition (also known as the boundary condition of the *second kind* or the *natural* boundary condition) involves applying the value of the spatial derivative of the dependent variable, u, on some points or surface areas on the boundary of a domain.

In heat flow problems, it implies specifying the value of heat flow rate $\mathbf{q}$ passing a cross-sectional area A per unit time. It is applied with respect to an outward unit normal vector perpendicular to the surface. The Neumann boundary condition is given by

$$-(\boldsymbol{\lambda} \cdot A \nabla u) \cdot \mathbf{n} = \mathbf{q}$$

where $\mathbf{n}$ is the unit outward vector. Specifically, the Neumann boundary condition on the end boundaries of a 1D domain $0 \leq x \leq L$ (see Fig. 4.10) can be expressed as

$$-\lambda_x A_x \left.\frac{\partial u}{\partial x}\right|_{x=0} = q_1(t) \cdot n_x, \quad -\lambda_x A_x \left.\frac{\partial u}{\partial x}\right|_{x=L} = q_2 \cdot n_x$$

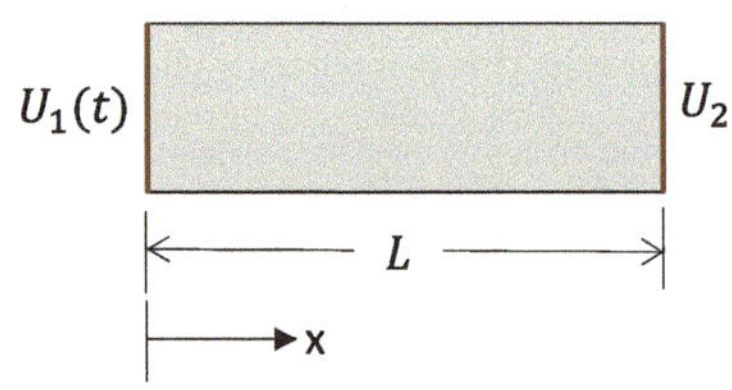

Fig. 4.9 A domain subjected to Dirichlet boundary conditions

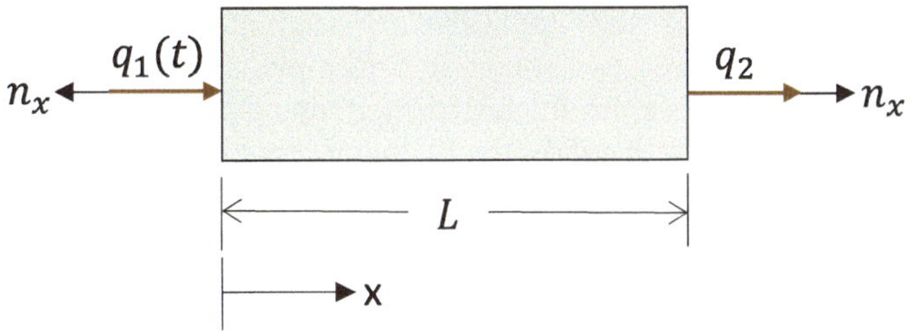

Fig. 4.10 A domain subjected to Neumann boundary conditions

where λ_x is the thermal conductivity of the domain in the x -direction. In this case, $q_1(t)$ is a time-dependent heat flux at $x = 0$, and q_2 is constant at $x = L$. The magnitude of the normal vector n_x on the left-hand side of the domain is -1, since its direction is opposite to that of the x-axis. And its magnitude on the right-hand side is $+1$. Thus, if a specified heat flux is set to enter the domain at $x = 0$ and another to leave the domain at $x = L$, the Neumann boundary conditions are expressed as

$$-\lambda_x A_x \left.\frac{\partial u}{\partial x}\right|_{x=0} = -q_1(t), \quad -\lambda_x A_x \left.\frac{\partial u}{\partial x}\right|_{x=L} = q_2$$

or

$$\lambda_x A_x \left.\frac{\partial u}{\partial x}\right|_{x=0} = q_1(t), \quad \lambda_x A_x \left.\frac{\partial u}{\partial x}\right|_{x=L} = -q_2.$$

If the surface area at $x = 0$ is effectively insulated that allows no heat flow, then $q = 0$, and the boundary condition on this location is given by

$$\left.\frac{\partial u}{\partial x}\right|_{x=0} = 0$$

Robin Boundary Condition: Newton's Law of Cooling

The Robin boundary condition (also known as the boundary condition of the *third kind*) corresponds to a linear combination of the value of the dependent variable and its derivative on the boundary. This corresponds to imposing a linear function combining Dirichlet and Neumann boundary conditions on a specific point or area of the boundary.

In heat flow problems, this kind of boundary conditions occurs when a boundary surface thermally interacts with a surrounding environment at a prescribed temperature U_∞. This boundary condition is commonly associated with Newton's law of cooling as given by (4.7). Yet, the Robin boundary condition is also applicable to heat interaction at the boundary between two solids. For example, if two solids are in contact and we are interested in studying the temperature distribution in both solids, the boundary condition at the contact surface between them can be expressed as

$$\lambda_1 \frac{\partial u_1}{\partial x}\bigg|_{x=x_0} = b\left[u_1(x_0,t) - u_2(x_0,t)\right]$$

$$\lambda_2 \frac{\partial u_2}{\partial x}\bigg|_{x=x_0} = b\left[u_2(x_0,t) - u_1(x_0,t)\right]$$

where b is a constant of proportionality representing the thermal interaction between the two domains. Note that this kind of boundary conditions is applicable only if the sum of the heat fluxes across the contact area is zero.

Homogeneous Boundary Conditions

The above given Dirichlet and Neumann boundary conditions are nonhomogeneous since they are not identically zero at the boundary. Should these boundary conditions be

$$u\,(x_i,t) = 0, \quad \frac{\partial u}{\partial x}\bigg|_{x=x_i} = 0$$

they are said to be homogeneous.

Similarly, the Robin boundary condition representing Newton's law of cooling defined as

$$\frac{\partial u}{\partial x}\bigg|_{x=x_i} = \bar{h}\left[\,u\,(x_i,t) - U_\infty\right]$$

is nonhomogeneous, but if we let the surrounding temperature $U_\infty = 0$, such that the boundary condition reads

$$\frac{\partial u}{\partial x}\bigg|_{x=x_i} - \bar{h}\,u\,(x_i,t) = 0$$

it becomes homogeneous at $x = x_i$.

▪ **Example 4.1 Application of Newton's Law of cooling in time domain**

A heat pipe goes through a fluid maintained at $T_f = 20$ °C. If the temperature of the heat pipe, T_p, drops from 50 to 40 °C in 10 min, calculate the time needed for the heat pipe to reach 25 °C.

▪ **Solution**

Newton's law of cooling in the temporal domain can be expressed as

$$\frac{dT_p}{dt} = \overline{h}\,(T_f - T_p)$$

where $\overline{h}$ is a constant of proportionality. This equation is a first-order differential equation with its solution given by

$$\int \frac{dT}{T_f - T_p} = \overline{h} \int dt \quad \Rightarrow \quad \ln\left|T_f - T_p\right| = \overline{h}t + c.$$

This outcome can be expressed as

$$T_p = T_f - ce^{\overline{h}\,t}.$$

The initial condition at $t = 0\,°\text{C}$ implies

$$50 = 20 - c, \Rightarrow \quad c = -30.$$

Hence

$$T_p = T_f + 30\,e^{\overline{h}\,t}.$$

To determine the constant of proportionality, we have at $t = 10$ min, $T_p = 40\,°\text{C}$, implying

$$40 = 20 + 30\,e^{10\overline{h}} \Rightarrow \quad \overline{h} = -0.04\,\text{units}.$$

Thus

$$T_p = 20 + 30e^{-0.04\,t}.$$

The time needed for the heat pipe to reach 25 °C is therefore

$$25 = 20 + 30\,e^{-0.04\,t} \Rightarrow \quad t = 44\,\text{min}.$$

Hence, the heat pipe will require another $44 - 10 = 34$ min to reach to 25 °C.

Note that for the heat pipe to reach equilibrium with the surrounding fluid, i.e. $T_p = 20\,°\text{C}$, this equation predicts that this will occur at $t \to \infty$. Of course this is not practical, but we understand that this model is empirical and does not quite stand to our physical intuition. However, it remains reasonable for estimating the temperature above the equilibrium limit.

Example 4.2: Application of Newton's law of cooling in spatial domain

A heat pipe goes through a medium maintained at $T_\infty = 12\,°\text{C}$. A fluid enters the heat pipe at $x = 0$ m with temperature $T = 5\,°\text{C}$. If we assume that the coefficient of thermal interaction between the pipe and the surrounding medium is $b = 0.1\,\text{m}^{-1}$, what is the required length of the pipe sufficient to increase the fluid temperature to 10 °C.

Solution

Newton's law of cooling in the spatial domain can be given by

$$\frac{dT}{dx} = b\,(T_\infty - T).$$

This is a first-order differential equation with its solution yields

$$T = T_\infty + ce^{-bx}.$$

At $x = 0$ m, it implies

$$5 = 12 + c \Rightarrow \quad c = -7.$$

Hence

$$T = 12 - 7e^{-0.1x}.$$

For the fluid in the heat pipe to reach to 10 °C, it entails that

$$10 = 12 - 7e^{-0.1x} \Rightarrow \quad x = 12\,\text{m}.$$

Thus, we need a heat pipe with at least 12 m in length to raise the fluid temperature from 5 to 10 °C.

Problems

1. Express the steady-state heat conduction-convection equation for a 3D domain.
2. Express the heat Eq. (4.16) in the cylindrical coordinate system.
3. How would the resulting heat equation in Problem 2 look like for an axial symmetric problem?
4. Modify your answer for Problem 3 for the case of a planar 2D domain.
5. A wall 10 cm thick with its two faces maintained at uniform temperatures of 5 and 20°C. The thermal conductivity of the wall material is 0.9W/m.K. Determine the steady-state heat transfer per unit area.
   Hint: use $q/A = Q = -\lambda \Delta T/\Delta x$.
6. On a sunny day the temperature at the surface of a soil mass reached 25°C. At night, the air temperature becomes 5°C. If the convective heat transfer coefficient at the surface is 12W/m^2.K, determine the heat transfer per unit area from the soil to the air.
   Hint: use $q/A = Q = \overline{h}(T - T_{\text{air}})$.
7. A hot fluid with $c_p \equiv c_h = 2.0\text{kJ/kg.K}$ enters a counterflow heat exchanger at 190 °C with a flow rate 0.5 kg/s and leaving at 60 °C. A cold fluid with $c_p \equiv c_c = 1.5\text{kJ/kg.K}$ exits the heat exchanger at a rate of 1 kg/s with 150 °C. Calculate the temperature of the cold fluid at the inlet, and also the amount of heat transfer between the two fluids.
   Hint: use $\dot{m}_c c_c(T_{co} - T_{ci}) = \dot{m}_h c_h(T_{hi} - T_{ho})$.

Further Reading

Carslaw HS, Jaeger JC (1959) Conduction of heat in solids, 2nd edn. Oxford University Press

Lienhard IV JH, Lienhard V JH (2019) A heat transfer textbook, 5th edn. Phlogiston Press

Pitts D, Sissom L (1998) Heat transfer, 2nd edn. McGraw-Hill, Schaum's Outlines

Sidebotham G (2015) Heat transfer modeling, an inductive approach. Springer, Berlin

Heat Conduction in Infinite Domains

R. Al Khoury and C. Kasbergen, *Analytical Subsurface Environment*,
https://doi.org/10.1007/978-3-031-93910-5_5

Heat conduction in infinite domains has extensive applications in engineering and applied sciences, particularly in the field of ground subsurface research and engineering. The first notable instance of interest in this area was in the late nineteenth century when electricity utility companies sought to understand the heat flow process surrounding buried electric cables. Lord Kelvin introduced the well-known Kelvin's line-source model as the earliest approach to calculating heat flow in such a domain. According to this model, the ground is considered an infinite medium, within which an infinite line heat source representing the electric cable is embedded.The utilization of Kelvin's line source model extends beyond its original application in electricity. In 1948, Ingersoll and Plass made a significant breakthrough by simulating a ground heat exchanger using Kelvin's concept of an infinite line heat source. In the 1950s, Hooper and Lepper employed Kelvin's line source to analyze thermal probes. Furthermore, Kelvin's line source, along with its derivatives, has proven instrumental in determining the thermal parameters of a wide range of materials.

In this chapter we focus on solving heat flow problems in infinite domains that are subjected to heat sources with varying configurations. We employ the Fourier method and spectral analysis for solving the governing boundary value problems.

5.1 Infinite Domain with a Point Heat Source

We begin by considering an infinite domain $-\infty \leq r \leq \infty$ subjected to a point heat source with a Neumann boundary condition, as schematically shown in Fig. 5.1. The boundary value problem of a transient heat transfer in such a geometry is conveniently given by the spherical coordinate system, such that

$$\begin{aligned}
&\frac{1}{\alpha}\frac{\partial u}{\partial t} = \frac{\partial^2 u}{\partial r^2} + \frac{2}{r}\frac{\partial u}{\partial r} \\
&\text{IC:} \quad u(z,0) = 0 \\
&\text{BC:} \quad -4\pi r^2 \lambda \left.\frac{\partial u}{\partial r}\right|_{r=0} = q_o(t) \\
&\qquad u(\infty,t) = 0
\end{aligned} \tag{5.1}$$

in which $\alpha = \lambda/\rho c\,(\mathrm{m^2 s^{-1}})$ is the thermal diffusivity, wherein $\lambda\ (\mathrm{Wm^{-1}K^{-1}})$ is the thermal conductivity, $\rho\ (\mathrm{kg\,m^{-3}})$ is the mass density and $c\ (\mathrm{J\,kg^{-1}K^{-1}})$ is the specific heat capacity.

General Solution

Referring to ▶ Chap. 3, it can be recalled that by utilizing the Fourier transform, the spectral representations of the temporal and spatial derivatives are

Fig. 5.1 Point heat source

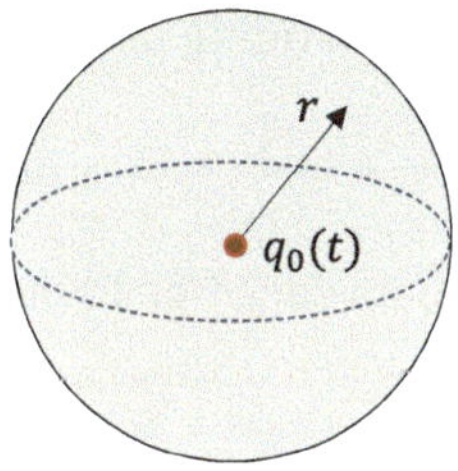

$$\frac{\partial u}{\partial t} \Rightarrow i\omega\, \hat{u}, \quad \frac{\partial u}{\partial x} \Rightarrow \frac{\partial \hat{u}}{\partial x}$$

where the hat $(\hat{\cdot})$ stands for a transformed variable. Hence, applying the Fourier transform to (5.1) results in

$$\begin{aligned} &\frac{d^2\hat{u}}{dr^2} + \frac{2}{r}\frac{d\hat{u}}{dr} - \frac{i\omega}{\alpha}\hat{u} = 0 \\ &\text{BC: } r^2\frac{d\hat{u}}{dr}\bigg|_{r=0} = -\frac{\hat{q}_o(\omega)}{4\pi\lambda} \\ &\qquad \hat{u}(\infty, \omega) = 0 \end{aligned} \tag{5.2}$$

where it is evident that through the transformation process the original partial differential equation has been reduced to an ordinary differential equation.

The first two terms in the differential equation in (5.2) can be written as

$$\frac{d^2\hat{u}}{dr^2} + \frac{2}{r}\frac{d\hat{u}}{dr} \equiv \frac{1}{r}\frac{d^2}{dr^2}(r\hat{u})$$

since

$$\frac{1}{r}\frac{d^2}{dr^2}(r\hat{u}) = \frac{1}{r}\frac{d}{dr}(r\frac{d\hat{u}}{dr} + \hat{u}) = \frac{d^2\hat{u}}{dr^2} + \frac{2}{r}\frac{d\hat{u}}{dr}$$

Letting $v = r\hat{u}$, we obtain

$$\frac{d^2 v}{dr^2} - \frac{i\omega}{\alpha} v = 0 \tag{5.3}$$

Recall Table 2.1 the general solution of this differential equation is given by

$$v = a_1 e^{\sqrt{i\omega/\alpha}\, r} + a_2 e^{-\sqrt{i\omega/\alpha}\, r}$$

Substituting back for $\hat{u}$, we obtain

$$\hat{u} = \frac{1}{r}(Ae^{\kappa r} + Be^{-\kappa r}), \quad \kappa = \sqrt{i\omega/\alpha}$$

Boundary Conditions

The second boundary condition in (5.2) implies $A = 0$, since otherwise it will lead to infinity at relatively far distances. Hence, the general solution becomes

$$\hat{u} = \frac{1}{r}Be^{-\kappa r}$$

Taking this equation with respect to r, substituting it into the first boundary condition in (5.2) and carrying out the limit as $r \to 0$ leads to

$$B = \frac{\hat{q}_o}{4\pi\lambda}$$

BVP Solution

Having determined the coefficients, the solution of boundary value problem (5.2) for frequency ω_n is thus

$$\hat{u}(r, \omega_n) = \frac{\hat{q}_o}{4\pi r\lambda} e^{-\kappa r} \tag{5.4}$$

To reconstruct the solution back to the time domain, we use the inverse fast Fourier transform (IFFT) algorithm, leading to

$$u(z, t) = \sum_n \hat{u}(z, \omega_n)\, e^{i\omega_n t}$$

- **Example 5.1: Point heat source**

A point heat source releases 50 W of heat in an infinite domain with a time-dependent rate shown in Fig. 5.2. The domain is characterized by thermal conductivity 2 $Wm^{-1}K^{-1}$, mass density 1500 kg m^{-3}, and specific heat capacity 800 J $kg^{-1}K^{-1}$. Calculate the temperature in space and at different times.

- **Solution**

To calculate the temperature distribution in this domain, we follow these steps:

1. Transform the input pulse to the frequency domain using FFT. For this, we used 4096 (2^{12}) FFT samples with a 300 s time step.
2. Calculate the temperature at different distances from the heat source using (5.4) for every frequency.
3. Use the inverse FFT (IFFT) to restore the time domain.

Figure 5.3 shows the evolution of temperature with time versus distance from the heat source. The figure clearly depicts

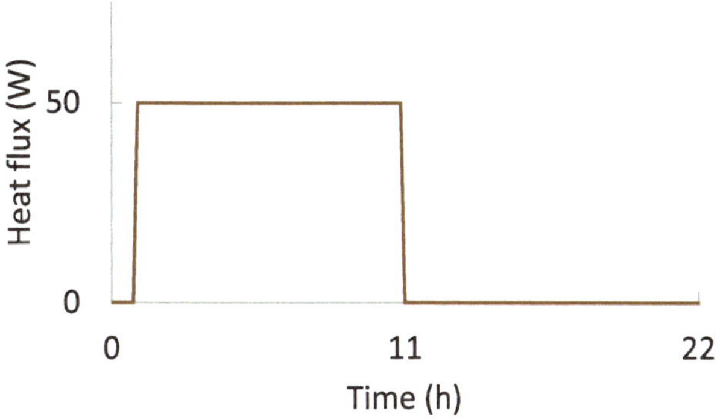

Fig. 5.2 Heat flux released from the point heat source

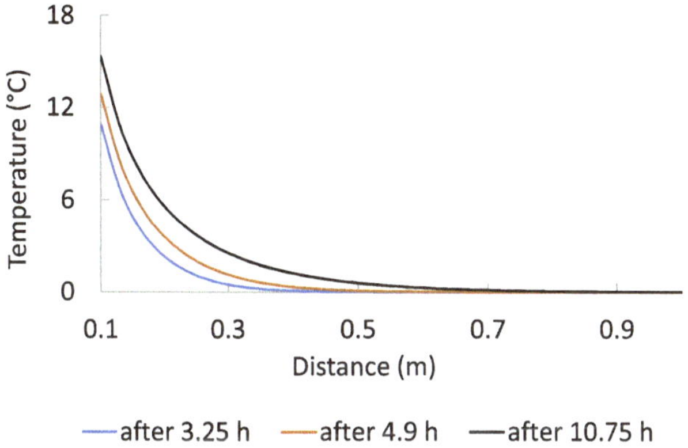

Fig. 5.3 Temperature evolution with time versus distance

the decay of temperature with distance and highlights the increase of the affected domain with time.

5.2 Infinite Line Heat Source (ILS)

Heat flow due to an infinite line heat source is in effect heat flow generated by a point source in a planar 2D domain with an axial (radial) symmetric effect, as schematically represented in Fig. 5.4. The heat equation and the relevant initial and boundary conditions of this problem are best described in the polar coordinate system, as

$$
\begin{aligned}
&\frac{1}{\alpha}\frac{\partial u}{\partial t} = \frac{\partial^2 u}{\partial r^2} + \frac{1}{r}\frac{\partial u}{\partial r} \\
&\text{IC}: \quad u(r,0) = 0 \\
&\text{BC}: \quad u(r \to \infty, t) = 0 \\
&\qquad\quad -\lambda \frac{\partial u}{\partial r} \cdot 2\pi r \bigg|_{r \to 0} = q(t)
\end{aligned}
\tag{5.5}
$$

Fig. 5.4 Infinite line heat source

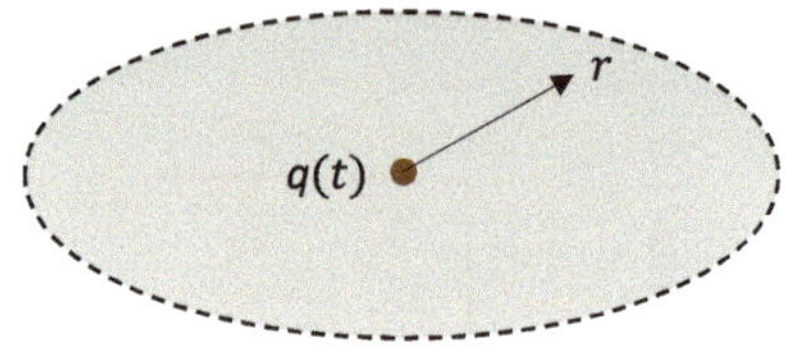

General Solution

Applying the Fourier transform to (5.5) results in

$$\begin{aligned} &\frac{d^2\hat{u}}{dr^2} + \frac{1}{r}\frac{d\hat{u}}{dr} - \frac{i\omega}{\alpha}\hat{u} = 0 \\ &\text{BC}: \quad \hat{u}(r \to \infty, \omega) = 0 \\ &\qquad -\lambda\frac{d\hat{u}}{\partial r} \cdot 2\pi r\bigg|_{r\to 0} = \hat{q}(\omega) \end{aligned} \tag{5.6}$$

The differential equation in (5.6) is referred to as a parametric Bessel equation. One way of approaching it involves transforming it into the standard form of the Bessel equation. To achieve this, we introduce the substitution $s = \sqrt{i\omega/\alpha}r = \kappa r$. Then, by applying the chain rule, we obtain

$$\begin{aligned} &\frac{1}{r} = \frac{\kappa}{s} \\ &\frac{d\hat{u}}{dr} = \frac{d\hat{u}}{ds}\frac{ds}{dr} = \kappa\frac{d\hat{u}}{ds} \\ &\frac{d^2\hat{u}}{dr^2} = \frac{d^2\hat{u}}{ds^2}\frac{ds^2}{dr^2} = \frac{d^2\hat{u}}{ds^2}(\frac{ds}{dr})^2 = \kappa^2\frac{d^2\hat{u}}{ds^2} \end{aligned}$$

Substituting these identities into (5.6), gives

$$\frac{d^2\hat{u}}{\partial s^2} + \frac{1}{s}\frac{d\hat{u}}{\partial s} - \hat{u} = 0$$

which is now a typical form of the modified Bessel equation, with its general solution given by

$$\hat{u} = AK_0(s) + BI_0(s)$$

where I_0 and K_0 are the modified Bessel functions of zero order of the first and second kind, respectively. A and B are arbitrary constants which need to be determined from the boundary conditions.

Substituting back for $s = \kappa r = \sqrt{i\omega/\alpha}r$, we obtain

$$\hat{u}(r, \omega) = AK_0(\kappa r) + BI_0(\kappa r) \tag{5.7}$$

Boundary Conditions

As $r \to \infty$, I_0 becomes unbounded, and accordingly the first boundary condition in (5.6) implies $B = 0.$. This leads (5.7) to become

$$\hat{u}(r, \omega) = AK_0(\kappa r) \tag{5.8}$$

Taking the derivative of (5.8) with respect to r yields

$$\frac{d\hat{u}}{dr} = -A\kappa K_1(\kappa r)$$

where K_1 is the modified Bessel function of the first order.

Knowing that (Abramowitz and Stegun, 1972):

$$\lim_{z \to 0} K_v(z) = \frac{1/2 \cdot \Gamma(v)}{\left(1/2 \cdot z\right)^v}$$

it entails that as $r \to 0$,

$$\lim_{r \to 0} K_1(\kappa r) \cong \frac{1/2 \cdot \Gamma(1)}{\left(1/2 \cdot \kappa r\right)^1} = \frac{1}{\kappa r} \tag{5.9}$$

Thus, the second boundary condition in (5.6), together with (5.8), implies

$$-\lambda 2\pi r \frac{d\hat{u}}{dr}\bigg|_{r \to 0} = -\lambda 2\pi r \left(-A\frac{1}{\kappa r}\kappa\right) = \hat{q}\,(\omega)$$

This readily leads to

$$A = \frac{\hat{q}\,(\omega)}{2\pi\,\lambda} \tag{5.10}$$

BVP Solution

Substituting (5.10) into (5.8), the solution of this boundary value problem for a given frequency ω_n can now be expressed as

$$\hat{u}(r, \omega_n) = \frac{\hat{q}\,(\omega_n)}{2\pi\,\lambda} K_0(\kappa r), \quad \kappa = \sqrt{i\omega_n/\alpha} \tag{5.11}$$

Having solved the problem in the frequency domain, the solution of this boundary value problem (5.5) in the time domain can then be obtained using the inverse fast Fourier transform (IFFT) algorithm, as

$$u(r, t) = \sum\nolimits_n \hat{u}(r, \omega_n)\, e^{i\omega_n t}$$

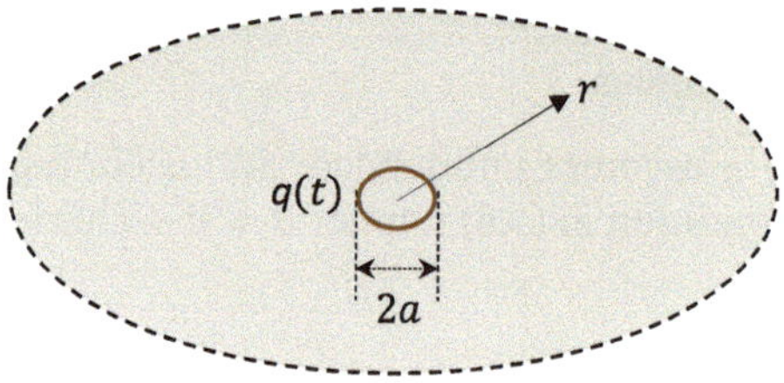

Fig. 5.5 Infinite cylindrical heat source

5.3 Infinite Cylindrical Heat Source (ICS)

Similar to the infinite line heat source, heat flow due to an infinite cylindrical heat source is in effect heat flow generated by a circular source in a planar 2D domain with an axial (radial) symmetric effect, as sketched in Fig. 5.5.

The boundary value problem of this case is given by

$$\begin{aligned}
&\frac{1}{\alpha}\frac{\partial u}{\partial t} = \frac{\partial^2 u}{\partial r^2} + \frac{1}{r}\frac{\partial u}{\partial r}, \quad r > a \\
&\text{IC}: \quad u(r,0) = 0 \\
&\text{BC}: \quad u(r \to \infty, t) = 0 \\
&\qquad\quad -\lambda \frac{\partial u}{\partial r} \cdot 2\pi a \bigg|_{r=a} = q(t)
\end{aligned} \tag{5.12}$$

where a is the radius of the heat cylinder.

General Solution

Applying the Fourier transform to (5.12) gives

$$\begin{aligned}
&\frac{d^2\hat{u}}{dr^2} + \frac{1}{r}\frac{d\hat{u}}{dr} - \frac{i\omega}{\alpha}\hat{u} = 0 \\
&\text{BC}: \quad \hat{u}(r \to \infty, \omega) = 0 \\
&\qquad\quad -\lambda \frac{d\hat{u}}{dr} \cdot 2\pi a \bigg|_{r=a} = \hat{q}_0(\omega)
\end{aligned} \tag{5.13}$$

Following the mathematical procedure for solving (5.6), the general solution of (5.13) can be expressed as

$$\hat{u}(r,\omega) = AK_0(\kappa r), \quad \kappa = \sqrt{i\omega/\alpha} \tag{5.14}$$

Boundary Conditions

Taking the derivative of (5.14) with respect to r, and solving at $r = a$, the second boundary condition in (5.13) implies

$$A = \frac{\hat{q}(\omega)}{2\pi a \lambda \kappa K_1(\kappa a)}$$

BVP Solution

Substituting A into (5.14), the solution of this boundary value problem in the frequency domain yields

$$\hat{u}(r,\omega_n) = \frac{\hat{q}(\omega_n)}{2\pi a\lambda\kappa\, K_1(\kappa a)} K_0(\kappa r), \quad \kappa = \sqrt{i\omega_n/\alpha} \tag{5.15}$$

The solution in the time domain can then be obtained using the inverse fast Fourier transform (IFFT) algorithm, as

$$u(r,t) = \sum\nolimits_n \hat{u}(r,\omega_n) e^{i\omega_n t} \tag{5.16}$$

Special Case: Line heat source

In the special case where $a \to 0$, the cylindrical heat source becomes a line heat source. Following (5.9), we can readily find that

$$\lim_{a\to 0} aK_1(\kappa a) \cong \frac{1}{\kappa}$$

Applying this limiting case to (5.15), we obtain

$$\hat{u}(r,\omega_n) = \frac{\hat{q}(\omega_n)}{2\pi\lambda} K_0(\kappa r), \quad \kappa = \sqrt{i\omega_n/\alpha}$$

which is identical to (5.11).

5.4 Cylindrical Heat Source in an Infinite Domain with θ-Dependence: Dirichlet Problem

In this section, we address heat flow in a planar domain encompassing a circular heat source of radius a located at its center. A Dirichlet boundary condition, dependent on θ and time, is prescribed at the heat source, see Fig. 5.6.

The relevant boundary value problem of this case is

$$\begin{aligned}
&\frac{\partial^2 u}{\partial r^2} + \frac{1}{r}\frac{\partial u}{\partial r} + \frac{1}{r^2}\frac{\partial^2 u}{\partial \theta^2} - \frac{1}{\alpha}\frac{\partial u}{\partial t} = 0 \\
&\text{IC}: \quad u(r,\theta,0) = 0 \\
&\text{BC}: \quad u(r\to\infty,\theta,t) = 0 \\
&\qquad\quad u(r,\theta,t) = u(r,\theta+2\pi,t) \\
&\qquad\quad u(a,\theta,t) = U(\theta,t)
\end{aligned} \tag{5.17}$$

General Solution

Applying the Fourier transform to the partial differential equation in (5.17) results in

$$\frac{\partial^2 \hat{u}}{\partial r^2} + \frac{1}{r}\frac{\partial \hat{u}}{\partial r} + \frac{1}{r^2}\frac{\partial^2 \hat{u}}{\partial \theta^2} - \frac{i\omega}{\alpha}\hat{u} = 0 \tag{5.18}$$

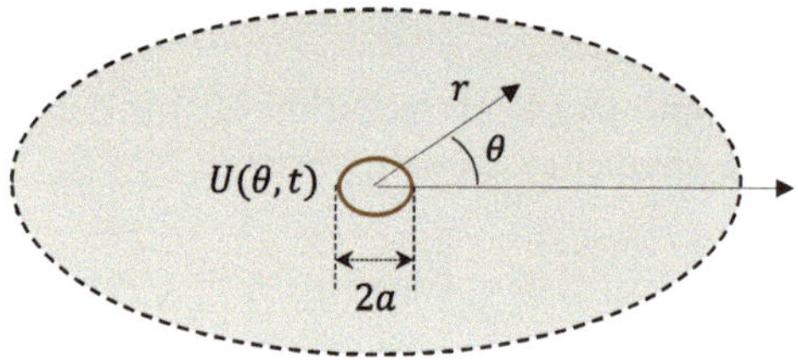

Fig. 5.6 Infinite cylindrical heat source with θ–dependence: Dirichlet problem

5

Using the method of separation of variables, we first assume

$$\hat{u}(r,\theta) = R(r)\,F(\theta) = RF \tag{5.19}$$

in which $R(r)$ is a function of r only and $F(\theta)$ is a function of θ only. Substituting (5.19) into (5.18) gives

$$R''F + \frac{1}{r}R'F + \frac{1}{r^2}RF'' - \frac{i\omega}{\alpha}RF = 0 \tag{5.20}$$

Multiplying this equation with r^2 and dividing by RF yields

$$r^2\frac{R''}{R} + r\frac{R'}{R} - \frac{i\omega}{\alpha}r^2 = -\frac{F''}{F} \tag{5.21}$$

Since the left hand side is a function of r only and the right hand side is a function of θ only, both sides must be equal a constant. Equating both sides of the equation by some arbitrary separation constant, say η, gives

$$r^2\frac{R''}{R} + r\frac{R'}{R} - \frac{i\omega}{\alpha}r^2 = -\frac{F''}{F} = \eta \tag{5.22}$$

This equations can be split into two separate equations, such that

$$\frac{d^2F}{d\theta^2} + \eta F = 0 \tag{5.23}$$

$$r^2\frac{d^2R}{dr^2} + r\frac{dR}{dr} - \left(\kappa^2 r^2 + \eta\right)R = 0 \tag{5.24}$$

where $\kappa^2 = i\omega/\alpha$ (see (5.21)).

We begin with solving the F boundary value problem:

$$F'' + \eta F = 0, \quad F(\theta) = F(\theta + 2\pi) \tag{5.25}$$

which is evidently 2π-periodic. We seek a solution that leads to eigenfunction expansion. Although this problem is not a regular Sturm-Liouville, its periodicity enables such an expansion. There are three possible general solutions to (5.25), namely:

$$\begin{aligned} F(\theta) &= c_1 + c_2\theta, & \eta &= 0 \\ F(\theta) &= c_1 \cosh(m\theta) + c_2 \sinh(m\theta), & \eta &= -m^2 < 0 \\ F(\theta) &= c_1 \cos(m\theta) + c_2 \sin(m\theta), & \eta &= m^2 > 0 \end{aligned} \tag{5.26}$$

Apparently the second solution is not periodic and thus can be dismissed. The first solution is periodic if $c_2 = 0$, with the remaining solution $F(\theta) = c_1, c_1 \neq 0$. Hence, $\eta = 0$ is an eigenvalue. Finally, the third solution in (5.26) can be 2π -periodic if we take $= 1, 2, \ldots$. The eigenvalues are then $\eta_m = m^2, m = 1, 2, \ldots$, leading the solution to be expressed as

$$\begin{aligned} F(\theta) &= c_1, & m &= 0 \\ F(\theta) &= \sum\nolimits_m c_1 \cos(m\theta) + c_2 \sin(m\theta), & m &= 1, 2, \cdots \end{aligned} \tag{5.27}$$

Setting $\eta = m^2$, the differential Eq. (5.24) can be written as

$$r^2 R'' + rR' - \left(\kappa^2 r^2 + m^2\right) R = 0 \tag{5.28}$$

which is close to the form of the modified Bessel equation, but not an exact match. A typical form of the modified Bessel equation is

$$x^2 y'' + xy' - \left(x^2 + m^2\right) y = 0 \tag{5.29}$$

To reduce (5.28) to a form identical to (5.29), we set $s = kr$, which gives

$$\begin{aligned} \frac{dR}{dr} &= \frac{dR}{ds}\frac{ds}{dr} = \kappa \frac{dR}{ds} \\ \frac{d^2R}{dr^2} &= \frac{d^2R}{ds^2}\frac{ds^2}{dr^2} = \kappa^2 \frac{d^2R}{ds^2} \end{aligned} \tag{5.30}$$

Inserting these derivatives into (5.28) leads to

$$s^2 \frac{d^2R}{ds^2} + s\frac{dR}{ds} - \left(s^2 + m^2\right) R = 0 \tag{5.31}$$

which is a modified Bessel equation of order m, with its general solution expressed by

$$R(s) = c_3 I_m(s) + c_4 K_m(s) \tag{5.32}$$

where I_m and K_m are the modified Bessel functions of m order of the first and second kind, respectively.

Boundary Conditions

As I_m is not bounded at far distances, c_3 must be zero, and the solution becomes

$$R(r) = c_4 K_m(\kappa r) \tag{5.33}$$

where we re-substituted $s = \kappa r$.

Collecting the two solutions in (5.27) and (5.33), and inserting into (5.19), it leads to

$$\hat{u}\,(r, \theta, \omega_n) = \sum\nolimits_m [A_m \cos(m\theta) + B_m \sin(m\theta)] K_m(\kappa r), \quad m = 0, 1, 2, \cdots \tag{5.34}$$

with $\kappa = \sqrt{i\omega_n/\alpha}$, and $A_m = c_1 c_4$ and $B_m = c_2 c_4$.

At $r = a$, the third boundary condition in (5.17) (in the frequency domain) implies

$$\sum\nolimits_m [A_m \cos(m\theta) + B_m \sin(m\theta)] K_m(\kappa a) = \hat{U}(\theta, \omega_n) \tag{5.35}$$

This is a trigonometric Fourier series with its coefficients given by

$$\begin{aligned} A_0 &= \frac{1}{2\pi K_0(\kappa a)} \int_0^{2\pi} \hat{U}\,(\theta, \omega_n)\, d\theta \\ A_m &= \frac{1}{\pi K_m(\kappa a)} \int_0^{2\pi} \hat{U}\,(\theta, \omega_n) \cos(m\theta)\, d\theta, \quad m = 1, 2, 3, \cdots \\ B_m &= \frac{1}{\pi K_m(\kappa a)} \int_0^{2\pi} \hat{U}\,(\theta, \omega_n) \sin(m\theta)\, d\theta, \quad m = 1, 2, 3, \cdots \end{aligned} \tag{5.36}$$

BVP Solution

Solving (5.36) and substituting the resulting values of A_m and B_m into (5.34), give the required solution for this boundary value problem in the frequency domain.

Then, using the inverse FFT, we finally obtain the solution in the time domain, such that

$$u\,(r, \theta, t) = \sum\nolimits_n \hat{u}\,(r, \theta, \omega_n)\, e^{i\omega_n t} \tag{5.37}$$

- **Example 5.2: Dirichlet problem**

An infinite planar domain is subjected at its center to a θ- and time-dependent cylindrical heat source with radius 0.1 m. The material properties of the domain are: thermal conductivity 2 $\mathrm{Wm^{-1}K^{-1}}$, mass density 1500 $\mathrm{kg\,m^{-3}}$, and specific heat capacity 800 $\mathrm{J\,kg^{-1}K^{-1}}$. The heat source imposes a Dirichlet boundary condition that is a linear function of θ, such that $U = \theta U_0(t)$, where $U_0(t)$ varies with time as shown in Fig. 5.7.

- **Solution**

As outlined in Example 5.1, this problem can be solved by first transforming the input signal in Fig. 5.7 into the frequency domain, then solving it for every frequency using (5.34) together with (5.36), followed by adding the resulted solutions for all involved frequencies. After that we transform the outcome back to the time domain using the IFFT algorithm.

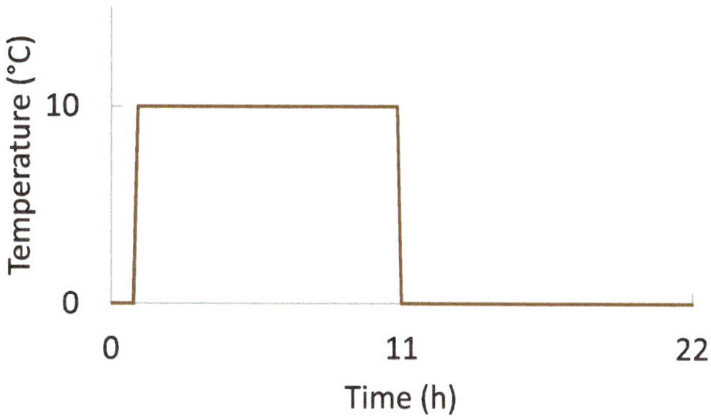

Fig. 5.7 Time variation of $U_0(t)$

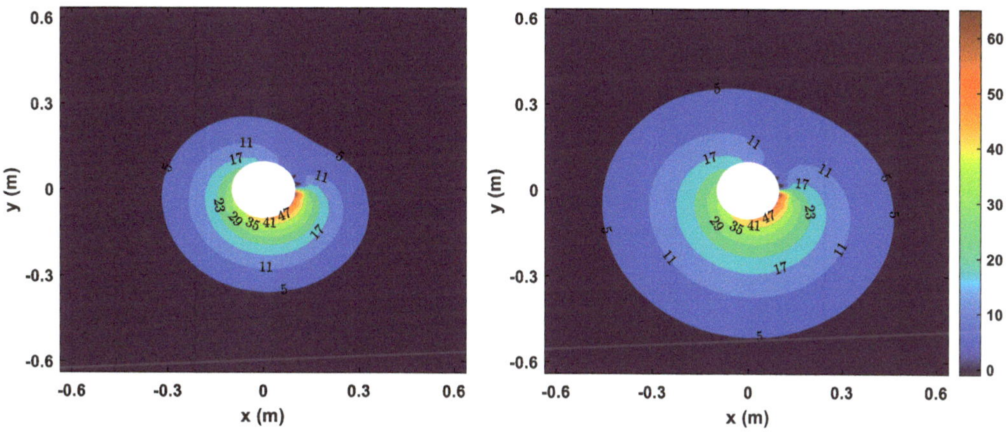

Fig. 5.8 Contour plots after 3.3 h (left), and 10 h (right) from the start of heating

Figure 5.8 shows the contour plots at two time intervals: after 3.3 and 10 h. The results clearly display the evolution of heat with time. This kind of analysis highlights the elegance of spectral analysis to simulate this relatively complex geometry and boundary condition using a systematic solution procedure.

5.5 Cylindrical Heat Source in an Infinite Domain with θ-Dependence: Neumann Problem

Here, we address the previous case but with a Neumann boundary condition applied at the heat source. In this context we study heat flow in a planar domain encompassing in its center a circular heat source with radius a, as depicted in Fig. 5.9. The heat source imposes a θ- and time-dependent heat flux.

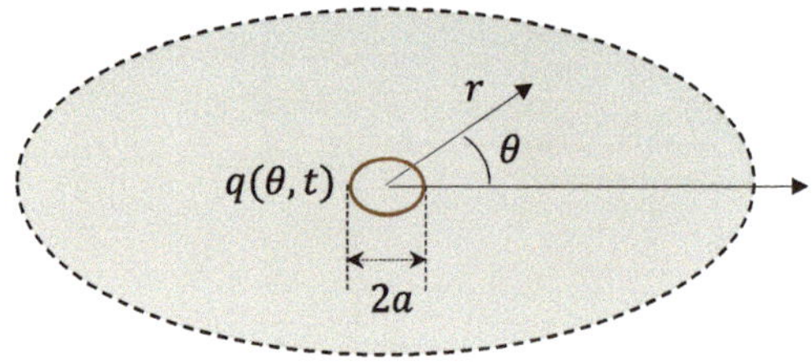

Fig. 5.9 Infinite cylindrical heat source with θ-dependence: Neumann problem

5

The boundary value problem of this case is

$$\begin{aligned} &\frac{\partial^2 u}{\partial r^2}+\frac{1}{r}\frac{\partial u}{\partial r}+\frac{1}{r^2}\frac{\partial^2 u}{\partial \theta^2}-\frac{1}{\alpha}\frac{\partial u}{\partial t}=0\\ &\text{IC}: \quad u(r,\theta,0)=0\\ &\text{BC}: \quad u(r\to\infty,\theta,t)=0\\ &\qquad\quad u(r,\theta,t)=u(r,\theta+2\pi,t)\\ &\qquad\quad -2\pi a\lambda\frac{\partial u}{\partial r}\bigg|_{r=a}=q(\theta,t) \end{aligned} \tag{5.38}$$

General Solution

The general solution of the differential equation in (5.38) in the frequency domain was given in detail in the previous case. It leads to

$$\hat{u}(r,\theta,\omega_n)=\sum\nolimits_m [A_m\cos(m\theta)+B_m\sin(m\theta)]K_m(\kappa r),\quad m=0,1,2,\cdots \tag{5.39}$$

with $\kappa=\sqrt{i\omega_n/\alpha}$.

Boundary Condition

Taking the derivative of (5.39) with respect to r gives

$$\frac{d\hat{u}}{dr}=\sum\nolimits_m [A_m\cos(m\theta)+B_m\sin(m\theta)]\left[-\kappa K_{m+1}(\kappa r)+\frac{m}{r}K_m(\kappa r)\right] \tag{5.40}$$

At $r=a$, the third boundary condition in (5.38) (in the frequency domain) implies

$$\sum\nolimits_m -\lambda 2\pi a[A_m\cos(m\theta)+B_m\sin(m\theta)]\left[-\kappa K_{m+1}(\kappa a)+\frac{m}{a}K_m(\kappa a)\right]=\hat{q}(\theta) \tag{5.41}$$

which is a trigonometric Fourier series with its coefficients determined by

$$
\begin{aligned}
A_0 &= \frac{1}{4\lambda\pi^2 a\kappa K_1(\kappa a)} \int_0^{2\pi} \hat{q}(\theta)\, d\theta, && m = 0 \\
A_m &= \frac{-1}{2\lambda\pi^2 a\left[-\kappa K_{m+1}(\kappa a) + \frac{m}{a} K_m(\kappa a)\right]} \int_0^{2\pi} \hat{q}(\theta)\cos(m\theta)\, d\theta, && m = 1, 2, 3, \cdots \\
B_m &= \frac{-1}{2\lambda\pi^2 a\left[-\kappa K_{m+1}(\kappa a) + \frac{m}{a} K_m(\kappa a)\right]} \int_0^{2\pi} \hat{q}(\theta)\sin(m\theta)\, d\theta, && m = 1, 2, 3, \cdots
\end{aligned}
\tag{5.42}
$$

BVP Solution

Solving (5.42) and substituting the resulting values of A_m and B_m into (5.39) yields the required solution for this Neumann boundary value problem in the frequency domain.

Special Case 1: Constant heat flux

In the special case where the prescribed heat flux is constant with magnitude $\hat{q}_0$, A_m and B_m in (5.42) become zero, and

$$
A_0 = \frac{\hat{q}_0}{2\lambda\pi a\kappa K_1(\kappa a)} \tag{5.43}
$$

Giving

$$
\hat{u}(r, \omega_n) = \frac{\hat{q}_0}{2\pi\lambda a\kappa K_1(\kappa a)} K_0(\kappa r) \tag{5.44}
$$

which is identical to the axial symmetric case given in ▶ Sect. 5.3, Eq. (5.15).

Special Case 2: Linear heat flux

In the special case where the prescribed heat flux is given by $\hat{q}_0\theta$, where $\hat{q}_0$ is constant, the Fourier coefficients in (5.42) become

$$
\begin{aligned}
A_0 &= \frac{\hat{q}_0}{4\lambda\pi^2 a\kappa K_1(\kappa a)} \int_0^{2\pi} \theta\, d\theta \\
&= \frac{\hat{q}_0}{2\lambda\pi a\kappa K_1(\kappa a)} \\
A_m &= \frac{-\hat{q}_0}{2\lambda\pi^2 a\left[-\kappa K_{m+1}(\kappa a) + \frac{m}{a} K_m(\kappa a)\right]} \int_0^{2\pi} \theta\cos(m\theta)\, d\theta, \quad m = 1, 2, 3, \cdots \\
&= 0 \\
B_m &= \frac{-\hat{q}_0}{2\lambda\pi^2 a\left[-\kappa K_{m+1}(\kappa a) + \frac{m}{a} K_m(\kappa a)\right]} \int_0^{2\pi} \theta\sin(m\theta)\, d\theta, \quad m = 1, 2, 3, \cdots \\
&= \frac{\hat{q}_0}{\lambda\pi a m\left[\frac{m}{a} K_m(\kappa a) - \kappa K_{m+1}(\kappa a)\right]}
\end{aligned}
\tag{5.45}
$$

where we had

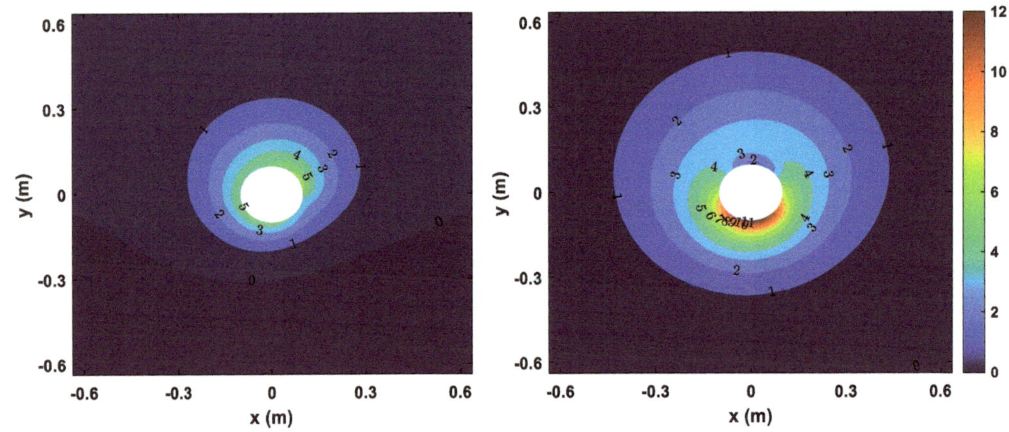

Fig. 5.10 Contour plots after 3.3 h (left), and 10 h (right) from the start of heating

$$\int_0^{2\pi} \theta \sin(m\theta)\, d\theta = -\frac{2\pi}{m} \tag{5.46}$$

Example 5.3: Neumann problem

Repeat Example 5.2, but for a Neumann boundary condition. The heat flux is assumed to vary linearly with θ, such that $q(t) = \theta q_0(t)$, where $q_0 = 20$ W, with a time variation similar to that given in Fig. 5.7

Solution

Following the solution steps outlined in the previous examples, and upon using (5.39) together with (5.45), we can readily obtain the solution of this problem. Figure 5.10 depicts the contour plots at two time intervals; 3.3 h and 10 h. The results clearly show the evolution of heat with time.

Problems

1. Compare the Fourier method solution of the point heat source model to the well-known Carslaw and Jaeger (1959) solution:

   $$T(r,t) = \frac{q}{2\pi r\lambda} \operatorname{erfc}\left(r \big/ 2\sqrt{\alpha t}\right)$$

   at several r distances. Also, go through the detailed derivation of the model using Laplace transform in Al-Khoury (2012), pp. 97–99, and discuss the difference in the derivation complexity between the Fourier transform and the Laplace transform.
2. Show that there is no steady-state solution to the infinite line heat source model (5.11).

Hint; Take the limit as $\omega \to 0$.

3. Solve the temperature distribution in the 2D planar of ▸ Sect. 5.4 for the special case where the prescribed temperature $U(t)$ at the heat source is constant (not a function of θ).
4. Compare the Fourier method solution of the infinite line heat source model to the well-known solution of Carslaw and Jaeger (1959) at several r distances.
   Hint: Use Hart and Couvillion (1986) approximation:

$$T(r,t) = \frac{q}{2\pi\lambda}\left[\ln\frac{r_\infty}{r} - 0.9818 + \frac{2r^2}{r_\infty^2} - \frac{1}{8}(\frac{4r^2}{r_\infty^2})\right]$$

   in which $r_\infty = 4\sqrt{\alpha t}$.
5. Also, go through the detailed derivation of the model using Laplace transform in Al-Khoury (2012), pp. 102–105, and discuss the difference between the Fourier transform and the Laplace transform. Use $q = 50\text{W/m}$; $\lambda = 2\text{W/m.K}$; $\rho = 1500\,\text{kg/m}^3$; $c = 800\text{J/kg.K}$.

References

Abramowitz M, Stegun IA (1972) Handbook of mathematical functions. Dover Publications, Inc

Al-Khoury R (2012) Computational modeling of shallow geothermal systems. CRC Press

Carslaw HS, Jaeger JC (1959) Conduction of heat in solids, 2nd ed. Oxford University Press

Hart P, Couvillion R (1986) Earth coupled heat transfer. National Water Well Association, pp 372–379

6

Heat Conduction in Finite Domains

R. Al Khoury and C. Kasbergen, *Analytical Subsurface Environment*,
https://doi.org/10.1007/978-3-031-93910-5_6

In the preceding chapter we addressed problems involving heat conduction in infinite domains. While these problems hold substantial significance in engineering and applied physics, practical cases often involve finite geometries. Finite domains relevant to engineering encompass a wide spectrum of applications, ranging from microchips to geothermal heat exchangers. Analyzing transient heat conduction in finite domains shares fundamental similarities with approaches used for infinite domains. However, the finite nature of the geometry introduces additional boundary conditions, thereby leading to certain mathematical and numerical complexities. In this chapter, we tackle the challenges posed by heat flow in finite domains, featured with various heat sources and boundary conditions. We will examine one-dimensional, planar, and cylindrical geometries, addressing both axially symmetric and non-symmetric heat transfer problems.

6.1 Finite 1D Solid Domain: Dirichlet Problem

The boundary value problem of a transient heat flow in a finite 1D solid domain, $0 \le z \le h$, subjected to Dirichlet boundary conditions, is given by

$$\begin{aligned} &\frac{1}{\alpha}\frac{\partial u}{\partial t} = \frac{\partial^2 u}{\partial z^2} \\ &\text{IC:} \quad u(z,0) = 0 \\ &\text{BC:} \quad u(0,t) = U(t) \\ &\qquad\;\; u(h,t) = 0 \end{aligned} \tag{6.1}$$

in which $\alpha = \lambda/\rho c$ is the thermal diffusivity $\left(\mathrm{m^2 s^{-1}}\right)$, $U(t)$ is a prescribed temperature (K) at the top surface, and h is the length (m) of the domain. The bottom surface is maintained at zero temperature, see Fig. 6.1.

General Solution

Applying the Fourier transform to (6.1) gives

$$\begin{aligned} &\frac{d^2\hat{u}}{dz^2} - \frac{i\omega}{\alpha}\hat{u} = 0 \\ &\text{BC:} \quad \hat{u}(0,\omega) = \hat{U}(\omega) \\ &\qquad\;\; \hat{u}(h,\omega) = 0 \end{aligned} \tag{6.2}$$

where it can be noticed that the original partial differential equation has been transformed to an ordinary differential equation. The general solution of the differential equation in (6.2) is given by

$$\hat{u}(z,\omega_n) = a_1 e^{\kappa z} + a_2 e^{-\kappa z}, \; \kappa = \sqrt{i\omega/\alpha} \tag{6.3}$$

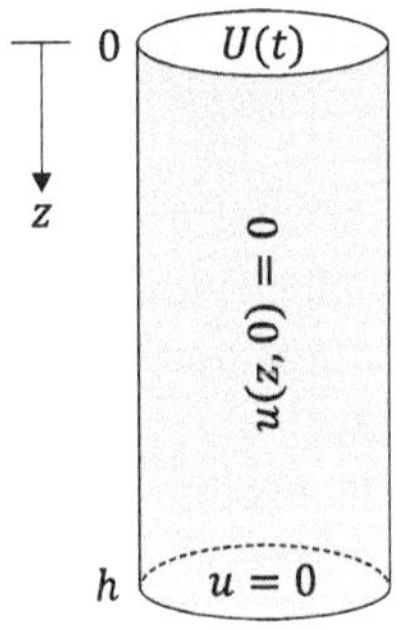

Fig. 6.1 Finite 1D solid domain

This solution can also be approached by a trial function of the form

$$\hat{u} = ae^{-i\eta z}$$

Substituting this expression into the differential equation in (6.2) gives the characteristic equation

$$\left(\eta^2 + \frac{i\omega}{\alpha}\right) a = 0 \tag{6.4}$$

where the term between the brackets must be zero, otherwise the solution will be trivial, rendering $\hat{u} = 0$. Thus

$$\eta = \mp\sqrt{-\frac{i\omega}{\alpha}}$$

which represent the two eigenvalues of the system:

$$\eta_1 = \sqrt{-\frac{i\omega}{\alpha}}; \quad \eta_2 = -\sqrt{-\frac{i\omega}{\alpha}}$$

Since $i^2 = -1$, η can be expressed in a more elegant form, such that

$$\eta_1 = i\sqrt{\frac{i\omega}{\alpha}}; \quad \eta_2 = -i\sqrt{\frac{i\omega}{\alpha}}$$

The solution of (6.2) can thus be expressed by the corresponding eigenfunctions, as

$$\begin{aligned} \hat{u}(z,\omega) &= a_1 e^{-i\eta_1 z} + a_2 e^{-i\eta_2 z} \\ &= a_1 e^{\sqrt{\frac{i\omega}{\alpha}} z} + a_2 e^{-\sqrt{\frac{i\omega}{\alpha}} z} \end{aligned} \tag{6.5}$$

Let $\kappa = \sqrt{i\omega_n/\alpha}$, then the second equation in (6.5) can be written as

$$\hat{u}(z, \omega_n) = a_1 e^{\kappa z} + a_2 e^{-\kappa z} \tag{6.6}$$

which is identical to (6.3).

Physically, the first term in (6.6) represents heat flow in the positive z-direction, and the second term denotes heat flow in the negative z-direction. Accordingly, Eq. (6.6) can also be expressed as

$$\begin{aligned}\hat{u}(z, \omega_n) &= A_1 e^{-i\kappa_1 z} + A_2 e^{-i\kappa_2 (z-h)} \\ &= A_1 e^{-i\kappa_1 z} + A_2 e^{-i\kappa_1 (h-z)} \\ &= A_1 e^{\kappa z} + A_2 e^{\kappa (h-z)}\end{aligned} \tag{6.7}$$

Although these formulations are mathematically valid general solutions to the problem, their suitability and robustness can vary depending on the specific characteristics and boundary conditions of the boundary value problems they address.

Boundary Conditions

At $z = 0$, the first boundary condition in (6.2) implies

$$a_1 + a_2 = \hat{U}$$

At $z = h$, the second boundary condition implies

$$a_1 e^{\kappa h} + a_2 e^{-\kappa h} = 0$$

Putting these two equations in a matrix form, gives

$$\begin{pmatrix} 1 & 1 \\ e^{\kappa h} & e^{-\kappa h} \end{pmatrix} \begin{pmatrix} a_1 \\ a_2 \end{pmatrix} = \begin{pmatrix} \hat{U} \\ 0 \end{pmatrix}$$

Solving for the vector $\{a_1 a_2\}^T$, yields

$$\begin{pmatrix} a_1 \\ a_2 \end{pmatrix} = \frac{1}{e^{-\kappa h} - e^{\kappa h}} \begin{pmatrix} e^{-\kappa h} & -1 \\ -e^{\kappa h} & 1 \end{pmatrix} \begin{pmatrix} \hat{U} \\ 0 \end{pmatrix}$$

With some algebraic rearrangements, it results in

$$a_1 = \frac{\hat{U}}{1 - e^{2\kappa h}}, \quad a_2 = \frac{\hat{U}}{1 - e^{-2kh}} \tag{6.8}$$

BVP Solution

Substituting (6.8) into (6.6), the solution of the boundary value problem (6.2) for any given frequency ω_n becomes

$$\hat{u}(z, \omega_n) = \left(\frac{e^{\kappa z}}{1 - e^{2\kappa h}} + \frac{e^{-\kappa z}}{1 - e^{-2\kappa h}} \right) \hat{U}(\omega_n) \tag{6.9}$$

The reader can verify that using any of the formulations in (6.7) would also lead to (6.9) (see Problem 1).

Notice that applying this model to a large system may result in inf (infinity) due to the positive exponent $e^{\kappa z}$ in the first term. To address this issue, we multiply both the nominator and the denominator of the first term by $e^{-2\kappa h}$, and with some algebraic manipulations, it leads to

$$\hat{u}(z, \omega_n) = \frac{\hat{U}(\omega_n)}{1 - e^{-2\kappa h}} \left[e^{-\kappa z} - e^{-\kappa(2h - z)} \right] \tag{6.10}$$

Having solved the problem in the frequency domain, the solution in the time domain can readily be obtained via the inverse fast Fourier transform (IFFT) algorithm, such that

$$u(z, t) = \sum_n \hat{u}(z, \omega_n) \, e^{i\omega_n t}$$

Special Case: Semi-infinite Domain

The model in (6.10) can be extended to a semi-infinite domain, simply by taking $h \to \infty$, leading to

$$\hat{u}(z, \omega_n) = \hat{U}(\omega_n) \, e^{-\kappa z} \tag{6.11}$$

The reader might find it intriguing to compare the temperature computed by this model with the conventional model commonly discussed in literature. The conventional model states that for a continuous boundary condition, U, the temperature distribution in a semi-infinite domain is given by

$$u(z, t) = \left[[1 - \text{erf} \left(\frac{z}{2\sqrt{\alpha t}} \right) \right] U \tag{6.12}$$

where erfXE "Erf" is the *error function*, see Problem 2. It is worth noting that the mathematical derivation leading to (6.12) is not trivial. It demands an in-depth knowledge of the Fourier integrals and their manipulations. See for example Al-Khoury (2012) and Xie (2010)

6.2 Finite 1D Solid Domain: Double Dirichlet Boundaries

Here we extend the previous boundary value problem to a case where the Dirichlet boundary condition at the bottom surface is not identically zero, see ◘ Fig. 6.2. This boundary value problem is expressed as

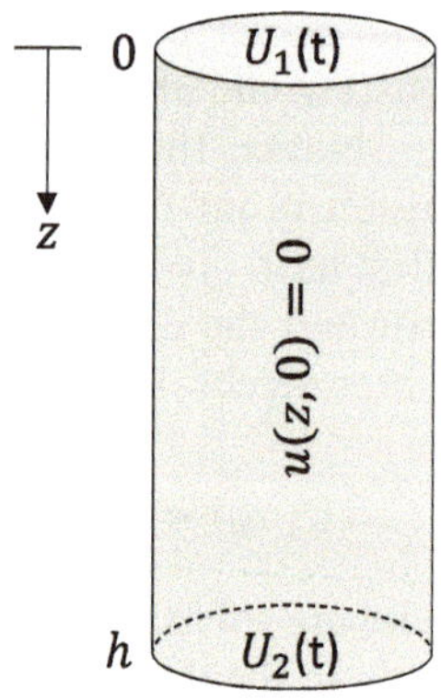

Fig. 6.2 Double Dirichlet boundary problem

$$
\begin{aligned}
&\frac{1}{\alpha}\frac{\partial u}{\partial t} = \frac{\partial^2 u}{\partial z^2} \\
&\quad \text{IC:} \quad u(z,0) = 0 \\
&\quad \text{BC:} \quad u(0,t) = U_1(t) \\
&\qquad\qquad u(h,t) = U_2(t).
\end{aligned}
\tag{6.13}
$$

General Solution

Applying the Fourier transform to (6.13) yields

$$
\begin{aligned}
&\frac{d^2\hat{u}}{dz^2} - \frac{i\omega}{\alpha}\hat{u} = 0 \\
&\text{BC:} \quad \hat{u}(0,\omega) = \hat{U}_1(\omega) \\
&\qquad\quad \hat{u}(h,\omega) = \hat{U}_2(\omega)
\end{aligned}
\tag{6.14}
$$

Following the procedure outlined in ▶ Sect. 6.1, the solution of the DE in (6.14) can be expressed as

$$
\hat{u}(z,\omega_n) = a_1 e^{-\kappa z} + a_2 e^{-\kappa(h-z)}, \qquad \kappa = \sqrt{\frac{i\omega}{\alpha}}. \tag{6.15}
$$

Boundary Conditions

At $z = 0$, the first boundary condition in (6.14) implies

$$
\hat{U}_1(\omega_n) = a_1 + a_2 e^{-\kappa h}
$$

At $z = h$, the second boundary condition implies

$$
\hat{U}_2(\omega_n) = a_1 e^{-\kappa h} + a_2
$$

Solving these two equations simultaneously gives

$$a_1 = \frac{1}{1 - e^{-2\kappa h}} (\hat{U}_1 - e^{-\kappa h} \hat{U}_2), \quad a_2 = \frac{1}{1 - e^{-2\kappa h}} (\hat{U}_2 - e^{-\kappa h} \hat{U}_1)$$

BVP Solution

Following this, the solution of this boundary value problem for a given frequency ω_n is then

$$\hat{u}(z, \omega_n) = \frac{1}{1 - e^{-2\kappa h}} \left[(\hat{U}_1 - e^{-\kappa h} \hat{U}_2) e^{-\kappa z} + (\hat{U}_2 - e^{-\kappa h} \hat{U}_1) e^{-\kappa (h-z)} \right]$$

with

$$\kappa = \sqrt{\frac{i\omega}{\alpha}} \tag{6.16}$$

The solution in the time domain can then be obtained using the IFFT algorithm, as

$$u(z, t) = \sum_n \hat{u}(z, \omega_n) e^{i\omega_n t}.$$

Alternative Solution Method

As the boundary conditions are both nonhomogeneous, this BVP can also be solved by splitting the solution into two sub-solutions, such that

$$u(z, t) = u_1(z, t) + u_2(z, t) \tag{6.17}$$

Substituting this equation into (6.13) gives

$$\begin{aligned} &\frac{1}{\alpha}\frac{\partial u_1}{\partial t} + \frac{1}{\alpha}\frac{\partial u_2}{\partial t} - \frac{\partial^2 u_1}{\partial z^2} - \frac{\partial^2 u_2}{\partial z^2} = 0 \\ &\text{BC:} \quad u_1(0, t) + u_2(0, t) = U_1(t) \\ &\qquad\quad u_1(h, t) + u_2(h, t) = U_2(t) \end{aligned}$$

This boundary value problem can be divided into two sub-problems:

- **Sub-problem 1**

$$\begin{aligned} &\frac{\partial^2 u_1}{\partial z^2} - \frac{1}{\alpha}\frac{\partial u_1}{\partial t} = 0 \\ &\text{IC:} \quad u_1(r, z, 0) = 0 \\ &\text{BC:} \quad u_1(z = 0, t) = U_1(t) \\ &\qquad\quad u_1(z = h, t) = 0 \end{aligned} \tag{6.18}$$

- **Sub-problem 2**

$$\begin{aligned}
&\frac{\partial^2 u_2}{\partial z^2} - \frac{1}{\alpha}\frac{\partial u_2}{\partial t} = 0 \\
&\text{IC:} \quad u_2(r, z, 0) = 0 \\
&\text{BC:} \quad u_2(z = 0, t) = 0 \\
&\qquad\quad u_2(z = h, t) = U_2(t)
\end{aligned} \tag{6.19}$$

Following the solution procedure given in ▶ Sect. 6.1, the solutions, in the frequency domain, of these two sub-problems would lead to

$$\hat{u}(z, \omega_n) = \frac{1}{1 - e^{-2\kappa h}}\left\{\hat{U}_1(\omega_n)\left[e^{-\kappa z} - e^{-\kappa(2h-z)}\right] + \hat{U}_2(\omega_n)\left[e^{-\kappa(h-z)} - e^{-\kappa(h+z)}\right]\right\}$$

with

$$\kappa = \sqrt{\frac{i\omega}{\alpha}} \tag{6.20}$$

which is identical to (6.16).

6.3 Finite 1D Solid Domain: Robin Boundary Problem

The boundary value problem of a transient heat flow in a finite 1D solid domain, $0 \le z \le h$, subjected to a Robin boundary condition at its surface, is given by

$$\begin{aligned}
&\frac{1}{\alpha}\frac{\partial u}{\partial t} = \frac{\partial^2 u}{\partial z^2} \\
&\text{IC:} \quad u(z, 0) = 0 \\
&\text{BC:} \quad \lambda \left.\frac{\partial u}{\partial z}\right|_{z=0} = b_s[u - U_\infty(t)] \\
&\qquad\quad u(h, t) = 0
\end{aligned} \tag{6.21}$$

in which b_s $\left(\mathrm{Wm^{-2}K^{-1}}\right)$ is a thermal interaction coefficient at the surface, and $U_\infty(t)$ is the temperature of the surrounding environment at $z = 0$, which might be air temperature. The boundary at $z = h$ is maintained at zero temperature, as shown in ◘ Fig. 6.3. Note that the Robin boundary condition is a direct application of Newton's law of cooling. In some literatures, this boundary condition is written as

$$-A\lambda \left.\frac{\partial u}{\partial z}\right|_{z=0} = b\left[u - U_\infty(t)\right]$$

where A is the cross-sectional area where the heat exchange between the domain and the surrounding environment takes place. Consequently, the thermal interaction coefficient, b, will have different units and values.

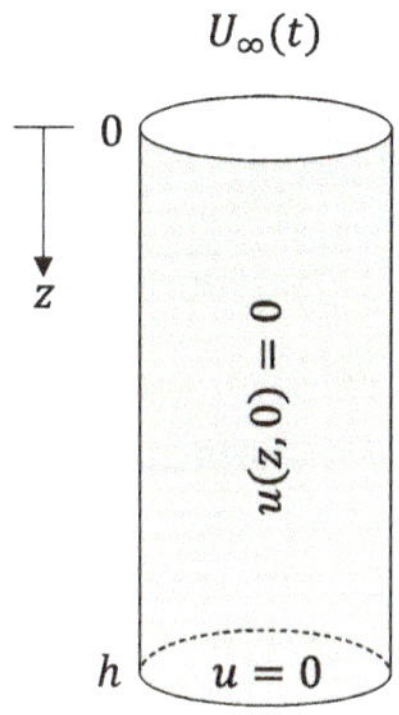

Fig. 6.3 Robin boundary problem

General Solution

Applying the Fourier transform to (6.21) yields

$$\begin{aligned} &\frac{d^2\hat{u}}{dz^2} - \frac{i\omega}{\alpha}\hat{u} = 0 \\ &\text{BC:} \quad -\lambda\frac{d\hat{u}}{dz}\bigg|_{z=0} = b_s\left[\hat{u} - \hat{U}_\infty(\omega)\right] \\ &\qquad \hat{u}(h,\omega) \;= 0 \end{aligned} \tag{6.22}$$

The general solution of this ordinary differential equation is given in detail in ▶ Sect. 6.1, yielding

$$\hat{u}\,(z,\omega_n) = a_1 e^{\kappa z} + a_2 e^{-\kappa z}, \quad \kappa = \sqrt{i\omega/\alpha}. \tag{6.23}$$

Boundary Conditions

Taking the first derivative of (6.23) with respect to z gives

$$\frac{d\hat{u}}{dz} = \kappa\,(a_1 e^{\kappa z} - a_2 e^{-\kappa z})$$

At $z = 0$, the first boundary condition in (6.22) implies

$$a_1(-\lambda\kappa - b_s) + a_2(\lambda\kappa - b_s) = -b_s\hat{U}_\infty$$

At $z = h$, the second boundary condition implies

$$a_1 e^{\kappa h} + a_2 e^{-\kappa h} = 0$$

Putting these two equations in a matrix form and solving for the vector $\{a_1 a_2\}^T$, yields

$$a_1 = \frac{-b_s \hat{U}_\infty}{(-\lambda\kappa - b_s) - (\lambda\kappa - b_s)\, e^{2\kappa h}}$$

$$a_2 = \frac{b_s \hat{U}_\infty}{(-\lambda\kappa - b_s)\, e^{-2\kappa h} - (\lambda\kappa - b_s)}.$$

BVP Solution

Having determined the two constants, the solution of the boundary value problem (6.22) for a given frequency ω_n is thus

$$\begin{aligned} \hat{u}\,(z, \omega_n) = & \frac{-b_s \hat{U}_\infty}{-\lambda\kappa\,(1 + e^{2\kappa h}) - b_s(1 - e^{2\kappa h})} e^{\kappa z} \\ & + \frac{b_s \hat{U}_\infty}{-\lambda\kappa\,(1 + e^{-2\kappa h}) + b_s(1 - e^{-2\kappa h})} e^{-\kappa z} \end{aligned} \tag{6.24}$$

Similar to the previous case, applying this model for a large system might lead to inf (infinity) due to the presence of the positive exponent $e^{\kappa z}$ in the first term. To address this issue, we multiply the nominator and the denominator of the first term by $e^{-2\kappa h}$, that, with some algebraic manipulations, gives

$$\hat{u}\,(z, \omega_n) = \frac{b_s \hat{U}_\infty}{\lambda\kappa\,(1 + e^{-2\kappa h}) - b_s(1 - e^{-2\kappa h})} \left[e^{-\kappa\,(2h - z)} - e^{-\kappa z} \right] \tag{6.25}$$

Having solved the problem in the frequency domain, using the inverse FFT algorithm yields the solution in the time domain, such that

$$u\,(z, t) = \sum_n \hat{u}\,(z, \omega_n)\, e^{i\omega_n t}.$$

Special Case: Semi-infinite Domain

The solution in (6.25) can readily be extended to a semi-infinite domain, simply by taking the limit as $h \to \infty$, leading to

$$\hat{u}\,(z, \omega_n) = -\frac{b_s \hat{U}_\infty}{\lambda\kappa - b_s} e^{-\kappa z}. \tag{6.26}$$

- **Example 6.1: Robin Problem**

The average monthly air temperature in a region is given in Table. 6.1. Assuming one-dimensional heat flow, calculate the temperature distribution in the ground down to a depth of 20 m during January, February, March, July and November. The material parameters of the ground are: thermal conductivity 1 $Wm^{-1}K^{-1}$, mass density 1500 kg m^{-3}, and specific heat

Table. 6.1 Average monthly air temperature

January	February	March	April	May	June
0	-5	0	5	7	10

July	August	September	October	November	December
15	20	15	10	7	5

capacity 800 J kg^{-1}K^{-1}. The thermal interaction coefficient between the air and the ground surface, b_S, is assumed -5 Wm^{-2}K^{-1}. The ground initial temperature is assumed 0 °C. Note that we assigned a negative value to b_s because the air temperature is generally higher than or equal to the initial temperature in the ground. As a result, heat flow typically occurs in the direction opposite to that of the normal vector $\boldsymbol{n}$, as depicted in Fig. 6.4.

Solution

Determining the temperature distribution in the ground can be achieved via:

a) Transforming the input pulse provided in Table 6.1 into the frequency domain using the FFT algorithm. Here, we employed 4096 (2^{12}) FFT samples with a time step of 1 month (2,592,000 s).
b) Evaluating the temperature at various distances from the ground surface using (6.25) for each frequency, followed by summing the results for all involved frequencies.
c) Transforming the calculated results back to the time domain, using the inverse FFT (IFFT) algorithm.

Fig. 6.4 Heat flow direction

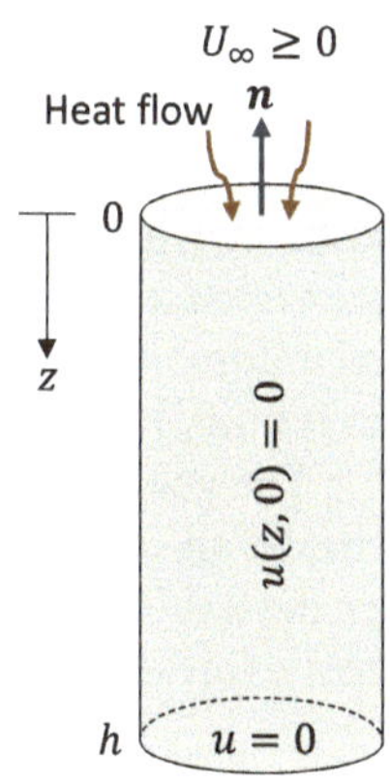

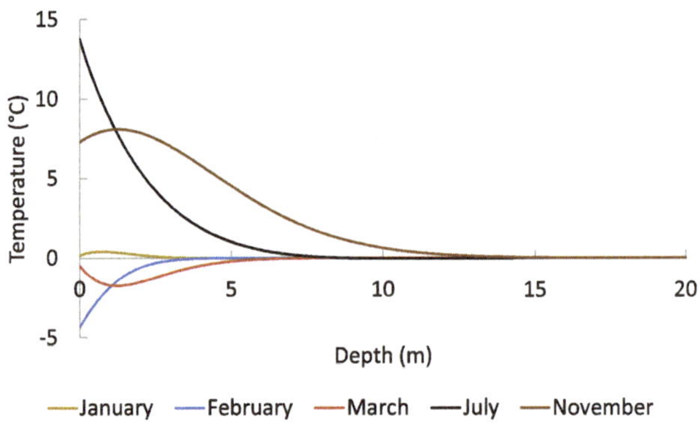

Fig. 6.5 Computed temperature distribution in the ground

Figure 6.5 shows the temperature distribution along the depth of the domain during January, February, March, July and November. Notably, the figure highlights how the seasonal fluctuations in air temperature affect only the upper 10–12 m of the ground. Below this depth, the ground temperature remains nearly constant, influenced by the natural heat transfer between the earth core and the air. This phenomenon bears significant relevance for shallow geothermal systems, as borehole heat exchangers are often designed to reach depths of up to 250 m, benefiting from the consistent temperature at deeper layers throughout the year.

6.4 Finite Line Heat Source (FLS): Constant Heat Flux

The Finite Line Source (FLS) is a mathematical model describing heat transfer from a line heat source to a surrounding medium. It is commonly applied in various engineering fields, such as thermal analysis of electronic components, underground cables, and geothermal systems. The model is tailored specifically to cases where a constant heat flux $q(t)$ radiates from a line source positioned along the axis of symmetry of a domain bounded in the vertical z direction and extending to infinity in the radial r direction, as illustrated in Fig. 6.6.

The geometry and the boundary condition in this case give rise to an axial-symmetric heat flow, with the z-axis standing as the axis of symmetry. Thus, the boundary value problem is best expressed in the cylindrical coordinate system, as

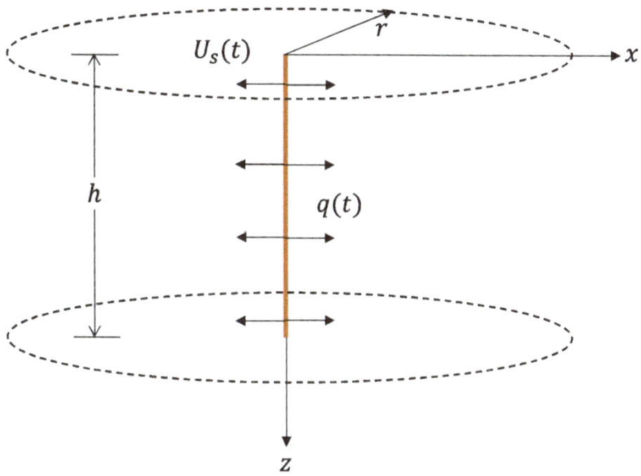

Fig. 6.6 Finite line heat source

$$
\begin{aligned}
&\frac{1}{\alpha}\frac{\partial u}{\partial t} - \frac{\partial^2 u}{\partial r^2} - \frac{1}{r}\frac{\partial u}{\partial r} - \frac{\partial^2 u}{\partial z^2} = 0 \\
&\text{IC:} \quad u\,(r,z,0) = 0 \\
&\text{BC:} \quad u\,(r \to \infty, z, t) = 0 \\
&\qquad\quad u\,(r,0,t) = U_s(t) \\
&\qquad\quad u\,(r,h,t) = 0 \\
&\qquad\quad -\lambda\frac{\partial u}{\partial r}\cdot 2\pi r\bigg|_{r\to 0} = q\,(t)
\end{aligned}
\tag{6.27}
$$

where h is the height of the domain, and U_s is the prescribed surface temperature. Other boundary conditions can also be applied on the upper and lower boundaries of the domain; however, the focus here is on the solution procedure.

▪ Solution

The partial differential equation in (6.27) is linear and homogeneous, but two of the boundary conditions are nonhomogeneous, since they are not identically zero. As the surface temperature $U_s(t)$ is not a function of r and the heat flux $q(t)$ is not a function of z, the problem in (6.27) can be broken into two sub-problems, with one addressing heat flow in the vertical z direction and the other in the radial r direction, such that

$$u\,(r,z,t) = u_1(z,t) + u_2(r,t) \tag{6.28}$$

Substituting (6.28) into (6.27) gives

$$\frac{1}{\alpha}\frac{\partial u_1}{\partial t} + \frac{1}{\alpha}\frac{\partial u_2}{\partial t} - \frac{\partial^2 u_2}{\partial r^2} - \frac{1}{r}\frac{\partial u_2}{\partial r} - \frac{\partial^2 u_1}{\partial z^2} = 0 \tag{6.29}$$

This equation can be split into two equations, expressed with their relevant boundary conditions as.

- **Sub-problem 1**

$$\begin{aligned}
&\frac{\partial^2 u_1}{\partial z^2} - \frac{1}{\alpha}\frac{\partial u_1}{\partial t} = 0 \\
&\text{IC:} \quad u_1(z,0) = 0 \\
&\text{BC:} \quad u_1(0,t) = U_s(t) \\
&\qquad\quad u_1(h,t) = 0
\end{aligned} \tag{6.30}$$

- **Sub-problem 2**

$$\begin{aligned}
&\frac{\partial^2 u_2}{\partial r^2} + \frac{1}{r}\frac{\partial u_2}{\partial r} - \frac{1}{\alpha}\frac{\partial u_2}{\partial t} = 0 \\
&\text{IC:} \quad u_2(r,0) = 0 \\
&\text{BC:} \quad u_2(r \to \infty, t) = 0 \\
&\qquad\quad -\lambda \frac{\partial u_2}{\partial r} \cdot 2\pi r \Big|_{r \to 0} = q\,(t).
\end{aligned} \tag{6.31}$$

BVP Solution

The solution of the BVP in (6.30) is given in ▶ Sect. 6.1, and the solution of the BVP in (6.31) is given in Chap. 5, ▶ Sect. 5.2. Thus, the solution of the boundary value problem (6.27) in the frequency domain can be expressed as

$$\hat{u}\,(r,z,\omega_n) = \underbrace{\frac{\hat{U}_s(\omega_n)}{1 - e^{-2\kappa h}}[e^{-\kappa z} - e^{-\kappa(2h-z)}]}_{\hat{u}_1(z,\omega)} + \underbrace{\frac{\hat{q}_0(\omega_n)}{2\pi\lambda}K_0(\kappa r)}_{\hat{u}_2(r,\omega)} \tag{6.32}$$

with $\kappa = \surd(i\omega_n/\alpha)$.

Having the solution in the frequency domain, the general solution in the time domain can readily be obtained by the IFFT algorithm, as

$$u(r,z,t) = \sum_n \hat{u}(r,z,\omega_n)e^{i\omega_n t}. \tag{6.33}$$

6.5 Finite Line Heat Source (FLS): Heat Flux Function of *z*

In the previous section we addressed a boundary value problem for a finite line heat source that extends across the entire thickness of the layer. Here, we expand our analysis to encompass a case where the heat source spans to depth shorter than the layer thickness, thereby rendering the prescribed heat flux a function of z, as depicted in ◘ Fig. 6.7.

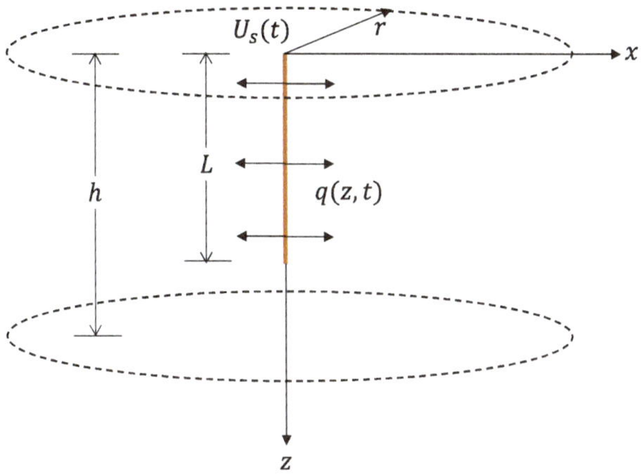

Fig. 6.7 Finite line heat source with the heat source a function of z

A relevant boundary value problem of this case can be

$$\begin{aligned}
&\frac{1}{\alpha}\frac{\partial u}{\partial t}-\frac{\partial^2 u}{\partial r^2}-\frac{1}{r}\frac{\partial u}{\partial r}-\frac{\partial^2 u}{\partial z^2}=0\\
&\text{IC:}\quad u(r,z,0)=0\\
&\text{BC:}\quad u(r\to\infty,t)=0\\
&\qquad u(r,0,t)=U_s(t)\\
&\qquad u(r,h,t)=0\\
&\qquad -\lambda\frac{\partial u}{\partial r}\cdot 2\pi r\bigg|_{r\to 0}=q(z,t).
\end{aligned}\tag{6.34}$$

where h is the height of the domain that is larger than the length L of the heat source.

▪ Solution

As for the previous case, this problem is nonhomogeneous, and the solution can best be attained by splitting the solution into two sub-solutions, such that

$$u(r,z,t)=u_1(z,t)+u_2(r,z,t)\tag{6.35}$$

where, in contrast to the previous case, u_2 is made a function of r and z. This disparity arises from that the heat flux $q(z,t)$ generates heat flow in both r and z directions. Inserting (6.35) into the partial differential equation in (6.34) gives

$$\frac{1}{\alpha}\frac{\partial u_1}{\partial t}+\frac{1}{\alpha}\frac{\partial u_2}{\partial t}-\frac{\partial^2 u_1}{\partial z^2}-\frac{\partial^2 u_2}{\partial r^2}-\frac{1}{r}\frac{\partial u_2}{\partial r}-\frac{\partial^2 u_2}{\partial z^2}=0\tag{6.36}$$

This equation can be divided into two equations, which can be expressed with their relevant boundary conditions as

6

- **Sub-problems 1**

$$
\begin{aligned}
&\frac{\partial^2 u_1}{\partial z^2} - \frac{1}{\alpha}\frac{\partial u_1}{\partial t} = 0 \\
&\text{IC:} \quad u_1(z,0) = 0 \\
&\text{BC:} \quad u_1(0,t) = U_s(t) \\
&\qquad\quad u_1(h,t) = 0.
\end{aligned}
\tag{6.37}
$$

- **Sub-problem 2**

$$
\begin{aligned}
&\frac{\partial^2 u_2}{\partial r^2} + \frac{1}{r}\frac{\partial u_2}{\partial r} + \frac{\partial^2 u_2}{\partial z^2} - \frac{1}{\alpha}\frac{\partial u_2}{\partial t} = 0 \\
&\text{IC:} \quad u_2(r,z,0) = 0 \\
&\text{BC:} \quad u_2(r \to \infty, t) = 0 \\
&\qquad\quad \left.\frac{\partial u_2}{\partial z}\right|_{z=0} = 0 \\
&\qquad\quad \left.\frac{\partial u_2}{\partial z}\right|_{z=h} = 0 \\
&\qquad\quad \left.-\lambda\frac{\partial u_2}{\partial r}\cdot 2\pi r\right|_{r\to 0} = q\,(z,t)
\end{aligned}
\tag{6.38}
$$

We note here that by adding the second and third boundary conditions in (6.38) to the first and second boundary conditions in (6.37) we obtain the original boundary conditions in (6.34). Alternatively, depending on the physics of the problem, we can set $u_2(r,0,t) = 0$ and $u_2(r,h,t) = 0$. See the "Alternative formulation" below.

- **Solving Sub-problem 1**

The solution of the boundary value problem (6.37) is given in detail in ▶ Sect. 6.1. In the frequency domain, it yields

$$
\hat{u}_1(z,\omega_n) = \frac{\hat{U}_s(\omega_n)}{1 - e^{-2\kappa h}}[e^{-\kappa z} - e^{-\kappa(2h-z)}], \quad \kappa = \sqrt{\frac{i\omega_n}{\alpha}}.
\tag{6.39}
$$

- **Solving Sub-problem 2**

Applying the Fourier transform to the BVP (6.38) gives

$$
\begin{aligned}
&\frac{\partial^2 \hat{u}_2}{\partial r^2} + \frac{1}{r}\frac{\partial \hat{u}_2}{\partial r} + \frac{\partial^2 \hat{u}_2}{\partial z^2} - \frac{i\omega}{\alpha}\hat{u}_2 = 0 \\
&\text{BC:} \quad \hat{u}_2(r \to \infty, z, \omega) = 0 \\
&\qquad\quad \left.\frac{\partial \hat{u}_2}{\partial z}\right|_{z=0} = 0 \\
&\qquad\quad \left.\frac{\partial \hat{u}_2}{\partial z}\right|_{z=h} = 0 \\
&\qquad\quad \left.-\lambda\frac{\partial \hat{u}_2}{\partial z}\cdot 2\pi r\right|_{r\to 0} = \hat{q}\,(z,\omega)
\end{aligned}
\tag{6.40}
$$

The solution of the partial differential equation in (6.40) can be carried out using the method of separation of variables. Assume

$$\hat{u}_2(r,z) = R(r)\,Z(z) = RZ \tag{6.41}$$

in which $R(r)$ is a function of r only and $Z(z)$ is a function of z only. Substituting (6.41) into the PDE in (6.40), gives

$$R''Z + \frac{1}{r}R'Z + RZ'' - \frac{i\omega}{\alpha}RZ = 0 \tag{6.42}$$

where the accents ("and") are the first and second derivative operators. Dividing through by RZ

$$\frac{R''}{R} + \frac{1}{r}\frac{R'}{R} - \frac{i\omega}{\alpha} = -\frac{Z''}{Z}$$

and equating both sides of this equation by some arbitrary constant, say η, we obtain

$$\frac{R''}{R} + \frac{1}{r}\frac{R'}{R} - \frac{i\omega}{\alpha} = -\frac{Z''}{Z} = \eta.$$

This separation yields two independent ordinary differential equations, such that

$$\frac{d^2Z}{dz^2} + \eta Z = 0 \tag{6.43}$$

$$\frac{d^2R}{dr^2} + \frac{1}{r}\frac{dR}{dr} - \left(\eta + \frac{i\omega}{\alpha}\right)R = 0 \tag{6.44}$$

We begin with solving $Z(z)$ equation. Depending on η, the three possible general solutions of (6.43) are

$$\begin{aligned} Z(z) &= c_1 + c_2 z, & \eta &= 0 \\ Z(z) &= c_1 \cosh(\beta z) + c_2 \sinh(\beta z), & \eta &= -\beta^2 < 0 \\ Z(z) &= c_1 \cos(\beta z) + c_2 \sin(\beta z), & \eta &= \beta^2 > 0 \end{aligned} \tag{6.45}$$

where c_1 and c_2 are arbitrary constants, need to be determined from the boundary conditions.

To satisfy the homogeneous boundary conditions, we seek a solution that is periodic. Apparently the second solution in (6.45) is nonperiodic and thus can be discarded. The first solution is periodic if $c_2 = 0$, with the remaining solution $Z(z) = c_1, c_1 \neq 0$. Hence, $\eta = 0$ is an eigenvalue. Finally, the third solution can be periodic, thereby rendering it suitable for solving the problem.

Taking the derivative of the third solution in (6.45) with respect to z gives

$$Z' = -c_1\beta \sin(\beta z) + c_2\beta \cos(\beta z)$$

At $z = 0$, the second boundary condition in (6.40) implies $c_2 = 0$, leading to

$$Z = c_1 \cos(\beta z)$$

Taking the first derivative of this equation with respect to z and then solving for $z = h$, the third boundary condition in (6.40) implies

$$-c_1 \beta \sin(\beta h) = 0$$

As c_1 or β cannot be zero, since it would lead to a trivial solution, it renders

$$\sin(\beta h) = 0$$

which is valid if and only if

$$\beta h = m\pi \quad \text{or} \quad \beta_m = \frac{m\pi}{h}, \qquad m = 1, 2, \ldots$$

6

This leads the solution to become

$$Z(z) = c_1 \cos(\beta_m z) \tag{6.46}$$

Having solved $Z(z)$, we now proceed to address $R(r)$. Equation (6.44) can be written as

$$R'' + \frac{1}{r} R' - \kappa_r^2 R = 0$$

with

$$\begin{aligned} \kappa_r^2 &= \frac{i\omega}{\alpha} + \beta_m^2 \\ &= \kappa^2 + \beta_m^2 \end{aligned}$$

Multiplying this differential equation by r^2 gives

$$r^2 R'' + rR' - \kappa_r^2 r^2 R = 0$$

which is a parametric form of the modified Bessel equation of order 0, with its solution given by

$$R(r) = c_3 I_0(\kappa_r r) + c_4 K_0(\kappa_r r)$$

in which I_0 and K_0 are the modified Bessel functions of zero order of the first and second kind, respectively. c_3 and c_4 are arbitrary constants, need to be determined from the boundary conditions.

The first boundary condition in (6.40) implies $c_3 = 0$, since I_0 is unbounded at far distances. Hence

$$R(r) = c_4 K_0(\kappa_r r) \tag{6.47}$$

Combining both solutions for $Z(z)$, (6.46), and $R(r)$, (6.47), yields the general solution for every m, as

$$\hat{u}_{2m} = R(r)\, Z(z) = A_m K_0(\kappa_r r) \cos(\beta_m z)$$

where $A_m = c_1.c_4$. Since the boundary value problem is linear, the solution can be expressed as an infinite summation over m eigenfunctions, leading to

$$\hat{u}_2(r, z, \omega) = \sum_m A_m K_0(\kappa_r r) \cos(\beta_m z), \qquad m = 0, 1, 2, \ldots \tag{6.48}$$

where you note that we also considered $m = 0$, since, as indicated above, it also generates a periodic solution.

Taking the first derivative of (6.48) with respect to r gives

$$\frac{\partial \hat{u}_2}{\partial r} = -A_m \cos(\beta_m z) \cdot \kappa_r K_1(\kappa_r r)$$

In the limit as $r \to 0$, we know (refer to ▶ Chap. 5) that

$$K_1(\kappa_r r)|_{r \to 0} \cong \frac{1}{\kappa_r r}$$

Hence, the fourth boundary condition in (6.40) implies

$$\sum_m A_m \cos(\beta_m z) \cdot \kappa_r \frac{1}{\kappa_r r} 2\pi r \lambda = \hat{q}(z, \omega)$$

Working out the algebra yields

$$\sum_m A_m \cos(\beta_m z) = \frac{\hat{q}(z, \omega)}{2\pi\lambda}$$

This is a Fourier cosine series with its coefficients expressed as

$$\begin{aligned} A_0 &= \frac{1}{h} \int_0^L \frac{\hat{q}(z, \omega)}{2\pi\lambda} dz, & m &= 0 \\ A_m &= \frac{2}{h} \int_0^L \frac{\hat{q}(z, \omega)}{2\pi\lambda} \cos(\beta_m z)\, dz, & m &= 1, 2, \ldots \end{aligned} \tag{6.49}$$

in which L is the length of the heat source. Notice that the coefficients (6.49) are those for a *half range cosine Fourier series* with period $2h$, see Zill and Wright (2011).

Following this, the solution of $\hat{u}_2$ for a given frequency ω_n becomes

$$\hat{u}_2(r, z, \omega_n) = \sum_m A_m \cos(\beta_m z)\, K_0(\kappa_r r), \quad m = 0, 1, 2, \ldots$$

with

$$\begin{aligned} A_0 &= \frac{1}{2\pi\lambda h} \int_0^L \hat{q}(z, \omega_n)\, dz, & m &= 0 \\ A_m &= \frac{1}{\pi\lambda h} \int_0^L \hat{q}(z, \omega_n) \cos(\beta_m z)\, dz, & m &= 1, 2, \ldots \\ \beta_m &= \frac{m\pi}{h}. \end{aligned} \tag{6.50}$$

Special Case 1: Constant Flux

In the special case where $\hat{q}$ is constant (not function of z), the Fourier coefficients become

$$A_0 = \frac{\hat{q}(\omega)L}{2\pi\lambda h}$$

$$A_m = \frac{\hat{q}(\omega)}{\pi\lambda h}\int_0^L \cos(\beta_m z)\,dz$$

$$= \frac{\hat{q}(\omega)}{\pi\lambda h\beta_m}\sin(\beta_m L), \qquad m = 1, 2, ...$$

6

Special Case 2: Constant Flux and $L = h$

In the special case where $\hat{q}$ is constant and $L = h$, the Fourier coefficients become

$$A_0 = \frac{\hat{q}(\omega)}{2\pi\lambda}$$

$$A_m = \frac{\hat{q}(\omega_n)}{\pi^2\lambda m}\sin(m\pi), \quad m = 1, 2, 3, \ldots \tag{6.51}$$

$$= 0.$$

BVP Solution

Collecting the two solutions, the solution of the boundary value problem (6.34) in the frequency domain entails

$$\hat{u}(r, z, \omega_n) = \hat{u}_1(z, \omega_n) + \hat{u}_2(r, z, \omega_n) \tag{6.52}$$

Substituting the solutions of $\hat{u}_1$ (6.39) and $\hat{u}_2$ (6.50) into (6.52) yields

$$\hat{u}(r, z, \omega_n) = \frac{\hat{U}_s(\omega_n)}{1 - e^{-2\kappa h}}[e^{-\kappa z} - e^{-\kappa(2h-z)}] + \sum_m A_m K_0(\kappa_r r)\cos(\beta_m z), \qquad m = 0, 1, 2, \ldots$$

with

$$A_0 = \frac{1}{2\pi\lambda h}\int_0^L \hat{q}(z, \omega_n)\,dz, \qquad m = 0$$

$$A_m = \frac{1}{\pi\lambda h}\int_0^L \hat{q}(z, \omega_n)\cos(\beta_m z)\,dz, \qquad m = 1, 2, \ldots \tag{6.53}$$

$$\kappa_r = \sqrt{\kappa^2 + \beta^2}$$

$$\kappa^2 = \frac{i\omega_n}{\alpha}$$

$$\beta_m = \frac{m\pi}{h}$$

For the special case where $\hat{q}$ is constant and $L = h$, employing (6.51), the solution becomes

$$\hat{u}\left(r,z,\omega_n\right) = \hat{U}_s(\omega_n)\left[\frac{e^{\kappa z}}{1-e^{2\kappa h}} + \frac{e^{-\kappa z}}{1-e^{-2\kappa h}}\right] + \frac{\hat{q}\left(\omega\right)}{2\pi\lambda}K_0(\kappa_r r) \tag{6.54}$$

which is identical to (6.32) (see Problem 3).

Applying the inverse fast Fourier transform gives the solution of this boundary value problem in the time domain, as

$$u\left(r,z,t\right) = \sum_n \hat{u}\left(r,z,\omega_n\right)e^{i\omega_n t}.$$

Alternative Formulation

In dividing the problem into two sub-problems in (6.35), we imposed an insulated boundary condition at $z = 0$ in Sub-problem 2, as given in (6.38). This implies that the prescribed surface temperature U_s will be affected by the heat source, which might be natural in many cases. If, however, our interest is on keeping the ground surface prescribed at all time, we ought to implement the following formulation:

- **Sub-problems 1**

$$\begin{aligned} &\frac{\partial^2 u_1}{\partial z^2} - \frac{1}{\alpha}\frac{\partial u_1}{\partial t} = 0 \\ &\text{IC:}\quad u_1(z,0) = 0 \\ &\text{BC:}\quad u_1(0,t) = U_s(t) \\ &\qquad\;\; u_1(h,t) = 0. \end{aligned} \tag{6.55}$$

- **Sub-problem 2**

$$\begin{aligned} &\frac{\partial^2 u_2}{\partial r^2} + \frac{1}{r}\frac{\partial u_2}{\partial r} + \frac{\partial^2 u_2}{\partial z^2} - \frac{1}{\alpha}\frac{\partial u_2}{\partial t} = 0 \\ &\text{IC:}\quad u_2(r,z,0) = 0 \\ &\text{BC:}\quad u_2(r\to\infty,t) = 0 \\ &\qquad\;\; u_2(r,0,t) = 0 \\ &\qquad\;\; u_2(r,h,t) = 0 \\ &\qquad\;\; -\lambda\frac{\partial u_2}{\partial r}\cdot 2\pi r\bigg|_{r\to 0} = q(z,t) \end{aligned} \tag{6.56}$$

Following the solution procedure outlined in the previous formulation, we can readily attain to

$$\hat{u}\left(r,z,\omega_n\right) = \hat{u}_1(z,\omega_n) + \hat{u}_2(r,z,\omega_n)$$

which results in

$$\hat{u}\,(r,z,\omega_n) = \hat{U}_s(\omega_n)\left[\frac{e^{\kappa z}}{1-e^{2\kappa h}} + \frac{e^{-\kappa z}}{1-e^{-2\kappa h}}\right] +$$

$$\sum_m \left[\frac{1}{\pi\lambda h}\int_0^L \hat{q}\,(z,\omega_n)\sin(\beta_m z)dz\right].$$

$$\sin(\beta_m z)K_0(\kappa_r r) \quad m = 1, 2, \ldots$$

with

$$\kappa_r = \sqrt{\kappa^2 + \beta_m^2}$$
$$\kappa^2 = \frac{i\omega_n}{\alpha} \qquad (6.57)$$
$$\beta_m = \frac{m\pi}{h}$$

The reader is encouraged to derive this solution (see Problem 4).

- **Example 6.2: Finite Line Heat Source**

A 10 m long line heat pipe is embedded in a 15 m deep soil layer. The thermal parameters of the soil mass are: thermal conductivity 2 $Wm^{-1}K^{-1}$, mass density 1500 kg m^{-3}, and specific heat capacity 800 J $kg^{-1}K^{-1}$. The surface of the soil layer is subjected to a Dirichlet boundary condition with a prescribed temperature varying with time, as shown in Fig. 6.8. At nearly 1600 h, the heat source operated and started to release heat flux for 240 h with a time variation depicted in the same figure. Compute the temperature distribution in the soil mass using both formulations addressed in ▸ Sect. 6.5.

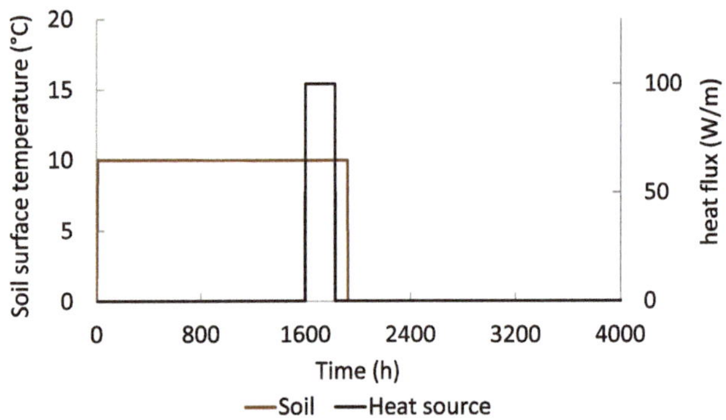

Fig. 6.8 Prescribed temperature at soil surface and released heat flux from heat pipe

▪ Solution

As outlined in Example 6.1, we start by transforming the input signals from the time domain to the frequency domain using the forward FFT algorithm. In this case, we used 4096 (2^{12}) FFT samples with 1 h (3600 s) time step. Then, upon using (6.53) or (6.57) with $M = 20$ and summing for all involved frequencies, we obtain the solution in the frequency domain. Finally, by applying the IFFT algorithm, we retrieve the solution in the time domain.

Employing the solution presented in (6.53), the contours of the temperature distributions in the domain after 72 and 240 h of operating the heat pipe are plotted in Fig. 6.9.

When employing the alternative formulation (6.57), the temperature in the domain takes on the distribution shown in Fig. 6.10. Comparing the results from the two formula-

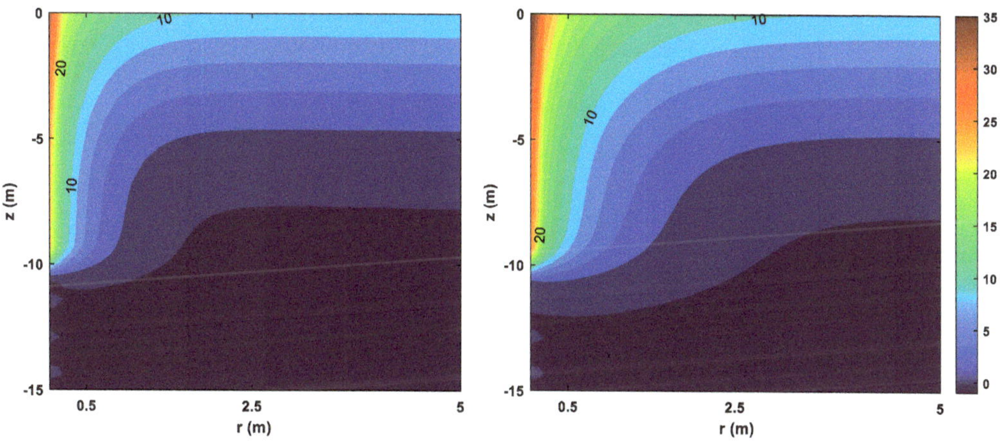

Fig. 6.9 Contour plots after 72 h (left), and 240 h (right) of operation

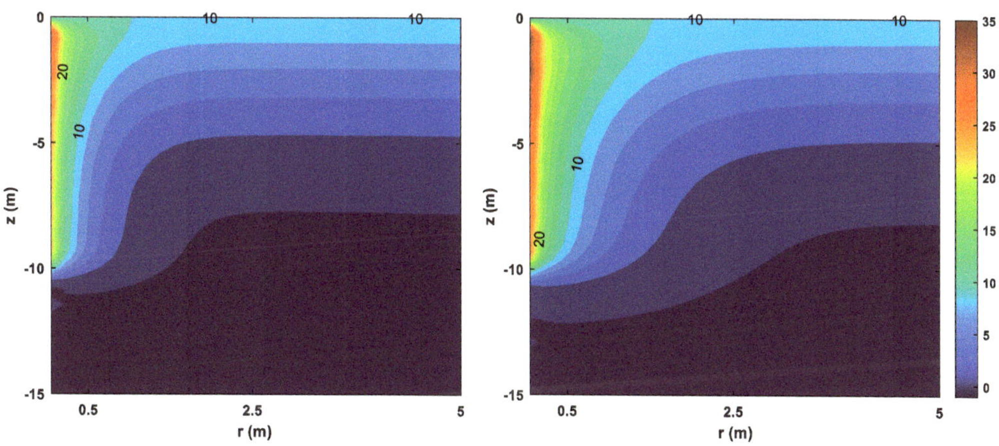

Fig. 6.10 Alternative formulation. Contour plots after 72 h (left), and 240 h (right) of operation

tions reveals that, even though the temperature range in the two cases is close, the distribution of temperature in the domain differs significantly, especially at the upper surface and close to the heat source.

6.6 Finite Cylindrical Heat Source (FCS): Constant Heat Flux

The concept of the finite cylindrical heat source (FCS) is specifically designed for scenarios in which the domain is bounded in the vertical z direction, while extending to infinity in the radial r direction. In contrast to the line heat source, this heat source is cylindrical, resembling real-world applications like heat pipes and heat exchangers, as shown schematically in Fig. 6.11. In this section, we address a case where the length of the heat source extends to the entire depth of the layer, and the emitted heat flux is only a function of time.

The boundary value problem for this FCS case can be described in the cylindrical coordinate system, as

$$\begin{aligned}
&\frac{1}{\alpha}\frac{\partial u}{\partial t} - \frac{\partial^2 u}{\partial r^2} - \frac{1}{r}\frac{\partial u}{\partial r} - \frac{\partial^2 u}{\partial z^2} = 0 \\
&\text{IC:} \quad u\,(r, z, 0) = 0 \\
&\text{BC:} \quad u\,(r \to \infty, z, t) = 0 \\
&\qquad\quad u\,(r, 0, t) = U_s(t) \\
&\qquad\quad u\,(r, h, t) = 0 \\
&\qquad\quad -\lambda 2\pi a \left.\frac{\partial u}{\partial r}\right|_{r=a} = q\,(t).
\end{aligned} \tag{6.58}$$

where a is the radius of the cylinder.

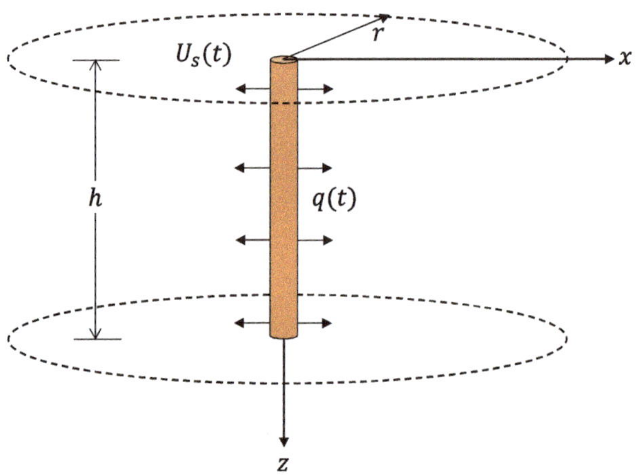

Fig. 6.11 Finite cylindrical heat source with a constant heat flux

BVP Solution

Similar to the solution of the boundary value problem in (6.27), the solution of this BVP can be attained by breaking the problem into two sub-problems: one addressing heat flow in the vertical z direction and the other in the radial r direction, such that

$$u(r, z, t) = u_1(z, t) + u_2(r, t)$$

The solution of u_1 is given in ▶ Sect. 6.1, and the solution of u_2 is given in Chap. 5, ▶ Sect. 5.3. Thus, the solution of the BVP (6.58) in the frequency domain can be expressed as

$$\hat{u}(r, z, \omega_n) = \underbrace{\frac{\hat{U}_s(\omega_n)}{1 - e^{-2\kappa h}}[e^{-\kappa z} - e^{-\kappa(2h-z)}]}_{\hat{u}_1(z,\omega)} + \underbrace{\frac{\hat{q}(\omega_n)}{2\pi a \lambda \kappa K_1(\kappa a)} K_0(\kappa r)}_{\hat{u}_2(r,\omega)}$$

with

$$\kappa = \sqrt{i\omega_n / \alpha} \tag{6.59}$$

Using the IFFT algorithm, the solution in the time domain can be attained via

$$u(r, z, t) = \sum_n \hat{u}(r, z, \omega_n) e^{i\omega_n t}.$$

6.7 Finite Cylindrical Heat Source (FCS): Heat Flux Function of *z*

In the previous section we addressed a boundary value problem for a finite cylindrical heat source (FCS) that extends through the entire thickness of the layer. Here, we expand on this problem to the case where the heat source is shorter than the layer thickness, making the prescribed heat flux a function of z and time, see Fig. 6.12.

The boundary value problem of this case is given by

$$\begin{aligned}
&\frac{1}{\alpha}\frac{\partial u}{\partial t} - \frac{\partial^2 u}{\partial r^2} - \frac{1}{r}\frac{\partial u}{\partial r} - \frac{\partial^2 u}{\partial z^2} = 0 \\
&\text{IC:} \quad u(r, z, 0) = 0 \\
&\text{BC:} \quad u(r \to \infty, z, t) = 0 \\
&\qquad\;\; u(r, 0, t) = U_s(t) \\
&\qquad\;\; u(r, h, t) = 0 \\
&\qquad\;\; -\lambda 2\pi a \left.\frac{\partial u}{\partial r}\right|_{r=a} = q(z, t), \quad 0 < z < L
\end{aligned} \tag{6.60}$$

where h is the height of the domain that is larger than the length L of the heat source.

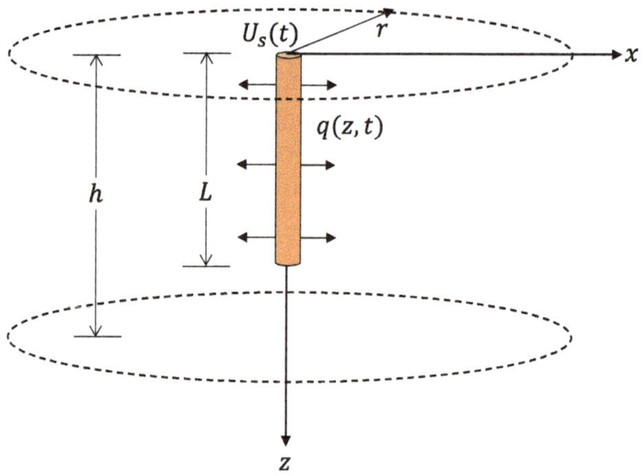

Fig. 6.12 Finite cylindrical heat source with the heat flux function of z

6

Solution

As for the line heat source case, which we have addressed in ▶ Sect. 6.5, we solve this problem by dividing it into two sub-problems, such that

$$u(r,z,t) = u_1(z,t) + u_2(r,z,t) \tag{6.61}$$

where, in contrast to the previous case, u_2 is made a function of r and z. This disparity arises from that the heat flux $q(z,t)$ here generates heat flow in both r and z directions.

Inserting (6.61) into the partial differential equation in (6.60) gives

$$\frac{1}{\alpha}\frac{\partial u_1}{\partial t} + \frac{1}{\alpha}\frac{\partial u_2}{\partial t} - \frac{\partial^2 u_1}{\partial z^2} - \frac{\partial^2 u_2}{\partial r^2} - \frac{1}{r}\frac{\partial u_2}{\partial r} - \frac{\partial^2 u_2}{\partial z^2} = 0$$

This equation can be split into two equations, which can be expressed with their relevant boundary conditions, as

Sub-problems 1

$$\begin{aligned} &\frac{\partial^2 u_1}{\partial z^2} - \frac{1}{\alpha}\frac{\partial u_1}{\partial t} = 0 \\ &\text{IC:} \quad u_1(z,0) = 0 \\ &\text{BC:} \quad u_1(0,t) = U_s(t) \\ &\qquad\;\; u_1(h,t) = 0. \end{aligned} \tag{6.62}$$

- **Sub-problem 2**

$$
\begin{aligned}
&\frac{\partial^2 u_2}{\partial r^2} + \frac{1}{r}\frac{\partial u_2}{\partial r} + \frac{\partial^2 u_2}{\partial z^2} - \frac{1}{\alpha}\frac{\partial u_2}{\partial t} = 0 \\
&\text{IC:} \quad u_2(r,z,0) = 0 \\
&\text{BC:} \quad u_2(r \to \infty, t) = 0 \\
&\qquad \left.\frac{\partial u_2}{\partial z}\right|_{z=0} = 0 \\
&\qquad \left.\frac{\partial u_2}{\partial z}\right|_{z=h} = 0 \\
&\qquad \left.-\lambda \frac{\partial u_2}{\partial r} \cdot 2\pi r\right|_{r=a} = q(z,t)
\end{aligned}
\tag{6.63}
$$

where we notice that adding the second and third boundary conditions in (6.63) to the first and second boundary conditions in (6.62) we obtain the original boundary conditions in (6.60).

- **Solving Sub-problem 1**

The solution of this sub-problem is given in detail in ► Sect. 6.1. In the frequency domain, the solution yields

$$
\hat{u}_1(z,\omega_n) = \frac{\hat{U}_s(\omega_n)}{1 - e^{-2\kappa h}}[e^{-\kappa z} - e^{-\kappa(2h-z)}], \quad \kappa = \sqrt{\frac{i\omega_n}{\alpha}}
\tag{6.64}
$$

- **Solving Sub-problem 2**

The solution of $u_2(r,z,t)$ is given in detail in ► Sect. 6.5. In the frequency domain, the solution yields

$$
\hat{u}_2(r,z,\omega) = \sum_m A_m K_0(\kappa_r r)\cos(\beta_m z)
$$

with

$$
\begin{aligned}
\kappa_r &= \sqrt{\kappa^2 + \beta^2} \\
\kappa^2 &= \frac{i\omega}{\alpha} \\
\beta_m &= \frac{m\pi}{h}
\end{aligned}
\tag{6.65}
$$

Taking the derivative of (6.65) with respect to r gives

$$
\frac{\partial \hat{u}_2}{\partial r} = -\sum_m A_m \cos\left(\frac{m\pi z}{h}\right)\kappa_r K_1(\kappa_r r)
$$

Then, at $r = a$, the fourth boundary condition in (6.63) implies

$$2\pi a\lambda \sum_{m} A_m \cos\left(\frac{m\pi z}{h}\right)\kappa_r K_1(\kappa_r a) = \hat{q}(z, \omega_n)$$

Re-arranging gives

$$\sum_{m} A_m \cos\left(\frac{m\pi z}{h}\right) = \frac{\hat{q}(z, \omega_n)}{2\pi a\lambda\kappa_r K_1(\kappa_r a)}$$

This is a half range Fourier cosine series with its coefficients A_m expressed as

$$A_0 = \frac{1}{h}\int_0^L \frac{\hat{q}(z, \omega_n)}{2\pi a\lambda\kappa_r K_1(\kappa_r a)} dz, \qquad m = 0$$

$$A_m = \frac{2}{h}\int_0^L \frac{\hat{q}(z, \omega_n)}{2\pi a\lambda\kappa_r K_1(\kappa_r a)} \cos(\beta_m z)\, dz, \quad m = 1, 2, \ldots \tag{6.66}$$

Following this, the solution of $\hat{u}_2$ can be expressed as

$$\hat{u}_2(r, z, \omega) = \sum_{m} A_m \cos(\beta_m z) K_0(\kappa_r r), \quad m = 0, 1, 2, \ldots \tag{6.67}$$

Special Case 1: Constant Heat Flux

In the special case where $\hat{q}$ is constant, the Fourier coefficients become

$$A_0 = \frac{\hat{q}(\omega_n) L}{2\pi a\lambda\, h\, \kappa_r K_1(\kappa_r a)}$$

$$A_m = \frac{\hat{q}(\omega_n)}{m\pi^2 a\lambda\, \kappa_r K_1(\kappa_r a)} \sin\left(\frac{m\pi L}{h}\right).$$

Special Case 2: Constant Heat Flux and $L = h$.

In the special case where $\hat{q}$ is constant and $L = h$, the Fourier coefficients become

$$A_0 = \frac{\hat{q}(\omega_n)}{2\pi a\lambda\, \kappa_r K_1(\kappa_r a)}$$

$$A_m = 0, \qquad m = 1, 2, \ldots \tag{6.68}$$

BVP Solution

Recalling (6.61) and expressing it in the frequency domain gives

$$\hat{u}(r, z, \omega_n) = \hat{u}_1(z, \omega_n) + \hat{u}_2(r, z, \omega_n)$$

Then, substituting the solutions of $\hat{u}_1$ (6.64) and $\hat{u}_2$ (6.67) into this identity yields

$$\hat{u}\,(r,z,\omega_n) = \frac{\hat{U}_s(\omega_n)}{1 - e^{-2\kappa h}}[\,e^{-\kappa z} - e^{-\kappa(2h-z)}] + \sum_m A_m K_0(\kappa_r r)\cos(\beta_m z), \qquad m = 0,1,2,\ldots \tag{6.69}$$

For the special case where $\hat{q}$ is constant and $L = h$, employing (6.68), the solution becomes

$$\hat{u}\,(r,z,\omega_n) = \frac{\hat{U}_s(\omega_n)}{1 - e^{-2\kappa h}}[\,e^{-\kappa z} - e^{-\kappa(2h-z)}] + \frac{\hat{q}\,(\omega_n)}{2\pi a\lambda\kappa_r K_1(\kappa_r a)} K_0(\kappa_r r) \tag{6.70}$$

which is identical to (6.59). See Problem 6.

Employing the inverse fast Fourier transform (IFFT) algorithm gives the solution in the time domain, such that

$$u\,(r,z,t) = \sum_n \hat{u}\,(r,z,\omega_n)\,e^{i\omega_n t}.$$

Alternative Formulation

When partitioning this BVP into two sub-problems in (6.61), we imposed an insulated boundary condition at $z = 0$ in Sub-problem 2, as given in (6.63). Yet, if our aim is to maintain the ground surface consistently prescribed, we must adopt the following formulation:

- **Sub-problems 1**

$$\frac{\partial^2 u_1}{\partial z^2} - \frac{1}{\alpha}\frac{\partial u_1}{\partial t} = 0$$

$$\begin{aligned} &\text{IC:} \quad u_1(z,0) = 0 \\ &\text{BC:} \quad u_1(0,t) = U_s(t) \\ &\qquad\;\; u_1(h,t) = 0. \end{aligned}$$

- **Sub-problem 2**

$$\frac{\partial^2 u_2}{\partial r^2} + \frac{1}{r}\frac{\partial u_2}{\partial r} + \frac{\partial^2 u_2}{\partial z^2} - \frac{1}{\alpha}\frac{\partial u_2}{\partial t} = 0$$

$$\begin{aligned} &\text{IC:} \quad u_2(r,z,0) = 0 \\ &\text{BC:} \quad u_2(r \to \infty, t) = 0 \\ &\qquad\;\; u_2(r,0,t) = 0 \\ &\qquad\;\; u_2(r,h,t) = 0 \\ &\qquad\;\; -\lambda\frac{\partial u_2}{\partial r}\cdot 2\pi a\bigg|_{r=a} = q\,(z,t) \end{aligned}$$

Following the solution procedure outlined above, we can readily solve this set of boundary value problems in the frequency domain, yielding

$$\hat{u}\left(r,z,\omega_n\right) = \frac{\hat{U}_s(\omega_n)}{1 - e^{-2\kappa h}}[e^{-\kappa z} - e^{-\kappa(2h-z)}]+$$
$$\sum_m A_m \sin\left(\frac{m\pi z}{h}\right) K_0(\kappa_r r), \quad m = 1,2,\ldots$$

with

$$A_m = \frac{1}{\pi a\lambda h\kappa_r K_1(\kappa_r a)} \int_0^L \hat{q}\left(z,\omega_n\right) \sin\left(\frac{m\pi z}{h}\right) dz. \tag{6.71}$$

6.8 Disk with Inner Circle: θ-Dependent Dirichlet Boundary Conditions

This case deals with a 2D planar domain represented by a disk with radius a_2 encompassing in its center an inner circle with radius a_1. This setup is commonly encountered in various engineering and scientific applications, such as in the study of heat dissipation in electronic components, the analysis of temperature gradients in cylindrical reactors, or understanding thermal behavior in geological formations. The θ dependence introduces an additional complexity to the problem, as heat distribution can vary non-uniformly inside the disk. Accurate modeling and analysis of such systems are essential for optimizing design, predicting temperature profiles, and ensuring efficient thermal management.

In this section, we address a Dirichlet boundary problem where the prescribed temperature on the inner circle is assumed a function of time only, and that on the outer circles is a function of time and θ, see Fig. 6.13.

The boundary value problem of this case is

$$\begin{aligned}
&\frac{\partial^2 u}{\partial r^2} + \frac{1}{r}\frac{\partial u}{\partial r} + \frac{1}{r^2}\frac{\partial^2 u}{\partial \theta^2} - \frac{1}{\alpha}\frac{\partial u}{\partial t} = 0 \\
&\text{IC:} \quad u\left(r,\theta,0\right) = 0 \\
&\text{BC:} \quad u\left(r,\theta,t\right) = u\left(r,\theta+2\pi,t\right) \\
&\qquad\quad u\left(a_1,\theta,t\right) = U_{in}(t) \\
&\qquad\quad u\left(a_2,\theta,t\right) = U_{out}(\theta,t).
\end{aligned} \tag{6.72}$$

▪ Solution

As usual, as the boundary conditions are not homogeneous, we split the problem into two sub-problems, such that

$$u\left(r,\theta,t\right) = u_1(r,t) + u_2(r,\theta,t) \tag{6.73}$$

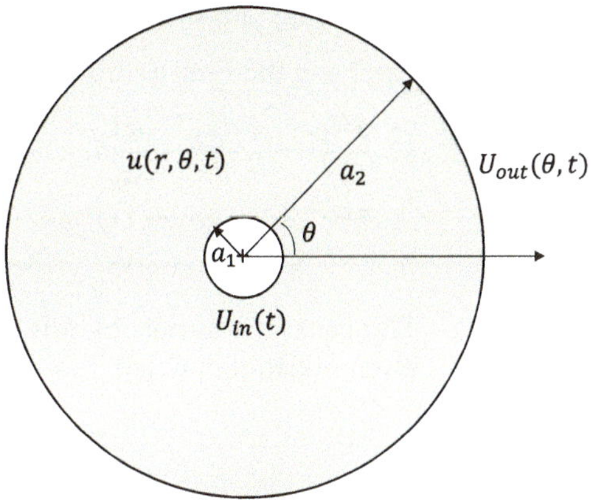

Fig. 6.13 Bounded 2D planar domain with Dirichlet boundaries

where we chose u_1 to be a function of r only, since U_{in} on the inner boundary is a function of time only. Substituting (6.73) into (6.72) gives

$$\frac{\partial^2 u_2}{\partial r^2} + \frac{\partial^2 u_1}{\partial r^2} + \frac{1}{r}\frac{\partial u_2}{\partial r} + \frac{1}{r}\frac{\partial u_1}{\partial r} + \frac{1}{r^2}\frac{\partial^2 u_2}{\partial \theta^2} - \frac{1}{\alpha}\frac{\partial u_2}{\partial t} - \frac{1}{\alpha}\frac{\partial u_1}{\partial t} = 0$$

This equation can be divided into two equations, such that

$$\frac{\partial^2 u_1}{\partial r^2} + \frac{1}{r}\frac{\partial u_1}{\partial r} - \frac{1}{\alpha}\frac{\partial u_1}{\partial t} = 0$$
$$\frac{\partial^2 u_2}{\partial r^2} + \frac{1}{r}\frac{\partial u_2}{\partial r} + \frac{1}{r^2}\frac{\partial^2 u_2}{\partial \theta^2} - \frac{1}{\alpha}\frac{\partial u_2}{\partial t} = 0$$

Expressing these two equations with their relevant boundary conditions, we obtain.

- **Sub-problem 1**

$$\begin{aligned} &\frac{\partial^2 u_1}{\partial r^2} + \frac{1}{r}\frac{\partial u_1}{\partial r} - \frac{1}{\alpha}\frac{\partial u_1}{\partial t} = 0 \\ &\text{IC:} \quad u_1(r,\theta,0) = 0 \\ &\text{BC:} \quad u_1(a_1,t) = U_{in}(t) \\ &\qquad\;\; u_1(a_2,t) = 0. \end{aligned} \tag{6.74}$$

- **Sub-problem 2**

$$\begin{aligned} &\frac{\partial^2 u_2}{\partial r^2} + \frac{1}{r}\frac{\partial u_2}{\partial r} + \frac{1}{r^2}\frac{\partial^2 u_2}{\partial \theta^2} - \frac{1}{\alpha}\frac{\partial u_2}{\partial t} = 0 \\ &\text{IC:} \quad u_2(r,\theta,0) = 0 \\ &\text{BC:} \quad u_2(r,\theta,t) = u_2(r,\theta+2\pi,t) \\ &\qquad\;\; u_2(a_1,\theta,t) = 0 \\ &\qquad\;\; u_2(a_2,\theta,t) = U_{out}(\theta,t) \end{aligned} \tag{6.75}$$

- **Solving Sub-problem 1**

Applying the Fourier transform to (6.74) yields

$$\begin{aligned}&\frac{d^2\hat{u}_1}{dr^2}+\frac{1}{r}\frac{d\hat{u}_1}{dr}-\frac{i\omega}{\alpha}\hat{u}_1=0\\&\text{BC:}\quad \hat{u}_1(a_1,\omega)=\hat{U}_{in}(\omega)\\&\qquad\quad \hat{u}_1(a_2,\omega)=0\end{aligned}\tag{6.76}$$

The general solution of this ordinary differential equation is given in detail in Chap. 5, ▶ Sect. 5.4. It leads to

$$\hat{u}_1(r,\omega)=c_1 I_0(\kappa r)+c_2 K_0(\kappa r),\quad \kappa=\sqrt{i\omega/\alpha}\tag{6.77}$$

At $r=a_1$, the first boundary condition in (6.76) implies

$$c_1 I_0(\kappa a_1)+c_2 K_0(\kappa a_1)=\hat{U}_{in}$$

At $r=a_2$, the second boundary condition implies

$$c_1 I_0(\kappa a_2)+c_2 K_0(\kappa a_2)=0$$

Putting these two equations in a matrix format gives

$$\begin{pmatrix} I_0(\kappa a_1) & K_0(\kappa a_1)\\ I_0(\kappa a_2) & K_0(\kappa a_2)\end{pmatrix}\begin{pmatrix}c_1\\c_2\end{pmatrix}=\begin{pmatrix}\hat{U}_{in}\\0\end{pmatrix}$$

Solving for c_1 and c_2 yields

$$\begin{aligned}&\begin{pmatrix}c_1\\c_2\end{pmatrix}=\frac{1}{\Delta}\begin{pmatrix} K_0(\kappa a_2) & -K_0(\kappa a_1)\\ -I_0(\kappa a_2) & I_0(\kappa a_1)\end{pmatrix}\begin{pmatrix}\hat{U}_{in}\\0\end{pmatrix}\\&\Delta=I_0(\kappa a_1)K_0(\kappa a_2)-K_0(\kappa a_1)I_0(\kappa a_2)\end{aligned}$$

This leads to

$$c_1=\frac{1}{\Delta}K_0(\kappa a_2)\,\hat{U}_{in},\quad c_2=-\frac{1}{\Delta}I_0(\kappa a_2)\,\hat{U}_{in}$$

Thus, the solution of the BVP (6.76) can be expressed as

$$\hat{u}_1(r,\omega)=\frac{\hat{U}_{in}}{\Delta}[K_0(\kappa a_2)I_0(\kappa r)-I_0(\kappa a_2)K_0(\kappa r)]$$

with

$$\begin{aligned}&\kappa=\sqrt{i\omega/\alpha}\\&\Delta=I_0(\kappa a_1)K_0(\kappa a_2)-K_0(\kappa a_1)I_0(\kappa a_2)\end{aligned}\tag{6.78}$$

- **Solving Sub-problem 2**

Applying the Fourier transform to (6.75) results in

$$\begin{aligned}&\frac{\partial^2\hat{u}_2}{\partial r^2}+\frac{1}{r}\frac{\partial\hat{u}_2}{\partial r}+\frac{1}{r^2}\frac{\partial^2\hat{u}_2}{\partial\theta^2}-\frac{i\omega}{\alpha}\hat{u}_2=0\\&\text{BC:}\quad \hat{u}_2(r,\theta,\omega)=\hat{u}_2(r,\theta+2\pi,\omega)\\&\qquad\quad \hat{u}_2(a_1,\theta,\omega)=0\\&\qquad\quad \hat{u}_2(a_2,\theta,\omega)=\hat{U}_{out}(\theta,\omega)\end{aligned}\tag{6.79}$$

Solution of (6.79) can be carried out using the method of separation of variables. Assume

$$\hat{u}_2(r,z) = R(r)\,F(\theta) = RF \tag{6.80}$$

in which $R(r)$ is a function of r only and $F(\theta)$ is a function of θ only. Substituting (6.80) into (6.79) gives

$$R''F + \frac{1}{r}R'F + \frac{1}{r^2}RF'' - \frac{i\omega}{\alpha}RF = 0$$

Dividing by RF and rearranging yields

$$r^2\frac{R''}{R} + r\frac{R'}{R} - \frac{i\omega}{\alpha}r^2 = -\frac{F''}{F}$$

Since the left hand side is a function of r only and the right hand side is a function of θ only, both sides must be equal a constant. Equating both sides of the equation by some arbitrary separation constant, say η, we get

$$r^2\frac{R''}{R} + r\frac{R'}{R} - \frac{i\omega}{\alpha}r^2 = -\frac{F''}{F} = \eta$$

The separated equations are then

$$\frac{d^2F}{d\theta^2} + \eta F = 0 \tag{6.81}$$

$$r^2\frac{d^2R}{dr^2} + r\,\frac{dR}{dr} - (\kappa^2 r^2 + \eta)\,R = 0 \tag{6.82}$$

where $\kappa^2 = i\omega/\alpha$ (see (6.77)).

We begin with solving $F(\theta)$ equation. The relevant boundary value problem of this equation is

$$F'' + \eta F = 0, \quad F(\theta) = F(\theta + 2\pi) \tag{6.83}$$

which is evidently 2π-periodic. This problem is not a regular Sturm-Liouville, yet, due to periodicity, it generates eigenvalues and eigenfunctions [Zill, p. 730]. The three possible general solutions of (6.83) are

$$\begin{aligned} F(\theta) &= c_1 + c_2\theta, & \eta &= 0 \\ F(\theta) &= c_1\cosh(m\theta) + c_2\sinh(m\theta), & \eta &= -m^2 < 0 \\ F(\theta) &= c_1\cos(m\theta) + c_2\sin(m\theta), & \eta &= m^2 > 0 \end{aligned} \tag{6.84}$$

As explained earlier, the second solution in (6.84) is not periodic and thus can be dismissed. The first solution is periodic if $c_2 = 0$, with the remaining solution $F(\theta) = c_1, c_1 \neq 0$. Hence, $\eta = 0$ is an eigenvalue. Finally, the third solution can be 2π-periodic if we take $= 1, 2, \ldots$. Considering that $\eta = 0$ is also an eigenvalue, the eigenvalues of the system are then $\eta_m = m^2, m = 0, 1, 2, \ldots$, leading the solution to be expressed as

$$F(\theta) = \sum_m c_3\cos(m\theta) + c_4\sin(m\theta), \quad m = 0, 1, 2, \ldots \tag{6.85}$$

The solution of $R(r)$ has been discussed in Chap. 5, ▶ Sect. 5.4, where it was shown that the differential Eq. (6.82) can be put into a standard form of the modified Bessel function of order m with the general solution expressed as

$$R(r) = c_5 I_m(\kappa r) + c_6 K_m(\kappa r) \tag{6.86}$$

Collecting both solutions, Eqs. (6.85) and (6.86), and substituting them into (6.80), leads to

$$\begin{aligned}\hat{u}_2(r,\theta,\omega) &= \sum_m c_3 \cos(m\theta)[c_5 I_m(\kappa r) + c_6 K_m(\kappa r)] \\ &+ \sum_m c_4 \sin(m\theta)[c_5 I_m(\kappa r) + c_6 K_m(\kappa r)].\end{aligned} \tag{6.87}$$

Boundary Conditions

At $r = a_1$, the second boundary condition in (6.75) implies

$$\begin{aligned}&\sum_m c_3 \cos(m\theta)\,[c_5 I_m(\kappa a_1) + c_6 K_m(\kappa a_1)] \\ &+ \sum_m c_4 \sin(m\theta)\,[c_5 I_m(\kappa a_1) + c_6 K_m(\kappa a_1)] = 0\end{aligned} \tag{6.88}$$

Rearranging gives

$$\sum_m [c_5 I_m(\kappa a_1) + c_6 K_m(\kappa a_1)]\,[c_3 \cos(m\theta) + c_4 \sin(m\theta)] = 0$$

This will be satisfied for every value of θ, if

$$c_5 I_m(\kappa a_1) + c_6 K_m(\kappa a_1) = 0$$

which leads to

$$c_6 = -\frac{c_5 I_m(\kappa a_1)}{K_m(\kappa a_1)}$$

Substituting this relationship into (6.87) gives

$$\begin{aligned}\hat{u}_2(r,\theta,\omega) &= \sum_m A_m \cos(m\theta)[I_m(\kappa r) - \frac{I_m(\kappa a_1)}{K_m(\kappa a_1)} K_m(\kappa r)] \\ &+ \sum_m B_m \sin(m\theta)[I_m(\kappa r) - \frac{I_m(\kappa a_1)}{K_m(\kappa a_1)} K_m(\kappa r)]\end{aligned} \tag{6.89}$$

where $A_m = c_3.c_5$ and $B_m = c_4.c_5$.

At $r = a_2$, the third boundary condition in (6.75) implies

$$\begin{aligned}&\sum_m A_{mn}[I_m(\kappa a_2) - \frac{I_m(\kappa a_1)}{K_m(\kappa a_1)} K_m(\kappa a_2)] \cos(m\theta) \\ &+ \sum_m B_{mn}[I_m(\kappa a_2) - \frac{I_m(\kappa a_1)}{K_m(\kappa a_1)} K_m(\kappa a_2)] \sin(m\theta) = \hat{U}_{out}(\theta)\end{aligned} \tag{6.90}$$

which is a trigonometric Fourier series with its coefficients

$$A_m = \begin{cases} \frac{1}{2\pi}\int\limits_0^{2\pi} \dfrac{\hat{U}_{out}(\theta)}{I_m(\kappa a_2) - \dfrac{I_m(\kappa a_1)}{K_m(\kappa a_1)} K_m(\kappa a_2)} d\theta, & m = 0 \\ \frac{1}{\pi}\int\limits_0^{2\pi} \dfrac{\hat{U}_{out}(\theta)}{I_m(\kappa a_2) - \dfrac{I_m(\kappa a_1)}{K_m(\kappa a_1)} K_m(\kappa a_2)} \cos(m\theta)\, d\theta, & m = 1, 2, \ldots \end{cases}$$

$$B_m = \frac{1}{\pi}\int\limits_0^{2\pi} \frac{\hat{U}_{out}(\theta)}{I_m(\kappa a_2) - \dfrac{I_m(\kappa a_1)}{K_m(\kappa a_1)} K_m(\kappa a_2)} \sin(m\theta)\, d\theta, \quad m = 1, 2, \ldots \tag{6.91}$$

Substituting (6.91) into (6.87) gives the required solution for $\hat{u}_2$.

BVP Solution

The solution of the boundary value problem (6.72) in the frequency domain, for a given frequency ω_n, is therefore

$$\hat{u}(r, \theta, \omega_n) = \hat{u}_1(r, \omega_n) + \hat{u}_2(r, \theta, \omega_n) \tag{6.92}$$

where $\hat{u}_1(r, \omega_n)$ is given in (6.78) and $\hat{u}_2(r, \theta, \omega_n)$ is given in (6.89) and (6.91).

Having the solution in the frequency domain, the solution in the time domain can then be obtained by applying the IFFT algorithm, such that

$$u(r, \theta, t) = \sum_n \hat{u}(r, \theta, \omega_n)\, e^{i\omega_n t}.$$

Special Case 1: U_{out} is Constant

In the special case where U_{out} is not a function of θ, the Fourier coefficients become

$$A_0 = \frac{\hat{U}_{out}(\omega_n)}{I_0(\kappa a_2) - \frac{I_0(\kappa a_1)}{K_0(\kappa a_1)} K_0(\kappa a_2)}$$

$$A_m = 0, \quad m = 1, 2, \ldots$$

$$B_m = 0, \quad m = 1, 2, \ldots$$

Leading the solution to be

$$\hat{u}(r, \omega_n) = \hat{u}_1(r, \omega_n) + \hat{u}_2(r, \omega_n)$$

where $\hat{u}_1$ is given in (6.78), and

$$\hat{u}_2(r, \omega_n) = A_0[I_0(\kappa r) - \frac{I_0(\kappa a_1)}{K_0(\kappa a_1)} K_0(\kappa r)].$$

Special Case 2: U_{out} is Linear

In the special case where

$$\hat{U}_{out}(\theta,\omega) = \hat{U}_0(\omega)\,\theta$$

the Fourier coefficients become

$$A_0 = \frac{\pi \hat{U}_o(\omega_n)}{I_0(\kappa a_2) - \dfrac{I_0(\kappa a_1)}{K_0(\kappa a_1)} K_0(\kappa a_2)}, \qquad m = 0$$

$$A_m = 0. \qquad m = 1, 2, \ldots$$

$$B_m = \frac{-2\hat{U}_0(\omega_n)}{m[I_m(\kappa a_2) - \dfrac{I_m(\kappa a_1)}{K_m(\kappa a_1)} K_m(\kappa a_2)]}, \quad m = 1, 2, \ldots$$

leading the solution to be

$$\hat{u}\,(r,\theta,\omega_n) = \hat{u}_1(r,\omega_n) + \hat{u}_2(r,\theta,\omega_n)$$

where $\hat{u}_1$ is given in (6.78), and

$$\hat{u}_2(r,\theta) = A_0[I_0(\kappa r) - \frac{I_0(\kappa a_1)}{K_0(\kappa a_1)} K_0(\kappa r)] + \sum_m B_m[I_m(\kappa r) - \frac{I_m(\kappa a_1)}{K_m(\kappa a_1)} K_m(\kappa r)] \sin(m\theta).$$

6.9 Disk with Inner Circle: θ-Dependent Dirichlet–Neumann Boundary Conditions

Consider a 2D planar domain featured by a disk with radius a_2 encompassing in its center an inner circle with radius a_1. The inner circle acts as a Dirichlet heat source, function of time only, and the outer boundary is a Neumann type boundary condition, function of θ and time, as sketched in ◘ Fig. 6.14.

The boundary value problem of this case is

$$\begin{aligned}
&\frac{\partial^2 u}{\partial r^2} + \frac{1}{r}\frac{\partial u}{\partial r} + \frac{1}{r^2}\frac{\partial^2 u}{\partial \theta^2} - \frac{1}{\alpha}\frac{\partial u}{\partial t} = 0\\
&\text{IC:} \quad u\,(r,\theta,0) = 0\\
&\text{BC:} \quad u\,(r,\theta,t) = u\,(r,\theta+2\pi,t)\\
&\qquad u\,(a_1,\theta,t) = U_{in}(t)\\
&\qquad \left.\frac{\partial u\,(r,\theta,t)}{\partial r}\right|_{r=a_2} = \frac{-q_{out}(\theta,t)}{2\pi a_2 \lambda}.
\end{aligned} \tag{6.93}$$

■ **Solution**

Since the boundary conditions are not homogeneous, the BVP problem (6.93) is best addressed by splitting it into two sub-problems, such that

Fig. 6.14 Bounded 2D planner domain with Dirichlet–Neumann boundaries

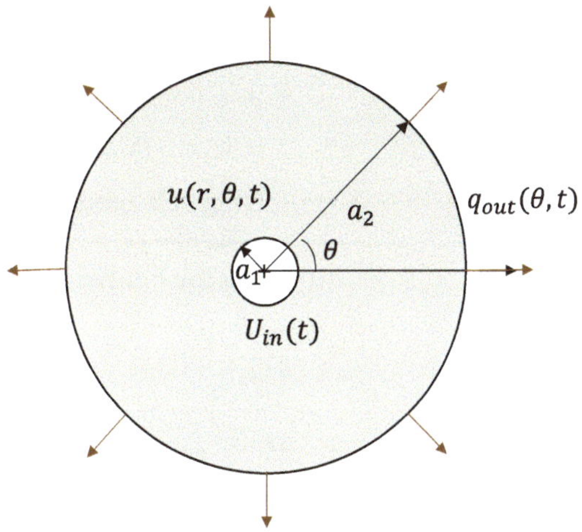

$$u(r,\theta,t) = u_1(r,t) + u_2(r,\theta,t) \tag{6.94}$$

Substituting (6.94) into (6.93) and splitting the resulting equation into two sub-problems, leads to.

- **Sub-problem 1**

$$\begin{aligned}
&\frac{\partial^2 u_1}{\partial r^2} + \frac{1}{r}\frac{\partial u_1}{\partial r} - \frac{1}{\alpha}\frac{\partial u_1}{\partial t} = 0 \\
&\text{IC:} \quad u_1(r,0) = 0 \\
&\text{BC:} \quad u_1(a_1,t) = U_{in}(t) \\
&\qquad\;\; u_1(a_2,t) = 0
\end{aligned} \tag{6.95}$$

- **Sub-problem 2**

$$\begin{aligned}
&\frac{\partial^2 u_2}{\partial r^2} + \frac{1}{r}\frac{\partial u_2}{\partial r} + \frac{1}{r^2}\frac{\partial^2 u_2}{\partial \theta^2} - \frac{1}{\alpha}\frac{\partial u_2}{\partial t} = 0 \\
&\text{IC:} \quad u_2(r,\theta,0) = 0 \\
&\text{BC:} \quad u_2(r,\theta,t) = u_2(r,\theta+2\pi,t) \\
&\qquad u_2(a_1,\theta,t) = 0 \\
&\qquad \left.\frac{\partial u_2}{\partial r}\right|_{r=a_2} = \frac{-q_{out}(\theta,t)}{2\pi a_2 \lambda} = Q_{out}(\theta,t)
\end{aligned} \tag{6.96}$$

- **Solving Sub-problem 1**

The boundary value problem in (6.95) is identical to (6.74), and thus its solution in the frequency domain can be expressed as

$$\hat{u}_1(r,\omega) = \frac{\hat{U}_{in}}{\Delta}[K_0(\kappa a_2)I_0(\kappa r) - I_0(\kappa a_2)K_0(\kappa r)]$$

6

with

$$\begin{aligned} &\kappa = \sqrt{i\omega/\alpha} \\ &\Delta = I_0(\kappa a_1)K_0(\kappa a_2) - K_0(\kappa a_1)I_0(\kappa a_2). \end{aligned} \tag{6.97}$$

- **Solving Sub-problem 2**

Applying the Fourier transform to (6.96) gives

$$\begin{aligned} &\frac{\partial^2 \hat{u}_2}{\partial r^2} + \frac{1}{r}\frac{\partial \hat{u}_2}{\partial r} + \frac{1}{r^2}\frac{\partial^2 \hat{u}_2}{\partial \theta^2} - \frac{i\omega}{\alpha}\hat{u}_2 = 0 \\ &\text{BC:} \quad \hat{u}_2(r,\theta,\omega) = \hat{u}_2(r,\theta+2\pi,\omega) \\ &\qquad \hat{u}_2(a_1,\theta,\omega) = 0 \\ &\qquad \left.\frac{\partial \hat{u}_2}{\partial r}\right|_{r=a_2} = \frac{-\hat{q}_{out}(\theta,\omega)}{2\pi a_2 \lambda} = \hat{Q}_{out}(\theta,\omega) \end{aligned} \tag{6.98}$$

The solution of the differential equation in (6.98), together with the first and second boundary conditions, was addressed in the previous section, yielding

$$\begin{aligned} \hat{u}_2(r,\theta,\omega) = &\sum_m A_m \cos(m\theta)[I_m(\kappa r) - \frac{I_m(\kappa a_1)}{K_m(\kappa a_1)} K_m(\kappa r)] \\ &+ \sum_m B_m \sin(m\theta)[I_m(\kappa r) - \frac{I_m(\kappa a_1)}{K_m(\kappa a_1)} K_m(\kappa r)] \end{aligned} \tag{6.99}$$

Taking the derivative of (6.99) with respect to r gives

$$\begin{aligned} \frac{\partial \hat{u}_2}{\partial r} = &\sum_m A_m\kappa \cos(m\theta)\cdot\{I_{m+1}(\kappa r) + m\frac{I_m(\kappa r)}{\kappa r} \\ &- \frac{I_m(\kappa a_1)}{K_m(\kappa a_1)}[-K_{m+1}(\kappa r) + m\frac{K_m(\kappa r)}{\kappa r}]\} \\ &+ \sum_m B_m\kappa \sin(m\theta)\cdot\{I_{m+1}(\kappa r) + m\frac{I_m(\kappa r)}{\kappa r} - \frac{I_m(\kappa a_1)}{K_m(\kappa a_1)}[-K_{m+1}(\kappa r) \\ &+ m\frac{K_m(\kappa r)}{\kappa r}]\} \end{aligned}$$

At $r = a_2$, the third boundary condition in (6.96) implies

$$\sum_m W_m \cos(m\theta) + X_m \sin(m\theta) = \hat{Q}_{out}(\theta, \omega$$

with

$$
\begin{aligned}
W_m &= A_m G_m \\
X_m &= B_m G_m \\
G_m &= \kappa \left\{ I_{m+1}(\kappa a_2) + m \frac{I_m(\kappa a_2)}{\kappa a_2} - \frac{I_m(\kappa a_1)}{K_m(\kappa a_1)} \right. \\
&\quad \left. \left[-K_{m+1}(\kappa a_2) + m \frac{K_m(\kappa a_2)}{\kappa a_2} \right] \right\}
\end{aligned}
\tag{6.100}
$$

which is a trigonometric Fourier series, with its coefficients determined by

$$
W_m = \begin{cases} \dfrac{1}{2\pi} \displaystyle\int_0^{2\pi} \hat{Q}_{out}(\theta)\, d\theta, & m = 0 \\ \dfrac{1}{\pi} \displaystyle\int_0^{2\pi} \hat{Q}_{out}(\theta) \cos(m\theta)\, d\theta, & m = 1, 2, \ldots \end{cases} \tag{6.101}
$$

$$
X_m = \frac{1}{\pi} \int_0^{2\pi} \hat{Q}_{out}(\theta) \sin(m\theta)\, d\theta, \qquad m = 1, 2, \ldots
$$

Upon determining the Fourier coefficients W_m and X_m, the constants A_m and B_m can be obtained from (6.100), and by substituting them into (6.99), we attain the required solution for $\hat{u}_2$.

BVP Solution

Based on the above derivations, the solution of the boundary value problem (6.93) in the frequency domain can be summarized as

$$
\hat{u}(r, \theta, \omega_n) = \hat{u}_1(r, \omega_n) + \hat{u}_2(r, \theta, \omega_n) \tag{6.102}
$$

where

$$
\hat{u}_1(r, \omega_n) = \frac{\hat{U}_{in}}{\Delta} [K_0(\kappa a_2) I_0(\kappa r) - I_0(\kappa a_2) K_0(\kappa r)]
$$

with

$$
\begin{aligned}
\kappa &= \sqrt{i\omega_n / \alpha} \\
\Delta &= I_0(\kappa a_1) K_0(\kappa a_2) - K_0(\kappa a_1) I_0(\kappa a_2)
\end{aligned}
$$

and

$$
\begin{aligned}
\hat{u}_2(r, \theta, \omega_n) &= \sum_m A_m \cos(m\theta) \left[I_m(\kappa r) - \frac{I_m(\kappa a_1)}{K_m(\kappa a_1)} K_m(\kappa r) \right] \\
&\quad + \sum_m B_m \sin(m\theta) \left[I_m(\kappa r) - \frac{I_m(\kappa a_1)}{K_m(\kappa a_1)} K_m(\kappa r) \right]
\end{aligned}
$$

in which the Fourier coefficients are determined by

$$A_m = \frac{W_m}{G_m}, \quad B_m = \frac{X_m}{G_m}$$

with

$$G_m = \kappa\{I_{m+1}(\kappa a_2) + m\frac{I_m(\kappa a_2)}{\kappa a_2} - \frac{I_m(\kappa a_1)}{K_m(\kappa a_1)}$$
$$[-K_{m+1}(\kappa a_2) + m\frac{K_m(\kappa a_2)}{\kappa a_2}]\}$$

and

$$W_m = \begin{cases} \frac{1}{2\pi}\int\limits_0^{2\pi} \hat{Q}_{out}(\theta)\, d\theta, & m = 0 \\ \frac{1}{\pi}\int\limits_0^{2\pi} \hat{Q}_{out}(\theta)\cos(m\theta)\, d\theta, & m = 1, 2, \ldots \end{cases}$$

$$X_m = \frac{1}{\pi}\int\limits_0^{2\pi} \hat{Q}_{out}(\theta)\sin(m\theta)\, d\theta, \qquad m = 1, 2, \ldots$$

where

$$\hat{Q}_{out}(\theta) = \frac{-\hat{q}_{out}(\theta)}{2\pi a_2 \lambda}$$

Using the IFFT algorithm, the solution in the time domain can readily be retrieved via

$$u(r, \theta, t) = \sum_n \hat{u}(r, \theta, \omega_n)\, e^{i\omega_n t}.$$

▪ Example 6.3: Heat Flow in a Disc

A disk, 1 m in radius, encompassing in its center an inner circle with radius 0.2 m. The thermal parameters of the domain are: thermal conductivity 1 $\mathrm{Wm^{-1}K^{-1}}$, mass density 1500 $\mathrm{kg\,m^{-3}}$, and specific heat capacity 800 $\mathrm{J\,kg^{-1}K^{-1}}$. The inner circle is subjected to a Dirichlet boundary condition and the outer circle is subjected to a Neumann boundary condition, both with prescribed quantities varying with time as shown in ◘ Fig. 6.15. The Dirichlet boundary condition is assumed constant in space and the heat flux is linearly varying with θ, i.e. $q = q_0(t)\theta$. Determine the temperature distribution in the disk.

▪ Solution

As outlined in Example 6.1, we begin by transforming the input signals from the time domain to the frequency domain using the FFT algorithm. Here, we utilized 4096 (2^{12}) FFT samples with 600 s per time step. Then, upon employing (6.102), for $M = 20$,

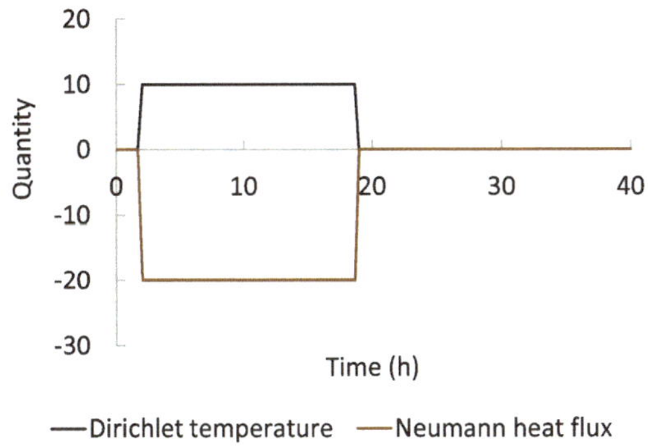

Fig. 6.15 Prescribed temperature at inner circle and imposed heat flux at outer circle

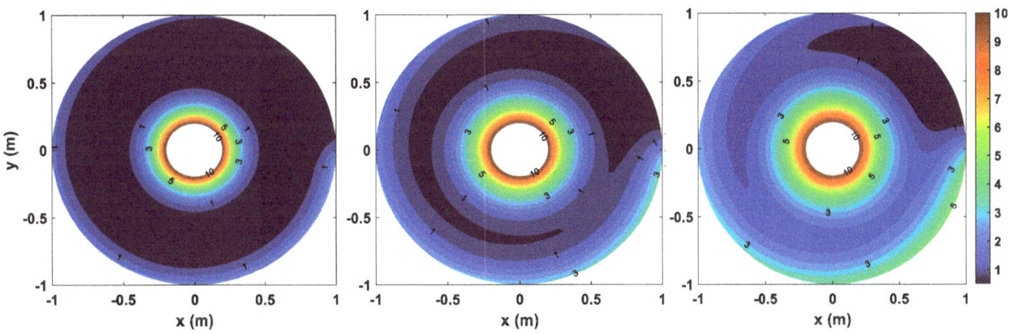

Fig. 6.16 Contour plots of temperature distributions after 3, 9 and 17 h

and summing the results for all involved frequencies, we obtain the solution in the frequency domain. Finally, applying the IFFT algorithm, we obtain the solution in the time domain.

Figure 6.16 shows the computed temperature in the domain after 3, 9 and 17 h of operating the system. The figure clearly shows the effect of the θ-dependent Neumann boundary condition on the temperature distribution in the domain.

6.10 Steady-State Temperature in a Disk with Inner Circle: θ-Dependent Neumann and Robin Boundary Conditions

Although the focus of this book is on solving transient problems, it is useful to solve some important steady-state problems with relatively complicated geometry and boundary conditions. Here, we solve heat flow in a disk with a Neumann boundary condition on the inner boundary and a Robin boundary condition on the outer, as depicted in Fig. 6.17.

Fig. 6.17 A disk with a steady-state Neumann-Robin boundary conditions

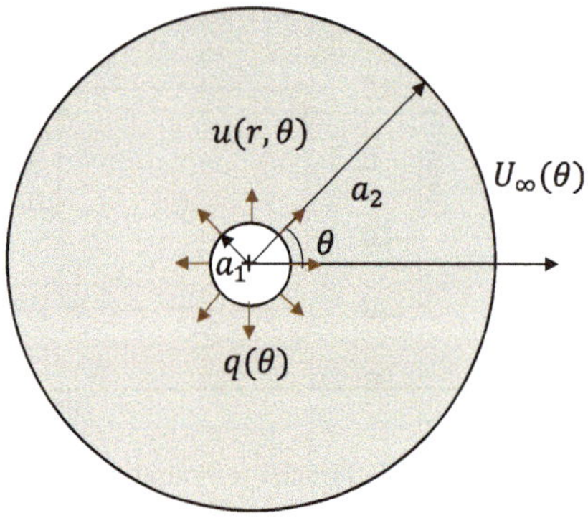

The relevant boundary value problem of this case is

$$\begin{aligned} &\frac{\partial^2 u}{\partial r^2}+\frac{1}{r}\frac{\partial u}{\partial r}+\frac{1}{r^2}\frac{\partial^2 u}{\partial \theta^2}=0 \\ &\text{BC:}\quad u(r,\theta)=u(r,\theta+2\pi) \\ &\qquad \left.\frac{\partial u(r,\theta)}{\partial r}\right|_{r=a_1}=-\frac{q(\theta)}{2\pi a_1\lambda}=Q(\theta) \\ &\qquad \left.\frac{\partial u(r,\theta)}{\partial r}\right|_{r=a_2}=\overline{h}\left[U_\infty(\theta)-u(a_2)\right] \end{aligned} \tag{6.103}$$

where U_∞ is the temperature of the surrounding environment.

- **Solution**

Since the boundary conditions in this boundary value problem are non-homogeneous, this problem can be addressed by dividing it into two sub-problems, such that

$$u(r,\theta,t)=u_1(r,\theta)+u_2(r,\theta) \tag{6.104}$$

Substituting (6.104) into (6.103) and separating the outcome into two equations, it leads to having two sub-BVPs, such that.

- **Sup-problem 1**

$$\begin{aligned} &\frac{\partial^2 u_1}{\partial r^2}+\frac{1}{r^2}\frac{\partial^2 u_1}{\partial \theta}+\frac{1}{r}\frac{\partial u_1}{\partial r}=0 \\ &\text{BC:}\quad \left.\frac{\partial u_1}{\partial r}\right|_{r=a_1}=\frac{-q(\theta)}{2\pi a_1\lambda}=Q(\theta) \\ &\qquad \left.\frac{\partial u_1}{\partial r}\right|_{r=a_2}=-\overline{h}u_1(a_2). \end{aligned} \tag{6.105}$$

Sub-problem 2

$$\frac{\partial^2 u_1}{\partial r^2} + \frac{1}{r^2}\frac{\partial^2 u_1}{\partial \theta} + \frac{1}{r}\frac{\partial u_1}{\partial r} = 0$$
$$\text{BC:} \quad \left.\frac{\partial u_1}{\partial r}\right|_{r=a_1} = \frac{-q(\theta)}{2\pi a_1 \lambda} = Q(\theta) \tag{6.106}$$
$$\left.\frac{\partial u_1}{\partial r}\right|_{r=a_2} = -\bar{h}u_1(a_2).$$

Note how we delineated the Robin boundary condition between the two sub-problems.

Solving Sub-problem 1

Using the separation of variables, we set

$$u_1 = R(r)F(\theta) \tag{6.107}$$

in which $F(\theta)$ is 2π-periodic, such that $F(\theta) = F(\theta + 2\pi)$. Substituting (6.107) into the differential equation in (6.105), and setting the separation constant to η, we obtain

$$\frac{r^2R'' + rR'}{R} = -\frac{F''}{F} = \eta$$

The separated equations are then

$$r^2R'' + rR' - \eta R = 0 \tag{6.108}$$

$$F'' + \eta F = 0 \tag{6.109}$$

The associated BVP of $F(\theta)$ is

$$F'' + \eta F = 0, \quad F(\theta) = F(\theta + 2\pi)$$

As explained in ▶ Sect. 6.8, the periodicity of this problem generates eigenvalues and eigenfunctions, and the solution leads to $\eta \equiv \eta_m = m^2, m = 0, 1, 2, \ldots$ with

$$F(\theta) = \sum_m c_1 \cos(m\theta) + c_2 \sin(m\theta), \quad m = 0, 1, 2, \ldots \tag{6.110}$$

The Eq. (6.108) is a typical *Cauchy-Euler differential equation*, for which its solution can be conducted by assuming a trial solution of the form

$$R = r^\kappa$$

where κ is to be determined. The first and second derivatives of this trial solution are

$$R' = \kappa r^{\kappa-1}, \quad R'' = \kappa(\kappa-1)r^{\kappa-2}$$

Substituting these derivatives into (6.108) and exchanging m^2 for η, gives

$$r^2\kappa(\kappa-1)r^{\kappa-2} + r\kappa r^{\kappa-1} - m^2r^\kappa = 0$$

Dividing by r^κ yields

$$\kappa(\kappa-1)+\kappa-m^2=0$$

This implies $\kappa^2=m^2$, that is $\kappa=\pm m$ with $m=0,1,2,\ldots$. For $m=0$, κ would imply having two repeated roots ($\kappa=+0$ and $\kappa=-0$, which are identical). Therefore, the solution yields (see Zill and Wright 2011)

$$R=c_3+c_4\ln r,\quad m=0 \tag{6.111}$$

For all other m, the solution is given by

$$R=c_5r^m+c_6r^{-m},\quad m=1,2,\ldots \tag{6.112}$$

Combining the solutions in (6.110)–(6.112), we obtain

$$u_1=R(r)F(\theta)$$
$$=\begin{cases}(c_3+c_4\ln r)\,c_1, & m=0\\ (c_5r^m+c_6r^{-m})\,[c_1\cos(m\theta)+c_2\sin(m\theta)], & m=1,2,\ldots\end{cases}$$

This outcome can be written as

$$u_1=\begin{cases}A_{01}+C_{01}\ln r, & m=0\\ r^m[A_{m1}\cos(m\theta)+B_{m1}\sin(m\theta)]+r^{-m}[C_{m1}\cos(m\theta) & \\ +D_{m1}\sin(m\theta)], & m=1,2,\ldots\end{cases}$$

where, for $m=0$, we replaced c_1c_3 by A_{01}, and c_1c_4 by C_{01}; and for $m=1,2,\ldots$, we replaced c_1c_5 by A_{m1}, c_2c_5 by B_{m1}, c_1c_6 by C_{m1} and c_2c_6 by D_{m1}.

The superposition principle then entails

$$\begin{aligned}u_1(r,\theta)&=(A_{01}+C_{01}\ln r)\\&+\sum_m r^m[A_{m1}\cos(m\theta)+B_{m1}\sin(m\theta)]\\&+r^{-m}[C_{m1}\cos(m\theta)+D_{m1}\sin(m\theta)],\qquad m=1,2,\ldots\end{aligned} \tag{6.113}$$

Boundary Conditions

Taking the derivative of (6.113) with respect to r, yields

$$\frac{\partial u_1}{\partial r}=\frac{1}{r}C_{01}+\sum_m \begin{matrix}mr^{m-1}[A_{m1}\cos(m\theta)+B_{m1}\sin(m\theta)]-\\ mr^{-m-1}[C_{m1}\cos(m\theta)+D_{m1}\sin(m\theta)]\end{matrix}$$

At $r=a_2$, the second boundary condition in (6.105) implies

$$\frac{1}{a_2}C_{01}+\sum_m \begin{matrix}ma_2^{m-1}[A_{m1}\cos(m\theta)+B_{m1}\sin(m\theta)]-\\ ma_2^{-m-1}[C_{m1}\cos(m\theta)+D_{m1}\sin(m\theta)]\end{matrix}$$
$$=-\bar{h}\{(A_{01}+C_{01}\ln a_2)+\sum_m \begin{matrix}a_2^m[A_{m1}\cos(m\theta)+B_{m1}\sin(m\theta)]+\\ a_2^{-m}[C_{m1}\cos(m\theta)+D_{m1}\sin(m\theta)]\}\end{matrix}$$

Collecting terms yields

$$\left(\frac{1}{a_2}+\bar{h}\ln a_2\right)C_{01}+\bar{h}A_{01}+\sum_m(ma_2^{m-1}+\bar{h}a_2^m)[A_{m1}\cos(m\theta)$$
$$+B_{m1}\sin(m\theta)]+(-ma_2^{-m-1}+\bar{h}a_2^{-m})[C_{m1}\cos(m\theta)+D_{m1}\sin(m\theta)]=0$$

which can be split into two equations

$$\left(\frac{1}{a_2}+\bar{h}\ln a_2\right)C_{01}+\bar{h}A_{01}=0$$

and

$$\sum_m(ma_2^{m-1}+\bar{h}a_2^m)[A_{m1}\cos(m\theta)+B_{m1}\sin(m\theta)]=$$
$$-\sum_m(-ma_2^{-m-1}+\bar{h}a_2^{-m})[C_{m1}\cos(m\theta)$$
$$+D_{m1}\sin(m\theta)]$$

This leads to

$$C_{01}=-\frac{\bar{h}A_{01}}{\frac{1}{a_2}+\bar{h}\ln a_2}$$
$$\sum_m C_{m1}\cos(m\theta)+D_{m1}\sin(m\theta)=$$
$$-\sum_m\frac{ma_2^{m-1}+\bar{h}a_2^m}{-ma_2^{-m-1}+\bar{h}a_2^{-m}}[A_{m1}\cos(m\theta)+B_{m1}\sin(m\theta)]$$

Thus, the solution so far can be expressed as

$$u_1(r,\theta)=A_{01}\left[1-\frac{\bar{h}\ln r}{\frac{1}{a_2}+\bar{h}\ln a_2}\right]$$
$$+\sum_m\left(r^m-r^{-m}\frac{ma_2^{m-1}+\bar{h}a_2^m}{-ma_2^{-m-1}+\bar{h}a_2^{-m}}\right) \tag{6.114}$$
$$[A_{m1}\cos(m\theta)+B_{m1}\sin(m\theta)]$$

6

Taking the derivative of (6.114) with respect to r yields

$$\frac{\partial u_1}{\partial r} = -\frac{\bar{h}}{r\left(\frac{1}{a_2} + \bar{h}\ln a_2\right)} A_{01} + \sum_m \left(m r^{m-1} + m r^{-m-1} \frac{m a_2^{m-1} + \bar{h} a_2^m}{-m a_2^{-m-1} + \bar{h} a_2^{-m}} \right) [A_{m1}\cos(m\theta) + B_{m1}\sin(m\theta)]$$

At $r = a_1$, the first boundary condition in (6.105) implies

$$-\frac{\bar{h}}{a_1\left(\frac{1}{a_2} + \bar{h}\ln a_2\right)} A_{01} + \sum_m \left(m a_1^{m-1} + m a_1^{-m-1} \frac{m a_2^{m-1} + \bar{h} a_2^m}{-m a_2^{-m-1} + \bar{h} a_2^{-m}} \right) [A_{m1}\cos(m\theta) + B_{m1}\sin(m\theta)] = Q(\theta)$$

This equation can be written as

$$W_{01} + \sum_m W_{m1}\cos(m\theta) + X_{m1}\sin(m\theta) = Q(\theta), \qquad m = 1, 2, \ldots \tag{6.115}$$

in which

$$\begin{aligned} W_{01} &= -\frac{\bar{h}}{a_1(\frac{1}{a_2} + \bar{h}\ln a_2)} A_{01} \\ W_{m1} &= \left(m a_1^{m-1} + m a_1^{-m-1} \frac{m a_2^{m-1} + \bar{h} a_2^m}{-m a_2^{-m-1} + \bar{h} a_2^{-m}} \right) A_{m1} \\ X_{m1} &= \left(m a_1^{m-1} + m a_1^{-m-1} \frac{m a_2^{m-1} + \bar{h} a_2^m}{-m a_2^{-m-1} + \bar{h} a_2^{-m}} \right) B_{m1} \\ Q(\theta) &= -\frac{q(\theta)}{2\pi a \lambda} \end{aligned} \tag{6.116}$$

Equation (6.115) is a typical trigonometric Fourier series with its coefficients given by

$$\begin{aligned} W_{01} &= \frac{1}{2\pi}\int_0^{2\pi} Q(\theta)\, d\theta \\ W_{m1} &= \frac{1}{\pi}\int_0^{2\pi} Q(\theta)\cos(m\theta)\, d\theta \\ X_{m1} &= \frac{1}{\pi}\int_0^{2\pi} Q(\theta)\sin(m\theta)\, d\theta \end{aligned} \tag{6.117}$$

Thus, the solution of u_1 can be expressed as

$$u_1(r,\theta) = A_{01}\left(1 - \frac{\bar{h}\ln r}{\dfrac{1}{a_2} + \bar{h}\ln a_2}\right) + \sum_m \left(r^m - r^{-m}\frac{ma_2^{m-1} + \bar{h}a_2^m}{-ma_2^{-m-1} + \bar{h}a_2^{-m}}\right)$$
$$[A_{m1}\cos(m\theta) + B_{m1}\sin(m\theta)],\ m = 1, 2, \ldots \tag{6.118}$$

with the coefficients given by (6.116) and (6.117).

Solving Sub-problem 2

The general solution of the differential equation in (6.106) has been derived in detail in solving Sub-problem 1, that yielded

$$u_2(r,\theta) = (A_{02} + C_{02}\ln r) + \sum_m \begin{matrix} r^m[A_{m2}\cos(m\theta) + B_{m2}\sin(m\theta)]+ \\ r^{-m}[C_{m2}\cos(m\theta) + D_{m2}\sin(m\theta)] \end{matrix},\quad m = 1, 2, \ldots \tag{6.119}$$

in which we amended the constants to suit this sub-problem.

Boundary Conditions

Taking the derivative of (6.119) with respect to r and solving for the first boundary condition in (6.106) at $r = a_1$ yields

$$C_{02} = 0,\ \sum_m C_{m2}\cos(m\theta) + D_{m2}\sin(m\theta) = \sum_m a_1^{2m}[A_{m2}\cos(m\theta) + B_{m2}\sin(m\theta)]$$

According to this result, Eq. (6.119) can be reduced to

$$u_2(r,\theta) = A_{02} + \sum_m (r^m + r^{-m}a_1^{2m})\,[A_{m2}\cos(m\theta) + B_{m2}\sin(m\theta)] \tag{6.120}$$

Taking the derivative of (6.120) with respect to r and solving for the second boundary condition in (6.106) at $r = a_2$ yields

$$A_{02}\bar{h} + \sum_m (ma_2^{m-1} - ma_2^{-m-1}a_1^{2m} + \bar{h}a_2^m + \bar{h}a_2^{-m}a_1^{2m})\cdot [A_{m2}\cos(m\theta) + B_{m2}\sin(m\theta)] = \bar{h}U_\infty(\theta)$$

which is a trigonometric Fourier series with its coefficients given by

$$A_{02} = \frac{1}{2\pi} \int_0^{2\pi} U_\infty(\theta)\, d\theta$$

$$A_{m2} = \frac{1}{\pi} \int_0^{2\pi} \frac{\bar{h} U_\infty(\theta)}{m a_2^{m-1} - m a_2^{-m-1} a_1^{2m} + \bar{h} a_2^m + \bar{h} a_2^{-m} a_1^{2m}} \cos(m\theta)\, d\theta$$

$$B_{m2} = \frac{1}{\pi} \int_0^{2\pi} \frac{\bar{h} U_\infty(\theta)}{m a_2^{m-1} - m a_2^{-m-1} a_1^{2m} + \bar{h} a_2^m + \bar{h} a_2^{-m} a_1^{2m}} \sin(m\theta)\, d\theta \tag{6.121}$$

Thus the solution of u_2 can be expressed as

$$u_2(r,\theta) = A_{02} + \sum_{m=1}^{\infty} \left(r^m + r^{-m} a_1^{2m} \right) \left[A_{m2} \cos(m\theta) + B_{m2} \sin(m\theta) \right] \tag{6.122}$$

with the Fourier coefficients given in (6.121)

BVP Solution

Having solved the two sub-problems, the final solution of this boundary value problem can be summarized as the following:

$$u(r,\theta) = u_1(r,\theta) + u_2(r,\theta) \tag{6.123}$$

with

$$u_1(r,\theta) = A_{01} \left(1 - \frac{\bar{h} \ln r}{\frac{1}{a_2} + \bar{h} \ln a_2} \right) + \sum_m \left[r^m - r^{-m} \frac{(m a_2^{m-1} + \bar{h} a_2^m)}{-m a_2^{-m-1} + \bar{h} a_2^{-m}} \right]$$

$$[A_{m1} \cos(m\theta) + B_{m1} \sin(m\theta)], \quad m = 1, 2, \ldots$$

in which

$$A_{01} = -\frac{a_1}{\bar{h}} \left(\frac{1}{a_2} + \bar{h} \ln a_2 \right) W_{01}$$

$$A_{m1} = \frac{-m a_2^{-m-1} + \bar{h} a_2^{-m}}{m a_1^{m-1}(-m a_2^{-m-1} + \bar{h} a_2^{-m}) + m a_1^{-m-1}(m a_2^{m-1} + \bar{h} a_2^m)} W_{m1}$$

$$B_{m1} = \frac{-m a_2^{-m-1} + \bar{h} a_2^{-m}}{m a_1^{m-1}(-m a_2^{-m-1} + \bar{h} a_2^{-m}) + m a_1^{-m-1}(m a_2^{m-1} + \bar{h} a_2^m)} X_{m1}$$

with

$$W_{01} = \frac{1}{2\pi} \int_0^{2\pi} Q(\theta)\, d\theta$$

$$W_{m1} = \frac{1}{\pi} \int_0^{2\pi} Q(\theta) \cos(m\theta)\, d\theta$$

$$X_{m1} = \frac{1}{\pi} \int_0^{2\pi} Q(\theta) \sin(m\theta)\, d\theta$$

with

$$Q(\theta) = -\frac{q(\theta)}{2\pi a_1 \lambda}$$

and

$$u_2(r,\theta) = A_{02} + \sum_m (r^m + r^{-m} a_1^{2m})$$
$$[A_{m2} \cos(m\theta) + B_{m2} \sin(m\theta)], \quad m = 1, 2, \ldots$$

with

$$A_{02} = \frac{1}{2\pi} \int_0^{2\pi} U_\infty(\theta)\, d\theta$$

$$A_{m2} = \frac{1}{\pi} \int_0^{2\pi} \frac{\bar{h} U_\infty(\theta)}{m\,(a_2^{m-1} - a_2^{-m-1} a_1^{2m}) + \bar{h}(a_2^m + a_2^{-m} a_1^{2m})}$$
$$\cos(m\theta)\, d\theta$$

$$B_{m2} = \frac{1}{\pi} \int_0^{2\pi} \frac{\bar{h} U_\infty(\theta)}{m\,(a_2^{m-1} - a_2^{-m-1} a_1^{2m}) + \bar{h}(a_2^m + a_2^{-m} a_1^{2m})}$$
$$\sin(m\theta)\, d\theta$$

Special Case 1: Constant Boundary Conditions

In the special case where the prescribed heat flux q and the temperature at the surrounding environment U_∞ are not function of θ, the solution becomes

$$u(r) = u_1(r) + u_2(r)$$

where

$$u_1(r) = A_{01} \left(1 - \frac{\bar{h} \ln r}{\frac{1}{a_2} + \bar{h} \ln a_2}\right)$$

with

$$A_{01} = -\frac{a_1}{\bar{h}} \left(\frac{1}{a_2} + \bar{h} \ln a_2\right) W_{01}, \quad \text{in which} \quad W_{01} = Q, \quad Q = -\frac{q}{2\pi a_1 \lambda}$$

and

$$u_2(r) = A_{02}$$

with

$$A_{02} = U_\infty.$$

Special Case 2: Linear Boundary Conditions

In the special case where the prescribed heat flux q and the temperature at the surrounding environment U_∞ are function of θ, with $q = q_0\theta$ and $U_\infty = U_{\infty 0}\theta$, the solution becomes

$$u(r,\theta) = u_1(r,\theta) + u_2(r,\theta)$$

where

$$u_1(r,\theta) = A_{01}\left(1 - \frac{\bar{h}\ln r}{\frac{1}{a_2} + \bar{h}\ln a_2}\right) + \sum_m B_{m1}\left(r^m - r^{-m}\frac{(m a_2^{m-1} + \bar{h}a_2^m)}{-m a_2^{-m-1} + \bar{h}a_2^{-m}}\right)\sin(m\theta), \quad m = 1,2,3,\ldots \tag{6.124}$$

with

$$A_{01} = -\frac{a_1}{\bar{h}}\left(\frac{1}{a_2} + \bar{h}\ln a_2\right)W_{01} \quad \text{where} \quad W_{01} = \frac{Q\pi}{\theta}$$
$$= -\frac{q_0}{2a_1\lambda}, \quad Q = -\frac{q_0\theta}{2\pi a_1\lambda}$$
$$B_{m1} = \frac{-m a_2^{-m-1} + \bar{h}a_2^{-m}}{m a_1^{m-1}(-m a_2^{-m-1} + \bar{h}a_2^{-m}) + m a_1^{-m-1}(m a_2^{m-1} + \bar{h}a_2^m)}X_{m1}$$

with

$$X_{m1} = \frac{q_0}{m\pi a_1\lambda}$$

and

$$u_2(r,\theta) = A_{02} + \sum_m B_{m2}(r^m + r^{-m}a_1^{2m}) \sin(m\theta), \quad m = 1,2,3,\ldots \tag{6.125}$$

where

$$A_{02} = \pi U_{\infty 0}$$
$$B_{m2} = \frac{2\bar{h}\,a_2^{m+1}\,U_{\infty 0}}{m\,[-\bar{h}a_2^{2m+1} + (m - \bar{h}a_2)\,a_1^{2m} - m a_2^{2m}]}.$$

6.11 Disk with Inner Circle: Transient θ-Dependent Neumann and Robin Boundary Conditions

In the previous section we solved a steady-state heat flow boundary value problem for a disk subjected to Neumann-Robin boundary conditions. Here, we extend the solution to the transient condition. The inner boundary condition is assumed solely a function of time, while the outer boundary condition can be a function of time and θ, as shown schematically in Fig. 6.18.

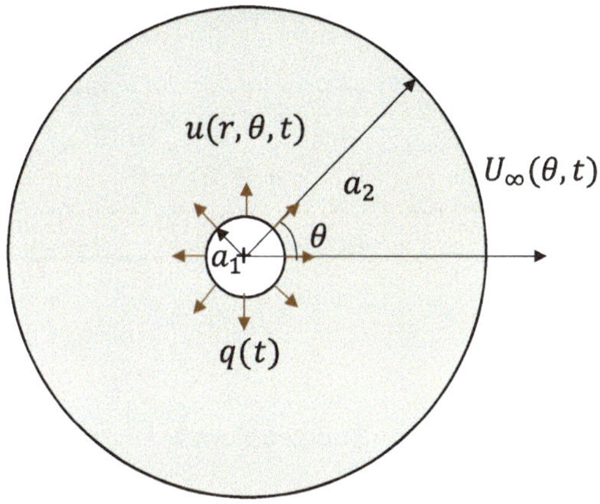

Fig. 6.18 Transient Neumann-Robin boundary problem

The relevant boundary value problem of this case is

$$
\begin{aligned}
&\frac{\partial^2 u}{\partial r^2}+\frac{1}{r}\frac{\partial u}{\partial r}+\frac{1}{r^2}\frac{\partial^2 u}{\partial \theta^2}-\frac{1}{\alpha}\frac{\partial u}{\partial t}=0\\
&\text{IC:}\quad u(r,\theta,0)=0\\
&\text{BC:}\quad u(r,\theta,t)=u(r,\theta+2\pi,t)\\
&\qquad \left.\frac{\partial u(r,\theta,t)}{\partial r}\right|_{r=a_1}=\frac{-q(t)}{2\pi a_1\lambda}\\
&\qquad \left.\frac{\partial u(r,\theta,t)}{\partial r}\right|_{r=a_2}=\bar{h}\left[U_\infty(\theta,t)-u(a_2)\right].
\end{aligned}
\tag{6.126}
$$

Solution

Similar to the previous case, the general solution of the differential equation in (6.126) requires splitting the problem into two sub-problems, such that

$$u(r,\theta,t)=u_1(r,t)+u_2(r,\theta,t) \tag{6.127}$$

where we chose u_1 not to be a function of θ since q is a function of time only. Inserting (6.127) into (6.126) leads to two boundary value sub-problems, such that

■ **Sub-problem 1**

$$
\begin{aligned}
&\frac{\partial^2 u_1}{\partial r^2} + \frac{1}{r}\frac{\partial u_1}{\partial r} - \frac{1}{\alpha}\frac{\partial u_1}{\partial t} = 0 \\
&\text{IC:} \quad u_1(r,0) = 0 \\
&\text{BC:} \quad \left.\frac{\partial u_1(r,t)}{\partial r}\right|_{r=a_1} = \frac{-q_{in}(t)}{2\pi a_1 \lambda} = Q(t) \\
&\qquad\quad \left.\frac{\partial u_1(r,t)}{\partial r}\right|_{r=a_2} = -\overline{h} u_1(a_2).
\end{aligned}
\tag{6.128}
$$

■ **Sub-problem 2**

$$
\begin{aligned}
&\frac{\partial^2 u_2}{\partial r^2} + \frac{1}{r}\frac{\partial u_2}{\partial r} + \frac{1}{r^2}\frac{\partial^2 u_2}{\partial \theta^2} - \frac{1}{\alpha}\frac{\partial u_2}{\partial t} = 0 \\
&\text{IC:} \quad u_2(r,\theta,0) = 0 \\
&\text{BC:} \quad u_2(r,\theta,t) = u_2(r,\theta + 2\pi, t) \\
&\qquad\quad \left.\frac{\partial u_2(r,\theta,t)}{\partial r}\right|_{r=a_1} = 0 \\
&\qquad\quad \left.\frac{\partial u_2(r,\theta,t)}{\partial r}\right|_{r=a_2} = \overline{h}\,[U_\infty(\theta) - u_2(a_2)]
\end{aligned}
\tag{6.129}
$$

Note the delineation of the Robin boundary condition between the two sub-problems.

■ **Solving Sub-problem 1**

Applying the Fourier transform to (6.128) gives

$$
\begin{aligned}
&\frac{d^2 \hat{u}_1}{dr^2} + \frac{1}{r}\frac{d\hat{u}_1}{dr} - \frac{i\omega}{\alpha}\hat{u}_1 = 0 \\
&\text{BC:} \quad \left.\frac{d\hat{u}_1(r,\omega)}{dr}\right|_{r=a_1} = \frac{-\hat{q}(\omega)}{2\pi a_1 \lambda} = \hat{Q}(\omega) \\
&\qquad\quad \left.\frac{d\hat{u}_1(r,\omega)}{dr}\right|_{r=a_2} = -\overline{h}\hat{u}_1(a_2,\omega)
\end{aligned}
\tag{6.130}
$$

The differential equation in (6.130) is a form of the modified Bessel equation, with its general solution given by

$$
\hat{u}_1(r,\omega) = c_1 I_0(\kappa r) + c_2 K_0(\kappa r), \quad \kappa = \sqrt{i\omega/\alpha} \tag{6.131}
$$

At $r = a_1$, the first boundary condition in (6.130) implies

$$
c_1 \kappa I_1(\kappa a_1) - c_2 \kappa K_1(\kappa a_1) = \hat{Q}(\omega)
$$

At $r = a_2$, the second boundary condition implies

$$
c_1[\kappa I_1(\kappa a_2) + \overline{h} I_0(\kappa a_2)] - c_2[\kappa K_1(\kappa a_2) - \overline{h} K_0(\kappa a_2)] = 0
$$

Putting these two equations in a matrix form yields

$$\begin{pmatrix} \kappa I_1(\kappa a_1) & -\kappa K_1(\kappa a_1) \\ \kappa I_1(\kappa a_2) + \bar{h} I_0(\kappa a_2) & -\kappa K_1(\kappa a_2) + \bar{h} K_0(\kappa a_2) \end{pmatrix} \begin{pmatrix} c_1 \\ c_2 \end{pmatrix} = \begin{pmatrix} \hat{Q} \\ 0 \end{pmatrix}$$

Solving for c_1 and c_2 results in

$$\begin{pmatrix} c_1 \\ c_2 \end{pmatrix} = \frac{1}{\Delta} \begin{pmatrix} -\kappa K_1(\kappa a_2) + \bar{h} K_0(\kappa a_2) & \kappa K_1(\kappa a_1) \\ -\kappa I_1(\kappa a_2) - \bar{h} I_0(\kappa a_2) & \kappa I_1(\kappa a_1) \end{pmatrix} \begin{pmatrix} \hat{Q} \\ 0 \end{pmatrix}$$

$$\Delta = \kappa I_1(\kappa a_1)[-\kappa K_1(\kappa a_2) + \bar{h} K_0(\kappa a_2)] + \kappa K_1(\kappa a_1)[\kappa I_1(\kappa a_2) + \bar{h} I_0(\kappa a_2)]$$

This gives

$$\begin{aligned} c_1 &= \frac{1}{\Delta}[-\kappa K_1(\kappa a_2) + \bar{h} K_0(\kappa a_2)]\, \hat{Q} \\ c_2 &= -\frac{1}{\Delta}[\kappa I_1(\kappa a_2) + \bar{h} I_0(\kappa a_2)]\, \hat{Q} \end{aligned}$$

Hence, the solution of the BVP (6.130) is

$$\begin{aligned} \hat{u}_1(r,\omega) = \frac{\hat{Q}_{in}}{\Delta} \big\{ &\big[-\kappa K_1(\kappa a_2) + \bar{h} K_0(\kappa a_2)\big] I_0(\kappa r) \\ &- \big[\kappa I_1(\kappa a_2) + \bar{h} I_0(\kappa a_2)\big] K_0(\kappa r) \big\} \end{aligned}$$

with

$$\begin{aligned} \kappa &= \sqrt{i\omega/\alpha} \\ \Delta &= \kappa I_1(\kappa a_1)\big[-\kappa K_1(\kappa a_2) + \bar{h} K_0(\kappa a_2)\big] \\ &\quad + \kappa K_1(\kappa a_1)\big[\kappa I_1(\kappa a_2) + \bar{h} I_0(\kappa a_2)\big]. \end{aligned} \tag{6.132}$$

- **Solving Sub-problem 2**

Applying the Fourier transform to (6.129) gives

$$\begin{aligned} &\frac{\partial^2 \hat{u}_2}{\partial r^2} + \frac{1}{r}\frac{\partial \hat{u}_2}{\partial r} + \frac{1}{r^2}\frac{\partial^2 \hat{u}_2}{\partial \theta^2} - \frac{i\omega}{\alpha}\hat{u}_2 = 0 \\ &\text{BC:} \quad \hat{u}_2(r,\theta,\omega) = \hat{u}_2(r,\theta+2\pi,\omega) \\ &\qquad \left.\frac{\partial \hat{u}_2(r,\theta,\omega)}{\partial r}\right|_{r=a_1} = 0 \\ &\qquad \left.\frac{\partial \hat{u}_2(r,\theta,\omega)}{\partial r}\right|_{r=a_2} = \bar{h}\,[\hat{U}_\infty(\theta,\omega) - \hat{u}_2(a_2)] \end{aligned} \tag{6.133}$$

The general solution of the differential equation in (6.133) can be carried out using the method of separation of variables. Assume

$$\hat{u}_2(r,\theta) = R(r)\,F(\theta) = RF \tag{6.134}$$

in which $R(r)$ is a function of r only and $F(\theta)$ is a function of θ only. Substituting (6.134) into (6.133) and following the procedure in (6.80) it leads to

$$\frac{d^2F}{d\theta^2} + m^2F = 0 \tag{6.135}$$

$$r^2\frac{d^2R}{dr^2} + r\frac{dR}{dr} - (\kappa^2 r^2 + m^2)R = 0 \tag{6.136}$$

with $\kappa^2 = i\omega/\alpha$.

As discussed earlier, the solution of (6.135) with its associated boundary condition $F(\theta) = F(\theta + 2\pi)$ is

$$F = \sum_m c_3 \cos(m\theta) + c_4 \sin(m\theta), \quad m = 0, 1, 2, \ldots \tag{6.137}$$

where c_3 and c_4 are constants to be determined.

The solution of (6.136) is

$$R(r) = c_5 I_m(\kappa r) + c_6 K_m(\kappa r) \tag{6.138}$$

Taking the derivative of (6.138) with respect to r gives

$$R'(r) = c_5\Big[\kappa I_{m+1}(\kappa r) + \frac{m}{r}I_m(\kappa r)\Big] + c_6\Big[-\kappa K_{m+1}(\kappa r) + \frac{m}{r}K_m(\kappa r)\Big]$$

At $r = a_1$, the second boundary condition in (6.133) implies

$$c_6 = -\frac{\kappa I_{m+1}(\kappa a_1) + \frac{m}{a_1}I_m(\kappa a_1)}{-\kappa K_{m+1}(\kappa a_1) + \frac{m}{a_1}K_m(\kappa a_1)}c_5$$

This leads to

$$R(r) = c_5[I_m(\kappa r) - YK_m(\kappa r)]$$

with

$$Y = \frac{\kappa I_{m+1}(\kappa a_1) + \frac{m}{a_1}I_m(\kappa a_1)}{-\kappa K_{m+1}(\kappa a_1) + \frac{m}{a_1}K_m(\kappa a_1)} \tag{6.139}$$

Collecting both solutions, (6.137) and (6.139), yields

$$\hat{u}_2(r,\theta) = \sum_m A_m \cos(m\theta)\,[I_m(\kappa r) - YK_m(\kappa r)] + \sum_m B_m \sin(m\theta)\,[I_m(\kappa r) - YK_m(\kappa r)] \tag{6.140}$$

where $A_m = c_3 \cdot c_5$ and $B_m = c_4 \cdot c_5$.

Taking the derivative of (6.140) with respect to r gives

$$\frac{\partial \hat{u}_2(r,\theta)}{\partial r} = \sum_m A_m \cos(m\theta)\Big\{\kappa I_{m+1}(\kappa r) + \frac{m}{r}I_m(\kappa r) - Y\Big[[-\kappa K_{m+1}(\kappa r) + \frac{m}{r}K_m(\kappa r)\Big]\Big\}$$
$$+ \sum_m B_m \sin(m\theta)\Big\{\kappa I_{m+1}(\kappa r) + \frac{m}{r}I_m(\kappa r) - Y\Big[-\kappa K_{m+1}(\kappa r) + \frac{m}{r}K_m(\kappa r)\Big]\Big\}$$

At $r = a_2$, the third boundary condition in (6.133) implies

$$\sum_m A_m \cos m\theta \left\{ \kappa I_{m+1}(\kappa a_2) + \frac{m}{a_2} I_m(\kappa a_2) \right.$$
$$\left. -Y \left[-\kappa K_{m+1}(\kappa a_2) + \frac{m}{a_2} K_m(\kappa a_2) \right] \right\}$$
$$+ \sum_m B_m \sin(m\theta) \left\{ \kappa I_{m+1}(\kappa a_2) + \frac{m}{a_2} I_m(\kappa a_2) \right.$$
$$\left. -Y \left[-\kappa K_{m+1}(\kappa a_2) + \frac{m}{a_2} K_m(\kappa a_2) \right] \right\}$$
$$= \bar{h}\hat{U}_\infty(\theta) - \bar{h} \sum_m A_m \cos(m\theta) \left[I_m(\kappa a_2) - Y K_m(\kappa a_2) \right]$$
$$- \bar{h} \sum_m B_m \sin(m\theta) \left[I_m(\kappa a_2) - Y K_m(\kappa a_2) \right]$$

Collecting terms, we obtain

$$\sum_m \left[A_m \cos(m\theta)(G_m + F_m) + B_m \sin(m\theta)(G_m + F_m) \right]$$
$$= \bar{h}\hat{U}_\infty(\theta)$$

with

$$G_m = \kappa I_{m+1}(\kappa a_2) + \frac{m}{a_2} I_m(\kappa a_2)$$
$$- Y \left[-\kappa K_{m+1}(\kappa a_2) + \frac{m}{a_2} K_m(\kappa a_2) \right]$$
$$F_m = \bar{h} \left[I_m(\kappa a_2) - Y K_m(\kappa a_2) \right]$$

Rearranging gives

$$\sum_m W_m \cos m\theta + X_m \sin m\theta = \bar{h}\hat{U}_\infty(\theta)$$

with

$$\begin{aligned} W_m &= A_m(G_m + F_m) \\ X_m &= B_m(G_m + F_m) \end{aligned} \tag{6.141}$$

which is a typical trigonometric Fourier series with its coefficients given by

$$W_m = \begin{cases} \frac{\bar{h}}{2\pi} \int_0^{2\pi} \hat{U}_\infty(\theta)\, d\theta, & m = 0 \\ \frac{\bar{h}}{\pi} \int_0^{2\pi} \hat{U}_\infty(\theta) \cos(m\theta)\, d\theta, & m = 1, 2, \ldots \end{cases} \tag{6.142}$$
$$X_m = \frac{\bar{h}}{\pi} \int_0^{2\pi} \hat{U}_\infty(\theta) \sin(m\theta)\, d\theta, \qquad m = 1, 2, \ldots$$

Upon determining the Fourier coefficients W_m and X_m, the constants A_m and B_m can be obtained, which, upon substituting into (6.140), gives the required solution for $\hat{u}_2$.

6

BVP Solution

The final solution of the boundary value problem (6.126) in the frequency domain is therefore

$$\hat{u}\,(r,\theta,\omega_n) = \hat{u}_1(r,\omega_n) + \hat{u}_2(r,\theta,\omega_n) \tag{6.143}$$

where

$$\hat{u}_1(r,\omega_n) = \frac{\hat{Q}}{\Delta}\left\{ \begin{array}{l} \left[-\kappa K_1(\kappa a_2) + \overline{h}K_0(\kappa a_2)\right] I_0(\kappa r) \\ -\left[\kappa I_1(\kappa a_2) + \overline{h}I_0(\kappa a_2)\right] K_0(\kappa r) \end{array} \right\}$$

with

$$\begin{aligned} \kappa &= \sqrt{i\omega/\alpha} \\ \Delta &= \kappa I_1(\kappa a_1)\left[-\kappa K_1(\kappa a_2) + \overline{h}K_0(\kappa a_2)\right] + \\ &\qquad \kappa K_1(\kappa a_1)\left[\kappa I_1(\kappa a_2) + \overline{h}I_0(\kappa a_2)\right] \end{aligned} \tag{6.144}$$

and

$$\begin{aligned} \hat{u}_2(r,\theta,\omega_n) &= \sum_m A_m \cos{(m\theta)}\left[I_m(\kappa r) - YK_m(\kappa r)\right] + \\ &\qquad \sum_m B_m \sin{(m\theta)}\left[I_m(\kappa r) - YK_m(\kappa r)\right] \end{aligned} \tag{6.145}$$

where

$$Y = \frac{\kappa I_{m+1}(\kappa a_1) + \frac{m}{a_1} I_m(\kappa a_1)}{-\kappa K_{m+1}(\kappa a_1) + \frac{m}{a_1} K_m(\kappa a_1)}$$

with the Fourier coefficients obtained via

$$A_m = \frac{W_m}{G_m + F_m}, \quad B_m = \frac{X_m}{G_m + F_m}$$

with

$$\begin{aligned} G_m &= \kappa I_{m+1}(\kappa a_2) + \frac{m}{a_2} I_m(\kappa a_2) - Y\left[-\kappa K_{m+1}(\kappa a_2) + \frac{m}{a_2} K_m(\kappa a_2)\right] \\ F_m &= \overline{h}\left[I_m(\kappa a_2) - YK_m(\kappa a_2)\right] \end{aligned} \tag{6.146}$$

and

$$\begin{aligned} W_m &= \left\{ \begin{array}{ll} \frac{\overline{h}}{2\pi} \int_0^{2\pi} \hat{U}_\infty(\theta)\, d\theta, & m = 0 \\ \frac{\overline{h}}{\pi} \int_0^{2\pi} \hat{U}_\infty(\theta) \cos{(m\theta)}\, d\theta, & m = 1, 2, \ldots \end{array} \right. \\ X_m &= \frac{\overline{h}}{\pi} \int_0^{2\pi} \hat{U}_\infty(\theta) \sin{(m\theta)}\, d\theta, \quad m = 1, 2, \ldots \end{aligned} \tag{6.147}$$

Having solved the problem in the frequency domain, using the IFFT algorithm, the solution in the time domain can be restored, such that

$$u(r,\theta,t) = \sum_n \hat{u}(r,\theta,\omega_n)\, e^{i\omega_n t}.$$

Special Case: Steady-State

In the particular case where the radial frequency ω approaches zero, the transient solution (6.143) converges to the steady-state solution (6.123). In the following, we carry on a detailed derivation to demonstrate the validity of this assertion.

Recall the solution of $\hat{u}_1(r,\omega)$ in (6.144), where we had

$$\begin{aligned}\hat{u}_1(r,\omega) =& \frac{\hat{Q}}{\Delta}\{\left[-\kappa K_1(\kappa a_2) + \bar{h}K_0(\kappa a_2)\right] I_0(\kappa r)\\ &-\left[\kappa I_1(\kappa a_2) + \bar{h} I_0(\kappa a_2)\right] K_0(\kappa r)\}\end{aligned}$$

with

$$\begin{aligned}\kappa &= \sqrt{i\omega/\alpha}\\ \Delta &= \kappa I_1(\kappa a_1)\left[-\kappa K_1(\kappa a_2) + \bar{h}K_0(\kappa a_2)\right]\\ &\quad + \kappa K_1(\kappa a_1)\left[\kappa I_1(\kappa a_2) + \bar{h} I_0(\kappa a_2)\right]\end{aligned}$$

Taking the limit as $\omega \to 0$, i.e. $\kappa \to 0$, the denominator will lead to

$$\Delta = \frac{\bar{h}}{a_1}$$

where we used the following limit approximations

$$\begin{aligned}&\lim_{\kappa\to 0} I_0(a_2\kappa) = 1\\ &\lim_{\kappa\to 0} I_1(\kappa) = 0\\ &\lim_{\kappa\to 0} \kappa K_1(a_1\kappa) \approx \frac{1}{a_1}\end{aligned}$$

In the same way, if we take the limit as $\kappa \to 0$, the nominator will lead to

$$Q\left\{-\frac{1}{a_2} - \bar{h}\ln(\kappa a_2) - [-\bar{h}\ln(\kappa r)]\right\} \quad \text{or} \quad Q\left[-\frac{1}{a_2} - \bar{h}(\ln a_2 - \ln r)\right]$$

where we used the following limit approximations

$$\begin{aligned}
&\lim_{\kappa\to 0} I_0(\kappa r) = 1\\
&\lim_{\kappa\to 0} I_1(\kappa) = 0\\
&\lim_{\kappa\to 0} \kappa K_1(a_2\kappa) \approx \frac{1}{a_2}\\
&\lim_{\kappa\to 0} K_0(a_2\kappa) \approx -\ln(a_2\kappa) = -\ln a_2 - \ln\kappa
\end{aligned} \tag{6.148}$$

Following these limit approximations, $\hat{u}_1(r,\omega)$ becomes

$$\hat{u}_1(r,\omega)\Big|_{\omega\to 0} = u_1(r) = -\frac{Q}{\bar{h}}\left[\frac{a_1}{a_2} + a_1\bar{h}\,(\ln a_2 - \ln r)\right] \tag{6.149}$$

6

Recall the solution of $\hat{u}_2(r,\theta,\omega)$ in (6.145):

$$\begin{aligned}
\hat{u}_2(r,\theta,\omega) &= \sum_m A_m \cos(m\theta)[I_m(\kappa r) - YK_m(\kappa r)]\\
&\quad + \sum_m B_m \sin(m\theta)[I_m(\kappa r) - YK_m(\kappa r)]
\end{aligned}$$

with

$$Y = \frac{\kappa I_{m+1}(\kappa a_1) + \frac{m}{a_1} I_m(\kappa a_1)}{-\kappa K_{m+1}(\kappa a_1) + \frac{m}{a_1} K_m(\kappa a_1)} \tag{6.150}$$

We proceed to solve the limiting case for the special scenario where $\hat{U}_\infty = \hat{U}_{\infty 0}\theta$. For $m = 0$, using the limit approximations derived earlier, we can readily find that as $\kappa \to 0$, Y in (6.150) approaches zero. Therefore, the corresponding Fourier coefficient in (6.147) simplifies to

$$W_0 = \pi\bar{h}U_{\infty 0} \Rightarrow A_0 = \pi U_{\infty 0}$$

where it can be figured out from (6.146) that for $m = 0$, $G_0 = 0$ and $F_0 = \bar{h}$.

Thus, using the first limit approximation in (6.148), we can readily find out that as $\kappa \to 0$, the solution (6.150) for $m = 0$ becomes

$$u_{20}(r,\theta) = \pi U_{\infty 0}$$

For $m \geq 1$, the relevant limits lead to

$$\begin{aligned}
W_m &= 0\\
X_m &= -\frac{2}{m}\bar{h}U_{\infty 0}
\end{aligned}$$

This yields

$$\begin{aligned}
A_m &= 0\\
B_m &= -\lim_{\kappa\to 0} \frac{2\bar{h}U_{\infty 0}}{m\left[I_{m+1}(\kappa a_2)\kappa + I_m(\kappa a_2)(\frac{m}{a_2} + \bar{h})\right]}
\end{aligned}$$

We know that $\lim_{z\to 0} I_m(z) = 0, m \geq 1$, hence, if we take the limit of B_m in its current form as $\kappa \to 0$, it will lead to infinity. To overcome this problem, we proceed to the limiting form of u_{2m}, which now reads

$$u_{2m}(r,\theta) = -\lim_{\kappa\to 0} \frac{2\bar{h}U_{\infty 0} \sin(m\theta) I_m(\kappa r)}{m\left[I_{m+1}(\kappa a_2)\kappa + I_m(\kappa a_2)(\frac{m}{a_2} + \bar{h})\right]}$$

Dividing the nominator and the denominator by $I_m(\kappa r)$ leads to

$$u_{2m}(r,\theta) = -\lim_{\kappa\to 0} \frac{2\bar{h}U_{\infty 0} \sin(m\theta)}{m\left[\dfrac{I_{m+1}(\kappa a_2)}{I_m(\kappa r)}\kappa + \dfrac{I_m(\kappa a_2)}{I_m(\kappa r)}(\dfrac{m}{a_2} + \bar{h})\right]}$$

Applying the limit approximations in (6.148), and knowing that

$$\lim_{\kappa\to 0} \frac{I_{m+1}(\kappa a_2)}{I_m(\kappa r)} = 0$$
$$\lim_{\kappa\to 0} \frac{I_m(\kappa a_2)}{I_m(\kappa r)} \approx \left(\frac{a_2}{r}\right)^m$$

we obtain

$$u_{2m}(r,\theta) = -\frac{2\bar{h}U_{\infty 0} \sin(m\theta)}{m(\frac{a_2}{r})^m(\frac{m}{a_2} + \bar{h})}$$

Thus

$$\begin{aligned} \hat{u}_2(r,\theta,\omega)\Big|_{\omega\to 0} &= u_2(r,\theta) \\ &= \pi U_{\infty 0} - \sum_m \frac{2a_2\bar{h}U_{\infty 0}}{(m^2 + a_2 m\bar{h})\left(\frac{a_2}{r}\right)^m} \sin(m\theta) \end{aligned} \tag{6.151}$$

Hence, the solution of this boundary value problem for the case when $\omega \to 0$ becomes

$$\begin{aligned} \hat{u}(r,\theta,\omega)\Big|_{\omega\to 0} &= u(r,\theta) \\ &= u_1(r) + u_2(r,\theta) \end{aligned} \tag{6.152}$$

where $u_1(r)$ is given in (6.149) and $u_2(r,\theta)$ is given in (6.151).

An interested reader can compare (6.152) to the steady-state solution given in ▶ Sect. 6.10. The comparison should result in a good matching between the two solution methods. See Problem 9.

6

▪ Problems

1. Use the three solutions in (6.7) and compare the results among themselves and with (6.10).
2. Looking in literature, use Laplace transform or Fourier integrals and try to solve the BVP in ▶ Sect. 6.1 for a semi-infinite domain, and check whether you get (6.12). Compare this solution to (6.11) after transforming it to the time domain using the IFFT algorithm.
3. Prove that the solution for a constant q and for $L = h$ in (6.54) gives nearly identical results to those obtained via (6.32).
4. Show that solving the set of PDEs in (6.55-and (6.56) would lead to (6.57).
5. Derive the solution in (6.59).
6. Prove that the solution for a constant q and for $L = h$ in (6.70) gives nearly identical results to those obtained via (6.59).
7. Prove the derivation of Special Case 1 in ▶ Sect. 6.10.
8. As for the previous problem, prove the derivation of Special Case 2 in ▶ Sect. 6.10.
9. For a disk with $a_1 = 0.1$, $a_2 = 1$, $\overline{h} = 1$, $q = 10$, and $U_\infty = 20\theta$ units, compare the computed temperature of a steady state solution given in ▶ Sect. 6.10 to the solution given in the Special Case of ▶ Sect. 6.11, for different r and θ values.

References

Al-Khoury R (2012) Computational modeling of shallow geothermal systems. CRC Press

Xie W-C (2010) Differential equations for engineers. Cambridge University Press

Zill, Wright (2011) Advanced engineering mathematics, 4th edn. Jones and Bartlett Publishers. Hint: See p. 133

Heat Flow in Porous Media

R. Al Khoury and C. Kasbergen, *Analytical Subsurface Environment*,
https://doi.org/10.1007/978-3-031-93910-5_7

Heat flow in porous medium domains is largely dominated by heat conduction and convection mechanisms. Modelling heat flow in such media is essential for various engineering and applied science fields, including geothermal energy and environmental studies of subsurface formations. In both fields, the groundwater plays a vital role in transporting heat from the heat source to distant regions. Understanding the movement of heat through subsurface formations allows engineers and scientists to evaluate the feasibility of geothermal energy systems and the impact on the subsurface environment. It helps designing geothermal wells, predicting temperature profiles, and optimizing the efficiency of heat exchange processes. It also helps analysing subsurface thermal plumes, studying the effects of industrial discharges and assessing the impact of temperature variations on the ground ecosystem and human health.

In this chapter we address analytical models for heat flow in subsurface formations in the presence of groundwater flow. Our focus lies on modelling heat flow due to infinite cylindrical and line heat sources. The solution is carried out using the spectral analysis in combination with the moving heat source technique.

7.1 Heat Conduction–Convection in Porous Media

A porous medium domain is commonly referred to domains encompassing pores in its material structure, and in most cases constituting multiple phases with different physical properties. Subsurface formations are typical examples of multiphase, porous medium domains, where the soil mass normally consists of three phases: solid, water and air. The water and the air are enclosed in the pores of the material, whereas the solid phase constitutes the skeleton of the material, often called the solid matrix.

Thermal parameters of multiphase constituents may significantly differ from each other. For instance, for a fully saturated sand, the thermal conductivity of the solid sand particles reaches up to 4 $\mathrm{Wm^{-1}K^{-1}}$, whereas for water it is 0.56 $\mathrm{Wm^{-1}K^{-1}}$. On the contrary, the specific heat of the sand particles is on the order of 1 $\mathrm{kJ\,kg^{-1}K^{-1}}$, but for water it is 4.18 $\mathrm{kJ\,kg^{-1}K^{-1}}$. Therefore, at the microscopic level, it is likely that there is a temperature gradient between the solid phase and the water phase. The significance of this gradient is influenced by the physics of the problem and the boundary conditions. For a highly transient case, or where heat generation exists in one of the phases, the temperature gradient can be significant and must be taken into consideration. In this case, the medium is said to be in *local thermal nonequilibrium*. However, in most practical conditions, the microscopic temperature gradient is generally much smaller than that at the macroscopic level and hence can be ignored. In this case the medium is said to be in *local thermal equilibrium*. In this book we focus on the local thermal equilibrium condition.

The macroscopic energy equations of the solid phase and water phase of a fully saturated soil material can be expressed, respectively, as

$$\begin{aligned}(1-\varphi)(\rho c)_s\frac{\partial T_s}{\partial t}+(1-\varphi)(\rho c)_s\boldsymbol{v}_s\cdot\nabla T_s+\nabla\cdot(1-\varphi)\boldsymbol{\lambda}_s\cdot\nabla T_s=0\\ \varphi(\rho c)_w\frac{\partial T_w}{\partial t}+\varphi(\rho c)_w\boldsymbol{v}_w\cdot\nabla T_w+\nabla\cdot\varphi\boldsymbol{\lambda}_w\cdot\nabla T_w=0,\end{aligned} \tag{7.1}$$

where the subscript s stands for the solid phase and w stands for the water phase, $\boldsymbol{v}_s$ and $\boldsymbol{v}_w$ are the solid phase and water phase velocity vectors, respectively, and φ is the porosity. For both phases, $\boldsymbol{\lambda}$ is the thermal conductivity ($\mathrm{Wm^{-1}K^{-1}}$), ρ is the mass density ($\mathrm{kgm^{-3}}$) and c is the specific heat capacity ($\mathrm{J\,kg^{-1}K^{-1}}$).

The basic assumption in the local thermal equilibrium is that the solid and fluid phases within a representative elementary volume (REV) are in equilibrium. This means that the average temperatures of the two phases are equal, such that

$$T_s=T_w=T.$$

Thus, the energy Eq. (7.1) can be modified to be written as

$$\begin{aligned}(1-\varphi)(\rho c)_s\frac{\partial T}{\partial t}+(1-\varphi)(\rho c)_s\boldsymbol{v}_s\cdot\nabla T+\nabla\cdot(1-\varphi)\boldsymbol{\lambda}_s\cdot\nabla T=0\\ \varphi(\rho c)_w\frac{\partial T}{\partial t}+\varphi(\rho c)_w\boldsymbol{v}_w\cdot\nabla T+\nabla\cdot\varphi\boldsymbol{\lambda}_w\cdot\nabla T=0.\end{aligned} \tag{7.2}$$

Adding these two equations, and considering that the solid phase is stationary, i.e. $\boldsymbol{v}_s=0$, the system macroscopic energy equation can be expressed as

$$(\rho c)_{\text{eff}}\frac{\partial T}{\partial t}+(\rho c)_w\varphi\boldsymbol{v}_w\cdot\nabla T+\nabla\cdot\boldsymbol{\lambda}_{\text{eff}}\cdot\nabla T=0, \tag{7.3}$$

where

$$\begin{aligned}(\rho c)_{\text{eff}}&=\varphi(\rho c)_w+(1-\varphi)(\rho c)_s\\ \boldsymbol{\lambda}_{\text{eff}}&=\varphi\boldsymbol{\lambda}_w+(1-\varphi)\boldsymbol{\lambda}_s.\end{aligned} \tag{7.4}$$

These parameters are commonly denoted as the *effective thermal parameters* of the porous medium domain, which are volume averaged over the representative elementary volume.

For a homogeneous and isotropic material with constant parameters, the heat Eq. (7.3) can be conveniently written as

$$\frac{1}{\alpha}\frac{\partial T}{\partial t}+\frac{U}{\alpha}\nabla T-\nabla\cdot\nabla T=0$$

in which

$$\alpha = \frac{\lambda_{\text{eff}}}{(\rho c)_{\text{eff}}}$$
$$U = \frac{(\rho c)_w}{(\rho c)_{\text{eff}}} \varphi v_w,$$

where we also considered a one-dimensional fluid flow velocity.

7.2 Moving Cylindrical Heat Source: Dirichlet Problem

7

The boundary value problem for a conduction–convection heat flow generated by a cylindrical heat source in a homogeneous and isotropic infinite domain with groundwater moving with velocity $-U$ along the x-axis, can be expressed as

$$\begin{aligned} &\frac{\partial^2 T}{\partial x^2} + \frac{\partial^2 T}{\partial y^2} + \frac{U}{\alpha}\frac{\partial T}{\partial x} - \frac{1}{\alpha}\frac{\partial T}{\partial t} = 0 \\ &\text{IC:} \quad T(x,y,t) = T(r,\theta,t) = 0 \\ &\text{BC:} \quad T(x,y,t)|_{(x^2+y^2=a^2)} = T_{\text{in}}(t) \\ &\qquad T(x,y,t)|_{(a,\theta)} = T(x,y,t)|_{(a,\theta+2\pi)} \end{aligned} \tag{7.5}$$

in which $T \equiv T(x,y,t)$ is the temperature (K) in the xy-plane, α is the material thermal diffusivity (m^2s^{-1}), and $T_{\text{in}}(t)$ is a prescribed temperature. The domain is initially at zero temperature, and for times $t > 0$ it is subjected to a Dirichlet boundary condition at $x^2 + y^2 = a^2$, with a the radius of the heat source. ◘ Figure 7.1 shows a schematic top view of the domain with the expected temperature distributions at different time intervals. Due to the circular shape of the heat source, the temperature on its boundary is periodic, imposed by the second boundary condition.

The heat equation in (7.5) is formulated in the Cartesian coordinate system. However, since the heat source is circular,

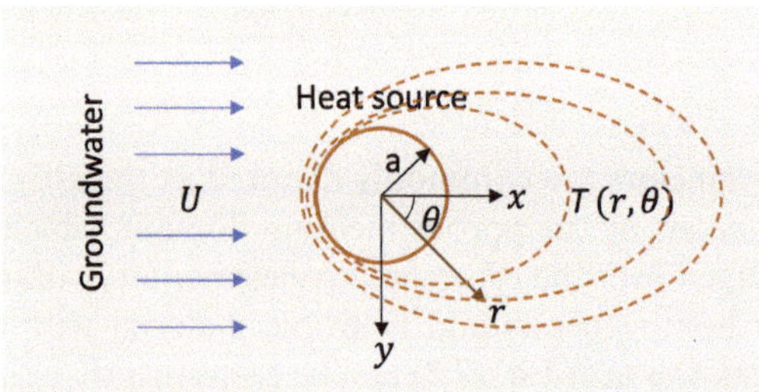

◘ **Fig. 7.1** Schematic representation of heat flow due to groundwater

it is more appropriate to express this equation in the polar coordinate system. Nonetheless, the presence of the first–order derivative of T in the third term of Eq. (7.5) introduces complexity when attempting to convert the equation to the polar system. To eliminate this term, we adopt a formulation similar to that proposed by Carslaw and Jaeger (1959). Specifically, we transform T into a new state variable, which can be expressed as

$$T(x, y, t) = u\,(x, y, t)\, e^{\eta x}, \tag{7.6}$$

where $u(x, y, t)$ is some function in space and time, and η is a constant to be determined.

The derivatives of (7.6) with respect to x and y are

$$\frac{\partial T}{\partial x} = u\eta e^{\eta x} + \frac{\partial u}{\partial x} e^{\eta x}$$
$$\frac{\partial^2 T}{\partial x^2} = \frac{\partial^2 u}{\partial x^2} e^{\eta x} + 2\frac{\partial u}{\partial x}\eta e^{\eta x} + u\eta^2 e^{\eta x}$$
$$\frac{\partial^2 T}{\partial y^2} = \frac{\partial^2 u}{\partial y^2} e^{\eta x}.$$

Substituting these identities into (7.5) gives

$$\frac{\partial^2 u}{\partial x^2} + \frac{\partial^2 u}{\partial y^2} + \left(2\eta + \frac{U}{\alpha}\right)\frac{\partial u}{\partial x} + \left(\eta^2 + \frac{U}{\alpha}\eta\right) u - \frac{1}{\alpha}\frac{\partial u}{\partial t} = 0.$$

The third term of this equation can be eliminated if we let $\eta = -U/2\alpha$, leading to

$$T(x, y, t) = u\,(x, y, t)\, e^{-Ux/\,2\alpha} \tag{7.7}$$

and

$$\frac{\partial^2 u}{\partial x^2} + \frac{\partial^2 u}{\partial y^2} - \frac{U^2}{4\alpha^2} u - \frac{1}{\alpha}\frac{\partial u}{\partial t} = 0. \tag{7.8}$$

Converting this equation to the polar coordinate system in the xy-plane yields

$$\frac{\partial^2 u}{\partial r^2} + \frac{1}{r}\frac{\partial u}{\partial r} + \frac{1}{r^2}\frac{\partial^2 u}{\partial \theta^2} - \frac{U^2}{4\alpha^2} u - \frac{1}{\alpha}\frac{\partial u}{\partial t} = 0 \tag{7.9}$$

for which we employed $x = r\cos\theta$, $y = r\sin\theta$, and $r = \sqrt{x^2 + y^2}$.

Based on (7.7), the relevant Dirichlet boundary condition for (7.9) at $x = a\cos\theta$ is therefore

$$u\,(a, \theta, t)\, e^{-Ua\cos(\theta)/\,2\alpha} = T_{\text{in}}(t). \tag{7.10}$$

Similarly, the periodic boundary condition in (7.5) implies

$$u\,(a, \theta, t) = u\,(a, \theta + 2\pi, t). \tag{7.11}$$

7

General Solution

Applying the Fourier transform to (7.9) yields

$$\frac{\partial^2 \hat{u}}{\partial r^2} + \frac{1}{r}\frac{\partial \hat{u}}{\partial r} + \frac{1}{r^2}\frac{\partial^2 \hat{u}}{\partial \theta^2} - \kappa^2 \hat{u} = 0, \quad \kappa^2 = \frac{U^2}{4\alpha^2} + \frac{i\omega}{\alpha}, \quad (7.12)$$

where $\hat{u} \equiv \hat{u}(r, \theta, \omega)$, with ω the angular frequency, $i = \sqrt{-1}$, and the transform is represented by

$$u \Rightarrow \hat{u}$$
$$\frac{\partial u}{\partial t} \Rightarrow i\omega\hat{u}$$
$$\frac{\partial^p u}{\partial \boldsymbol{x}^p} \Rightarrow \frac{\partial^p \hat{u}}{\partial \boldsymbol{x}^p}.$$

Equation (7.12) can be solved using the separation of variables, which can be expressed as

$$\hat{u}(r, \theta, \omega) = R(r, \omega) F(\theta) \equiv RF. \quad (7.13)$$

Substituting this Formula into (7.12), dividing by $R(r,\omega)F(\theta)$ and equating both terms to a constant, say η, gives

$$\frac{r^2}{R}\frac{d^2R}{dr^2} + \frac{r}{R}\frac{dR}{dr} - r^2\kappa^2 = -\frac{1}{F}\frac{d^2F}{d\theta^2} = \eta. \quad (7.14)$$

The relevant F–dependent boundary value problem in (7.14) is

$$F'' + \eta F = 0, \quad F(\theta) = F(\theta + 2\pi) \quad (7.15)$$

which is evidently 2π-periodic. This problem is not a regular Sturm–Liouville, but due to the periodicity, it generates eigenvalues and eigenfunctions. The three possible general solutions of (7.15) are

$$\begin{aligned} F(\theta) &= c_1 + c_2\theta, & \eta &= 0 \\ F(\theta) &= c_1 \cosh(m\theta) + c_2 \sinh(m\theta), & \eta &= -m^2 < 0 \\ F(\theta) &= c_1 \cos(m\theta) + c_2 \sin(m\theta), & \eta &= m^2 > 0. \end{aligned} \quad (7.16)$$

Apparently the second solution is not periodic and thus can be dismissed. The first solution in (7.16) is periodic if $c_2 = 0$, with the remaining solution $F(\theta) = c_1, c_1 \neq 0$. Hence, $\eta = 0$ is an eigenvalue. Finally, the third solution can be 2π-periodic if we take $m = 1, 2, \ldots$. Hence, considering $\eta = 0$, the eigenvalues of this BVP are $\eta_m = m^2, m = 0, 1, 2, \ldots$, leading the solution to be expressed as

$$F(\theta) = \sum\nolimits_m a_m \cos(m\theta) + b_m \sin(m\theta), \quad m = 0,\ 1,\ 2, \cdots. \quad (7.17)$$

Letting $\eta = m^2$, the R-dependent differential equation in (7.14) becomes

$$r^2\frac{d^2R}{dr^2}+r\frac{dR}{dr}-(r^2\kappa^2+m^2)R=0.$$

The solution of this ordinary differential equation is expressed in terms of the modified Bessel functions of first kind $I_m(\kappa r)$ and second kind $K_m(\kappa r)$. However, since $I_m(\kappa r)$ is unbounded at far distances, the solution reduces to

$$R\,(r,\omega)=K_m(\kappa r). \tag{7.18}$$

Substituting (7.17) and (7.18) into (7.13) leads to

$$\hat{u}\,(r,\theta,\omega)=\sum_m\left[a_m\cos\,(m\theta)+b_m\sin\,(m\theta)\right]K_m(\kappa r),\quad m=0,\ 1,\ 2,\cdots. \tag{7.19}$$

Boundary Condition

Applying the Fourier transform to the boundary condition in (7.10) gives

$$\hat{u}(a,\theta,\omega)=\hat{T}_{\text{in}}(\omega)\,e^{Ua\cos(\theta)/\,2\alpha}. \tag{7.20}$$

At $r=a$, (7.19) becomes

$$\hat{u}\,(a,\theta,\omega)=\sum\nolimits_m\left[a_m\cos(m\theta)+b_m\sin(m\theta)\right]K_m(\kappa a). \tag{7.21}$$

Equating (7.20) to (7.21) yields

$$\hat{T}_{\text{in}}(\omega)e^{U\,a\cos(\theta)/\,2\alpha}=\sum\nolimits_m\left[a_m\cos(m\theta)+b_m\sin(m\theta)\right]K_m(\kappa a). \tag{7.22}$$

This equation is a Fourier series with its coefficients expressed as

$$\begin{aligned}
a_0&=\frac{\hat{T}_{\text{in}}(\omega)}{K_0(\kappa a)}\frac{1}{2\pi}\int_0^{2\pi}e^{U\,a\cos(\theta)/\,2\alpha}d\theta, && m=0\\
a_m&=\frac{\hat{T}_{\text{in}}(\omega)}{K_m(\kappa a)}\frac{1}{\pi}\int_0^{2\pi}e^{U\,a\cos(\theta)/\,2\alpha}\cos(m\theta)d\theta, && m\geq 1\\
b_m&=\frac{\hat{T}_{\text{in}}(\omega)}{K_m(\kappa a)}\frac{1}{\pi}\int_0^{2\pi}e^{U\,a\cos(\theta)/\,2\alpha}\sin(m\theta)d\theta, && m\geq 1.\\
&=0
\end{aligned} \tag{7.23}$$

Since $\hat{T}_{\text{in}}$ is a function of frequency (time) only, the integral of the b_m coefficients is zero for all m. However, the integrals for the a_m coefficients have closed form expressions, given in terms of the modified Bessel function of the first kind (Abramowitz and Stegun 1972, p. 376), such that

$$\begin{aligned}
a_0&=\frac{\hat{T}_{\text{in}}(\omega)}{K_0(\kappa a)}I_0\left(\frac{Ua}{2\alpha}\right), && m=0\\
a_m&=\frac{2\hat{T}_{\text{in}}(\omega)}{K_m(\kappa a)}I_m\left(\frac{Ua}{2\alpha}\right), && m\geq 1
\end{aligned} \tag{7.24}$$

7

in which I_m is the modified Bessel function of the first kind of order m.

BVP Solution

Following the above derivation, the solution of $\hat{u}$ for a frequency ω_n can then be expressed as

$$\hat{u}(r,\theta,\omega_n) = \sum\nolimits_m \varepsilon_m \frac{\hat{T}_{\text{in}}(\omega_n)}{K_m(\kappa a)} I_m\left(\frac{Ua}{2\alpha}\right) \cos(m\theta) K_m(\kappa r)$$

with

$$\varepsilon_m = \begin{cases} 1 & m = 0 \\ 2 & m \geq 1 \end{cases}, \qquad \kappa = \sqrt{\frac{U^2}{4\alpha^2} + \frac{i\omega_n}{\alpha}}. \tag{7.25}$$

Applying the inverse Fourier transform, the solution in the time domain can be obtained via

$$u(r,\theta,t) = \sum\nolimits_n \hat{u}(r,\theta,\omega_n) e^{i\omega_n t}.$$

Having solved $u(r,\theta,t)$, employing (7.7), the temperature distribution in the domain can readily be reconstructed by

$$T(r,\theta,t) = u(r,\theta,t) e^{-Ur\cos(\theta)/2\alpha}. \tag{7.26}$$

Alternative Solution

The general solution of (7.15) can also be expressed as

$$F(\theta) = \sum\nolimits_m a_m e^{im\theta}, \qquad m = 0, \mp 1, \mp 2, \cdots. \tag{7.27}$$

Following this, Eq. (7.19) becomes

$$\hat{u}(r,\theta,\omega) = \sum\nolimits_m a_m e^{im\theta} K_m(\kappa r), \qquad m = 0, \mp 1, \mp 2, \cdots. \tag{7.28}$$

At $r = a$, (7.28) gives

$$\hat{u}(a,\theta,\omega) = \sum\nolimits_m a_m e^{im\theta} K_m(\kappa a), \qquad m = 0, \mp 1, \mp 2, \cdots.$$

Equating this equation to (7.20) yields

$$\hat{T}_{\text{in}}(\omega)\, e^{Ua\cos(\theta)/2\alpha} = \sum\nolimits_m a_m e^{im\theta} K_m(\kappa a).$$

Rearranging, we get

$$\frac{\hat{T}_{\text{in}}(\omega)\, e^{Ua\cos(\theta)/2\alpha}}{K_m(\kappa a)} = f(\theta) = \sum\nolimits_m a_m e^{im\theta}$$

which is a typical complex Fourier series, with its coefficients given by

$$
\begin{aligned}
a_m =& \frac{1}{2\pi}\int_0^{2\pi} f(\theta)e^{-im\theta}d\theta \\
=& \frac{\hat{T}_{\text{in}}(\omega)}{K_m(\kappa a)}\frac{1}{2\pi}\int_0^{2\pi} e^{U\,a\cos(\theta)/\,2\alpha}e^{-im\theta}d\theta.
\end{aligned}
\tag{7.29}
$$

Thus, the solution of the BVP for a given frequency ω_n becomes

$$
\hat{u}(r,\theta,\omega_n) = \sum\nolimits_m \frac{\hat{T}_{\text{in}}(\omega_n)}{K_m(\kappa a)}\left(\frac{1}{2\pi}\int_0^{2\pi} e^{U\,a\cos(\theta)/\,2\alpha}e^{-im\theta}d\theta\right)e^{im\theta}K_m(\kappa r),\quad m = 0,\ \mp 1,\ \mp 2,\cdots.
\tag{7.30}
$$

This integral can be solved by most programming tools, such as MATLAB. Nevertheless, it might be approximated by considering

$$
i^m J_m(z) = \frac{1}{2\pi}\int_0^{2\pi} e^{i\,z\cos(\theta)}e^{-im\theta}d\theta
\tag{7.31}
$$

in which J_m is the Bessel function of the first kind of order m. Comparing the integral in (7.29) to (7.31) we can deduce that

$$
z = \frac{Ua}{2i\alpha}.
$$

Hence

$$
a_m = \frac{\hat{T}_{\text{in}}(\omega)}{K_m(\kappa a)} i^m J_m\left(\frac{Ua}{2i\alpha}\right).
$$

And the solution becomes

$$
\hat{u}(r,\theta,\omega_n) = \sum\nolimits_m \frac{\hat{T}_{\text{in}}(\omega_n)}{K_m(\kappa a)} i^m J_m\left(\frac{Ua}{2i\alpha}\right)e^{im\theta}K_m(\kappa r),\qquad m = 0, \mp 1, \mp 2, \cdots.
\tag{7.32}
$$

This solution can also be written as

$$
\hat{u}(r,\theta,\omega_n) = \sum\nolimits_m \frac{\hat{T}_{\text{in}}(\omega_n)}{K_m(\kappa a)} i^{-m} J_m\left(\frac{iUa}{2\alpha}\right)e^{im\theta}K_m(\kappa r),\qquad m = 0, \mp 1, \mp 2, \cdots.
\tag{7.33}
$$

And, since

$$
I_m = i^{-m}J_m(iz),
$$

the solution can also be expressed as

$$
\hat{u}(r,\theta,\omega_n) = \sum\nolimits_m \frac{\hat{T}_{\text{in}}(\omega)}{K_m(\kappa a)} I_m\left(\frac{Ua}{2\alpha}\right)e^{im\theta}K_m(\kappa r),\qquad m = 0, \mp 1, \mp 2, \cdots.
\tag{7.34}
$$

The interested reader can implement these four formulations (7.30), (7.32), (7.33), and (7.34) in MATLAB or any other programming tool to find that they produce nearly identical results, see Problem 3.

7.3 Moving Cylindrical Heat Source: Neumann Problem

In sect.7.2 we solved a boundary value problem involving heat conduction and convection in a homogeneous, isotropic infinite domain subjected to a cylindrical heat source with a Dirichlet boundary condition. Here, we extend this solution to address the same problem, but with a Neumann boundary condition. The governing equations of this boundary value problem is

$$\frac{\partial^2 T}{\partial x^2} + \frac{\partial^2 T}{\partial y^2} + \frac{U}{\alpha}\frac{\partial T}{\partial x} - \frac{1}{\alpha}\frac{\partial T}{\partial t} = 0$$

$$\text{IC:} \quad T(x,y,t) = T(r,\theta,t) = 0 \tag{7.35}$$

$$\text{BC:} \quad \left.\frac{\partial T(r,\theta,t)}{\partial r}\right|_{r=a} = \frac{-q(t)}{2\pi a\lambda} = Q_{\text{in}}(t)$$

$$T(r,\theta,t)|_{r=a} = T(r,\theta+2\pi,t)|_{r=a},$$

where we expressed the boundary condition in the polar coordinate system since the heat source is circular.

General Solution

Following the mathematical procedure in ▶ Sect. 7.1, the general solution of the partial differential equation in (7.35) for a frequency ω_n is given by

$$\hat{T}(r,\theta,\omega_n) = \hat{u}(r,\theta,\omega_n)\, e^{-Ur\cos(\theta)/2\alpha} \tag{7.36}$$

in which

$$\hat{u}(r,\theta,\omega_n) = \sum\nolimits_m [a_m\cos(m\theta) + b_m\sin(m\theta)]K_m(\kappa r), \qquad m = 0,\ 1,\ 2,\cdots$$

with

$$\kappa = \sqrt{\frac{U^2}{4\alpha^2} + \frac{i\omega_n}{\alpha}}. \tag{7.37}$$

Boundary Conditions

Taking the derivative of $\hat{T}$ in (7.36) with respect to r yields

$$\frac{\partial \hat{T}}{\partial r} = \left(\frac{\partial \hat{u}}{\partial r} - \frac{U\cos(\theta)}{2\alpha}\hat{u}\right)e^{-Ur\cos(\theta)/2\alpha}, \tag{7.38}$$

where we removed the independent variables (r,θ,ω_n) for ease of notation.

At $r = a$, based on (7.38), the first boundary condition in (7.35), in the frequency domain, is thus

$$\left(\frac{\partial \hat{u}}{\partial r} - \frac{U\cos(\theta)}{2\alpha}\hat{u}\right)e^{-Ua\cos(\theta)/2\alpha} = \hat{Q}_{\text{in}}(\omega). \tag{7.39}$$

Taking the derivative of (7.37) with respect to r gives

$$\begin{aligned}\frac{\partial \hat{u}}{\partial r} =& a_m\kappa\cos(m\theta)\left[-K_{m+1}(\kappa r) + \frac{mK_m(\kappa r)}{\kappa r}\right] + \\ & b_m\kappa\sin(m\theta)\left[-K_{m+1}(\kappa r) + \frac{mK_m(\kappa r)}{\kappa r}\right].\end{aligned} \tag{7.40}$$

Solving this equation for $r = a$ and substituting the result into (7.39) yield

$$\begin{aligned}&\left\{a_m\kappa\cos(m\theta)\left[-K_{m+1}(\kappa a) + \frac{mK_m(\kappa a)}{\kappa a}\right] + \right.\\ &b_m\kappa\sin(m\theta)\left[-K_{m+1}(\kappa a) + \frac{mK_m(\kappa a)}{\kappa a}\right] - \\ &\left.\frac{U\cos(\theta)}{2\alpha}[a_m\cos(m\theta) + b_m\sin(m\theta)]K_m(\kappa a)\right\}e^{-Ua\cos(\theta)/2\alpha} = \hat{Q}_{\text{in}}(\omega).\end{aligned}$$

Rearranging gives

$$a_m\cos(m\theta) + b_m\sin(m\theta) = f(\theta)$$

with

$$f(\theta) = \frac{e^{Ua\cos(\theta)/2\alpha}}{-\kappa K_{m+1}(\kappa a) + (\frac{m}{a} - \frac{U\cos(\theta)}{2\alpha})K_m(\kappa a)}\hat{Q}_{\text{in}}(\omega).$$

This is a typical Fourier series with its coefficients given by

$$\begin{aligned} a_0 &= \frac{1}{2\pi}\int_0^{2\pi} f(\theta)d\theta, && m = 0 \\ a_m &= \frac{1}{\pi}\int_0^{2\pi} f(\theta)\cos(m\theta)d\theta, && m \geq 1 \\ b_m &= \frac{1}{\pi}\int_0^{2\pi} f(\theta)\sin(m\theta)d\theta, && m \geq 1. \end{aligned} \tag{7.41}$$

BVP Solution

Inserting the Fourier coefficients in (7.41) into (7.37) completes the solution of this boundary value problem in the frequency domain. Applying the inverse Fourier transform, the solution in the time domain can be obtained via

$$u(r,\theta,t) = \sum_n \hat{u}(r,\theta,\omega_n)e^{i\omega_n t}.$$

Having solved the problem for $u(r,\theta,t)$, the original state variable can then be determined via

$$T(r,\theta,t) = u(r,\theta,t)e^{-Ur\cos(\theta)/2\alpha}. \tag{7.42}$$

7.4 Multiple Heat Sources

The use of multiple cylindrical heat sources in a region is a common setup, especially for shallow geothermal energy systems. This kind of energy systems is often utilized for direct heating and cooling of buildings sharing a relatively small land. The thermal interaction among these heat sources can significantly influence the heat flow in the subsurface, thereby affecting the effectiveness of individual borehole heat exchangers. Factors such as the spacing between wells, the heat extraction rates of neighbouring sources, and the hydraulic properties of the aquifer collectively shape the temperature distribution in the region and hence impacting the energy yield per individual heat exchanger.

Modelling the thermal interaction between adjacent borehole heat exchangers can effectively be achieved via the superposition principle. For Neumann problems, this principle is readily applicable through a straightforward algebraic summation of temperatures induced by each heat exchanger. However, for Dirichlet problems, the superposition cannot directly be applied via simple algebraic methods. Hereafter, we discuss both cases in detail.

7.4.1 Multiple Heat Sources with Neumann Boundary Condition

The superposition principle for the Neumann boundary condition implies algebraic summation over temperatures induced by individual heat sources on a point in the domain (referred to as the reference point in ◘ Fig 7.2), such that

$$T(r,\theta,t) = \sum\nolimits_s \sum\nolimits_n \hat{u}_{sj}(r,\theta,\omega_n)e^{i\omega_n t}e^{-Ur_{sj}\cos(\theta_{sj})/2\alpha}, \tag{7.43}$$

where the index sj indicates the effect of heat source s on altering the temperature of point j, which can be a grid point; and $\hat{u}_{sj}(r,\theta,\omega_n)$ is as defined in (7.37), along with (7.41).

7.4.2 Multiple Heat Sources with Dirichlet Boundary Condition

The application of the algebraic sum in the Neumann problem does not translate directly to the Dirichlet boundary condition setup. The explicit summation of temperatures induced by the involved heat sources in the region can lead to unphysical increase or decrease of temperature at the reference point. For

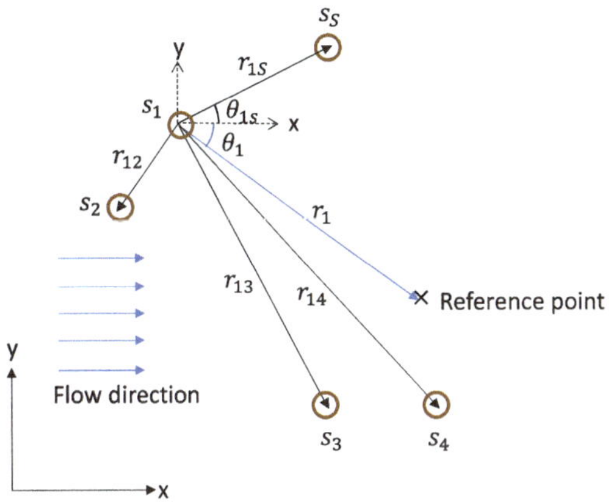

Fig. 7.2 Schematic representation of a multiple heat source system

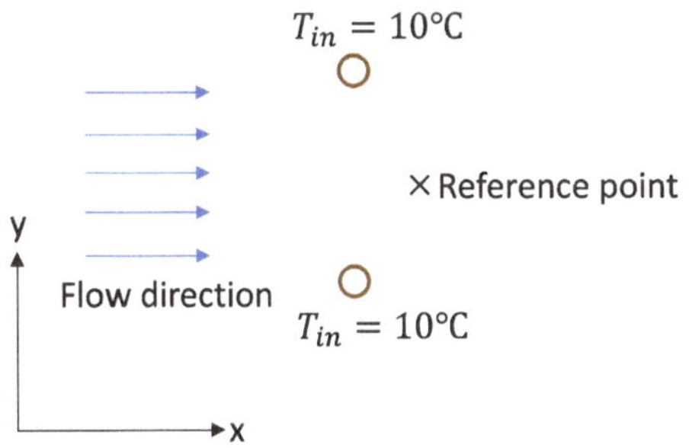

Fig. 7.3 Two heat sources with Dirichlet boundary conditions

instance, suppose that we have a setup with two heat sources, each with a prescribed temperature of 10 °C, as shown schematically in Fig 7.3. Also suppose that each of them, if acted alone, would elevate the temperature of the reference point to 7 °C. Then, if we apply the direct algebraic summation, similar to the approach used for the Neumann boundary condition setup, would erroneously yield 14 °C at this point. Such a result is obviously not correct. Thus, it is essential to couple the heat sources in a manner that ensures that the prescribed temperature at each source is maintained. This coupling involves adjusting the boundary temperatures of the heat sources so that the sum of all contributions leads to the desired prescribed value.

For a single heat source s, the temperature at a reference point j in the domain making a radial distance r_{sj} and an angle θ_{sj} from the heat source (see Fig 7.2) is given by (7.42). Then, employing (7.25) for $\hat{u}_{sj}(r, \theta, \omega_n)$, it implies that for a given frequency ω_n, the temperature distribution is expressed via

$$\hat{T}_{sj}(r,\theta,\omega_n) = \sum\nolimits_m \varepsilon_m \frac{\hat{T}_{\text{in}_s}(\omega_n)}{K_m(\kappa a_s)} I_m\left(\frac{Ua_s}{2\alpha}\right) \cos(m\theta_{sj}) K_m(\kappa r_{sj}) e^{-Ur_{sj}\cos(\theta_{sj})/2\alpha}$$

with

$$\varepsilon_m = \begin{cases} 1 & m = 0 \\ 2 & m \geq 1 \end{cases}, \qquad \kappa = \sqrt{\frac{U^2}{4\alpha^2} + \frac{i\omega_n}{\alpha}}, \tag{7.44}$$

where the index sj indicates the effect of heat source s on altering the temperature of point j, which can be a grid point, or a neighbouring heat source.

For multiple heat sources, as indicated above, it is important to tailor the boundary temperatures of each source to ensure that their collective contributions align with the desired prescribed values. These boundary temperatures effectively incorporate the temperatures induced by all involved heat sources, while maintaining the prescribed values. Let's denote the effective temperature amplitude at the boundary of heat source s as $\hat{T}_{\text{eff}_s}$. Consequently, (7.44) must be adjusted to read

$$\hat{T}_s(r,\theta,\omega_n) = \sum\nolimits_m \varepsilon_m \frac{\hat{T}_{\text{eff}_s}(\omega_n)}{K_m(\kappa a_s)} I_m\left(\frac{Ua_s}{2\alpha}\right) \cos(m\theta_s)\, K_m(\kappa r_{sj})\, e^{-Ur_{sj}\cos(\theta_{sj})/2\alpha}.$$

This equation can be expressed as (7.45)

$$\hat{T}_s(r,\theta,\omega_n) = \hat{T}_{\text{eff}_s}(\omega_n) R_{sj} e^{-Ur_{sj}\cos(\theta_{sj})/2\alpha} \tag{7.46}$$

in which

$$R_{sj} = \sum\nolimits_m \varepsilon_m \frac{1}{K_m(\kappa a_s)} I_m\left(\frac{Ua_s}{2\alpha}\right) \cos(m\theta_{sj})\, K_m(\kappa r_{sj}).$$

Solving (7.46) at the boundary of each heat source (omitting the independent variables (r,θ,ω) for simplicity) yields

$$\begin{aligned}
\hat{T}_{\text{in}_1} &= \hat{T}_{\text{eff}_1} + \hat{T}_{\text{eff}_2} R_{12} e^{-Ur_{12}\cos(\theta_{12})/2\alpha} + \cdots + \hat{T}_{\text{eff}_S} R_{1S} e^{-Ur_{1S}\cos(\theta_{1S})/2\alpha} \\
\hat{T}_{\text{in}_2} &= \hat{T}_{\text{eff}_1} R_{21} e^{-Ur_{21}\cos(\theta_{21})/2\alpha} + \hat{T}_{\text{eff}_2} + \cdots + \hat{T}_{\text{eff}_S} R_{2S} e^{-Ur_{2S}\cos(\theta_{2S})/2\alpha} \\
&\vdots \\
\hat{T}_{\text{in}_S} &= \hat{T}_{\text{eff}_1} R_{S1} e^{-Ur_{S1}\cos(\theta_{S1})/2\alpha} + \hat{T}_{\text{eff}_2} R_{S2} e^{-Ur_{S2}\cos(\theta_{S2})/2\alpha} + \cdots + \hat{T}_{\text{eff}_S}
\end{aligned} \tag{7.47}$$

in which $\hat{T}_{\text{in}1}$ is the prescribed temperature at heat source 1, etc., r_{12} represents the radial distance between heat source 1 and heat source 2, etc., and θ_{12} denotes the angle between heat source 1 and heat source 2, etc. (see Fig. 7.2). This set of equations indicates that the prescribed temperature at heat source 1, $\hat{T}_{\text{in}1}$, for instance, equals its effective temperature amplitude at its boundary a_1, plus temperatures induced by all neighbouring heat sources, which themselves undergo adjustment of their prescribed temperature.

Putting (7.47) in a matrix format, we obtain

$$\begin{pmatrix} \hat{T}_{\mathrm{in}_1} \\ \hat{T}_{\mathrm{in}_2} \\ \vdots \\ \hat{T}_{\mathrm{in}_S} \end{pmatrix} = \begin{pmatrix} 1 & R_{12}e^{-Ur_{12}\cos(\theta_{12})/2\alpha} & \dots & R_{1S}e^{-Ur_{1S}\cos(\theta_{1S})/2\alpha} \\ R_{21}e^{-Ur_{21}\cos(\theta_{21})/2\alpha} & 1 & \dots & R_{2S}e^{-Ur_{2S}\cos(\theta_{2S})/2\alpha} \\ \vdots & \vdots & \vdots & \vdots \\ R_{S1}e^{-Ur_{S1}\cos(\theta_{S1})/2\alpha} & R_{S2}e^{-Ur_{S2}\cos(\theta_{S2})/2\alpha} & \dots & 1 \end{pmatrix} \begin{pmatrix} \hat{T}_{\mathrm{eff}_1} \\ \hat{T}_{\mathrm{eff}_2} \\ \vdots \\ \hat{T}_{\mathrm{eff}_S} \end{pmatrix}.$$

Solving this equation for the effective temperatures vector yields

$$\begin{pmatrix} \hat{T}_{\mathrm{eff}_1} \\ \hat{T}_{\mathrm{eff}_2} \\ \vdots \\ \hat{T}_{\mathrm{eff}_S} \end{pmatrix} = \begin{pmatrix} 1 & R_{12}e^{-Ur_{12}\cos(\theta_{12})/2\alpha} & \dots & R_{1S}e^{-Ur_{1S}\cos(\theta_{1S})/2\alpha} \\ R_{21}e^{-Ur_{21}\cos(\theta_{21})/2\alpha} & 1 & \dots & R_{2S}e^{-Ur_{2S}\cos(\theta_{2S})/2\alpha} \\ \vdots & \vdots & \vdots & \vdots \\ R_{S1}e^{-Ur_{S1}\cos(\theta_{S1})/2\alpha} & R_{S2}e^{-Ur_{S2}\cos(\theta_{S2})/2\alpha} & \dots & 1 \end{pmatrix}^{-1} \begin{pmatrix} \hat{T}_{\mathrm{in}_1} \\ \hat{T}_{\mathrm{in}_2} \\ \vdots \\ \hat{T}_{\mathrm{in}_S} \end{pmatrix}. \tag{7.48}$$

Substituting the resulting effective temperature into (7.46), the solution in the time domain can then be obtained using the inverse fast Fourier transform (IFFT), as

$$T(r,\theta,t) = \sum_s \sum_n \hat{T}_s(r,\theta,\omega_n)\, e^{i\omega_n t}.$$

This equation gives the temperature at a point in the domain located at a distance r_s from each of the heat sources and positioned at an angle θ_s relative to them.

As it can be observed from the above formulation, using the matrix technique allows coupling heat sources with any arbitrary configurations and thermal loads. Unlike several existing models, there is no special treatment for corner or middle heat sources, the boreholes can have any diameter and flow rates, and the use of the FFT algorithm allows applying any thermal load scenarios.

▪ Example 7.1: Multiple Cylindrical Heat Sources

A two-dimensional, fully saturated porous medium domain, constituting a solid phase and a water phase, incorporates three cylindrical heat sources. The domain is in local thermal equilibrium, and its physical and thermal parameters are given in ◻ Table 7.1. The heat sources are distributed in the

Table 7.1 Physical and thermal parameters

Parameter	Magnitude	Unit
Solid mass density (ρ_s)	1500	kg/m^3
Solid specific heat (c_s)	800	J/kg.K
Solid thermal conductivity (λ_s)	2	W/m.K
Water mass density (ρ_w)	1000	kg/m^3
Water specific heat (c_w)	4180	J/kg.K
Water thermal conductivity (λ_w)	0.56	W/m.K
Porosity (φ)	0.3	–
Groundwater flow velocity	1e-5	m/s
Heat source radius (a)	0.15	m

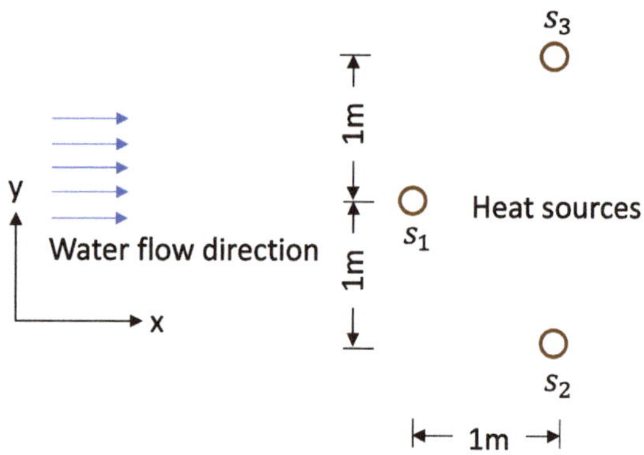

Fig. 7.4 Configuration of the heat sources

domain in a configuration depicted in Fig 7.4 (note that we choose these relatively short distances between the heat sources to highlight their thermal interactions). This geometry resembles a top view of a soil layer with a lateral groundwater flow passing through three neighbouring borehole heat exchangers.

The initial temperature in the porous domain and the heat sources is zero. Upon operating the system, the heat sources emit heat fluxes with a time distribution given as

$$q_{S_1} = \begin{cases} 0\,\text{W/m} - 150\,\text{W/m} & 10 \leq t \leq 12\,\text{h} \\ 150\,\text{W/m} & 12 < t \leq 112\,\text{h} \\ 0\,\text{W/m} & 112 < t \leq 114\,\text{h} \end{cases}$$

$$q_{S_2/S_3} = \begin{cases} 0\,\text{W/m} - 100\,\text{W/m} & 10 \leq t \leq 12\,\text{h} \\ 100\,\text{W/m} & 12 < t \leq 112\,\text{h}\,. \\ 0\,\text{W/m} & 112 < t \leq 114\,\text{h} \end{cases} \tag{7.49}$$

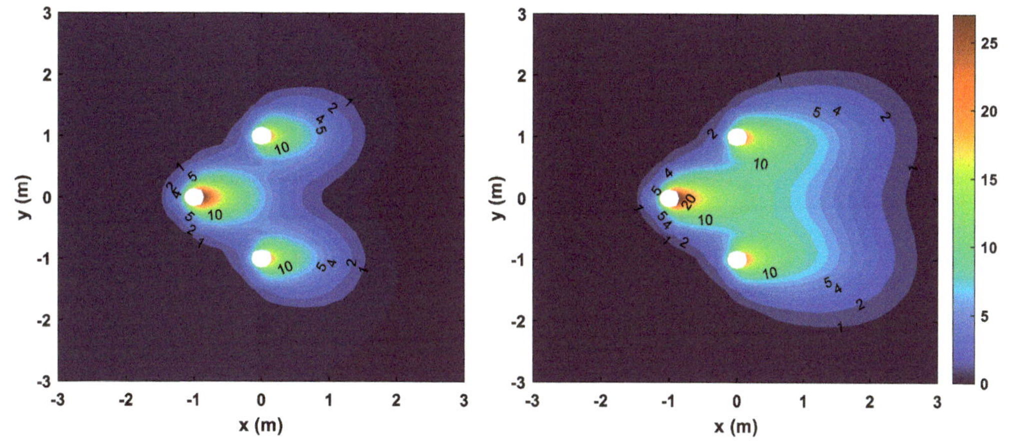

Fig. 7.5 Temperature contour plots after 50 h (left) and 100 h (right) of operation

Calculate the temperature distribution in the domain after 50 and 100 h of operation.

Solution

The temperature distribution in the domain can be determined through the following steps:

1. Transforming the input pulses given in (7.49) to the frequency domain using FFT. For this, we used 4096 (2^{12}) FFT samples with a 1 hr (3600 s) time step.
2. Calculating the transformed temperature $\hat{u}_s(r, \theta, \omega_n)$ generated by the three heat sources at selected grid points for every frequency using (7.37), along with (7.41).
3. Using the inverse fast Fourier transform (IFFT) to restore the time domain. The final temperature can then be calculated via (7.43).

   Figure 7.5 demonstrates the temperature distribution after 50 and 100 h of operating the heat sources. The figure clearly shows the temperature dissipation with elapsing time along the water flow direction. It also depicts the thermal interaction between the heat sources.

7.5 Moving Line Heat Source: Neumann Problem

As the radius of the heat source approaches zero, the infinite cylindrical heat source would resemble an infinite line heat source, where in a planar geometry, it acts as a point heat source.

In ▶ Sect. 7.3, we established that the solution of the governing boundary value problem due to a moving cylindrical heat source is given by

$$\hat{u}(r,\theta,\omega) = \sum_m [a_m \cos(m\theta) + b_m \sin(m\theta)]\, K_m(\kappa r), \qquad m = 0,\ 1,\ 2,\cdots$$

with

$$\begin{aligned}
a_0 &= \frac{1}{2\pi}\int_0^{2\pi} f(\theta)d\theta, && m = 0\\
a_m &= \frac{1}{\pi}\int_0^{2\pi} f(\theta)\cos(m\theta)d\theta, && m \geq 1\\
b_m &= \frac{1}{\pi}\int_0^{2\pi} f(\theta)\sin(m\theta)d\theta, && m \geq 1
\end{aligned} \tag{7.50}$$

in which

$$f(\theta) = -\frac{\hat{q}(\omega)e^{Ua\cos(\theta)/2\alpha}}{-2\pi a\lambda\kappa K_{m+1}(\kappa a) + 2\pi a\lambda(\frac{m}{a} - \frac{U\cos(\theta)}{2\alpha})K_m(\kappa a)}, \tag{7.51}$$

where we used

$$\hat{Q}_{\text{in}}(\omega) = \frac{-\hat{q}(\omega)}{2\pi a\lambda}.$$

Knowing that

$$\begin{aligned}
&\lim_{a\to 0} a\,K_0(\kappa a) = 0\\
&\lim_{a\to 0} a\,K_1(\kappa a) \simeq \frac{1}{\kappa}\\
&\lim_{a\to 0} a\,K_m(\kappa a) = \infty, \qquad m > 1\\
&\lim_{a\to 0} e^a \simeq 1.
\end{aligned}$$

Applying these limits to (7.51) reads for $m = 0$:

$$\begin{aligned}
\lim_{a\to 0} f(\theta) &= \left.\frac{-\hat{q}(\omega)\,e^{Ua\cos(\theta)/2\alpha}}{\left(-2\pi\lambda\kappa a\,K_1(\kappa a) - \pi\lambda\frac{Ua\cos(\theta)}{\alpha}K_0(\kappa a)\right)}\right|_{a\to 0}\\
&= \frac{\hat{q}(\omega)}{2\pi\lambda}.
\end{aligned}$$

Substituting this outcome into (7.50), the a_0 coefficient becomes

$$\begin{aligned}
a_0 &= \frac{\hat{q}(\omega)}{2\lambda\pi}\frac{1}{2\pi}\int_0^{2\pi} d\theta\\
&= \frac{\hat{q}(\omega)}{2\pi\lambda}.
\end{aligned} \tag{7.52}$$

In the same way, taking the limit as $a \to 0$ of $f(\theta)$ for $m > 0$, the Fourier coefficients in (7.50), for a constant $\hat{Q}_{\text{in}}$, become

$$a_m = b_m = 0, \qquad m = 1, 2, \cdots. \tag{7.53}$$

Having the Fourier coefficients (7.52) and (7.53), the solution of the moving infinite line heat source subjected to a Neumann boundary condition for a given frequency ω_n can then be expressed as

$$\hat{u}(r, \theta, \omega_n) = \frac{\hat{q}(\omega_n)}{2\lambda\pi} K_0(\kappa r). \tag{7.54}$$

Applying the IFFT algorithm gives the solution in the time domain, such that

$$u(r, \theta, t) = \sum_n \hat{u}(r, \theta, \omega_n) e^{i\omega_n t}.$$

Having this, the temperature distribution in the domain can then be calculated via

$$T(r, \theta, t) = u(r, \theta, t) e^{-Ur\cos(\theta)/2\alpha}. \tag{7.55}$$

For a multiple heat sources system, the temperature distribution can be calculated using the superposition principle, such that

$$T(r, \theta, t) = \sum_s u_s(r, \theta, t) e^{-Ur_s\cos(\theta_s)/2\alpha}, \tag{7.56}$$

where the subscript s stands for a line heat source.

Example 7.2: Multiple Line Heat Sources

A fully saturated porous medium domain, constituting a solid phase and a water phase, incorporates two line heat sources. The domain is in local thermal equilibrium, and its physical and thermal parameters are given in Table 7.2. The heat sources are set in the configuration shown in Fig. 7.6 (note that we choose this relatively short distance between the heat sources to highlight their thermal interactions). This geometry resembles a top view of a soil layer with a lateral groundwater flow passing through two neighbouring borehole heat exchangers.

Table 7.2 Physical and thermal parameters

Parameter	Magnitude	Unit
Solid mass density (ρ_s)	2650	kg/m^3
Solid specific heat (c_s)	900	J/kg · K
Solid thermal conductivity (λ_s)	2.5	W/m · K
Water mass density (ρ_w)	1000	kg/m^3
Water specific heat (c_w)	4180	J/kg · K
Water thermal conductivity (λ_w)	0.56	W/m · K
Porosity (φ)	0.3	–
Water flow velocity	1e-5	m/s
Heat source radius (a)	Line $a \to 0$	–

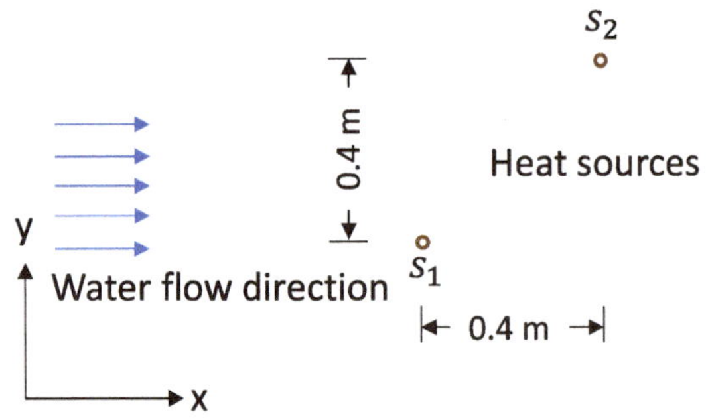

Fig. 7.6 Two line heat sources

7

The initial temperature in the porous medium domain and the heat sources is zero. Upon operating the system, the heat sources generate heat fluxes with a time distribution given as

$$
\begin{aligned}
q_{S_1} &= \begin{cases} 0\,\text{W/m} - 100\,\text{W/m} & 10 \le t \le 12\,\text{h} \\ 100\,\text{W/m} & 12 < t \le 112\,\text{h} \\ 0\,\text{W/m} & 112 < t \le 114\,\text{h} \end{cases} \\
q_{S_2} &= \begin{cases} 0\,\text{W/m} - 150\,\text{W/m} & 10 \le t \le 12\,\text{h} \\ 150\,\text{W/m} & 12 < t \le 112\,\text{h}\,. \\ 0\,\text{W/m} & 112 < t \le 114\,\text{h} \end{cases}
\end{aligned}
\tag{7.57}
$$

Calculate the temperature distribution in the domain after 50 and 100 h of operation.

Solution

Just like Example 7.1, we can tackle this example by initially transforming the input heat fluxes in (7.57) to the frequency domain. Then, upon utilizing (7.54) for each line heat source s allows us to determine $\hat{u}_s(r, \theta, \omega_n)$ induced by each heat source at selected grid points within the vicinity. Following this, applying the inverse fast Fourier transform (IFFT), we restore back the time domain. Finally, by using (7.56), we obtain the temperature distribution as a function of time.

Figure 7.7 shows the temperature distribution after 50 and 100 h of operating the heat sources. The figure clearly shows the temperature dissipation with elapsing time along the water flow direction, and the thermal interaction between the heat sources.

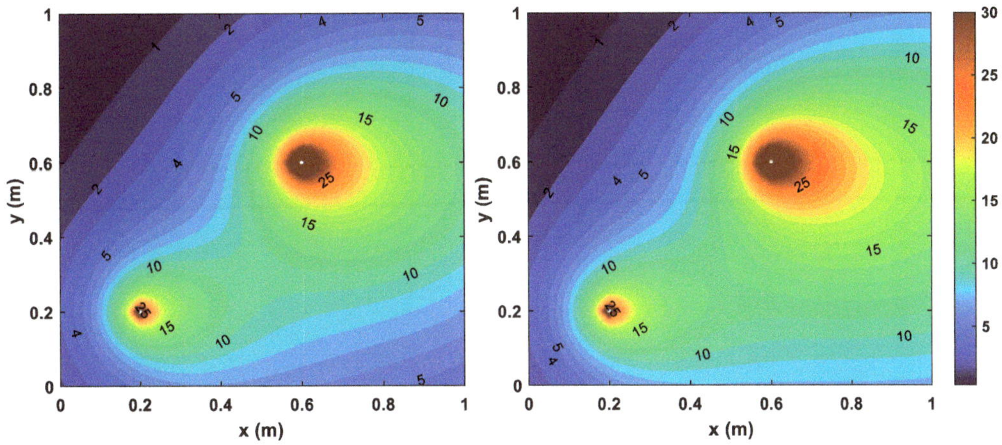

Fig. 7.7 Temperature contours due to two line heat sources after 50 h (left) and 100 h (right) of operating the system

Problems

1. Employing the transformations $x = r\cos\theta$, $y = r\sin\theta$ and $r = \sqrt{x^2 + y^2}$, demonstrate how the heat conduction–convection equation in its Cartesian form (7.8), can be converted into its polar form (7.9).
   Hint: See Zill and Cullen (2006).
2. Solve the Neumann problem in ▶ Sect. 7.3 using the exponential form (7.27).
3. Implement the four formulations (7.30), (7.32), (7.33) and (7.34) in MATLAB, Python, or any other programming tool, and compare the results.
4. Show that the temperature distribution due to line heat sources shown in Example 8.2 can be obtained using the solution given in ▶ Sect. 7.4 for cylindrical heat sources.
   Hint: Keep a small, say a is 0.005 m.
5. Implement the Dirichlet problem for multiple heat sources in a computer code and rerun Example 7.1 by replacing the prescribed heat fluxes emitted by the three heat sources with Dirichlet temperatures.
   Hint: See Al-Khoury et al. (2021).

References

Abramowitz M, Stegun IA (1972) Handbook of mathematical functions. Dover Publications, Inc.

Al-Khoury R, BniLam N, Arzanfudi M, Saeid S (2021) Analytical model for arbitrarily configured neighbouring shallow geothermal installations in the presence of groundwater flow. Geothermic, 93, Article 102063

BniLam N, Al-Khoury R (2016) Transient heat conduction in an infinite medium subjected to XE multiple cylindrical heat sources: an application to shallow geothermal systems. Renewe Energy 97:145–154

Carslaw HS, Jaeger JC (1959) Conduction of heat in solids, 2nd edn. Oxford University Press

Zill D, Cullen M (2006) Advanced engineering mathematics, 3rd edn. Jones and Bartlett Publishers

Heat Pipes and Heat Exchangers

R. Al Khoury and C. Kasbergen, *Analytical Subsurface Environment*,
https://doi.org/10.1007/978-3-031-93910-5_8

Heat pipes and heat exchangers are integral components in various engineering applications, spanning HVAC systems, industrial processes, refrigeration, and geothermal energy systems. In the context of this book, a heat pipe acts as a conduit for heat transfer, utilizing a working fluid to transport heat along its length and facilitate thermal interaction with the surrounding environment. The flow of the working fluid within heat pipes is unidirectional, entering from one end and exiting from the other. Similarly, a heat exchanger serves to transfer thermal energy between fluids and their surroundings. However, unlike heat pipes, the flow of the working fluid is multidirectional. The heat exchangers either employ several unidirectional channels carrying fluids moving in opposite directions or utilize the same fluid flowing through inner and outer pipes forming U-tubes.

In this chapter we will first address problems related to heat flow in one-dimensional heat pipes under varying boundary conditions. Then we delve into studying two prominent configurations of heat exchangers: the coaxial heat exchanger and the U-tube heat exchanger, with a specific emphasis on their application in geothermal energy systems.

8.1 1D Heat Pipe: Dirichlet Problem

Consider a heat pipe with length L carrying a working fluid is subjected to Dirichlet boundary conditions at both ends of the pipe, as depicted in Fig. 8.1. The boundary value problem of this case is

$$\begin{aligned}
&\lambda\frac{\partial^2 u}{\partial x^2} - \rho c v\frac{\partial u}{\partial x} - \rho c\frac{\partial u}{\partial t} = 0\\
&\text{IC:} \quad u(x,0) = 0\\
&\text{BC:} \quad u(0,t) = U_{in}(t)\\
&\qquad\;\; u(L,t) = U_{out}(t)
\end{aligned} \tag{8.1}$$

in which λ ($\mathrm{Wm^{-1}K^{-1}}$) is the thermal conductivity of the working fluid, ρ ($\mathrm{kg\,m^{-3}}$) is its mass density, c ($\mathrm{J\,kg^{-1}K^{-1}}$) is the specific heat capacity, v ($\mathrm{ms^{-1}}$) is the fluid velocity and L (m) is the length of the pipe.

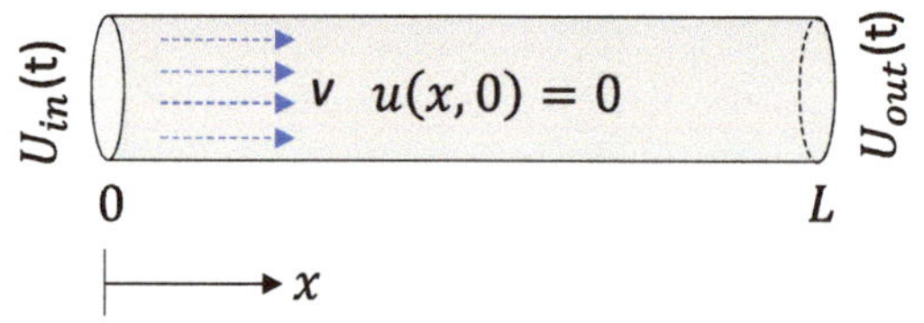

Fig. 8.1 A heat pipe under double Dirichlet boundary conditions

General Solution

Applying the Fourier transform to (8.1) yields

$$\begin{aligned} &\frac{d^2\hat{u}}{dx^2} - \frac{v}{\alpha}\frac{d\hat{u}}{dx} - \frac{i\omega}{\alpha}\hat{u} = 0 \\ &\text{BC:} \quad \hat{u}(0,\omega) = \hat{U}_{in}(\omega) \\ &\qquad\quad \hat{u}(L,\omega) = \hat{U}_{out}(\omega) \end{aligned} \tag{8.2}$$

in which $\alpha = \lambda/\rho c$ (m^2s^{-1}) is the thermal diffusivity. Apparently, the Fourier transform has converted the partial differential equation in (8.1) to an ordinary differential equation in (8.2), rendering the solution more feasible.

Since the coefficients in (8.2) are constant, the trial solution can be expressed as

$$\hat{u} = a\,e^{-\kappa x}$$

in which a and κ are constants which need to be determined. Substituting this trial solution into (8.2) and simplifying, leads to

$$\kappa^2 + \frac{v}{\alpha}\kappa - \frac{i\omega}{\alpha} = 0$$

Using the quadratic formula, gives

$$\kappa_1 = \frac{-v + \sqrt{v^2 + 4i\alpha\omega}}{2\alpha}, \qquad \kappa_2 = \frac{-v - \sqrt{v^2 + 4i\alpha\omega}}{2\alpha} \tag{8.3}$$

which represent the two eigenvalues of the system. In the special case where only heat conduction is involved, i.e. $v = 0$, these eigenvalues reduce to the conduction heat flow case, addressed in ▶ Sect. 6.1, ▶ Chap. 6.

The two eigenvalues in (8.3) have two corresponding eigenfunctions, leading the general solution to be

$$\hat{u} = a_1 e^{-\kappa_1 x} + a_2 e^{-\kappa_2 x}. \tag{8.4}$$

Boundary Conditions

Applying the boundary conditions in (8.2), implies:

at $x = 0$

$$a_1 + a_2 = \hat{U}_{in}$$

at $x = L$

$$a_1 e^{-\kappa_1 L} + a_2 e^{-\kappa_2 L} = \hat{U}_{out}$$

Solving these two equations simultaneously yields

$$a_1 = -\frac{\hat{U}_{in}e^{-\kappa_2 L} - \hat{U}_{out}}{e^{-\kappa_1 L} - e^{-\kappa_2 L}}, \qquad a_2 = \frac{\hat{U}_{in}e^{-\kappa_1 L} - \hat{U}_{out}}{e^{-\kappa_1 L} - e^{-\kappa_2 L}}. \tag{8.5}$$

BVP Solution

Substituting (8.5) into (8.4), the solution of the BVP (8.2) for a given frequency ω_n leads to

$$\hat{u}(x, \omega_n) = \frac{\hat{U}_{out} - \hat{U}_{in}e^{-\kappa_2 L}}{e^{-\kappa_1 L} - e^{-\kappa_2 L}} e^{-\kappa_1 x} + \frac{\hat{U}_{in}e^{-\kappa_1 L} - \hat{U}_{out}}{e^{-\kappa_1 L} - e^{-\kappa_2 L}} e^{-\kappa_2 x} \tag{8.6}$$

with the eigenvalues κ_1 and κ_2 given in (8.3).

For heat pipes with relatively large lengths, the second term in (8.6) becomes unbounded since κ_2 is always negative, leading to large positive exponent on that term. To tackle this, we multiply the nominator and denominator of both terms of the equation by $e^{\kappa_2 L}$, leading the solution to become

$$\hat{u}(x, \omega_n) = \frac{\hat{U}_{in} - \hat{U}_{out}e^{\kappa_2 L}}{1 - e^{-(\kappa_1 - \kappa_2)L}} e^{-\kappa_1 x} + \frac{\hat{U}_{out} - \hat{U}_{in}e^{-\kappa_1 L}}{1 - e^{-(\kappa_1 - \kappa_2)L}} e^{-\kappa_2 (x-L)} \tag{8.7}$$

8

Note that should we had considered the general solution in the form

$$\hat{u} = a_1 e^{-\kappa_1 x} + a_2 e^{\kappa_2 (L-x)}$$

instead of (8.4), we would attain the BVP solution given in (8.7) directly.

Having solved the problem in the frequency domain, the solution in the time domain can then be obtained via the inverse fast Fourier transform (IFFT) algorithm, such that

$$u(x, t) = \sum_n \hat{u}(x, \omega_n) e^{i\omega_n t}.$$

Special Case: Steady-State

As time goes to infinity, the system reaches a steady state and the frequency approaches zero. Thus, the eigenvalues become

$$\kappa_1 = 0, \quad \kappa_2 = -\frac{v}{\alpha}$$

Following this, the steady-state condition yields

$$u(x) = \frac{U_{in} - U_{out}}{1 - e^{\kappa_2 L}} \left[e^{\kappa_2 L} - e^{-\kappa_2 (x-L)}\right]. \tag{8.8}$$

8.2 1D Heat Pipe: Mixed Boundary Conditions

In addressing heat flow in a heat pipe under mixed boundary conditions, we effectively simulate an important type of heat exchangers known as the *straight-tube heat exchanger*. In this set-up, a single straight tube or a series of tubes carrying a working fluid are immersed within a shell filled with another fluid. At the inset, these two fluids have different temperatures, and as the system operates, the heat is transferred from the hotter fluid to the cooler fluid through the tube walls. In

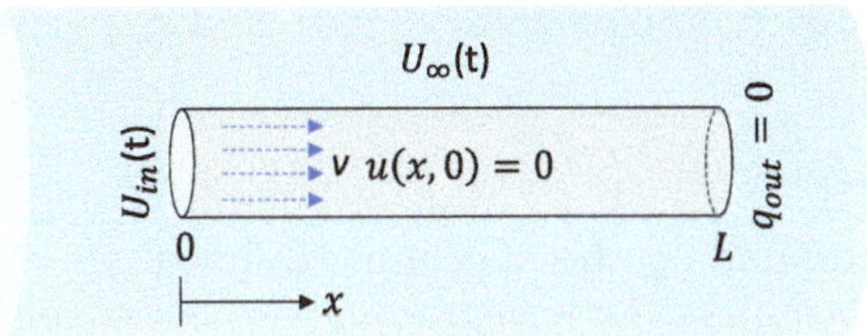

Fig. 8.2 A heat pipe immersed in an environment with temperature $\boldsymbol{U_\infty(t)}$

practical applications, a diverse array of shell-and-tube configurations exists, each carefully tailored to serve specific technological objectives. In this discussion, we focus on the heat transfer in a single straight-tube heat exchanger immersed in a surrounding environment with a temperature denoted as U_∞, which might be constant or varying with time, as schematically shown in Fig. 8.2.

The boundary value problem of a transient heat flow in a finite heat pipe subjected to a Dirichlet boundary condition at one end, an insulted (a Neumann-type) boundary condition at the other, and a convective (Robin-type) boundary condition along its length is given by

$$\begin{aligned} &\lambda \frac{\partial^2 u}{\partial x^2} A - \rho c v \frac{\partial u}{\partial x} A - \rho c \frac{\partial u}{\partial t} A = \overline{h}(U_\infty - u) P \\ &\text{IC:} \quad u(x, 0) = 0 \\ &\text{BC:} \quad u(0, t) = U_{in}(t) \\ &\qquad \left.\frac{\partial u}{\partial x}\right|_{x=L} = 0. \end{aligned} \tag{8.9}$$

in which A (m^2) is the cross sectional area of the heat pipe and P (m) is its circumference, $\overline{h}$ (Wm^{-2}K^{-1}) is a thermal interaction coefficient between the pipe and the surrounding environment with temperature U_∞(K). The definition of the other parameters are given in ▶ Sect. 8.1.

General Solution

Applying the Fourier transform to (8.9) gives

$$\begin{aligned} &\frac{d^2\hat{u}}{dx^2} - \frac{v}{\alpha}\frac{d\hat{u}}{dx} - \left(\frac{i\omega}{\alpha} - \frac{P\overline{h}}{\lambda A}\right)\hat{u} = \frac{P\overline{h}}{\lambda A}\hat{U}_\infty \\ &\text{BC:} \quad \hat{u}(0, \omega) = \hat{U}_{in}(\omega) \\ &\qquad \left.\frac{d\hat{u}}{dx}\right|_{x=L} = 0 \end{aligned} \tag{8.10}$$

This boundary value problem is nonhomogeneous due to the right-hand side term, which is spatially constant. Thus the solution can be attained by dividing the problem into a homogenous (complementary) part and a particular part, such that

$$\hat{u} = \hat{u}_h + \hat{u}_p \tag{8.11}$$

The corresponding homogenous differential equation of (8.10) is

$$\frac{d^2\hat{u}_h}{dx^2} - \frac{v}{\alpha}\frac{d\hat{u}_h}{dx} - \left(\frac{i\omega}{\alpha} - \mu\right)\hat{u}_h = 0, \quad \mu = \frac{P\overline{h}}{\lambda A} \tag{8.12}$$

Since the coefficients in this differential equation are constant, the trial solution can be expressed as

$$\hat{u}_h = a\, e^{-\kappa x}$$

Substituting this trial solution into (8.12), and simplifying, leads to

$$\kappa^2 + \frac{v}{\alpha}\kappa - \left(\frac{i\omega}{\alpha} - \mu\right) = 0$$

Using the quadratic formula yields, upon re-arrangement,

$$\kappa_1 = \frac{-v + \sqrt{4\alpha i\omega - 4\alpha^2\mu + v^2}}{2\alpha}, \quad \kappa_2 = \frac{-v - \sqrt{4\alpha i\omega - 4\alpha^2\mu + v^2}}{2\alpha} \tag{8.13}$$

Thus, the general solution of the homogeneous equation becomes

$$\hat{u}_h = a_1 e^{-\kappa_1 x} + a_2 e^{-\kappa_2 x} \tag{8.14}$$

The relevant particular differential equation of (8.10) is

$$\frac{d^2\hat{u}_p}{dx^2} - \frac{v}{\alpha}\frac{d\hat{u}_p}{dx} - \left(\frac{i\omega}{\alpha} - \mu\right)\hat{u}_p = \mu\hat{U}_\infty, \quad \mu = \frac{P\overline{h}}{\lambda A} \tag{8.15}$$

Since the right-hand side of this equation is constant in the x-direction, the trial solution can be expressed as a constant, such that

$$\hat{u}_p = b$$

Substituting this trial solution into (8.15) yields

$$-\left(\frac{i\omega}{\alpha} - \mu\right)b = \mu\hat{U}_\infty$$

From which

$$b = -\frac{\mu\hat{U}_\infty}{\frac{i\omega}{\alpha} - \mu}$$

Thus, the solution of the particular equation becomes

$$\hat{u}_p = \frac{\hat{U}_\infty}{1 - \frac{i\omega}{\mu\alpha}} \tag{8.16}$$

Substituting (8.14) and (8.16) into (8.11) gives

$$\hat{u} = a_1 e^{-\kappa_1 x} + a_2 e^{-\kappa_2 x} + \frac{\hat{U}_\infty}{1 - \frac{i\omega}{\mu\alpha}}. \tag{8.17}$$

Boundary Conditions

The first boundary condition in (8.10) implies

At $x = 0$

$$a_1 + a_2 = \hat{U}_{in} - \hat{u}_p \tag{8.18}$$

Taking the derivative of (8.17) with respect to x gives

$$\frac{d\hat{u}}{dx} = -\kappa_1 a_1 e^{-\kappa_1 x} - \kappa_2 a_2 e^{-\kappa_2 x}$$

Thus, at $x = L$, the second boundary condition in (8.10) implies

$$-\kappa_1 a_1 e^{-\kappa_1 L} - \kappa_2 a_2 e^{-\kappa_2 L} = 0 \tag{8.19}$$

Solving (8.18) and (8.19) simultaneously yields

$$a_1 = -\frac{\left(\hat{U}_{in} - \hat{u}_p\right)\kappa_2 e^{-\kappa_2 L}}{\kappa_1 e^{-\kappa_1 L} - \kappa_2 e^{-\kappa_2 L}}, \qquad a_2 = \frac{\left(\hat{U}_{in} - \hat{u}_p\right)\kappa_1 e^{-\kappa_1 L}}{\kappa_1 e^{-\kappa_1 L} - \kappa_2 e^{-\kappa_2 L}}. \tag{8.20}$$

BVP Solution

Substituting the constants in (8.20) into (8.17), the solution of the boundary value problem (8.10) for a frequency ω_n becomes

$$\hat{u}\left(x, \omega_n\right) = \frac{\hat{U}_{in} - \hat{u}_p}{\kappa_1 e^{-\kappa_1 L} - \kappa_2 e^{-\kappa_2 L}} \left[\kappa_1 e^{-(\kappa_1 L + \kappa_2 x)} - \kappa_2 e^{-(\kappa_2 L + \kappa_1 x)}\right] + \hat{u}_p \tag{8.21}$$

with

$$\hat{u}_p = \frac{\hat{U}_\infty}{1 - \frac{i\omega}{\mu\alpha}}, \quad \mu = \frac{P\bar{h}}{\lambda A}$$

and the eigenvalues κ_1 and κ_2 are given in (8.13).

To avoid the unbounded solutions for relatively large pipes, we multiply the nominator and denominator of both terms of the equation by $e^{\kappa_2 L}$, leading to

$$\hat{u}(x,\omega_n) = \frac{\hat{U}_{in} - \hat{u}_p}{\kappa_1 e^{-(\kappa_1-\kappa_2)L} - \kappa_2}[\kappa_1 e^{-L(\kappa_1-\kappa_2)} e^{-\kappa_2 x} - \kappa_2 e^{-\kappa_1 x}]$$

Having solved the boundary value problem in the frequency domain, the solution in the time domain can be restored via the IFFT algorithm, such that

$$u(x,t) = \sum_n \hat{u}(x,\omega_n)\, e^{i\omega_n t}.$$

8.3 Coaxial Heat Exchanger

Coaxial heat exchangers are devices designed for efficient heat transfer between two fluids, commonly utilized in heating, cooling, or refrigeration systems. They are typically configured in two main setups. One is open, comprising two concentric tubes, with one fluid flows through the inner tube while the other flows through the outer tube. The other setup is closed, wherein the same fluid circulates between the inner and outer tubes. Both configurations have widespread applications across diverse industries, including HVAC systems and geothermal energy extraction.

Here, we address heat flow in a closed coaxial heat exchanger with annular inlet (CXA) consisting of two channels: inner pipe, which we denote pipe-in, and outer pipe, which we denote pipe-out. Pipe-in acts as the intermediate medium that transfers heat between pipe-out and the outside environment, see Fig. 8.3.

The boundary value problem of a closed coaxial heat exchanger can be expressed as

$$\begin{aligned}
&\text{Pipe-in:} \quad \rho c \frac{\partial u_i}{\partial t} A_i + \rho c v \frac{\partial u_i}{\partial z} A_i - \lambda \frac{\partial^2 u_i}{\partial z^2} A_i = b_{io}(u_i - u_o)P_o \\
&\qquad + \bar{h}(u_i - U_\infty)P_i \\
&\text{Pipe-out:} \quad \rho c \frac{\partial u_o}{\partial t} A_o - \rho c v \frac{\partial u_o}{\partial z} A_o - \lambda \frac{\partial^2 u_o}{\partial z^2} A_o = b_{io}(u_o - u_i)P_o \\
&\text{IC:} \quad u_i(z,0) = u_o(z,0) = 0 \\
&\text{BC:} \quad u_i(0,t) = U_{in}(t), \qquad q_o(0,t) = 0 \\
&\qquad u_i(L,t) = u_o(L,t), \quad q_i(L,t) = q_o(L,t)
\end{aligned} \tag{8.22}$$

in which A_i and A_o (m^2) are the cross sectional area of pipe-in and pipe-out, respectively, P_i (m) is the circumference of pipe-in (representing the contact surface between pipe-in and the surrounding environment), P_o is the circumference of pipe-out (representing the contact surface between pipe-in and pipe-out), b_{io} ($Wm^{-2}K^{-1}$) is a thermal interaction coefficient at the pipe-out surface, $\bar{h}$ ($Wm^{-2}K^{-1}$) is a thermal interaction coefficient between the coaxial heat exchanger and a surrounding environment, and U_∞ is the temperature of the

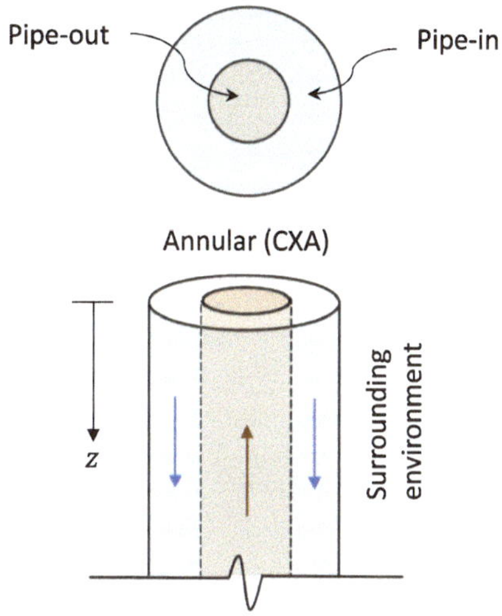

Fig. 8.3 A sketch of a coaxial heat exchanger (CXA)

surrounding environment which can be constant or function of time.

Note that the thermal interaction coefficients, b_{io} and $\overline{h}$, can be assigned positive or negative values, depending on the heat flow direction. Basically, they should be included in the heat equations as $b_{\text{io}} \cdot \boldsymbol{n}$ and $\overline{h} \cdot \boldsymbol{n}$, where $\boldsymbol{n}$ is a unit outward vector, which can be $+1$ or -1.

General Solution

This boundary value problem involves a coupled set of partial differential equations, where the coupling is made via the thermal interaction terms on the right-hand side of both heat equations.

Employing the Fourier transform on (8.22) yields

$$\begin{aligned}
&\frac{i\omega}{\alpha}A_i\hat{u}_i + \frac{v}{\alpha}A_i\frac{d\hat{u}_i}{dz} - A_i\frac{d^2\hat{u}_i}{dz^2} = \frac{P_o b_{io}}{\lambda}(\hat{u}_i - \hat{u}_o) + \frac{P_i\overline{h}}{\lambda}(\hat{u}_i - \hat{U}_\infty)\\
&\frac{i\omega}{\alpha}A_o\hat{u}_o - \frac{v}{\alpha}A_o\frac{d\hat{u}_o}{dz} - A_o\frac{d^2\hat{u}_o}{dz^2} = \frac{P_o b_{io}}{\lambda}(\hat{u}_o - \hat{u}_i)\\
&\text{BC:}\quad \hat{u}_i(0,\omega) = \hat{U}_{in}(\omega), \qquad \hat{q}_o(0,\omega) = 0\\
&\qquad\quad \hat{u}_i(L,\omega) = \hat{u}_o(L,\omega), \quad \hat{q}_i(L,t) = \hat{q}_o(L,t)
\end{aligned} \tag{8.23}$$

in which $\alpha = \lambda/\rho c$ (m^{2}s^{-1}) is the thermal diffusivity. This set of differential equations is nonhomogeneous due to the thermal interaction term between pipe-in and the surrounding environment. To effectively address such equations, it is essential to express the solution as the sum of the solutions of the corresponding homogeneous and particular equations, such that

$$\begin{aligned}\hat{u}_i &= \hat{u}_{ih} + \hat{u}_{ip} \\ \hat{u}_o &= \hat{u}_{oh} + \hat{u}_{op}\end{aligned} \tag{8.24}$$

The corresponding set of homogeneous equations of (8.23) are

$$\begin{aligned}&\frac{i\omega}{\alpha}A_i\hat{u}_{ih} + \frac{v}{\alpha}A_i\frac{d\hat{u}_{ih}}{dz} - A_i\frac{d^2\hat{u}_{ih}}{dz^2} - \frac{P_i\bar{h}}{\lambda}\hat{u}_{ih} = \frac{P_o b_{io}}{\lambda}(\hat{u}_{ih} - \hat{u}_{oh}) \\ &\frac{i\omega}{\alpha}A_o\hat{u}_{oh} - \frac{v}{\alpha}A_o\frac{d\hat{u}_{oh}}{dz} - A_o\frac{d^2\hat{u}_{oh}}{dz^2} = \frac{P_o b_{io}}{\lambda}(\hat{u}_{oh} - \hat{u}_{ih})\end{aligned} \tag{8.25}$$

8

where it can be noticed that the summation of these two equations leads to zero on the right-hand side, making them homogeneous. As the coefficients in (8.25) are constant, the trial solution of this system of ordinary differential equations can be expressed as

$$\hat{u}_{ih} = ae^{-\kappa z}, \quad \hat{u}_{oh} = ce^{-\kappa z}$$

in which a and c are constants, need to be determined from the boundary conditions, and κ is a constant, needs to be determined from solving the system's eigenvalues. Substituting these trial solutions into (8.25), rearranging and putting the results in a matrix form, yields

$$\begin{pmatrix} \frac{i\omega}{\alpha}A_i - \frac{v}{\alpha}A_i\kappa - \kappa^2 A_i - \frac{P_i\bar{h}}{\lambda} - \frac{P_o b_{io}}{\lambda} & \frac{P_o b_{io}}{\lambda} \\ \frac{P_o b_{io}}{\lambda} & \frac{i\omega}{\alpha}A_o + \frac{v}{\alpha}A_o\kappa - \kappa^2 A_o - \frac{P_o b_{io}}{\lambda} \end{pmatrix}\begin{pmatrix} a \\ c \end{pmatrix} = \begin{pmatrix} 0 \\ 0 \end{pmatrix} \tag{8.26}$$

The solution of this set of equations can only be attained by letting the determinant of the matrix equals to zero. The determinant of this matrix obviously yields a complex four degree polynomial of the form

$$m_4\kappa^4 + m_3\kappa^3 + m_2\kappa^2 + m_1\kappa + m_0 = 0$$

which involves the four eigenvalues of the coaxial heat exchanger. The general solution can then be expressed by the superposition of the corresponding four eigenfunctions, as

$$\begin{aligned}\hat{u}_{ih} &= a_1e^{-\kappa_1 z} + a_2e^{-\kappa_2 z} + a_3e^{\kappa_3(L-z)} + a_4e^{\kappa_4(L-z)} \\ \hat{u}_{oh} &= c_1e^{-\kappa_1 z} + c_2e^{-\kappa_2 z} + c_3e^{\kappa_3(L-z)} + c_4e^{\kappa_4(L-z)}\end{aligned} \tag{8.27}$$

Since the heat flow in pipe-in and pipe-out is coupled, the constants in (8.27) must be related to each other. From (8.26) the constants are related via

$$c_j = R_i a_j$$

where

$$R_j = -\frac{\frac{i\omega}{\alpha}A_i - \frac{v}{\alpha}A_i\kappa_j - \kappa_j^2 A_i - \frac{P_i\bar{h}}{\lambda} - \frac{P_o b_{io}}{\lambda}}{\frac{P_o b_{io}}{\lambda}} \tag{8.28}$$

with j denotes that for each eigenvalue κ_j there is a corresponding R_j. Following this, (8.27) can be expressed as

$$\begin{aligned}
\hat{u}_{ih} &= a_1 e^{-\kappa_1 z} + a_2 e^{-\kappa_2 z} + a_3 e^{\kappa_3(L-z)} + a_4 e^{\kappa_4(L-z)} \\
\hat{u}_{oh} &= R_1 a_1 e^{-\kappa_1 z} + R_2 a_2 e^{-\kappa_2 z} + R_3 a_3 e^{\kappa_3(L-z)} + R_4 a_4 e^{\kappa_4(L-z)}
\end{aligned} \tag{8.29}$$

Having solved the homogeneous equations, we now proceed to solving the particular equations. The corresponding set of particular equations of (8.23) are

$$\begin{aligned}
&\frac{i\omega}{\alpha}A_i\hat{u}_{ip} + \frac{v}{\alpha}A_i\frac{d\hat{u}_{ip}}{dz} - A_i\frac{d^2\hat{u}_{ip}}{dz^2} - \frac{P_i\bar{h}}{\lambda}\hat{u}_{ip} - \frac{P_o b_{io}}{\lambda}(\hat{u}_{ip} - \hat{u}_{op}) \\
&\quad = -\frac{P_i\bar{h}}{\lambda}\hat{U}_\infty \\
&\frac{i\omega}{\alpha}A_o\hat{u}_{op} - \frac{v}{\alpha}A_o\frac{d\hat{u}_{op}}{dz} - A_o\frac{d^2\hat{u}_{op}}{dz^2} - \frac{P_o b_{io}}{\lambda}(\hat{u}_{op} - \hat{u}_{ip}) = 0
\end{aligned} \tag{8.30}$$

Since the right-hand side of the first differential equation is constant along z, the trial solutions can be expressed as constants, such that

$$\hat{u}_{ip} = a_p, \quad \hat{u}_{op} = c_p$$

Substituting these trial solutions into (8.30) and putting the resulting equations in a matrix form gives

$$\begin{pmatrix} \frac{i\omega}{\alpha}A_i - \frac{P_i\bar{h}}{\lambda} - \frac{P_o b_{io}}{\lambda} & \frac{P_o b_{io}}{\lambda} \\ \frac{P_o b_{io}}{\lambda} & \frac{i\omega}{\alpha}A_o - \frac{P_o b_{io}}{\lambda} \end{pmatrix} \begin{pmatrix} a_p \\ c_p \end{pmatrix} = \begin{pmatrix} -\frac{P_i\bar{h}}{\lambda}\hat{U}_\infty \\ 0 \end{pmatrix}$$

Solving this set of equations for the constants vector, results in

$$\begin{pmatrix} a_p \\ c_p \end{pmatrix} = \frac{1}{\Delta}\begin{pmatrix} \frac{i\omega}{\alpha}A_o - \frac{P_o b_{io}}{\lambda} & -\frac{P_o b_{io}}{\lambda} \\ -\frac{P_o b_{io}}{\lambda} & \frac{i\omega}{\alpha}A_i - \frac{P_i\bar{h}}{\lambda} - \frac{P_o b_{io}}{\lambda} \end{pmatrix}\begin{pmatrix} -\frac{P_i\bar{h}}{\lambda}\hat{U}_\infty \\ 0 \end{pmatrix}$$

$$\Delta = \left(\frac{i\omega}{\alpha}A_i - \frac{P_i\bar{h}}{\lambda} - \frac{P_o b_{io}}{\lambda}\right)\left(\frac{i\omega}{\alpha}A_o - \frac{P_o b_{io}}{\lambda}\right) - \left(\frac{P_o b_{io}}{\lambda}\right)^2$$

from which

$$a_p = -\frac{1}{\Delta}\left(\frac{i\omega}{\alpha}A_o - \frac{P_o b_{io}}{\lambda}\right)\left(\frac{P_i \bar{h}}{\lambda}\hat{U}_\infty\right)$$

$$c_p = \frac{1}{\Delta}\frac{P_o P_i \bar{h} b_{io}}{\lambda^2}\hat{U}_\infty$$

Hence, the solutions of the particular equations are

$$\hat{u}_{ip} = -\frac{1}{\Delta}\left(\frac{i\omega}{\alpha}A_o - \frac{P_o b_{io}}{\lambda}\right)\left(\frac{P_i \bar{h}}{\lambda}\hat{U}_\infty\right)$$
$$\hat{u}_{op} = \frac{1}{\Delta}\frac{P_o P_i \bar{h}\, b_{io}}{\lambda^2}\hat{U}_\infty \tag{8.31}$$

Substituting (8.29) and (8.31) into (8.24) yields the general solution of the differential equations in (8.23), such that

8

$$\hat{u}_i(z,\omega) = a_1 e^{-\kappa_1 z} + a_2 e^{-\kappa_2 z} + a_3 e^{\kappa_3 (L-z)} + a_4 e^{\kappa_4 (L-z)} + \hat{u}_{ip}$$
$$\hat{u}_o(z,\omega) = R_1 a_1 e^{-\kappa_1 z} + R_2 a_2 e^{-\kappa_2 z} + R_3 a_3 e^{\kappa_3 (L-z)} + R_4 a_4 e^{\kappa_4 (L-z)} + \hat{u}_{op}$$

with

$$\hat{u}_{ip} = -\frac{1}{\Delta}\left(\frac{i\omega}{\alpha}A_o - \frac{P_o b_{io}}{\lambda}\right)\left(\frac{P_i \bar{h}}{\lambda}\hat{U}_\infty\right)$$
$$\hat{u}_{op} = \frac{1}{\Delta}\frac{P_o P_i \bar{h}\, b_{io}}{\lambda^2}\hat{U}_\infty. \tag{8.32}$$

Boundary Conditions

Applying the boundary conditions in (8.23) on (8.32) implies:

At $z = 0$:

$$a_1 + a_2 + a_3\, e^{\kappa_3 L} + a_4\, e^{\kappa_4 L} = \hat{U}_{in} - \hat{u}_{ip}$$

Taking the derivative of $\widehat{u}_o$ (second equation in (8.32)) with respect to z, and solving for $z = 0$, gives

$$-\kappa_1 R_1 a_1 - \kappa_2 R_2 a_2 - \kappa_3 R_3 a_3\, e^{\kappa_3 L} - \kappa_4 R_4 a_4\, e^{\kappa_4 L} = 0$$

At $z = L$:

$$a_1 e^{-\kappa_1 L}(1 - R_1) + a_2 e^{-\kappa_2 L}(1 - R_2) + a_3(1 - R_3) + a_4(1 - R_4) = \hat{u}_{op} - \hat{u}_{ip}$$

Taking the derivative of both equations in (8.32) with respect to z, and solving for $z = L$, gives

$$-\kappa_1 a_1(1 - R_1)e^{-\kappa_1 L} - \kappa_2 a_2(1 - R_2)e^{-\kappa_2 L} - \kappa_3 a_3(1 - R_3) - \kappa_4 a_4(1 - R_4) = 0$$

Putting these four equations in a matrix form, we obtain

$$\begin{pmatrix} 1 & 1 & e^{\kappa_3 L} & e^{\kappa_4 L} \\ \kappa_1 R_1 & \kappa_2 R_2 & \kappa_3 R_3 e^{\kappa_3 L} & \kappa_4 R_4 e^{\kappa_4 L} \\ (1-R_1)\,e^{-\kappa_1 L} & (1-R_2)\,e^{-\kappa_2 L} & (1-R_3) & (1-R_4) \\ \kappa_1(1-R_1)\,e^{-\kappa_1 L} & \kappa_2(1-R_2)\,e^{-\kappa_2 L} & \kappa_3(1-R_3) & \kappa_4(1-R_4) \end{pmatrix} \begin{pmatrix} a_1 \\ a_2 \\ a_3 \\ a_4 \end{pmatrix} = \begin{pmatrix} \hat{U}_{in} - \hat{u}_{ip} \\ 0 \\ \hat{u}_{op} - \hat{u}_{ip} \\ 0 \end{pmatrix}. \tag{8.33}$$

BVP Solution

Solving the system of Eq. (8.33) for vector $\{a_1 a_2 a_3 a_4\}$ and substituting into (8.32), yields, the solution of this BVP in the frequency domain.

Following this, the solution of the boundary value problem (8.22) in the time domain can then be obtained by the IFFT algorithm, that can be expressed as

$$u_i(z,t) = \sum_n \hat{u}_i(z,\omega_n)\,e^{i\omega_n t}$$
$$u_o(z,t) = \sum_n \hat{u}_o(z,\omega_n)\,e^{i\omega_n t}$$

Example 8.1: Heat Flow in Coaxial Borehole Heat Exchanger

A coaxial borehole heat exchanger is installed in ground to be utilized for direct space heating/cooling purposes. The pipe is 50 m long with pipe-in diameter 0.15 m and pipe-out diameter 0.075 m. It carries a mixture of water and Ethylene glycol with thermal conductivity 0.5 Wm^{-1}K^{-1}, mass density 1050 kg m^{-3} and specific heat capacity 3795 J kg^{-1}K^{-1}. The mixture flows at a rate 0.01 ms^{-1}. The thermal interaction coefficient between pipe-in and pipe-out, b_{io}, is assumed −2 Wm^{-2}K^{-1}, and the thermal interaction coefficient between pipe-in and the surrounding soil mass is assumed −10 Wm^{-2}K^{-1}. The heat exchanger is operated for 2 h during the summer season. At the onset, the water mixture enters the inner pipe at a temperature of 20 °C and undergoes thermal exchange with the adjacent soil, which maintains a temperature of 10 °C, (see Fig. 8.4). The thermal interaction is assumed to be active for the entire duration of the heat exchanger

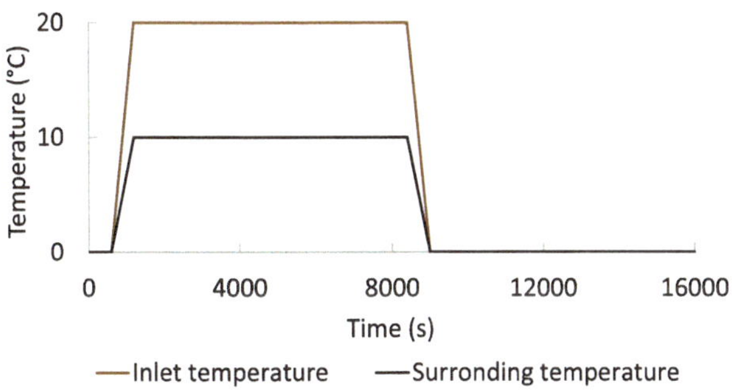

Fig. 8.4 Input temperatures at inlet of inner pipe and of surrounding soil

operation. Initially, both the borehole and the surrounding medium share an initial temperature of 0 °C.

Find the temperature distribution versus time and space along the heat exchanger. Discuss the result.

Solution

As usual, we begin by transforming the input temperature and the surrounding medium temperature from the time domain to the frequency domain using the FFT. For this example we used 16,384 (2^{14}) FFT samples with a time step of 120 s for each. Subsequently, upon solving the system of Eqs. (8.32) and (8.33), and transforming the results back to the time domain, we obtain the temperature distribution along pipe-in and pipe-out versus time, as shown in Fig. 8.5.

This figure reveals several noteworthy physical processes. First, it is evident that the heat exchanger is operating in cooling mode, as it can be noticed by the decrease of the mixture temperature along the borehole. Second, along the depth of pipe-in, and prior to the arrival of the pumped water mixture, the initially existing water in the pipes interacts with the surrounding soil medium. This interaction leads to a slight increase in temperature. Subsequently, as the pumped water arrives, the temperature rises accordingly. Third, a notable observation is manifested in the behavior of pipe-out before the arrival of the water mixture. The temperature in this pipe increases from its initial temperature merely due to the thermal interaction with pipe-in. This trend continues until the arrival of the pumped water, at which point the temperature starts to increase accordingly.

It is also worth noting that the temperature of the outer pipe at $z = 0$m (represented by the blue line on the right-hand side figure) exhibits a further temperature increase, primarily attributed to its thermal interaction with the inlet of pipe-in.

Figure 8.6 shows the temperature distribution along pipe-in and pipe-out after 2 h of operation. The figure shows that on this time the working fluid has reached to the mid length of pipe-out. The temperature in the upper half of this pipe has increased from its initial temperature of 0 °C due to conductive heat exchange with pipe-in and the surrounding environment.

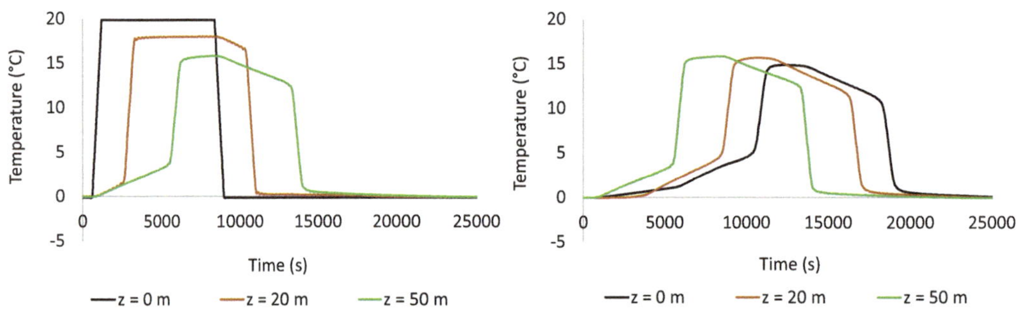

Fig. 8.5 Temperature distribution versus time in pipe-in (left) and pipe-out (right)

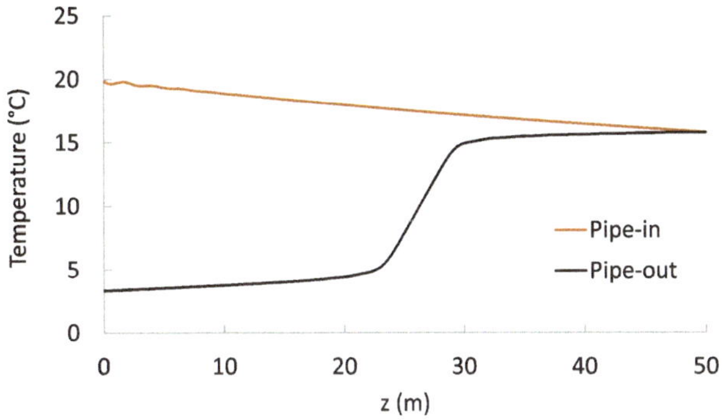

Fig. 8.6 Temperature distribution along pipe-in and pipe-out after 2 h of operation

Important Note for Computer Implementation

When using the fast Fourier transform (FFT) to convert signals from the time domain to the frequency domain, it is recommended to include a small time gap at the beginning, see Fig. 8.4. This practice helps identify any potential aliasing in the inverse-transformed signal. Additionally, it is advised to let the input signal gradually increase rather than making it a sharp, rectangular signal. For instance, in this case, we delay the input signal by 5 steps (equivalent to 10 min) and allow it to rise over an additional 5 steps before stabilizing at its desired value.

However, in some cases, such as the current example, the gradual rise might not be sufficient and can introduce oscillations into the calculated results. To mitigate this issue, it is recommended to incorporate damping into the system. This can be effectively accomplished by treating the angular frequency as a complex quantity, defined as

$$\omega \rightarrow \omega\,(1.0 - i\,\zeta)$$

where ζ is some damping coefficient. In this example we assigned $\zeta = 0.01$. Figure 8.7 demonstrates the computed results without any damping in the system. The figure re-

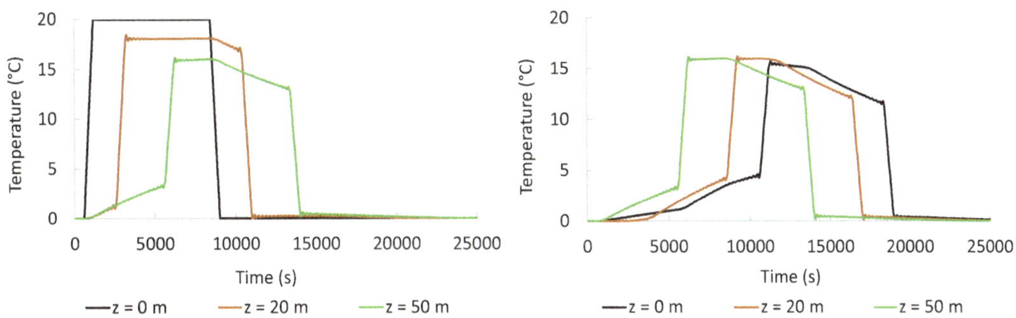

Fig. 8.7 Calculation results without damping

veals that, even though the results are accurate, they contain oscillations that are not quite physical. It is noteworthy that the damping coefficient is not totally for numerical purposes, rather, it can be attributed to the friction of the working fluid with the pipe tube.

8.4 U-tube Heat Exchanger

U-tube heat exchangers are important devices in the field of thermal engineering, widely employed for heat transfer between two fluids or mediums. Their design consists of a U-shaped tube bundle, typically made of metal, polyethylene or polypropylene materials, which facilitates the exchange of thermal energy between the two fluids or mediums. The distinctive closed form U-shaped configuration allows for compact and space-efficient installations, while maximizing heat transfer efficiency. This type of heat exchangers finds diverse applications in various industries, including HVAC systems, power generation, and geothermal energy. Here, we address U-tube heat exchangers commonly utilized in shallow geothermal systems, where heat is extracted from the ground for direct use in space heating and cooling. This technology is gaining popularity due to its environmentally friendly source of energy.

Two main configurations of U-tube borehole heat exchangers (BHE) are in use for direct heating: a single U-tube BHE, which consists of three channels: pipe-in, pipe-out and grout; and a double U-tube BHE, which consists of 2 pipes-in, 2 pipes-out and grout. ◘ Figure 8.8 shows schematically these two configurations. The heat flow mechanism in pipe-in and pipe-out is conductive-convective, while the grout channel operates solely through conductive heat transfer. The grout acts as the intermediate medium that transfers heat between pipe-in, pipe-out, and the surrounding soil mass.

In this section, we focus on heat flow in a single U-tube BHE embedded in a surrounding environment with a time-varying temperature $U_\infty(t)$, see ◘ Fig. 8.9. The surrounding environment acts as a source of heat that exhibits no thermal interaction with the borehole, i.e. its temperature remains constant.

This type of heat exchanger is highly slender because the ratio of the U-tube diameter and grout to the length of the borehole can be <1/1000. Given this physical characteristic, the BHE can be modeled as a 1D element while preserving the physical and thermal properties of the borehole components and their thermal interactions. Following this postulate, the boundary value problem for this scenario can be expressed as

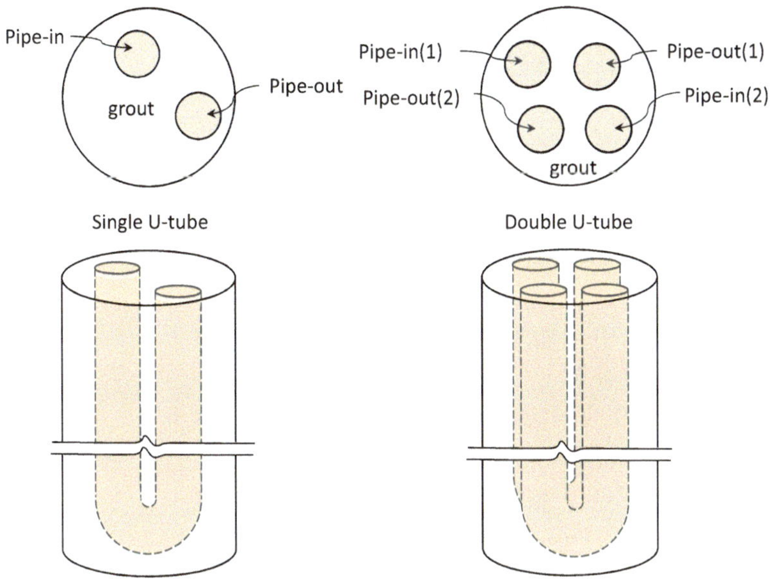

Fig. 8.8 A sketch of a single and a double borehole heat exchanger

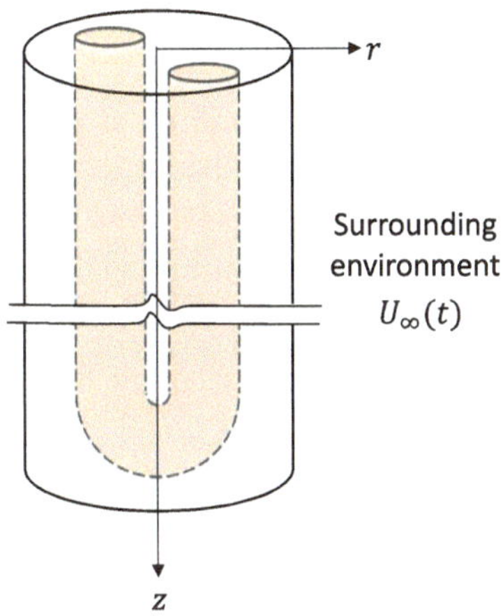

Fig. 8.9 A single U-tube BHE embedded in a surrounding environment with a time-varying temperature

$$
\begin{aligned}
&\text{Pipe - in:} && \frac{1}{\alpha_w}\frac{\partial u_i}{\partial t}A_i + \frac{v_w}{\alpha_w}\frac{\partial u_i}{\partial z}A_i - \frac{\partial^2 u_i}{\partial z^2}A_i = \frac{b_{ig}}{\lambda_w}(u_i - u_g)P_i \\
&\text{Pipe - out:} && \frac{1}{\alpha_w}\frac{\partial u_o}{\partial t}A_o - \frac{v_w}{\alpha_w}\frac{\partial u_o}{\partial z}A_o - \frac{\partial^2 u_o}{\partial z^2}A_o = \frac{b_{og}}{\lambda_w}(u_o - u_g)P_o \\
&\text{Grout:} && \frac{1}{\alpha_g}\frac{\partial u_g}{\partial t}A_g - \frac{\partial^2 u_g}{\partial z^2}A_g = \\
& && \frac{b_{ig}}{\lambda_g}(u_g - u_i)P_i + \frac{b_{og}}{\lambda_g}(u_g - u_o)P_o + \frac{\overline{h}}{\lambda_g}(u_g - U_\infty)P_g \\
&\text{IC:} && u_i(z,0) = u_o(z,0) = u_g(z,0) = 0 \\
&\text{BC:} && \text{at } z = 0: \\
& && u_i(0,t) = U_{in}(t), \quad q_o(0,t) = 0, \quad q_g(0,t) = 0 \\
& && \text{at } z = L: \\
& && u_i(L,t) = u_o(L,t), \quad q_i(L,t) = q_o(L,t) \\
& && -\lambda_g \left.\frac{\partial u_g}{\partial z}\right|_{z=L} = \overline{h}(u_g - U_\infty)
\end{aligned}
\tag{8.34}
$$

where the subscript w stands for the working fluid (water) running inside the U-tube, and the subscripts i, o and g stand for pipe-in, pipe-out and grout respectively; $\alpha_w = \lambda_w/\rho_w c_w$, $\alpha_g = \lambda_g/\rho_g c_g$; A_i is the cross-sectional area of pipe-in, etc.; P_i is the circumference of pipe-in, etc.; and b_{ig}, b_{og} and $\overline{h}$ are some thermal interaction coefficients, respectively between pipe-in and grout, pipe-out and grout, and grout and surrounding environment. As indicated in the previous section, the thermal interaction coefficients, b_{ig}, etc., can be assigned positive or negative values, depending on the heat flow direction. Basically, they should be included in the heat equations as $b_{\text{ig}} \cdot \boldsymbol{n}$, where $\boldsymbol{n}$ is a unit outward vector, which can be $+1$ or -1.

General Solution

Applying the Fourier transform to the boundary value problem (8.34) yields

$$
\begin{aligned}
&\frac{i\omega\hat{u}_i}{\alpha_w}A_i + \frac{v_w}{\alpha_w}\frac{d\hat{u}_i}{dz}A_i - \frac{d^2\hat{u}_i}{dz^2}A_i = \frac{b_{ig}}{\lambda_w}(\hat{u}_i - \hat{u}_g)P_i \\
&\frac{i\omega\hat{u}_o}{\alpha_w}A_o - \frac{v_w}{\alpha_w}\frac{d\hat{u}_o}{dz}A_o - \frac{d^2\hat{u}_o}{dz^2}A_o = \frac{b_{og}}{\lambda_w}(\hat{u}_o - \hat{u}_g)P_o \\
&\frac{i\omega\hat{u}_g}{\alpha_w}A_g - \frac{d^2\hat{u}_g}{dz^2}A_g = \frac{b_{ig}}{\lambda_g}(\hat{u}_g - \hat{u}_i)P_i + \frac{b_{og}}{\lambda_g}(\hat{u}_g - \hat{u}_o)P_o + \frac{\overline{h}}{\lambda_g}(\hat{u}_g - \hat{U}_\infty)P_g \\
&\text{BC:} \quad \text{at } z = 0: \\
&\quad \hat{u}_i(0,\omega) = \hat{U}_{in}(\omega), \quad \hat{q}_o(0,\omega) = 0, \quad \hat{q}_g(0,\omega) = 0 \\
&\quad \text{at } z = L: \\
&\quad \hat{u}_i(L,\omega) = \hat{u}_o(L,\omega), \quad \hat{q}_i(L,\omega) = \hat{q}_o(L,\omega) \\
&\quad -\lambda_g \left.\frac{d\hat{u}_g}{dz}\right|_{z=L} = \overline{h}(\hat{u}_g - \hat{U}_\infty)
\end{aligned}
\tag{8.35}
$$

This set of differential equations is nonhomogeneous due to the thermal interaction term at the right-hand side of the grout heat equation. The solution of this set of equations can thus be conducted by splitting it into solving the homogeneous and particular set of equations, such that

$$\begin{aligned}\hat{u}_i(z,\omega) &= \hat{u}_{ih} + \hat{u}_{ip}\\ \hat{u}_o(z,\omega) &= \hat{u}_{oh} + \hat{u}_{op}\\ \hat{u}_g(z,\omega) &= \hat{u}_{gh} + \hat{u}_{gp}\end{aligned} \tag{8.36}$$

The corresponding homogeneous equations of (8.35) are

$$\begin{aligned}&\frac{v_w}{\alpha_w}\frac{d\hat{u}_{ih}}{dz}A_i - \frac{d^2\hat{u}_{ih}}{dz^2}A_i - \left(\frac{b_{ig}}{\lambda_w}P_i - \frac{A_i}{\alpha_w}i\omega\right)\hat{u}_{ih} + \frac{b_{ig}P_i}{\lambda_w}\hat{u}_{gh}\\ &\quad = 0\\ &-\frac{v_w}{\alpha_w}\frac{d\hat{u}_{oh}}{dz}A_o - \frac{d^2\hat{u}_{oh}}{dz^2}A_o - \left(\frac{b_{og}}{\lambda_w}P_o - \frac{A_o}{\alpha_w}i\omega\right)\hat{u}_{oh}\\ &+\frac{b_{og}P_o}{\lambda_w}\hat{u}_{gh} = 0\\ &-\frac{d^2\hat{u}_{gh}}{dz^2}A_g - \left(\frac{b_{ig}P_i}{\lambda_g} + \frac{b_{og}P_o}{\lambda_g} + \frac{\bar{h}P_g}{\lambda_g} - \frac{A_g}{\alpha_g}i\omega\right)\hat{u}_{gh}\\ &+\frac{b_{ig}P_i}{\lambda_g}\hat{u}_{ih} + \frac{b_{og}P_o}{\lambda_g}\hat{u}_{oh} = 0\end{aligned} \tag{8.37}$$

The trial solutions of this system of ordinary differential equations can be expressed as

$$\hat{u}_{ih} = ae^{-\kappa z}, \quad \hat{u}_{oh} = be^{-\kappa z}, \quad \hat{u}_{gh} = ce^{-\kappa z}$$

in which a, b and c are constants to be determined from the boundary conditions, and κ is a constant to be determined from solving the system's eigenvalues. Substituting these trial solutions into (8.37) yields

$$\begin{aligned}&-\frac{v_w}{\alpha_w}\kappa A_i a e^{-\kappa z} - \kappa^2 A_i a e^{-\kappa z} - \left(\frac{b_{ig}P_i}{\lambda_w} - \frac{A_i}{\alpha_w}i\omega\right)ae^{-\kappa z} + \frac{b_{ig}P_i}{\lambda_w}ce^{-\kappa z} = 0\\ &\frac{v_w}{\alpha_w}\kappa A_o b e^{-\kappa z} - \kappa^2 A_o b e^{-kz} - \left(\frac{b_{og}P_o}{\lambda_w} - \frac{A_o}{\alpha_w}i\omega\right)be^{-\kappa z} + \frac{b_{og}P_o}{\lambda_w}ce^{-\kappa z} = 0\\ &-\kappa^2 A_g c e^{-\kappa z} - \left(\frac{b_{ig}P_i}{\lambda_g} + \frac{b_{og}P_o}{\lambda_g} + \frac{\bar{h}P_g}{\lambda_g} - \frac{A_g}{\alpha_g}i\omega\right)ce^{-\kappa z} + \frac{b_{ig}P_i}{\lambda_g}ae^{-\kappa z} + \frac{b_{og}P_o}{\lambda_g}be^{-\kappa z}\\ &= 0\end{aligned}$$

Dividing by $e^{-\kappa z}$ and putting the result in a matrix form, gives

$$\begin{pmatrix} x_{11} & 0 & x_{13}\\ 0 & x_{22} & x_{23}\\ x_{31} & x_{32} & x_{33}\end{pmatrix}\begin{pmatrix} a\\ b\\ c\end{pmatrix} = \begin{pmatrix} 0\\ 0\\ 0\end{pmatrix} \tag{8.38}$$

with

$$x_{11} = -\kappa^2 A_i - \frac{v_w}{\alpha_w}\kappa A_i + \frac{A_i}{\alpha_w} i\omega - \frac{b_{ig}P_i}{\lambda_w}, \qquad x_{13} = \frac{b_{ig}P_i}{\lambda_w}$$

$$x_{22} = -\kappa^2 A_o + \frac{v_w}{\alpha_w}\kappa A_o + \frac{A_o}{\alpha_w} i\omega - \frac{b_{og}P_o}{\lambda_w}, \qquad x_{23} = \frac{b_{og}P_o}{\lambda_w}$$

$$x_{31} = \frac{b_{ig}P_i}{\lambda_g}, \quad x_{32} = \frac{b_{og}P_o}{\lambda_g}$$

$$x_{33} = -\kappa^2 A_g + \frac{A_g}{\alpha_g} i\omega - \frac{b_{ig}P_i}{\lambda_g} - \frac{b_{og}P_o}{\lambda_g} - \frac{\overline{h}P_g}{\lambda_g}$$

The nontrivial solution of the system of Eq. (8.38) can be obtained if and only if the determinant of the matrix is zero. Obviously, the determinant of this matrix yields a complex polynomial of degree six, of the form

$$m_6\,\kappa^6 + m_5\,\kappa^5 + m_4\,\kappa^4 + m_3\,\kappa^3 + m_2\,\kappa^2 + m_1\,\kappa + m_0 = 0 \tag{8.39}$$

8

where upon solving for κ it reveals six eigenvalues in three pairs of complex numbers with opposite signs. Accordingly, the general solution of the homogeneous equations (8.37) can be expressed as

$$\begin{aligned}
\hat{u}_{ih} &= a_1 e^{-\kappa_1 z} + a_2 e^{-\kappa_2 z} + a_3 e^{-\kappa_3 z} + a_4 e^{\kappa_4 (L-z)} \\
&\quad + a_5 e^{\kappa_5 (L-z)} + a_6 e^{\kappa_6 (L-z)} \\
\hat{u}_{oh} &= b_1 e^{-\kappa_1 z} + b_2 e^{-\kappa_2 z} + b_3 e^{-\kappa_3 z} + b_4 e^{\kappa_4 (L-z)} \\
&\quad + b_5 e^{\kappa_5 (L-z)} + b_6 e^{\kappa_6 (L-z)} \\
\hat{u}_{gh} &= c_1 e^{-\kappa_1 z} + c_2 e^{-\kappa_2 z} + c_3 e^{-\kappa_3 z} + c_4 e^{\kappa_4 (L-z)} \\
&\quad + c_5 e^{\kappa_5 (L-z)} + c_6 e^{\kappa_6 (L-z)}
\end{aligned} \tag{8.40}$$

As the system of equations is coupled, the constants a, b and c are related. Equation (8.38) reveals that

$$a = \frac{-x_{13}}{x_{11}} c = R_{ac} c, \quad b = \frac{-x_{23}}{x_{22}} c = R_{bc} c \tag{8.41}$$

Substituting (8.41) into (8.40) gives

$$\begin{aligned}
\hat{u}_i &= R_{ac1} c_1 e^{-\kappa_1 z} + R_{ac2} c_2 e^{-\kappa_2 z} + R_{ac3} c_3 e^{-\kappa_3 z} \\
&\quad + R_{ac4} c_4 e^{\kappa_4 (L-z)} + R_{ac5} c_5 e^{\kappa_5 (L-z)} + R_{ac6} c_6 e^{\kappa_6 (L-z)} \\
\hat{u}_o &= R_{bc1} c_1 e^{-\kappa_1 z} + R_{bc2} c_2 e^{-\kappa_2 z} + R_{bc3} c_3 e^{-\kappa_3 z} \\
&\quad + R_{bc4} c_4 e^{\kappa_4 (L-z)} + R_{bc5} c_5 e^{\kappa_5 (L-z)} + R_{bc6} c_6 e^{\kappa_6 (L-z)} \\
\hat{u}_g &= c_1 e^{-\kappa_1 z} + c_2 e^{-\kappa_2 z} + c_3 e^{-\kappa_3 z} + c_4 e^{\kappa_4 (L-z)} \\
&\quad + c_5 e^{\kappa_5 (L-z)} + c_6 e^{\kappa_6 (L-z)}
\end{aligned}$$

Having solved the homogeneous equations, we now turn to solving the particular equations. The relevant particular equations of (8.35) are

$$\frac{v_w}{\alpha_w}\frac{d\hat{u}_{ip}}{dz}A_i - \frac{d^2\hat{u}_{ip}}{dz^2}A_i - \left(\frac{b_{ig}P_i}{\lambda_w} - \frac{A_i}{\alpha_w}i\omega\right)\hat{u}_{ip} + \frac{b_{ig}P_i}{\lambda_w}\hat{u}_{gp} = 0$$

$$-\frac{v_w}{\alpha_w}\frac{d\hat{u}_{op}}{dz}A_o - \frac{d^2\hat{u}_{op}}{dz^2}A_o - \left(\frac{b_{og}}{\lambda_w}P_o - \frac{A_o}{\alpha_w}i\omega\right)\hat{u}_{op} + \frac{b_{og}P_o}{\lambda_w}\hat{u}_{gp} = 0$$

$$-\frac{d^2\hat{u}_{gp}}{dz^2}A_g - \left(\frac{b_{ig}P_i}{\lambda_g} + \frac{b_{og}P_o}{\lambda_g} + \frac{\bar{h}P_g}{\lambda_g} - \frac{A_g}{\alpha_g}i\omega\right)\hat{u}_{gp} + \frac{b_{ig}P_i}{\lambda_g}\hat{u}_{ip} + \frac{b_{og}P_o}{\lambda_g}\hat{u}_{op} = -\frac{\bar{h}P_g}{\lambda_g}\hat{U}_\infty \tag{8.42}$$

As the nonhomogeneous term $\widehat{U}_\infty(\omega)$ is spatially constant, the solution can be expressed as a constant, such that

$$\hat{u}_{ip} = a_p, \quad \hat{u}_{op} = b_p, \quad \hat{u}_{gp} = c_p \tag{8.43}$$

Substituting (8.43) into (8.42) leads to

$$-\left(\frac{b_{ig}P_i}{\lambda_w} - \frac{A_i}{\alpha_w}i\omega\right)a_p + \frac{b_{ig}P_i}{\lambda_w}c_p = 0$$

$$-\left(\frac{b_{og}}{\lambda_w}P_o - \frac{A_o}{\alpha_w}i\omega\right)b_p + \frac{b_{og}P_o}{\lambda_w}c_p = 0$$

$$-\left(\frac{b_{ig}P_i}{\lambda_g} + \frac{b_{og}P_o}{\lambda_g} + \frac{\bar{h}P_g}{\lambda_g} - \frac{A_g}{\alpha_g}i\omega\right)c_p + \frac{b_{ig}P_i}{\lambda_g}a_p + \frac{b_{og}P_o}{\lambda_g}b_p = -\frac{\bar{h}P_g}{\lambda_g}\hat{U}_\infty$$

Putting these equations in a matrix form, gives

$$\begin{pmatrix} \frac{A_i}{\alpha_w}i\omega - \frac{b_{ig}P_i}{\lambda_w} & 0 & \frac{b_{ig}P_i}{\lambda_w} \\ 0 & \frac{A_o}{\alpha_w}i\omega - \frac{b_{og}}{\lambda_w}P_o & \frac{b_{og}P_o}{\lambda_w} \\ \frac{b_{ig}P_i}{\lambda_g} & \frac{b_{og}P_o}{\lambda_g} & \frac{A_g}{\alpha_g}i\omega - \frac{b_{ig}P_i}{\lambda_g} - \frac{b_{og}P_o}{\lambda_g} - \frac{\bar{h}P_g}{\lambda_g} \end{pmatrix}\begin{pmatrix} a_p \\ b_p \\ c_p \end{pmatrix} = \begin{pmatrix} 0 \\ 0 \\ -\frac{\bar{h}P_g}{\lambda_g}\hat{U}_\infty \end{pmatrix}$$

Solving this set of equations simultaneously gives the particular constants $\{a_p b_p c_p\}$, which upon substituting in (8.43) gives the required general solution of the particular part of the boundary value problem.

Accordingly, the general solution in the frequency domain for the three BHE components is given by

$$\begin{aligned}
\hat{u}_i &= R_{ac1}c_1e^{-\kappa_1 z} + R_{ac2}c_2e^{-\kappa_2 z} + R_{ac3}c_3e^{-\kappa_3 z} \\
&\quad + R_{ac4}c_4e^{\kappa_4(L-z)} + R_{ac5}c_5e^{\kappa_5(L-z)} + R_{ac6}c_6e^{\kappa_6(L-z)} + \hat{u}_{ip} \\
\hat{u}_o &= R_{bc1}c_1e^{-\kappa_1 z} + R_{bc2}c_2e^{-\kappa_2 z} + R_{bc3}c_3e^{-\kappa_3 z} \\
&\quad + R_{bc4}c_4e^{\kappa_4(L-z)} + R_{bc5}c_5e^{\kappa_5(L-z)} + R_{bc6}c_6e^{\kappa_6(L-z)} + \hat{u}_{op} \\
\hat{u}_g &= c_1e^{-\kappa_1 z} + c_2e^{-\kappa_2 z} + c_3e^{-\kappa_3 z} + c_4e^{\kappa_4(L-z)} + c_5e^{\kappa_5(L-z)} + c_6e^{\kappa_6(L-z)} + \hat{u}_{gp}.
\end{aligned} \tag{8.44}$$

Boundary Conditions

Taking the derivatives of (8.44) with respect to z, the boundary conditions in (8.35) imply:

At $z = 0$:

$$\begin{aligned}
&R_{ac1}c_1 + R_{ac2}c_2 + R_{ac3}c_3 + R_{ac4}c_4e^{\kappa_4 L} + \\
&\qquad R_{ac5}c_5e^{\kappa_5 L} + R_{ac6}c_6e^{\kappa_6 L} + \hat{u}_{ip} = \hat{U}_{in}(\omega) \\
&-\kappa_1 R_{bc1}c_1 - \kappa_2 R_{bc2}c_2 - \kappa_3 R_{bc3}c_3 - \\
&\qquad \kappa_4 R_{bc4}c_4e^{\kappa_4 L} - \kappa_5 R_{bc5}c_5e^{\kappa_5 L} - \kappa_6 R_{bc6}c_6e^{\kappa_6 L} = 0
\end{aligned}$$

and

$$\begin{aligned}
&-\kappa_1 c_1 - \kappa_2 c_2 - \kappa_3 c_3 - \\
&\qquad \kappa_4 c_4e^{\kappa_4 L} - \kappa_5 c_5e^{\kappa_5 L} - \kappa_6 c_6e^{\kappa_6 L} = 0
\end{aligned}$$

At $z = L$:

$$\begin{aligned}
&(R_{ac1} - R_{bc1})\, c_1e^{-\kappa_1 L} + (R_{ac2} - R_{bc2})\, c_2e^{-\kappa_2 L} \\
&\quad + (R_{ac3} - R_{bc3})\, c_3e^{-\kappa_3 L} + (R_{ac4} - R_{bc4})\, c_4 \\
&\quad + (R_{ac5} - R_{bc5})\, c_5 + (R_{ac6} - R_{bc6})\, c_6 = \hat{u}_{op} - \hat{u}_{ip}
\end{aligned}$$

and

$$\begin{aligned}
&(-R_{ac1} + R_{bc1})\, \kappa_1 c_1e^{-\kappa_1 L} + (-R_{ac2} + R_{bc2})\, \kappa_2 c_2e^{-\kappa_2 L} \\
&\quad + (-R_{ac3} + R_{bc3})\, \kappa_3 c_3e^{-\kappa_3 L} + (-R_{ac4} + R_{bc4})\, \kappa_4 c_4 + (-R_{ac5} + R_{bc5})\, \kappa_5 c_5 \\
&\quad + (-R_{ac6} + R_{bc6})\, \kappa_6 c_6 = 0
\end{aligned}$$

Additionally, the third boundary condition implies

$$\begin{aligned}
&c_1e^{-\kappa_1 L}\left(-\kappa_1 + \frac{\bar{h}}{\lambda_g}\right) + c_2e^{-\kappa_2 L}\left(-\kappa_2 + \frac{\bar{h}}{\lambda_g}\right) + c_3e^{-\kappa_3 L}\left(-\kappa_3 + \frac{\bar{h}}{\lambda_g}\right) \\
&+ c_4\left(-\kappa_4 + \frac{\bar{h}}{\lambda_g}\right) + c_5\left(-\kappa_5 + \frac{\bar{h}}{\lambda_g}\right) + c_6\left(-\kappa_6 + \frac{\bar{h}}{\lambda_g}\right) \\
&= -\frac{\bar{h}}{\lambda_g}(\hat{u}_{gp} - \hat{U}_\infty)
\end{aligned}$$

Putting these six equations in a matrix form, yields

$$\begin{pmatrix} a_{11} & a_{12} & a_{13} & a_{14} & a_{15} & a_{16} \\ a_{21} & a_{22} & a_{23} & a_{24} & a_{25} & a_{26} \\ a_{31} & a_{32} & a_{33} & a_{34} & a_{35} & a_{36} \\ a_{41} & a_{42} & a_{43} & a_{44} & a_{45} & a_{46} \\ a_{51} & a_{52} & a_{53} & a_{54} & a_{55} & a_{56} \\ a_{61} & a_{62} & a_{63} & a_{64} & a_{65} & a_{66} \end{pmatrix} \begin{pmatrix} c_1 \\ c_2 \\ c_3 \\ c_4 \\ c_5 \\ c_6 \end{pmatrix} = \begin{pmatrix} \hat{U}_{in}(\omega) - \hat{u}_{ip} \\ 0 \\ 0 \\ \hat{u}_{op} - \hat{u}_{ip} \\ 0 \\ -\frac{\bar{h}}{\lambda_g}(\hat{u}_{gp} - \hat{U}_\infty) \end{pmatrix} \tag{8.45}$$

where

$$a_{11} = R_{ac1} \quad a_{21} = -\kappa_1 R_{ac1} \quad a_{31} = -\kappa_1$$
$$a_{12} = R_{ac2} \quad a_{22} = -\kappa_2 R_{ac2} \quad a_{32} = -\kappa_2$$
$$a_{13} = R_{ac3} \quad a_{23} = -\kappa_3 R_{ac3} \quad a_{33} = -\kappa_3$$
$$a_{14} = R_{ac4} e^{\kappa_4 L} \quad a_{24} = -\kappa_4 R_{ac4} e^{\kappa_4 L} \quad a_{34} = -\kappa_4 e^{\kappa_4 L}$$
$$a_{15} = R_{ac5} e^{\kappa_5 L} \quad a_{25} = -\kappa_5 R_{ac5} e^{\kappa_5 L} \quad a_{35} = -\kappa_5 e^{\kappa_5 L}$$
$$a_{16} = R_{ac6} e^{\kappa_6 L} \quad a_{26} = -\kappa_6 R_{ac6} e^{\kappa_6 L} \quad a_{36} = -\kappa_6 e^{\kappa_6 L}$$

$$a_{41} = (R_{ac1} - R_{bc1}) e^{-\kappa_1 L} \quad a_{51} = (-R_{ac1} + R_{bc1}) \kappa_1 e^{-\kappa_1 L}$$
$$a_{61} = e^{-\kappa_1 L} \left(-\kappa_1 + \bar{h}/\lambda_g\right)$$
$$a_{42} = (R_{ac2} - R_{bc2}) e^{-\kappa_2 L} \quad a_{52} = (-R_{ac2} + R_{bc2}) \kappa_2 e^{-\kappa_2 L}$$
$$a_{62} = e^{-\kappa_2 L} \left(-\kappa_2 + \bar{h}/\lambda_g\right)$$
$$a_{43} = (R_{ac3} - R_{bc3}) e^{-\kappa_3 L} \quad a_{53} = (-R_{ac3} + R_{bc3}) \kappa_3 e^{-\kappa_3 L}$$
$$a_{63} = e^{-\kappa_3 L} \left(-\kappa_3 + \bar{h}/\lambda_g\right)$$
$$a_{44} = (R_{ac4} - R_{bc4}) \quad a_{54} = (-R_{ac4} + R_{bc4}) \kappa_4$$
$$a_{64} = \left(-\kappa_4 + \bar{h}/\lambda_g\right)$$
$$a_{45} = (R_{ac5} - R_{bc5}) \quad a_{55} = (-R_{ac5} + R_{bc5}) \kappa_5$$
$$a_{65} = \left(-\kappa_5 + \bar{h}/\lambda_g\right)$$
$$a_{46} = (R_{ac6} - R_{bc6}) \quad a_{56} = (-R_{ac6} + R_{bc6}) \kappa_6$$
$$a_{66} = \left(-\kappa_6 + \bar{h}/\lambda_g\right).$$

BVP Solution

Solving the system of Eqs. (8.45) simultaneously gives the six constants $c_1 \cdots c_6$, which upon substituting into (8.44) yields the required solution of the boundary value problem in the frequency domain.

Having solved the system in the frequency domain, the solution in the time domain can readily be obtained using the IFFT algorithm, such that

$$u_i(z,t) = \sum_n \hat{u}_i(z,\omega_n)\, e^{i\omega_n t}$$
$$u_o(z,t) = \sum_n \hat{u}_o(z,\omega_n)\, e^{i\omega_n t}$$
$$u_g(z,t) = \sum_n \hat{u}_g(z,\omega_n)\, e^{i\omega_n t}.$$

8.5 Single U-tube BHE Embedded in a Soil Mass

In the previous section we addressed a scenario where the borehole is installed in an environment that serves as a heat source with a constant temperature. Here, we expand this scenario to include a case where the borehole is embedded in a soil mass, engaging in a full thermal interaction process with the borehole. In this section we study heat flow in a single borehole heat exchanger, where its length is smaller than the thickness of the soil layer, as shown in ◘ Fig. 8.10.

This system is highly coupled since the surrounding soil can absorb heat from the borehole or release heat to it, and conversely, the borehole can absorb or supply heat from or to the soil mass. As indicated in the previous section, this type of heat exchanger is extremely slender because the ratio between the U-tube diameter and grout to the length of the borehole can be <1/1000. Given this physical characteristic, the BHE can be modeled as a 1D element while preserving the physical and thermal properties of the borehole components and their thermal interactions. This assumption makes the BHE-soil system axially symmetric, where the axis of symmetry goes through the borehole, as can be seen in ◘ Fig. 8.10. Thus, the

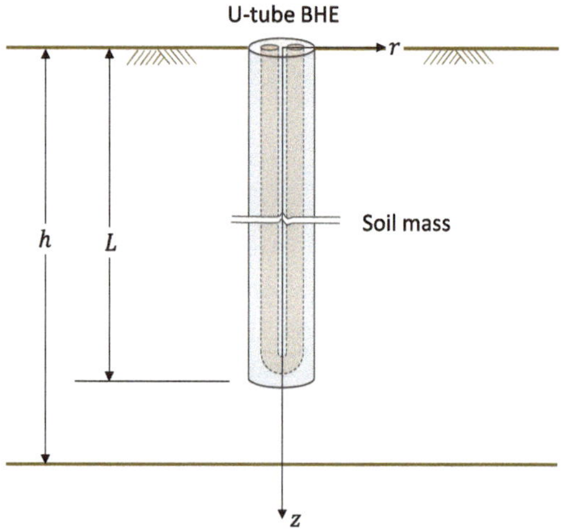

◘ **Fig. 8.10** A single U-tube BHE in full thermal interaction with a soil mass

coupled boundary value problem of this system can be symbolically expressed as

$$\begin{aligned} du_{\text{BHE}} &= du_b(z,t) + du_{bs} \\ du_{\text{soil}} &= du_s(r,z,t) + du_{sb} \end{aligned} \tag{8.46}$$

where d is some differential operator, subscripts b and s stand for the BHE and the soil mass respectively, and $du_{bs} \equiv du_{sb}$ are some thermal interaction terms coupling the borehole to the soil mass and vice versa. The soil mass interacts with the BHE along its boundary on the axis of symmetry (z–axis), while the BHE exchanges heat with the soil mass through the body of the grout.

Hereafter, we derive the solutions for the governing heat equations for both the soil mass and the borehole heat exchanger. Then, we couple these solutions using a staggered approach.

Heat Flow in the Soil Mass

The boundary value problem of heat flow in a soil layer of thickness h, initially at 0 °C, subjected to a varying temperature at its surface U_s, and exchanging heat with a cylindrical borehole heat exchanger with radius r_b along its z–axis, is given by

$$\begin{aligned} &\frac{1}{\alpha_s}\frac{\partial u_s}{\partial t} - \frac{\partial^2 u_s}{\partial r^2} - \frac{1}{r}\frac{\partial u_s}{\partial r} - \frac{\partial^2 u_s}{\partial z^2} = 0 \\ &\text{IC:}\quad u_s(r,z,0) = 0 \\ &\text{BC:}\quad u_s(r \to \infty, z, t) = 0 \\ &\qquad\quad u_s(r,0,t) = U_s(t) \\ &\qquad\quad \left.\frac{\partial u_s}{\partial z}\right|_{z=h} = 0 \\ &\qquad\quad -\left.\frac{\partial u_s}{\partial r}\right|_{r=r_b} = \frac{b_{gs}}{\lambda_s}[u_s(r_b,z) - u_g(z)], \quad 0 \le z \le L \end{aligned} \tag{8.47}$$

Note that other types of boundary conditions can be specified for the soil surface and its bottom. These may include Newton's law of cooling at the surface. See Problem 6.

The partial differential equation in (8.47) is linear and homogeneous, but two of the boundary conditions are nonhomogeneous because they are not identically zero. To solve this boundary value problem using the separation of variables and the eigenfunction expansion we divide the problem into two sub-problems, such that

$$u_s(r,z,t) = u_1(z,t) + u_2(r,z,t) \tag{8.48}$$

where it can be noticed that u_1 is a function of z and t, and u_2 is a function of r, z and t.

Substituting (8.48) into (8.47) yields

$$\frac{1}{\alpha_s}\frac{\partial u_1}{\partial t} - \frac{\partial^2 u_1}{\partial z^2} + \frac{1}{\alpha_s}\frac{\partial u_2}{\partial t} - \frac{\partial^2 u_2}{\partial r^2} - \frac{1}{r}\frac{\partial u_2}{\partial r} - \frac{\partial^2 u_2}{\partial z^2} = 0$$

Splitting this equation into two andwl expressing them together with their relevant boundary conditions, we obtain

- **Sub-problems 1**

$$\begin{aligned} &\frac{\partial^2 u_1}{\partial z^2} - \frac{1}{\alpha_s}\frac{\partial u_1}{\partial t} = 0 \\ &\text{IC:}\quad u_1(z,0) = 0 \\ &\text{BC:}\quad u_1(0,t) = U_s(t) \\ &\qquad\quad \left.\frac{\partial u_1}{\partial z}\right|_{z=h} = 0. \end{aligned} \tag{8.49}$$

- **Sub-problem 2**

$$\begin{aligned} &\frac{\partial^2 u_2}{\partial r^2} + \frac{1}{r}\frac{\partial u_2}{\partial r} + \frac{\partial^2 u_2}{\partial z^2} - \frac{1}{\alpha_s}\frac{\partial u_2}{\partial t} = 0 \\ &\text{IC:}\quad u_2(r,z,0) = 0 \\ &\text{BC:}\quad u_2(r \to \infty, t) = 0 \\ &\qquad\quad u_2(r,0,t) = 0 \\ &\qquad\quad \left.\frac{\partial u_2}{\partial z}\right|_{z=h} = 0 \\ &\qquad\quad -\left.\frac{\partial u_2}{\partial r}\right|_{r_b} = \frac{b_{gs}}{\lambda_s}\{[u_1(z) + u_2(r_b,z)] - u_g(z)\}. \end{aligned} \tag{8.50}$$

- **Solving Sub-problem 1**

Transforming the boundary value problem (8.49) to the frequency domain results in

$$\begin{aligned} &\frac{\partial^2 \hat{u}_1}{\partial z^2} - \frac{i\omega}{\alpha_s}\hat{u}_1 = 0 \\ &\text{BC:}\quad \hat{u}_1(0,\omega) = \hat{U}_s(\omega) \\ &\qquad\quad \left.\frac{\partial \hat{u}_1(z,\omega)}{\partial z}\right|_{z=h} = 0 \end{aligned} \tag{8.51}$$

Since the coefficients in this differential equation are constant, the trial solution can be expressed as

$$\hat{u}_1 = a\,e^{-\kappa z}$$

Substituting this trial solution into (8.51), and simplifying, leads to

$$\kappa^2 - \frac{i\omega_n}{\alpha_s} = 0$$

Using the quadratic formula yields, upon re-arrangement,

$$\kappa_1 = \kappa = \sqrt{\frac{i\omega_n}{\alpha_s}}, \quad \kappa_2 = -\kappa \tag{8.52}$$

Thus, the general solution of the homogeneous equation becomes

$$\hat{u}_1 = a_1 e^{-\kappa z} + a_2 e^{\kappa z}. \tag{8.53}$$

Boundary Conditions

The first boundary condition in (8.51) implies at $z = 0$

$$a_1 + a_2 = \hat{U}_s \tag{8.54}$$

Taking the derivative of (8.53) with respect to x gives

$$\frac{d\hat{u}_1}{dx} = -\kappa a_1 e^{-\kappa x} + \kappa a_2 e^{\kappa x}$$

Thus, at $z = h$, the second boundary condition in (8.51) implies

$$-\kappa a_1 e^{-\kappa h} + \kappa a_2 e^{\kappa h} = 0 \tag{8.55}$$

Solving (8.54) and (8.55) simultaneously yields

$$a_1 = \frac{\hat{U}_s\, e^{\kappa h}}{e^{\kappa h} + e^{-\kappa h}}, \qquad a_2 = \frac{\hat{U}_s\, e^{-\kappa h}}{e^{\kappa h} + e^{-\kappa h}}. \tag{8.56}$$

BVP Solution

Substituting the constants in (8.56) into (8.53), the solution of the boundary value problem (8.51) for a frequency ω_n becomes

$$\begin{aligned} \hat{u}_1\,(z, \omega_n) &= \frac{\hat{U}_s\, e^{\kappa h}}{e^{\kappa h} + e^{-\kappa h}} e^{-\kappa z} + \frac{\hat{U}_s\, e^{-\kappa h}}{e^{\kappa h} + e^{-\kappa h}} e^{\kappa z} \\ &= \frac{\hat{U}_s}{1 + e^{-2\kappa h}} [e^{-\kappa z} + e^{-\kappa\,(2h - z)}] \end{aligned} \tag{8.57}$$

with the eigenvalue κ given in (8.52).

- **Solving Sub-problem 2**

Applying the Fourier transform to the BVP (8.50) gives

$$\begin{aligned} &\frac{\partial^2 \hat{u}_2}{\partial r^2} + \frac{1}{r}\frac{\partial \hat{u}_2}{\partial r} + \frac{\partial^2 \hat{u}_2}{\partial z^2} - \frac{i\omega}{\alpha}\hat{u}_2 = 0 \\ &\text{BC:} \quad \hat{u}_2(r \to \infty, z, \omega) = 0 \\ &\qquad\quad \hat{u}_2(r, 0, \omega) = 0 \\ &\qquad\quad \left.\frac{\partial \hat{u}_2}{\partial z}\right|_{z=h} = 0 \\ &\qquad\quad -\left.\frac{\partial u_2}{\partial r}\right|_{r=r_b} = \frac{b_{gs}}{\lambda_s}\left[\hat{u}_1(z) + \hat{u}_2(r_b, z) - \hat{u}_g(z)\right]. \end{aligned} \tag{8.58}$$

Using the method of separation of variables, the solution of the partial differential equation in (8.58) can be undertaken by assuming

$$\hat{u}_2(r, z) = R(r)\, Z(z) \equiv RZ \tag{8.59}$$

in which $R(r)$ is a function of r only and $Z(z)$ is a function of z only. Substituting (8.59) into the partial differential equation in (8.58), and dividing through by RZ, results in

$$\frac{R''}{R} + \frac{1}{r}\frac{R'}{R} - \frac{i\omega}{\alpha} = -\frac{Z''}{Z}$$

Equating both sides of the equation by some arbitrary constant, say β^2, yields

$$\frac{R''}{R} + \frac{1}{r}\frac{R'}{R} - \frac{i\omega}{\alpha} = -\frac{Z''}{Z} = \beta^2$$

This equation can be split into two independent ordinary differential equations, such that

8

$$\frac{d^2Z}{dz^2} + \beta^2 Z = 0 \tag{8.60}$$

$$\frac{d^2R}{dr^2} + \frac{1}{r}\frac{dR}{dr} - \left(\frac{i\omega}{\alpha} + \beta^2\right) R = 0 \tag{8.61}$$

The solution of the second order ordinary differential Eq. (8.60) can readily be obtained via

$$Z(z) = c_1 \cos \beta z + c_2 \sin \beta z \tag{8.62}$$

where c_1 and c_2 are constants to be determined from the boundary conditions.

Multiplying (8.61) by r^2, it can be expressed as

$$r^2R'' + rR' - \kappa_r^2 r^2 R = 0, \quad \kappa_r = \sqrt{\kappa^2 + \beta^2}$$

This equation is the parametric form of the modified Bessel equation of order 0, with its solution given by

$$R(r) = c_3 I_0(\kappa_r r) + c_4 K_0(\kappa_r r)$$

in which I_0 and K_0 are the modified Bessel functions of zero order of the first and second kind, respectively, and c_3 and c_4 are arbitrary constants, need to be determined from the boundary conditions.

Following the first boundary condition in (8.58), c_3 must be zero since I_0 is unbounded at far distances, leading to

$$R(r) = c_4 K_0(\kappa_r r) \tag{8.63}$$

Substituting (8.62) and (8.63) into (8.59) the general solution finally entails

$$\hat{u}_2(r, z) = R(r)\, Z(z) = K_0(\kappa_r r)[c_5 \cos(\beta z) + c_6 \sin(\beta z)] \tag{8.64}$$

where $c_5 = c_1 \cdot c_4$ and $c_6 = c_2 \cdot c_4$.

Boundary Conditions

Applying the boundary conditions in (8.58) into (8.64) implies:

At $z = 0$

$$c_5 = 0 \tag{8.65}$$

At $z = h$

$$c_6 \beta \cos(\beta h) = 0 \tag{8.66}$$

Since c_6 and β cannot be zero, otherwise it leads to a nontrivial solution, it is therefore

$$\cos(\beta h) = 0 \tag{8.67}$$

which is true only if

$$\beta h = \left(m + \frac{1}{2}\right)\pi \Rightarrow \beta_m = \frac{\pi}{h}(m + \frac{1}{2}), \quad m = 0, 1, 2, \ldots$$

Following this, the solution for every m becomes

$$\hat{u}_{2m}(r, z) = c_m \sin(\beta_m z) K_0(\kappa_r r)$$

where we arbitrarily changed c_6 to c_m.

Since the boundary value problem is linear, an infinite sum over all solutions is also a solution, leading to

$$\hat{u}_2(r, z) = \sum_m c_m \sin(\beta_m z)\, K_0(\kappa_r r) \tag{8.68}$$

Taking the derivative of (8.68) with respect to r yields

$$\frac{\partial \hat{u}_2}{\partial r} = -\sum_m c_m \sin(\beta_m z)\, \kappa_r K_1(\kappa_r r)$$

At $r = r_b$, the third boundary condition in (8.58) implies

$$\begin{aligned}\sum_m \lambda_s c_m \sin(\beta_m z)\, \kappa_r K_1(\kappa_r r_b) = \\ \sum_m b_{gs} c_m \sin(\beta_m z) K_0(\kappa_r r_b) + b_{gs}\left[\hat{u}_1(z) - \hat{u}_g(z)\right]\end{aligned} \tag{8.69}$$

where $\widehat{u}_1(z)$ is given in (8.57), and $\widehat{u}_g(z)$ will be determined by solving the BVP of the borehole heat exchanger. Rearranging (8.69) gives

$$\sum_m c_m \sin(\beta_m z) = \hat{q}(z,\omega_n)$$
where
$$\hat{q}(z,\omega_n) = \frac{b_{gs}\left[\hat{u}_1(z) - \hat{u}_g(z)\right]}{\lambda_s \kappa_r K_1(\kappa_r r_b) - b_{gs} K_0(\kappa_r r_b)} \tag{8.70}$$

which is a half range sine Fourier series, with its coefficients given by

$$c_m = \frac{2}{h}\int_0^L \hat{q}(z,\omega_n)\sin(\beta_m z)\,dz \tag{8.71}$$

8

Inserting c_m into (8.68) yields the required solution of $\widehat{u}_2$.

Solution of Soil BVP

Substituting the solutions of u_1 and u_2 into the transformed form of (8.48), yields the solution of u_s in the frequency domain, as

$$\hat{u}_s(r,z,\omega_n) = \hat{u}_1(z,\omega_n) + \sum_m c_m \sin(\beta_m z) K_0(\kappa_r r)$$

where

$$\begin{aligned}
\hat{u}_1(z,\omega_n) &= \frac{\hat{U}_s(\omega_n)}{1+e^{-2\kappa h}}\left[e^{-\kappa z} + e^{-\kappa(2h-z)}\right] \\
c_m &= \frac{2}{h}\int_0^L \hat{q}(z,\omega_n)\sin(\beta_m z)\,dz \\
\hat{q}(z,\omega_n) &= \frac{b_{gs}\left[\hat{u}_1(z) - \hat{u}_g(z)\right]}{\lambda_s \kappa_r K_1(\kappa_r r_b) - b_{gs} K_0(\kappa_r r_b)} \\
\kappa &= \sqrt{\frac{i\omega_n}{\alpha}}, \quad \kappa_r = \sqrt{\frac{i\omega_n}{\alpha} + \beta_m^2}, \quad \beta_m = \frac{\pi}{h}\left(m + \frac{1}{2}\right), \\
& m = 0, 1, 2, \ldots
\end{aligned} \tag{8.72}$$

in which $\widehat{u}_g(z)$ will be determined by solving the BVP of the borehole heat exchanger. It will be incorporated into this equation sequentially using a staggered iterative scheme.

Heat Flow in the BHE

The boundary value problem of the BHE can be expressed as

$$
\begin{aligned}
&\text{Pipe-in:} && \frac{1}{\alpha_w}\frac{\partial u_i}{\partial t}A_i + \frac{v_w}{\alpha_w}\frac{\partial u_i}{\partial z}A_i - \frac{\partial^2 u_i}{\partial z^2}A_i = \frac{b_{ig}}{\lambda_w}(u_i - u_g)P_i \\
&\text{Pipe-out:} && \frac{1}{\alpha_w}\frac{\partial u_o}{\partial t}A_o - \frac{v_w}{\alpha_w}\frac{\partial u_o}{\partial z}A_o - \frac{\partial^2 u_o}{\partial z^2}A_o = \frac{b_{og}}{\lambda_w}(u_o - u_g)P_o \\
&\text{Grout:} && \frac{1}{\alpha_g}\frac{\partial u_g}{\partial t}A_g - \frac{\partial^2 u_g}{\partial z^2}A_g \\
& && = \frac{b_{ig}}{\lambda_g}(u_g - u_i)P_i + \frac{b_{og}}{\lambda_g}(u_g - u_o)P_o + \frac{b_{gs}}{\lambda_g}\left[u_g - u_s(r_b, z)\right]P_g \\
&\text{IC:} && u_i(z,0) = u_o(z,0) = u_g(z,0) = 0 \\
&\text{BC:} && \text{at } z = 0: \\
& && u_i(0,t) = U_{in}(t), \quad q_o(0,t) = 0, \quad q_g(0,t) = 0 \\
& && \text{at } z = L: \\
& && u_i(L,t) = u_o(L,t), \quad q_i(L,t) = q_o(L,t) \\
& && q_g(L,t) = 0
\end{aligned}
\tag{8.73}
$$

where the subscripts i, o, and g stand for pipe-in, pipe-out, and grout, respectively, and A_i is the cross-sectional area of pipe-in, etc., and P_i is the circumference of pipe-in, etc. $u_s(r_b, z)$, with r_b representing the radius of the borehole, is the soil temperature along its boundary with the BHE, to be incorporated sequentially from solving the soil mass heat Eq. (8.72). Note that while we assume the proposed boundary conditions reasonably reflect the physics of the problem, other boundary conditions may also be applied.

This set of differential equations is nonhomogeneous due to the thermal interaction between the grout and the surrounding soil mass. The solution for this set of equations can thus be conducted by splitting it into solving the homogeneous set of equations and the particular set, such that

$$
\begin{aligned}
\hat{u}_i(z,\omega) &= \hat{u}_{ih} + \hat{u}_{ip} \\
\hat{u}_o(z,\omega) &= \hat{u}_{oh} + \hat{u}_{op} \\
\hat{u}_g(z,\omega) &= \hat{u}_{gh} + \hat{u}_{gp}.
\end{aligned}
\tag{8.74}
$$

Homogeneous Equations

Applying the Fourier transform to the associated homogeneous equations of (8.73) yields

$$
\begin{aligned}
&\frac{v_w}{\alpha_w}\frac{d\hat{u}_{ih}}{dz}A_i - \frac{d^2\hat{u}_{ih}}{dz^2}A_i - \left(\frac{b_{ig}}{\lambda_w}P_i - \frac{A_i}{\alpha_w}i\omega\right)\hat{u}_{ih} + \frac{b_{ig}P_i}{\lambda_w}\hat{u}_{gh} \\
&\quad = 0 \\
&-\frac{v_w}{\alpha_w}\frac{d\hat{u}_{oh}}{dz}A_o - \frac{d^2\hat{u}_{oh}}{dz^2}A_o - \left(\frac{b_{og}}{\lambda_w}P_o - \frac{A_o}{\alpha_w}i\omega\right)\hat{u}_{oh} \\
&\quad + \frac{b_{og}P_o}{\lambda_w}\hat{u}_{gh} = 0 \\
&-\frac{d^2\hat{u}_{gh}}{dz^2}A_g - \left(\frac{b_{ig}P_i}{\lambda_g} + \frac{b_{og}P_o}{\lambda_g} + \frac{b_{gs}P_g}{\lambda_g} - \frac{A_g}{\alpha_g}i\omega\right)\hat{u}_{gh} \\
&\quad + \frac{b_{ig}P_i}{\lambda_g}\hat{u}_{ih} + \frac{b_{og}P_o}{\lambda_g}\hat{u}_{oh} = 0
\end{aligned}
\tag{8.75}
$$

The trial solutions of this system of ordinary differential equations can be expressed as

$$\hat{u}_{ih} = ae^{-\kappa z}, \quad \hat{u}_{oh} = be^{-\kappa z}, \quad \hat{u}_{gh} = ce^{-\kappa z} \tag{8.76}$$

in which a, b and c are the integration constants, and κ is a contact to be determined from solving the system's eigenvalues. Substituting (8.76) into (8.75) results in

$$-\frac{v_w}{\alpha_w}\kappa A_i ae^{-\kappa z} - \kappa^2 A_i ae^{-\kappa z} - \left(\frac{b_{ig}P_i}{\lambda_w} - \frac{A_i}{\alpha_w}i\omega\right)ae^{-\kappa z} + \frac{b_{ig}P_i}{\lambda_w}ce^{-\kappa z} = 0$$

$$\frac{v_w}{\alpha_w}\kappa A_o be^{-\kappa z} - \kappa^2 A_o be^{-kz} - \left(\frac{b_{og}P_o}{\lambda_w} - \frac{A_o}{\alpha_w}i\omega\right)be^{-\kappa z} + \frac{b_{og}P_o}{\lambda_w}ce^{-\kappa z} = 0$$

$$-\kappa^2 A_g ce^{-\kappa z} - \left(\frac{b_{ig}P_i}{\lambda_g} + \frac{b_{og}P_o}{\lambda_g} + \frac{b_{gs}P_g}{\lambda_g} - \frac{A_g}{\alpha_g}i\omega\right)ce^{-\kappa z} + \frac{b_{ig}P_i}{\lambda_g}ae^{-\kappa z}$$
$$+ \frac{b_{og}P_o}{\lambda_g}be^{-\kappa z} = 0$$

Dividing by $e^{-\kappa z}$ and putting the result in a matrix form gives

$$\begin{pmatrix} x_{11} & 0 & x_{13} \\ 0 & x_{22} & x_{23} \\ x_{31} & x_{32} & x_{33} \end{pmatrix}\begin{pmatrix} a \\ b \\ c \end{pmatrix} = \begin{pmatrix} 0 \\ 0 \\ 0 \end{pmatrix} \tag{8.77}$$

with

$$x_{11} = -\kappa^2 A_i - \frac{v_w}{\alpha_w}\kappa A_i + \frac{A_i}{\alpha_w}i\omega - \frac{b_{ig}P_i}{\lambda_w},$$

$$x_{13} = \frac{b_{ig}P_i}{\lambda_w}$$

$$x_{22} = -\kappa^2 A_o + \frac{v_w}{\alpha_w}\kappa A_o + \frac{A_o}{\alpha_w}i\omega - \frac{b_{og}P_o}{\lambda_w},$$

$$x_{23} = \frac{b_{og}P_o}{\lambda_w}, \qquad x_{31} = \frac{b_{ig}P_i}{\lambda_g}$$

$$x_{32} = \frac{b_{og}P_o}{\lambda_g},$$

$$x_{33} = -\kappa^2 A_g + \frac{A_g}{\alpha_g}i\omega - \frac{b_{ig}P_i}{\lambda_g} - \frac{b_{og}P_o}{\lambda_g} - \frac{b_{gs}P_g}{\lambda_g}$$

The solution of (8.77) can be obtained by setting the determinant of the matrix to zero. Obviously, evaluating this determinant yields a complex sixth degree polynomial, of the form

$$m_6\kappa^6 + m_5\kappa^5 + m_4\kappa^4 + m_3\kappa^3 + m_2\kappa^2 + m_1\kappa + m_0 = 0$$

where upon solving for κ it reveals six eigenvalues in three pairs of complex numbers with opposite signs. Accordingly, the solution of (8.75) can be expressed in terms of the corresponding eigenfunctions, such that

$$\begin{aligned}\hat{u}_{ih} &= a_1 e^{-\kappa_1 z} + a_2 e^{-\kappa_2 z} + a_3 e^{-\kappa_3 z} + a_4 e^{\kappa_4 (L-z)} + a_5 e^{\kappa_5 (L-z)} + a_6 e^{\kappa_6 (L-z)} \\ \hat{u}_{oh} &= b_1 e^{-\kappa_1 z} + b_2 e^{-\kappa_2 z} + b_3 e^{-\kappa_3 z} + b_4 e^{\kappa_4 (L-z)} + b_5 e^{\kappa_5 (L-z)} + b_6 e^{\kappa_6 (L-z)} \\ \hat{u}_{gh} &= c_1 e^{-\kappa_1 z} + c_2 e^{-\kappa_2 z} + c_3 e^{-\kappa_3 z} + c_4 e^{\kappa_4 (L-z)} + c_5 e^{\kappa_5 (L-z)} + c_6 e^{\kappa_6 (L-z)}\end{aligned} \tag{8.78}$$

As this system of equations is coupled, the constants a, b and c are related. Equation (8.77) reveals that

$$a = \frac{-x_{13}}{x_{11}} c = R_{ig} c, \quad \text{and} \quad b = \frac{-x_{23}}{x_{22}} c = R_{og} c \tag{8.79}$$

Substituting (8.79) into (8.78) gives

$$\begin{aligned}\hat{u}_{ih} &= R_{ig1} c_1 e^{-\kappa_1 z} + R_{ig2} c_2 e^{-\kappa_2 z} + R_{ig3} c_3 e^{-\kappa_3 z} \\ &\quad + R_{ig4} c_4 e^{\kappa_4 (L-z)} + R_{ig5} c_5 e^{\kappa_5 (L-z)} + R_{ig6} c_6 e^{\kappa_6 (L-z)} \\ \hat{u}_{oh} &= R_{og1} c_1 e^{-\kappa_1 z} + R_{og2} c_2 e^{-\kappa_2 z} + R_{og3} c_3 e^{-\kappa_3 z} \\ &\quad + R_{og4} c_4 e^{\kappa_4 (L-z)} + R_{og5} c_5 e^{\kappa_5 (L-z)} + R_{og6} c_6 e^{\kappa_6 (L-z)} \\ \hat{u}_{gh} &= c_1 e^{-\kappa_1 z} + c_2 e^{-\kappa_2 z} + c_3 e^{-\kappa_3 z} + c_4 e^{\kappa_4 (L-z)} + c_5 e^{\kappa_5 (L-z)} + c_6 e^{\kappa_6 (L-z)}.\end{aligned} \tag{8.80}$$

Particular Equations

Having solved the homogeneous equations, we now turn to solving the particular equations. The relevant particular equations of (8.73) are

$$\begin{aligned}&\frac{v_w}{\alpha_w}\frac{d\hat{u}_{ip}}{dz} A_i - \frac{d^2\hat{u}_{ip}}{dz^2} A_i - \left(\frac{b_{ig} P_i}{\lambda_w} - \frac{A_i}{\alpha_w} i\omega\right)\hat{u}_{ip} + \frac{b_{ig} P_i}{\lambda_w}\hat{u}_{gp} = 0 \\ &-\frac{v_w}{\alpha_w}\frac{d\hat{u}_{op}}{dz} A_o - \frac{d^2\hat{u}_{op}}{dz^2} A_o - \left(\frac{b_{og}}{\lambda_w} P_o - \frac{A_o}{\alpha_w} i\omega\right)\hat{u}_{op} \\ &+ \frac{b_{og} P_o}{\lambda_w}\hat{u}_{gp} = 0 \\ &-\frac{d^2\hat{u}_{gp}}{dz^2} A_g - \left(\frac{b_{ig} P_i}{\lambda_g} + \frac{b_{og} P_o}{\lambda_g} + \frac{b_{gs} P_g}{\lambda_g} - \frac{A_g}{\alpha_g} i\omega\right) u_{gp} \\ &+ \frac{b_{ig} P_i}{\lambda_g}\hat{u}_{ip} + \frac{b_{og} P_o}{\lambda_g}\hat{u}_{op} = -\frac{b_{gs} P_g}{\lambda_g} u_s(r_b, z)\end{aligned} \tag{8.81}$$

where

$$\hat{u}_s(r_b, z) = \hat{u}_1(z, \omega_n) + \sum_m c_m \sin(\beta_m z) K_0(\kappa_r r_b)$$

with

$$\hat{u}_1(z,\omega_n) = \frac{\hat{U}_s(\omega_n)}{1+e^{-2\kappa h}}\left[e^{-\kappa z} + e^{-\kappa(2h-z)}\right]$$

$$c_m = \frac{2}{h}\int_0^L \hat{q}\,(z,\omega_n)\sin(\beta_m z)\,dz$$

$$\hat{q}\,(z,\omega_n) = \frac{b_{gs}\left[\hat{u}_1(z) - \hat{u}_g(z)\right]}{\lambda_s\kappa_r K_1(\kappa_r r_b) - b_{gs}K_0(\kappa_r r_b)}$$

$$\kappa = \sqrt{\frac{i\omega_n}{\alpha}},\quad \kappa_r = \sqrt{\frac{i\omega_n}{\alpha} + \beta_m^2},\quad \beta_m = \frac{\pi}{h}\left(m + \frac{1}{2}\right),$$

$$m = 0, 1, 2, \ldots$$

Since the nonhomogeneous term $\widehat{u}_s(r_b, z)$ is given in terms of exponential and trigonometric functions along z, the trial solutions of (8.81) can be expressed as

8

$$\begin{aligned}
\hat{u}_{ip} &= \sum_m \left[a_{1m}\cos(\beta_m z) + a_{2m}\sin(\beta_m z)\right] + a_3 e^{-\kappa z} + a_4 e^{-\kappa(2h-z)}\\
\hat{u}_{op} &= \sum_m \left[b_{1m}\cos(\beta_m z) + b_{2m}\sin(\beta_m z)\right] + b_3 e^{-\kappa z} + b_4 e^{-\kappa(2h-z)}\\
\hat{u}_{gp} &= \sum_m \left[c_{1m}\cos(\beta_m z) + c_{2m}\sin(\beta_m z)\right] + c_3 e^{-\kappa z} + c_4 e^{-\kappa(2h-z)}
\end{aligned} \tag{8.82}$$

Taking the first and second derivatives of (8.82) with respect to z, and substituting the results into (8.81) leads for:

Pipe-In

$$\begin{aligned}
&\frac{v_w A_i}{\alpha_w}\left\{\sum_m \left[\beta_m\{-a_{1m}\sin(\beta_m z) + a_{2m}\cos(\beta_m z)\}\right] - \kappa a_3 e^{-\kappa z} + \kappa a_4 e^{-\kappa(2h-z)}\right\}\\
&- A_i\left\{\sum_m \left[\beta_m^2\{-a_{1m}\cos(\beta_m z) - a_{2m}\sin(\beta_m z)\}\right] + \kappa^2 a_3 e^{-\kappa z} + \kappa^2 a_4 e^{-\kappa(2h-z)}\right\}\\
&- \left(\frac{b_{ig}P_i}{\lambda_w} - \frac{A_i i\omega}{\alpha_w}\right)\left\{\sum_m \left[a_{1m}\cos(\beta_m z) + a_{2m}\sin(\beta_m z)\right] + a_3 e^{-\kappa z} + a_4 e^{-\kappa(2h-z)}\right\}\\
&+ \frac{b_{ig}P_i}{\lambda_w}\left\{\sum_m \left[c_{1m}\cos(\beta_m z) + c_{2m}\sin(\beta_m z)\right] + c_3 e^{-\kappa z} + c_4 e^{-\kappa(2h-z)}\right\} = 0.
\end{aligned} \tag{8.83}$$

Pipe-Out

$$\begin{aligned}
&-\frac{v_w A_o}{\alpha_w}\left\{\sum_m \left[\beta_m\{-b_{1m}\sin(\beta_m z) + b_{2m}\cos(\beta_m z)\}\right] - \kappa b_3 e^{-\kappa z} + \kappa b_4 e^{-\kappa(2h-z)}\right\}\\
&- A_o\left\{\sum_m \left[\beta_m^2\{-b_{1m}\cos(\beta_m z) - b_{2m}\sin(\beta_m z)\}\right] + \kappa^2 b_3 e^{-\kappa z} + \kappa^2 b_4 e^{-\kappa(2h-z)}\right\}\\
&- \left(\frac{b_{og}P_o}{\lambda_w} - \frac{A_o i\omega}{\alpha_w}\right)\left\{\sum_m \left[b_{1m}\cos(\beta_m z) + b_{2m}\sin(\beta_m z)\right] + b_3 e^{-\kappa z} + b_4 e^{-\kappa(2h-z)}\right\}\\
&+ \frac{b_{og}P_o}{\lambda_w}\left\{\sum_m \left[c_{1m}\cos(\beta_m z) + c_{2m}\sin(\beta_m z)\right] + c_3 e^{-\kappa z} + c_4 e^{-\kappa(2h-z)}\right\} = 0.
\end{aligned} \tag{8.84}$$

Grout

$$-A_g\left\{\begin{array}{c}\sum_m [\beta_m^2\{-c_{1m}\cos(\beta_m z) - c_{2m}\sin(\beta_m z)\}]+\\ \kappa^2 c_3 e^{-\kappa z} + \kappa^2 c_4 e^{-\kappa(2h-z)}\end{array}\right\}$$
$$-\left(\frac{b_{ig}P_i}{\lambda_g} + \frac{b_{og}P_o}{\lambda_g} + \frac{b_{gs}P_g}{\lambda_g} - \frac{A_g i\omega}{\alpha_g}\right)\left\{\begin{array}{c}\sum_m [c_{1m}\cos(\beta_m z) + c_{2m}\sin(\beta_m z)]+\\ c_3 e^{-\kappa z} + c_4 e^{-\kappa(2h-z)}\end{array}\right\}$$
$$+\frac{b_{ig}P_i}{\lambda_g}\left\{\begin{array}{c}\sum_m [a_{1m}\cos(\beta_m z) + a_{2m}\sin(\beta_m z)]+\\ a_3 e^{-\kappa z} + a_4 e^{-\kappa(2h-z)}\end{array}\right\}$$
$$+\frac{b_{og}P_o}{\lambda_g}\left\{\begin{array}{c}\sum_m [b_{1m}\cos(\beta_m z) + b_{2m}\sin(\beta_m z)]+\\ b_3 e^{-\kappa z} + b_4 e^{-\kappa(2h-z)}\end{array}\right\} = -\frac{b_{gs}P_g}{\lambda_g}\hat{u}_s(r_b, z) \tag{8.85}$$

in which

$$\hat{u}_s(r_b, z) = \frac{\hat{U}_s(\omega_n)}{1 + e^{-2\kappa h}}[e^{-\kappa z} + e^{-\kappa(2h-z)}] + \sum_m Z_m \sin(\beta_m z) K_0(\kappa_r r_b), \quad m = 0, 1, 2, \ldots$$

with

$$Z_m = \int_0^L \hat{q}(z, \omega_n)\sin(\beta_m z)\, dz$$

and

$$\hat{q}(z, \omega_n) = W\left\{\frac{\hat{U}_s(\omega_n)}{1 + e^{-2\kappa h}}[e^{-\kappa z} + e^{-\kappa(2h-z)}] - \hat{u}_g(z)\right\}$$

with

$$W = \frac{2}{h}\cdot\frac{b_{gs}}{\lambda_s\kappa_r K_1(\kappa_r r_b) - b_{gs}K_0(\kappa_r r_b)}$$
$$\hat{u}_g(z) = \hat{u}_{gp}(z) + \hat{u}_{gh}(z)$$

where

$$\hat{u}_{gh}(z) = \sum_{j=1}^{3} C_j e^{-\kappa_j z} + \sum_{j=4}^{6} C_j e^{-\kappa_j(L-z)}$$
$$\hat{u}_{gp}(z) = \sum_s [c_{1s}\cos(\beta_s z) + c_{2s}\sin(\beta_s z)] + c_3 e^{-\kappa z} + c_4 e^{-\kappa(2h-z)}, \quad s = 0, 1, 2, \ldots \tag{8.86}$$

Simplify the equation of $\hat{u}_s(r_b, z)$ further, it can be expressed as

$$\hat{u}_s(r_b, z) = Z_{01}\, e^{-\kappa z} + Z_{02}\, e^{-\kappa(2h-z)} + \sum_m Z_m \sin(\beta_m z) K_0(\kappa_r r_b), \quad m = 0, 1, 2, \ldots$$

with

$$Z_{01} = \frac{\hat{U}_s(\omega_n)}{1 + e^{-2\kappa h}}, \quad Z_{02} = Z_{01} \tag{8.87}$$

Collecting terms of $e^{-\kappa z}$, $e^{-\kappa(L-z)}$, $\cos(\beta_m z)$ and $\sin(\beta_m z)$ from (8.83) to (8.86), and then equating the coefficients of like terms, we obtain 12 equations: 6 for the exponential terms, and 6 for the trigonometric terms, such that:

Exponential Terms

$$\begin{aligned}
&-\frac{v_w A_i}{\alpha_w}\kappa a_3 - A_i\kappa^2 a_3 - \left(\frac{b_{ig}P_i}{\lambda_w} - \frac{A_i i\omega}{\alpha_w}\right) a_3 + \frac{b_{ig}P_i}{\lambda_w} c_3 \\
&= 0 \\
&\frac{v_w A_o}{\alpha_w}\kappa b_3 - A_o\kappa^2 b_3 - \left(\frac{b_{og}P_o}{\lambda_w} - \frac{A_o i\omega}{\alpha_w}\right) b_3 \\
&\qquad + \frac{b_{og}P_o}{\lambda_w} c_3 = 0 \\
&- A_g\kappa^2 c_3 - \left(\frac{b_{ig}P_i}{\lambda_g} + \frac{b_{og}P_o}{\lambda_g} + \frac{b_{gs}P_g}{\lambda_g} - \frac{A_g i\omega}{\alpha_g}\right) c_3 \\
&+ \frac{b_{ig}P_i}{\lambda_g} a_3 + \frac{b_{og}P_o}{\lambda_g} b_3 = -\frac{b_{gs}P_g}{\lambda_g} Z_{01} \\
&\frac{v_w A_i}{\alpha_w}\kappa a_4 - A_i\kappa^2 a_4 - \left(\frac{b_{ig}P_i}{\lambda_w} - \frac{A_i i\omega}{\alpha_w}\right) a_4 + \frac{b_{ig}P_i}{\lambda_w} c_4 \\
&= 0 \\
&- \frac{v_w A_o}{\alpha_w}\kappa b_4 - A_o\kappa^2 b_4 - \left(\frac{b_{og}P_o}{\lambda_w} - \frac{A_o i\omega}{\alpha_w}\right) b_4 \\
&+ \frac{b_{og}P_o}{\lambda_w} c_4 = 0 \\
&- A_g\kappa^2 c_4 - \left(\frac{b_{ig}P_i}{\lambda_g} + \frac{b_{og}P_o}{\lambda_g} + \frac{b_{gs}P_g}{\lambda_g} - \frac{A_g i\omega}{\alpha_g}\right) c_4 \\
&+ \frac{b_{ig}P_i}{\lambda_g} a_4 + \frac{b_{og}P_o}{\lambda_g} b_4 = -\frac{b_{gs}P_g}{\lambda_g} Z_{02}.
\end{aligned} \tag{8.88}$$

Trigonometric Terms

$$
\begin{aligned}
&\frac{v_w A_i}{\alpha_w}(\beta_m a_{2m}) - A_i[-\beta_m^2 a_{1m}] - \left(\frac{b_{ig}P_i}{\lambda_w} - \frac{A_i i\omega}{\alpha_w}\right) a_{1m} \\
&\quad + \frac{b_{ig}P_i}{\lambda_w} c_{1m} = 0 \\
&-\frac{v_w A_o}{\alpha_w}(\beta_m\, b_{2m}) - A_o[-\beta_m^2 b_{1m}] - \left(\frac{b_{og}P_o}{\lambda_w} - \frac{A_o i\omega}{\alpha_w}\right) b_{1m} \\
&+ \frac{b_{og}P_o}{\lambda_w} c_{1m} = 0 \\
&- A_g[-\beta_m^2 c_{1m}] - \left(\frac{b_{ig}P_i}{\lambda_g} + \frac{b_{og}P_o}{\lambda_g} + \frac{b_{gs}P_g}{\lambda_g} - \frac{A_g i\omega}{\alpha_g}\right) c_{1m} \\
&+ \frac{b_{ig}P_i}{\lambda_g} a_{1m} + \frac{b_{og}P_o}{\lambda_g} b_{1m} = 0 \\
&\frac{v_w A_i}{\alpha_w}(-\beta_m\, a_{1m}) - A_i[-\beta_m^2 a_{2m}] - \left(\frac{b_{ig}P_i}{\lambda_w} - \frac{A_i i\omega}{\alpha_w}\right) b_{2m} \\
&\quad + \frac{b_{ig}P_i}{\lambda_w} c_{2m} = 0 \\
&-\frac{v_w A_o}{\alpha_w}(-\beta_m\, b_{1m}) - A_o[-\beta_m^2 b_{2m}] - \left(\frac{b_{og}P_o}{\lambda_w} - \frac{A_o i\omega}{\alpha_w}\right) b_{2m} \\
&+ \frac{b_{og}P_o}{\lambda_w} c_{2m} = 0 \\
&- A_g[-\beta_m^2 c_{2m}] - \left(\frac{b_{ig}P_i}{\lambda_g} + \frac{b_{og}P_o}{\lambda_g} + \frac{b_{gs}P_g}{\lambda_g} - \frac{A_g i\omega}{\alpha_g}\right) c_{2m} \\
&+ \frac{b_{ig}P_i}{\lambda_g} a_{2m} + \frac{b_{og}P_o}{\lambda_g} b_{2m} = -\frac{b_{gs}P_g}{\lambda_g} Z_m K_0(\kappa_r r_b).
\end{aligned}
\tag{8.89}
$$

General Solution

Solving (8.88) and (8.89) simultaneously yields the particular constants $a_3, a_4, b_3, b_4, c_3, c_4$ and $a_{1m}, a_{2m}, b_{1m}, b_{2m}, c_{1m}, c_{2m}$, for $m = 0,1,2,\ldots$. Having done that, the general solution of the BHE boundary value problem in the frequency domain yields

$$
\begin{aligned}
\hat{u}_i(z) &= \sum_{j=1}^{3} C_j R_{igj}\, e^{-k_j z} + \sum_{j=4}^{6} C_j R_{igj}\, e^{-k_j(z-L)} + \hat{u}_{ip}(z) \\
\hat{u}_o(z) &= \sum_{j=1}^{3} C_j R_{ogj}\, e^{-k_j z} + \sum_{j=4}^{6} C_j R_{ogj}\, e^{-k_j(z-L)} + \hat{u}_{op}(z) \\
\hat{u}_g(z) &= \sum_{j=1}^{3} C_j\, e^{-k_j z} + \sum_{j=4}^{6} C_j\, e^{-k_j(z-L)} + \hat{u}_{gp}(z)
\end{aligned}
$$

with

$$
\hat{u}_s(r,z) = Z_{01}\, e^{-\kappa z} + Z_{02}\, e^{-\kappa(2h-z)} + \sum_m [Z_m \sin(\beta_m z)]\, K_0(\kappa_r r), \quad m = 0,1,2,\ldots \tag{8.90}
$$

where $\widehat{u}_{ip}(z)$, $\widehat{u}_{op}(z)$ and $\widehat{u}_{gp}(z)$ are given in (8.82), Z_{01} and Z_{o2} are given in (8.87), and C_j are the homogeneous constants, see (8.80).

Boundary Conditions

Taking the derivatives of (8.90) with respect to z gives

$$\frac{d\hat{u}_i}{dz} = \frac{d\hat{u}_{ip}}{dz} - \kappa_1 R_{ig1} C_1 e^{-\kappa_1 z} - \kappa_2 R_{ig2} C_2 e^{-\kappa_2 z} - \kappa_3 R_{ig3} C_3 e^{-\kappa_3 z} - \kappa_4 R_{ig4} C_4 e^{\kappa_4 (L-z)} - \kappa_5 R_{ig5} C_5 e^{\kappa_5 (L-z)} - \kappa_6 R_{ig6} C_6 e^{\kappa_6 (L-z)}$$

$$\frac{d\hat{u}_o}{dz} = \frac{d\hat{u}_{op}}{dz} - \kappa_1 R_{og1} C_1 e^{-\kappa_1 z} - \kappa_2 R_{og2} C_2 e^{-\kappa_2 z} - \kappa_3 R_{og3} C_3 e^{-\kappa_3 z} - \kappa_4 R_{og4} C_4 e^{\kappa_4 (L-z)} - \kappa_5 R_{og5} C_5 e^{\kappa_5 (L-z)} - \kappa_6 R_{og6} C_6 e^{\kappa_6 (L-z)}$$

$$\frac{d\hat{u}_g}{dz} = \frac{d\hat{u}_{gp}}{dz} - \kappa_1 C_1 e^{-\kappa_1 z} - \kappa_2 C_2 e^{-\kappa_2 z} - \kappa_3 C_3 e^{-\kappa_3 z} - \kappa_4 C_4 e^{\kappa_4 (L-z)} - \kappa_5 C_5 e^{\kappa_5 (L-z)} - \kappa_6 C_6 e^{\kappa_6 (L-z)}$$

with

$$\frac{d\hat{u}_{ip}}{dz} = \sum_m \{\beta_m[-a_{1m}\sin(\beta_m z) + a_{2m}\cos(\beta_m z)]\} - \kappa a_3 e^{-\kappa z} + \kappa a_4 e^{-\kappa(2h-z)}$$

$$\frac{d\hat{u}_{op}}{dz} = \sum_m \{\beta_m[-b_{1m}\sin(\beta_m z) + b_{2m}\cos(\beta_m z)]\} - \kappa b_3 e^{-\kappa z} + \kappa b_4 e^{-\kappa(2h-z)}$$

$$\frac{d\hat{u}_{gp}}{dz} = \sum_m \{\beta_m[-c_{1m}\sin(\beta_m z) + c_{2m}\cos(\beta_m z)]\} - \kappa c_3 e^{-\kappa z} + \kappa c_4 e^{-\kappa(2h-z)}$$

where

$$\beta_m = \frac{\pi}{h}\left(m + \frac{1}{2}\right) \tag{8.91}$$

The boundary conditions in (8.73), in the frequency domain, implies:

At $z = 0$:

$$\begin{gathered}\sum_{j=1}^{3} C_j R_{igj} + \sum_{j=4}^{6} C_j R_{igj} e^{k_j L} = \hat{U}_{in}(\omega) - \hat{u}_{ip}(0)\\ -\sum_{j=1}^{3} \kappa_j R_{ogj} C_j - \sum_{j=1}^{3} \kappa_j R_{ogj} C_j e^{\kappa_j L} = -\left.\frac{d\hat{u}_{op}}{dz}\right|_{z=0}\\ -\sum_{j=1}^{3} \kappa_j C_j - \sum_{j=1}^{3} \kappa_j C_j e^{\kappa_j L} = -\left.\frac{d\hat{u}_{gp}}{dz}\right|_{z=0}\end{gathered} \tag{8.92}$$

At $z = L$:

$$\begin{aligned}
&\sum_{j=1}^{3} C_j \left(R_{igj} - R_{ogj}\right) e^{-k_j L} + \sum_{j=4}^{6} C_j \left(R_{igj} - R_{ogj}\right) \\
&\quad = \hat{u}_{op}(L) - \hat{u}_{ip}(L) \\
&\quad - \sum_{j=1}^{3} C_j \left(R_{igj} - R_{ogj}\right) e^{-\kappa_j L} - \sum_{j=1}^{3} C_j \left(R_{igj} - R_{ogj}\right) \\
&\quad = \left.\frac{d\hat{u}_{op}}{dz}\right|_{z=L} - \left.\frac{d\hat{u}_{ip}}{dz}\right|_{z=L} \\
&\quad - \sum_{j=1}^{3} \kappa_j C_j \, e^{-k_j L} - \sum_{j=4}^{6} \kappa_j C_j = - \left.\frac{d\hat{u}_{gp}}{dz}\right|_{z=L}
\end{aligned} \tag{8.93}$$

Solving the system of equations in (8.92) and (8.93) simultaneously gives the six constants $C_1 \cdots C_6$, which upon substituting into (8.90) yields the required solution of the boundary value problem in the frequency domain.

Staggered Algorithm

A staggered algorithm can be utilized to solve the coupled boundary value problems for the soil mass and the borehole heat exchanger (BHE). This algorithm facilitates the sequential solution of equations governing the thermal interactions between the soil and the BHE by iteratively updating the temperature distributions in each domain until convergence is achieved. An outline of this computational technique is given below.

(a) **First iteration**: We begin with computing the soil temperature using (8.72), while assuming the grout temperature $\widehat{u}_g(z)$ to be equal to the initial temperature, which is set to zero. The soil temperature distribution is computed in all grid points, including at the boundary with the borehole $\widehat{u}_s(r_b, z)$. Subsequently, the computed soil temperature at the borehole boundary is used as the input to determine the temperature distribution in the BHE components: $\widehat{u}_i$, $\widehat{u}_o$ and$\widehat{u}_g$.

(b) **Iteration to convergence**: The computed $\widehat{u}_g(z)$ from the first iteration is then used as input to recalculate the soil temperature in the domain. The process of recalculating the soil and BHE temperatures is repeated iteratively. In each iteration, the updated temperature of the grout is used to adjust the soil temperature calculation, and vice versa. This iterative process continues until convergence is achieved, meaning the changes in temperature distributions between successive iterations fall below a specified threshold, defined by the user.

By using a staggered approach, this algorithm effectively couples the thermal interactions between the soil mass and the borehole heat exchanger, ensuring that the interdependent temperature fields of the two domains are accurately captured.

Solution of BVP

Having solved this coupled boundary value problem in the frequency domain, the solution in the time domain can be attained via the inverse fast Fourier transform algorithm, such that

$$\begin{aligned} u_i(z,t) &= \sum_n \hat{u}_i(z,\omega_n)\, e^{i\omega_n t} \\ u_o(z,t) &= \sum_n \hat{u}_o(z,\omega_n)\, e^{i\omega_n t} \\ u_g(z,t) &= \sum_n \hat{u}_g(z,\omega_n)\, e^{i\omega_n t} \\ u_s(r,z,t) &= \sum_n \hat{u}_s(r,z,\omega_n)\, e^{i\omega_n t} \end{aligned}$$

■ **Example 8.2: Heat Flow in a U-tube Borehole Heat Exchanger**

A single U-tube borehole heat exchanger is installed in the ground to be utilized for direct space heating/cooling purposes. The BHE is 80 m in length and 0.15 m in diameter and is embedded in a 100 m deep soil layer. The inner diameter of pipe-in and pipe-out is 0.025 m with pipe material thickness of 0.003 m. ◻ Figure 8.11 shows a schematic representation of this system.

The U-tube carries a mixture of water and Ethylene glycol with thermal conductivity 0.56 $Wm^{-1}K^{-1}$, mass density 1052 kg m^{-3} and specific heat capacity 3795 J $kg^{-1}K^{-1}$. The mixture flows at a rate of 0.1 ms^{-1}. The grout has thermal conductivity of 1.5 $Wm^{-1}K^{-1}$, mass density 1420 kg m^{-3} and specific heat capacity 800 J $kg^{-1}K^{-1}$; and the soil has thermal conductivity of 2.2 $Wm^{-1}K^{-1}$, mass density 1680 kg m^{-3} and specific heat capacity 400 J $kg^{-1}K^{-1}$. The thermal interaction coefficients between pipe-in and grout, b_{ig}, and between pipe-out and grout, b_{og}, are both −15 $Wm^{-2}K^{-1}$; and the thermal interaction coefficient between the grout and the surrounding soil mass is −10 $Wm^{-2}K^{-1}$. (Note that we assigned negative values for the thermal interaction coefficients because we are addressing a cooling mode. See the note in the ▶ Sect. 8.3.)

The heat exchanger is operated for 2000 h during the spring season. Initially, both the borehole and the surrounding soil mass share an initial temperature of 0 °C. At the onset, the water mixture enters the inner pipe at a temperature of 25 °C and undergoes thermal exchange with the adjacent soil mass with a surface temperature maintained at 10 °C. The

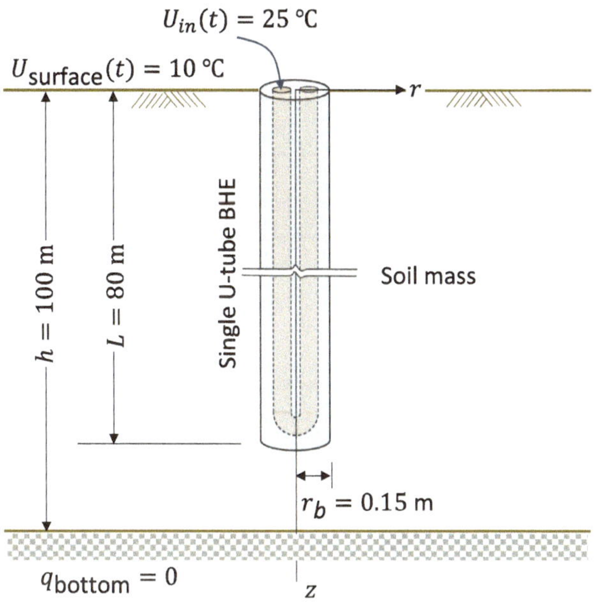

Fig. 8.11 A sketch of the geometry and boundary conditions for Example 9.2

time variations of these prescribed temperatures are shown in Fig. 8.12.

Find the temperature distribution versus time along the heat exchanger and in the soil mass at different times. Discuss the outcome.

Solution

As usual, we begin by transforming the input temperatures in Fig. 8.12 from the time domain to the frequency domain using the FFT algorithm. Here we used 4096 FFT samples with a time step of 3600 s. Subsequently, upon solving the system of Eq. (8.90) together with (8.92)–(8.93), and transforming the results back to the time domain, we obtain the temperature distribution along pipe-in, pipe-out, grout and soil mass.

Figure 8.13 shows the temperature variations over time in pipe-in and pipe-out along the depth of the BHE. In pipe-in, the temperature decreases from 25 °C at the top to approximately 19 °C at the bottom. This drop of temperature is due to its interaction with the soil mass, which is at deeper depths is still at 0 °C. The temperature at the bottom of pipe-in is then picked-up in pipe-out, where it continues to cool down along its height due to the continuous thermal interaction with pipe-in, grout and surrounding soil. The reached temperature at the top of pipe-out is approximately 15 °C, indicating that the working fluid has experienced a temperature drop of approximately 10 °C.

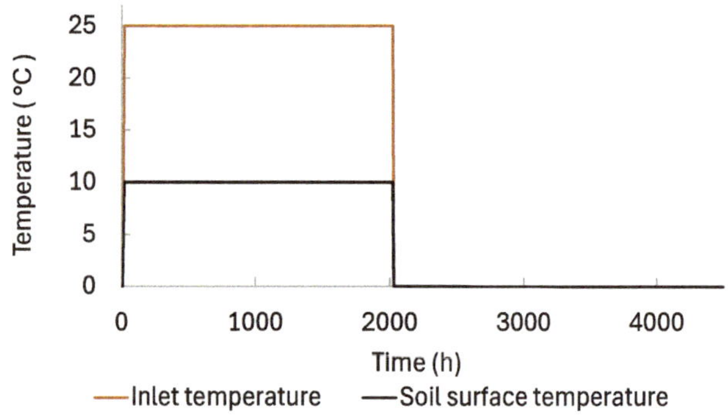

Fig. 8.12 Time variation of $\boldsymbol{U_{in}}$ and $\boldsymbol{U_{surface}}$

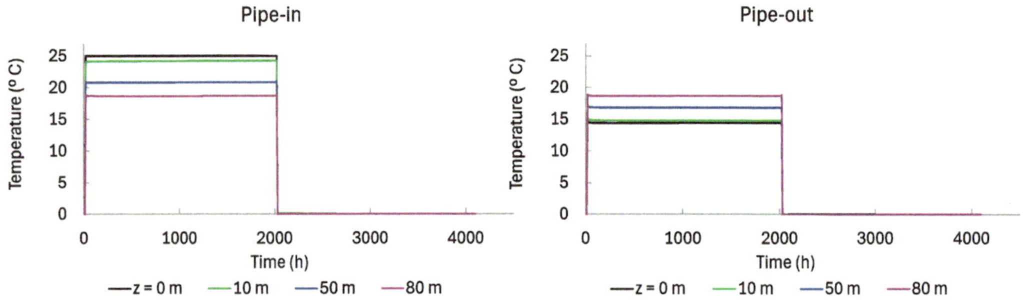

Fig. 8.13 Temperature variation in pipe-in and pipe out

Figure 8.14 shows the temperature variations over time in the grout and in the surrounding soil mass, 0.55 m away from the BHE boundary surface. At the top, the soil surface temperature is maintained at 10 °C. Since heat transfer in the soil mass is conductive and takes a relatively long time to reach deeper depths, the temperatures at these deeper levels are primarily influenced by the thermal interaction with the borehole. The grout is in direct contact with the soil mass and the heat pipes, and thus its temperature is influenced by a combination of all involved components. This interaction leads to a complex temperature profile within the grout, reflecting the heat exchange dynamics between the borehole and the surrounding soil. This interplay highlights the importance of considering the detailed geometry, materials and heat transfer mechanisms in the BHE system when analyzing its energy performance.

Figure 8.15 shows the contour plots of the soil mass temperature after 15 days and 80 days of operation. The figure depicts the temperature evolution over time, influenced by the thermal interplay between the prescribed surface temperature and the thermal interaction with the borehole.

Figure 8.16 presents the heat plume in the soil mass at 2 days and 10 days after system operation was halted and the surface temperature reached 0 °C.

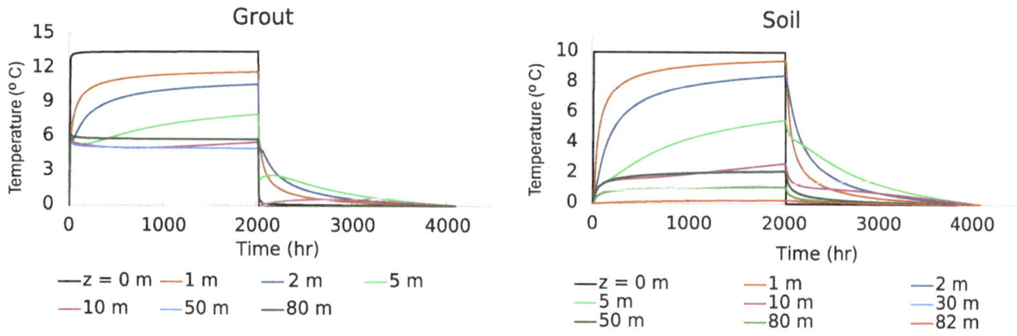

Fig. 8.14 Temperature variation in grout and soil

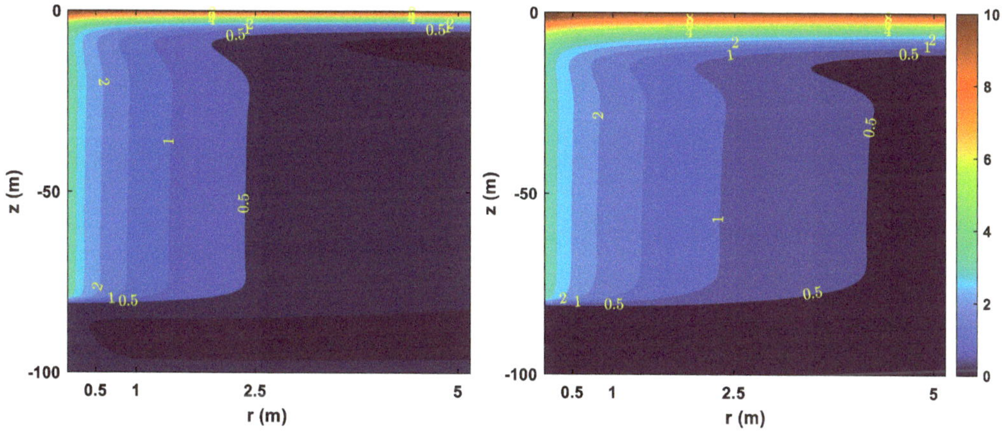

Fig. 8.15 Contour plots of soil temperature after 15 days (left) and 80 days (right) of operation

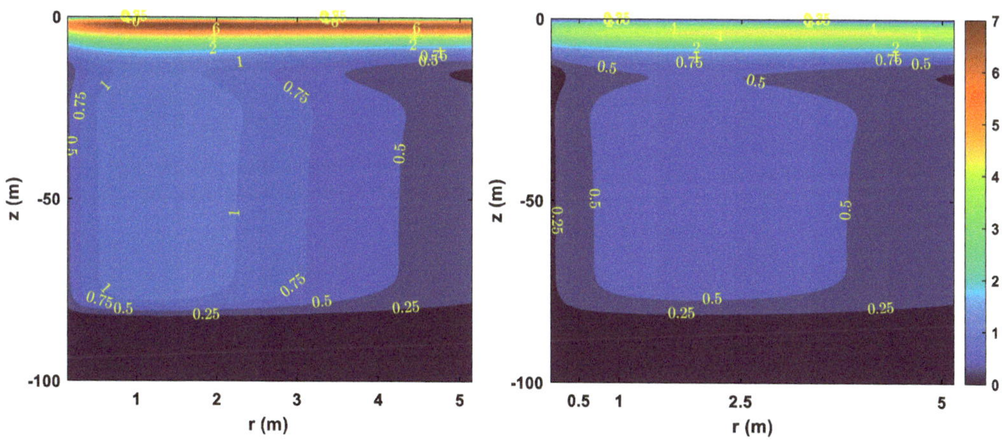

Fig. 8.16 Plume of heat after 2 days (left) and 10 days (right) of switching off the system

8.6 Double U-tube BHE Embedded in a Soil Mass

Consider a double U-tube BHE, consisting of five components (two pipes-in, denoted as $i1$ and $i2$; two pipes-out, denoted as $o1$ and $o2$; and grout, denoted as g), see Fig. 8.8. The governing heat equations of this borehole heat exchanger can be expressed as

Pipes-In

$$\frac{1}{\alpha_w}\frac{\partial u_{i1}}{\partial t}A_{i1} + \frac{v_w}{\alpha_w}\frac{\partial u_{i1}}{\partial z}A_{i1} - \frac{\partial^2 u_{i1}}{\partial z^2}A_{i1} = \frac{b_{ig1}}{\lambda_w}(u_{i1} - u_g)P_{i1}$$

$$\frac{1}{\alpha_w}\frac{\partial u_{i2}}{\partial t}A_{i2} + \frac{v_w}{\alpha_w}\frac{\partial u_{i2}}{\partial z}A_{i2} - \frac{\partial^2 u_{i2}}{\partial z^2}A_{i2} = \frac{b_{ig2}}{\lambda_w}(u_{i2} - u_g)P_{i2}.$$

Pipes-Out

$$\frac{1}{\alpha_w}\frac{\partial u_{o1}}{\partial t}A_{o1} - \frac{v_w}{\alpha_w}\frac{\partial u_{o1}}{\partial z}A_{o1} - \frac{\partial^2 u_{o1}}{\partial z^2}A_{o1} = \frac{b_{og1}}{\lambda_w}(u_{o1} - u_g)P_{o1}$$

$$\frac{1}{\alpha_w}\frac{\partial u_{o2}}{\partial t}A_{i2} - \frac{v_w}{\alpha_w}\frac{\partial u_{o2}}{\partial z}A_{o2} - \frac{\partial^2 u_{i2}}{\partial z^2}A_{o2} = \frac{b_{og2}}{\lambda_w}(u_{o2} - u_g)P_{o2}.$$

Grout

$$\frac{1}{\alpha_g}\frac{\partial u_g}{\partial t}A_g - \frac{\partial^2 u_g}{\partial z^2}A_g = \frac{b_{ig1}}{\lambda_g}(u_g - u_{i1})P_{i1} + \frac{b_{ig2}}{\lambda_g}(u_g - u_{i2})P_{i2}$$
$$+\frac{b_{og1}}{\lambda_g}(u_g - u_{o1})P_{o1} + \frac{b_{og2}}{\lambda_g}(u_g - u_{o2})P_{o2} + \frac{b_{gs}}{\lambda_g}[u_g - u_s(r_b)]P_g$$

where we observe that the grout works as an intermediate medium, facilitating the transfer of heat between the U-tubes and the soil mass. This general formulation describes heat transfer in boreholes with various U-tube sizes and materials. However, in practice, the U-tubes typically have the same size and are made of the same material. This means that the heat flow in pipe-in $i1$ is similar to pipe-in $i2$, and that of pipe-out $o1$ is similar to pipe-out $o2$. Considering this, we can simplify the governing heat equations to become

Pipes-In

$$\frac{1}{\alpha_w}\frac{\partial u_i}{\partial t}A_i + \frac{v_w}{\alpha_w}\frac{\partial u_i}{\partial z}A_i - \frac{\partial^2 u_i}{\partial z^2}A_i = \frac{b_{ig}}{\lambda_w}(u_i - u_g)P_i.$$

Pipes-Out

$$\frac{1}{\alpha_w}\frac{\partial u_o}{\partial t}A_o - \frac{v_w}{\alpha_w}\frac{\partial u_o}{\partial z}A_o - \frac{\partial^2 u_o}{\partial z^2}A_o = \frac{b_{og}}{\lambda_w}(u_o - u_g)P_o.$$

Grout

$$\frac{1}{\alpha_g}\frac{\partial u_g}{\partial t}A_g - \frac{\partial^2 u_g}{\partial z^2}A_g = 2\frac{b_{ig}}{\lambda_g}(u_g - u_i)P_i + 2\frac{b_{og}}{\lambda_g}(u_g - u_o)P_o + \frac{b_{gs}}{\lambda_g}[u_g - u_s(r_b)]P_g$$

where we notice that the grout is in thermal contact with two pipes-in and two pipes-out. An interested reader can solve and implement both formulations in a computer code and compare the validity of the second approach to the first.

Problems

1. Solve the boundary value problem in ▶ Sect. 8.1 using the general solution: $\hat{u}(x,\omega) = a_1 e^{-\kappa_1 x} + a_2 e^{\kappa_2 (L-x)}$.
2. Solve the boundary value problem in ▶ Sect. 8.1 by dividing it into two sub-problems, such that

   $u(z,t) = u_1(z,t) + u_2(z,t)$

   where

   Sub-problem 1

   $$\lambda\frac{\partial^2 u_1}{\partial z^2} - \rho c v\frac{\partial u_1}{\partial z} - \rho c\frac{\partial u_1}{\partial t} = 0$$

   IC: $u_1(z,0) = 0$

   BC: $u_1(0,t) = U_{in}(t)$

   $u_1(L,t) = 0.$

   Sub-problem 2

   $$\lambda\frac{\partial^2 u_2}{\partial z^2} - \rho c v\frac{\partial u_2}{\partial z} - \rho c\frac{\partial u_2}{\partial t} = 0$$

   IC: $u_2(z,0) = 0$

   BC: $u_2(0,t) = 0$

   $u_2(L,t) = U_{out}(t)$

   Then, compare the your solution with the one given in this section.
3. Solve the boundary value problem of ▶ Sect. 8.2 but for a Neumann boundary condition at $z = 0$.
4. Solve the following boundary value problem:

   $$\lambda\frac{\partial^2 u}{\partial x^2} - \rho c v\frac{\partial u}{\partial x} - \rho c\frac{\partial u}{\partial t} = b\left(U_0 e^{-x} - u\right)$$

   IC: $u(x,0) = 0$

   BC: $u(0,t) = U_{in}(t)$

   $$\left.\frac{\partial u}{\partial x}\right|_{x=L} = 0$$

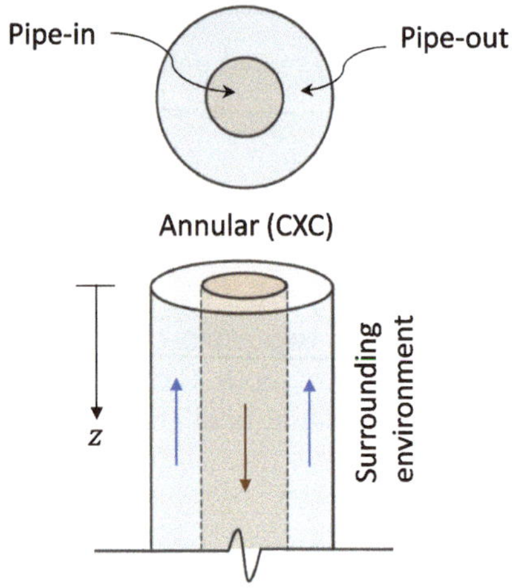

Fig. 8.17 Coaxial heat exchanger with centered inlet (CXC)

in which b is a thermal interaction coefficient, and U_0 is some temperature, a function of time.

5. Similar to the boundary value problem of ▶ Sect. 8.3, formulate the boundary value problem of a coaxial heat exchanger with centered inlet (CXC). This heat exchanger is similar to the CXA but in this type, pipe-out is the intermediate medium that transfers heat between pipe-in and the surrounding environment (see Fig. 8.17).
6. Similar to the boundary value problem of ▶ Sect. 8.5, formulate the boundary value problem of a U-tube borehole heat exchanger embedded in a soil mass and subjected to air temperature T_a. Then solve the resulting governing equations considering the parameters and input signals as those given in Example 8.2. Assume that the thermal interaction coefficient between the ground surface and the air is $b_{as} = -10$ units.
   Hint: Use the Newton's law of cooling to describe the soil surface boundary condition.

Further Readings

Al-Khoury R (2012) Computational modeling of shallow geothermal systems. CRC Press, Leiden, NL

BniLam N, Al-Khoury R (2018) A semi-analytical model for detailed 3D heat flow in shallow geothermal systems. Int J Heat Mass Transf 123:911–927

Erol S, Hashemi MA, François B (2015) Analytical solution of discontinuous heat extraction for sustainability and recovery aspects of borehole heat exchangers. Int J Therm Sci 88:47–58

Groundwater Flow

9 Groundwater Flow

R. Al Khoury and C. Kasbergen, *Analytical Subsurface Environment*,
https://doi.org/10.1007/978-3-031-93910-5_9

Studying groundwater flow is vital in civil and environmental engineering. It focuses on the movement and distribution of water in subsurface formations and its effect on the subsurface environment and engineering projects. Understanding groundwater movement plays a fundamental role in various engineering applications, including the design and management of water supply systems, groundwater remediation, geothermal energy extraction and assessment of environmental impacts. Basically, groundwater flow refers to the movement of water through porous geological formations such as soil, sand, and fractured rock. It primarily occurs as a result of the hydraulic gradient between regions in the ground and consolidation due to mechanical forces. These effects drive the groundwater to flow from areas of higher pore pressure to areas of lower pressure. The velocity of the groundwater flow is proportional to the hydraulic gradient and the property of the pores of the domain, commonly referred to as the permeability of the material. By studying the movement and behavior of groundwater, engineers can make informed decisions to optimize resource utilization, protect the environment, and ensure the safety and stability of infrastructure projects. In particular, understanding groundwater flow is essential for addressing geothermal heat transfer and contaminant transport, which are the core topics of this book.

In this chapter, we will provide a brief overview of groundwater flow and soil mechanics fundamentals. Our focus will primarily be on deriving the relevant conservation of mass equations. We will particularly emphasize linear problems that lend themselves to solution through the Fourier method.

9.1 Groundwater Flow in Subsurface Formations

The driving force for groundwater flow in subsurface formations is primarily the pore water pressure gradient. Pore water pressure refers to the pressure exerted by the water within the interconnected pore spaces of soil or rock. It plays a crucial role in determining the direction and velocity of groundwater movement. In the context of groundwater flow, *hydrostatic pressure* and *hydraulic pressure* are two related but distinct engineering concepts.

Hydrostatic pressure is the pressure exerted by a fluid at rest due to the force of gravity. In the context of groundwater, hydrostatic pore pressure, p_w (Pa), is determined by the depth and weight of the water column above a reference line or point as shown in ◘ Fig. 9.1. It is defined as $p_w = \rho g h$ in which ρ (kg m^{-3}) is the mass density of water, g (m s^{-2}) is the gravitational acceleration, and h (m) is the elevation of the water table from the reference line, also known as the *piezometric surface*. In this equation, ρg represents the weight (N) of water

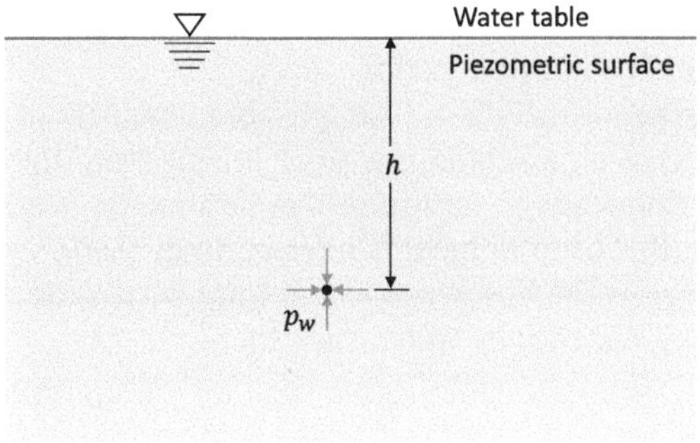

Fig. 9.1 Hydrostatic pressure

per unit volume. As the name suggests, hydrostatic pressure does not, by itself, cause the groundwater to flow.

The hydraulic pressure, on the other hand, refers to a pressure exerted by the fluid motion. It is related to the energy associated with the movement of water through the aquifer. Hydraulic pressure is influenced by factors such as the groundwater flow rate, the hydraulic conductivity of the aquifer, and the geometry.

In subsurface aquifers, the pore water pressure is measured by what is known as the *hydraulic head* or the *piezometric head* (m). The head is a measure that relates energy in an incompressible fluid to a height of a column of the fluid above some point in the aquifer. The *total head* h (m) is given by

$$h = h_{el} + h_p + h_v$$

in which:

h_{el}: the *elevation head*, a height above an arbitrary datum level;

h_p: the *pressure head*, a height of water in a scaled pipe, known as the *piezometer tube*. It is defined as $h_p = p/\rho g$, where p is the pore water pressure (we discarded the subscript w from p_w for simplicity of notation); and

h_v: the *velocity head*, a head of water generated due to its motion. It is defined as $h_v = v^2/2g$. However, since the groundwater flow is rather slow, this head is usually neglected in most groundwater flow studies. Thus, ignoring the velocity head and denoting the elevation head by z, the total head can then be expressed as

$$\begin{aligned} h &= h_{el} + h_p \\ &= z + \frac{p}{\rho g} \end{aligned} \tag{9.1}$$

which is typically the *piezometric head.* Figure 9.2 shows schematically this relationship.

The groundwater flow through pore channels in an aquifer is affected by a *retardation force* arising from the viscosity of water that leads to a drop in the piezometric head. Mathematically, this drop in head is referred to as the *hydraulic gradient*, which is the difference in the total head with respect to the traveled distance by water, such that

$$i = \frac{\Delta h}{\Delta L} \tag{9.2}$$

Darcy, in his well-known experiment, observed that the flow rate of water, Q ($m^3\,s^{-1}$), through a sand column with a cross-sectional area, A (m^2), is proportional to this hydraulic gradient, such that

$$q = \frac{Q}{A} \propto i \tag{9.3}$$

This quantity is referred to as *Darcy's velocity* ($m\,s^{-1}$). In a fully saturated porous medium, groundwater flow occurs within interconnected pore spaces. Since natural pores vary in geometry, the fluid flowing through them exhibits different velocities. As a result, the observed fluid velocity in an aquifer is in effect an average of the various velocities occurring within the pores. This averaged velocity is referred to as the *discharge velocity*, denoted as $q = \varphi v$, with φ the porosity.

The constant of proportionality in (9.3) is a parameter called the *coefficient of permeability*, K ($m\,s^{-1}$), also known as the *hydraulic conductivity*. It is defined as

$$K = \frac{\kappa \gamma_w}{\mu}$$

in which κ (m^2) is the *absolute permeability*, γ_w ($N\,m^{-3}$) is the unit weight of water, and μ (Pa . s) is its dynamic viscosity. Figure 9.3 depicts schematically Darcy's experimental setup.

Substituting (9.2) into (9.3) and considering the proportionality constant yields what is commonly referred to as *Darcy's law*, which states that

$$q = -K\frac{\Delta h}{\Delta L} \tag{9.4}$$

where the negative sign indicates that the flow occurs on the direction of decreasing total head.

Expressing (9.4) in its differential form in a three-dimensional domain reveals

$$\mathbf{q} = -\mathbf{K} \cdot \nabla h \tag{9.5}$$

where the quantitates **q** and **K** are tensors that may have different magnitudes on different directions. This entails that

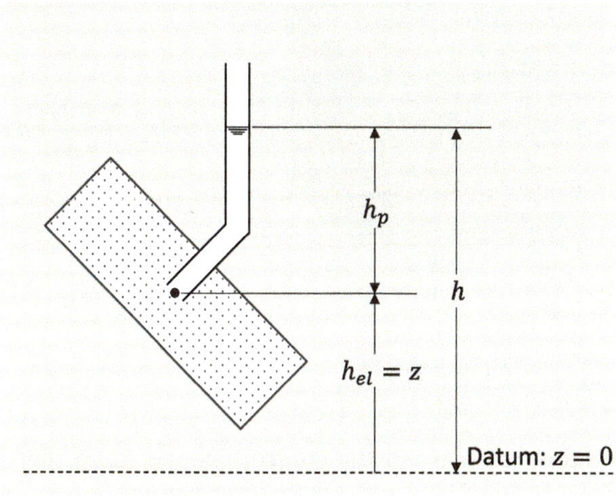

Fig. 9.2 Piezometric head

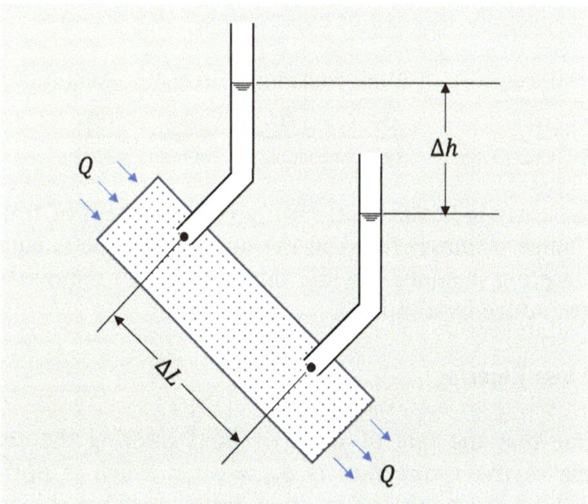

Fig. 9.3 Darcy's experimental setup

Darcy's law for an anisotropic medium can be expressed in the Cartesian coordinate system as

$$q_x = -K_x \frac{\partial h}{\partial x}, \quad q_y = -K_y \frac{\partial h}{\partial y}, \quad q_z = -K_z \frac{\partial h}{\partial z}$$

9.2 Conservation of Mass

Consider the control volume, *dxdydz*, shown in Fig. 9.4. Groundwater entering this control volume might accumulate, leave, and produced via a source or removed by a sink. The

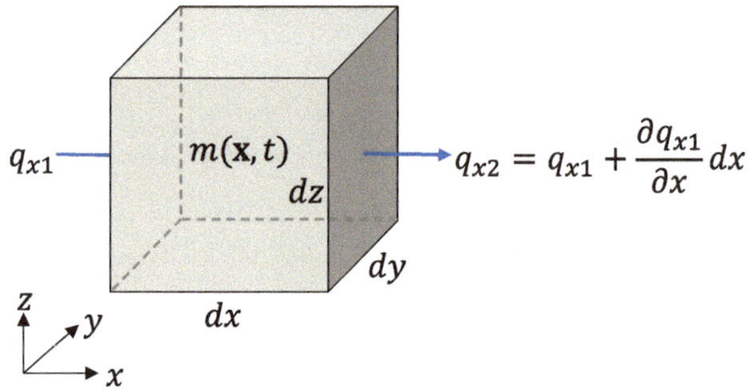

Fig. 9.4 Control volume

conservation of mass in the control volume states that there must be a balance between a fluid entering the control volume and that accumulated, generated and leaving it. This implies

Mass accumulation rate = net entering mass flow rate + net mass production

Mathematically, this can be expressed as

$$\underbrace{\frac{\Delta m_{\mathrm{ac}}}{\Delta t}}_{\text{mass accumulation}} = \underbrace{\frac{\Delta m_{\mathrm{in}}}{\Delta t} - \frac{\Delta m_{\mathrm{out}}}{\Delta t}}_{\text{net mass entering}} + \underbrace{\frac{\Delta m_{\mathrm{pr}}}{\Delta t} - \frac{\Delta m_{\mathrm{re}}}{\Delta t}}_{\text{net mass production}}$$

which clearly indicates that, for a given increment of time Δt, the change in mass stored in the control volume is equal to that entering it minus leaving, minus (or plus) removed by a sink (or added by a source).

Net Mass Entering

Assume that the rate of mass of fluid entering the control volume in the x–direction is $q_{x1} = \varphi v_{x1}$, which is normally referred to as the mass flux in x–direction. Then, using the Taylor series, the balance of mass in the control volume can be described as

$$\rho q_{x1} dydz = \rho \left(q_{x1} + \frac{\partial q_{x1}}{\partial x} dx \right) dydz \tag{9.6}$$

The net mass flux in the control volume in the x-direction is therefore

$$-\rho \frac{\partial q_{x1}}{\partial x} dxdydz \tag{9.7}$$

Similarly, the net mass flux crossing the other two directions are

$$-\rho \frac{\partial q_{y1}}{\partial y} dxdydz, \quad -\rho \frac{\partial q_{z1}}{\partial z} dxdydz \tag{9.8}$$

Summing (9.4.8) and (9.4.9), the net mass flux in the control volume is hence

$$-\rho\left(\frac{\partial q_{x1}}{\partial x}+\frac{\partial q_{y1}}{\partial y}+\frac{\partial q_{z1}}{\partial z}\right)dV, \quad \text{or} \quad -\rho\nabla\cdot\mathbf{q}dV \tag{9.9}$$

where $dV = dxdydz$.

Accumulated Mass

The accumulated mass in the control volume for a porous medium domain is given by

$$\frac{\partial(\varphi\rho)}{\partial t}dV$$

Expanding this derivative using the chain rule, we obtain

$$\frac{\partial(\varphi\rho)}{\partial t}=\rho\frac{\partial\varphi}{\partial t}+\varphi\frac{\partial\rho}{\partial t} \tag{9.10}$$

The first term on the right-hand side of this equation represents the mass rate of water generated by the compaction of the porous medium as its porosity φ changes in response to variations in pore pressure. This entails that

$$\frac{\partial\varphi}{\partial t}=\frac{\partial\varphi}{\partial p}\frac{\partial p}{\partial t} \tag{9.11}$$

where p (Pa) is the pore pressure. Note that, we here assumed that there is no change in the total stress on the porous medium, giving

$$d\sigma' = -dp$$

in which $\sigma\prime$ (Pa) is the effective stress. The first term on the right-hand side of (9.11) is related to the *compressibility* of the solid matrix m_v, also referred to as the *compaction coefficient*, defined as

$$m_v = -\frac{1}{V_s}\frac{\partial V_s}{\partial p} = -\frac{\partial\varphi}{\partial p}$$

Thus

$$\frac{\partial\varphi}{\partial t}=-m_v\frac{\partial p}{\partial t} \tag{9.12}$$

The second term on the right-hand side of (9.10) represents the mass rate of water produced by the expansion of the water under a change in its density ρ. This entails that

$$\frac{\partial\rho}{\partial t}=\frac{\partial\rho}{\partial p}\frac{\partial p}{\partial t} \tag{9.13}$$

where the first term on the right-hand side of this equation is related to the compressibility of the water β, defined as

$$\beta = -\frac{1}{\rho}\frac{\partial \rho}{\partial p}$$

Thus, (9.13) can be written as

$$\frac{\partial \rho}{\partial t} = -\rho\beta\frac{\partial p}{\partial t} \tag{9.14}$$

Substituting (9.12) and (9.14) into (9.10) yields

$$\begin{aligned}\frac{\partial (\varphi\rho)}{\partial t} &= -\rho m_v\frac{\partial p}{\partial t} - \varphi\rho\beta\frac{\partial p}{\partial t}\\ &= -\rho(m_v + \varphi\beta)\frac{\partial p}{\partial t}\end{aligned} \tag{9.15}$$

Referring to (9.1), we established that the total head is related to the pore pressure, thus allowing expressing (9.15) as

$$\frac{\partial (\varphi\rho)}{\partial t} = -\rho(m_v + \varphi\beta)\rho g\frac{\partial h}{\partial t}$$

In hydrology, this relationship is written as

9

$$\frac{\partial (\varphi\rho)}{\partial t} = -\rho S_s\frac{\partial h}{\partial t} \tag{9.16}$$

in which $S_s = \rho g(m_v + \varphi\beta)$ (m^{-1}) is the *specific storativity*, also known as the *specific storage*, for a saturated domain. It is defined as the amount of water added to, or removed from, a control volume per unit drop of piezometric head. It combines the effects of the compressibility of the solid matrix and the compressibility of water.

Net production

The net production represents mass rate that is added to or removed from the control volume, expressed as

$$\dot{Q} \tag{9.17}$$

The net production rate (kg s^{-1}) can be due to a source or a sink, but also can be a result of leakage from one or more boundary surfaces of the aquifer.

9.3 Groundwater Flow Balance Equation

Collecting the above three conserved quantities, the macroscopic mass balance equation for a fully saturated porous medium domain is given by

$$\underbrace{-\rho S_s \partial h/\partial t dV}_{\text{accumulation}} = \underbrace{-\rho\nabla\cdot\mathbf{q}dV}_{\text{net mass flux}} + \underbrace{\dot{Q}}_{\text{production rate}} \tag{9.18}$$

Using Darcy's law (9.5), the mass balance equation can then be expressed as

$$\underbrace{S_s \partial h/\partial t}_{\text{accumulation}} = \underbrace{\nabla \cdot (\mathbf{K} \cdot \nabla h)}_{\text{net mass flux}} - \underbrace{\dot{m}}_{\text{production rate}} \tag{9.19}$$

where we used $\dot{m} \equiv \dot{Q}/\rho dV$, ($s^{-1}$). It is important to note that the sign of the production rate may be either negative or positive, depending on whether there is a source or sink present.

The balance Eq. (9.19) is commonly used for describing groundwater flow in 3D saturated porous media domains. For a homogeneous domain, with $\mathbf{K} = K$, it is usually given by

$$\frac{1}{\alpha}\frac{\partial h}{\partial t} = \nabla \cdot \nabla h - \frac{\dot{m}}{K} \tag{9.20}$$

where $\alpha = K/S_s$ ($m^2\ s^{-1}$) is the *diffusivity* of the porous medium. In other contexts, it is also given by

$$\frac{S}{T}\frac{\partial h}{\partial t} = \nabla \cdot \nabla h - \frac{\dot{m}}{K} \tag{9.21}$$

in which $S = S_s B$, a dimensionless *storage coefficient* with B (m) being the thickness of the aquifer, and $T = KB$ ($m^2\ s^{-1}$) is the *transmissivity* of the aquifer.

Depending on the problem, it is sometime more convenient to express (9.21) in terms of the drawdown $s = h_0 - h$, with h_0 the initial piezometric head, such that

$$\frac{S}{T}\frac{\partial s}{\partial t} = \nabla \cdot \nabla s \tag{9.22}$$

where, here, we ignored the source term.

In (9.21), the operator $\nabla . \nabla = \nabla^2$ is the *Laplacian*, which is, in the Cartesian coordinate system, defined as

$$\nabla^2 \equiv \frac{\partial^2}{\partial x^2} + \frac{\partial^2}{\partial y^2} + \frac{\partial^2}{\partial z^2}$$

In the cylindrical coordinate system, the Laplacian is

$$\nabla^2 \equiv \frac{\partial^2}{\partial r^2} + \frac{1}{r}\frac{\partial}{\partial r} + \frac{1}{r^2}\frac{\partial^2}{\partial \theta^2} + \frac{\partial^2}{\partial z^2}, \quad \text{or}$$

$$\nabla^2 \equiv \frac{1}{r}\frac{\partial}{\partial r}\left(r\frac{\partial}{\partial r}\right) + \frac{1}{r^2}\frac{\partial^2}{\partial \theta^2} + \frac{\partial^2}{\partial z^2}$$

For an axial symmetric domain, the Laplacian is given by

$$\nabla^2 \equiv \frac{\partial^2}{\partial r^2} + \frac{1}{r}\frac{\partial}{\partial r} + \frac{\partial^2}{\partial z^2}, \quad \text{or}$$

$$\nabla^2 \equiv \frac{1}{r}\frac{\partial}{\partial r}\left(r\frac{\partial}{\partial r}\right) + \frac{\partial^2}{\partial z^2}$$

and for an axial symmetric, planar 2D domain, it is given by

$$\nabla^2 \equiv \frac{\partial^2}{\partial r^2} + \frac{1}{r}\frac{\partial}{\partial r}, \quad \text{or} \quad \nabla^2 \equiv \frac{1}{r}\frac{\partial}{\partial r}\left(r\frac{\partial}{\partial r}\right)$$

Equation (9.21) and its derivatives are commonly utilized for modeling groundwater flow problems in confined, unconfined, and leaky aquifers.

Confined Aquifer

A confined aquifer is a fully saturated subsurface layer bounded by impermeable formations, known as *aquicludes*. The confinement exerts a pressure higher than the atmospheric pressure such that when a well is installed in a confined aquifer, the water arises above the top boundary of the aquifer. Figure 9.5 shows schematically a confined aquifer, together with other kind of aquifers.

Unconfined Aquifer

An unconfined aquifer, also known as *water table aquifer*, is a sublayer where the water table is at atmospheric pressure. Depending on the weather conditions in terms of rainfalls or drought, and on the usage of the groundwater in the aquifer, the water table can rise or fall.

Leaky Aquifer

A leaky aquifer, also known as a *semi-confined aquifer*, is a sublayer bounded on the upper and lower boundaries by aquitards, or one boundary is an aquitard, and the other is an aquiclude. Leaky aquifers occur more frequently in nature as aquicludes are rarely completely impermeable, allowing some groundwater to leak to its bordering aquifers.

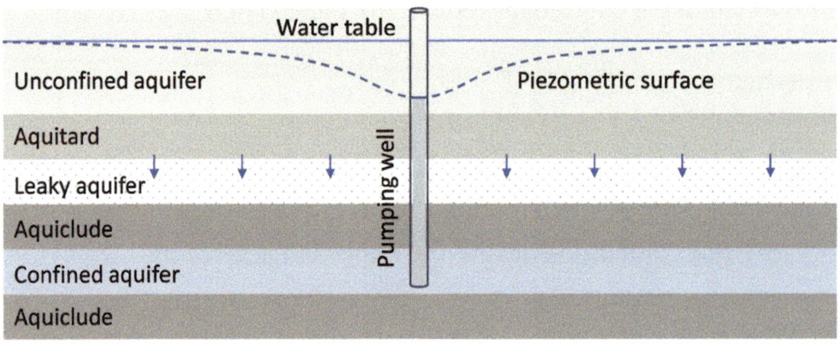

Fig. 9.5 Various types of subsurface aquifers

9.4 Initial and Boundary Conditions

The dependent variables used in formulating the groundwater flow problems are the piezometric head h, the drawdown head s, or the pore pressure. Commonly, formulations involving h or s with their relevant initial and boundary conditions are employed. The pressure formulation is rarely used because pressure at the boundaries is usually not known, except for boundaries in contact with water bodies. Yet, for a constant mass density, h can be readily calculated from the pressure.

Understanding and appropriately defining the initial and boundary conditions are essential for predicting the behavior of groundwater flow and its impact on various engineering applications. In the following discussion, we explicitly address initial and boundary conditions for the piezometric head h, but these considerations are equally applicable to the drawdown head s.

Initial Conditions

The initial condition refers to the state of the groundwater system at the beginning of the simulation or modeling period. It represents the starting point from which the groundwater flow evolves over time. Essentially, the initial condition implies prescribing h or s values at time $t = 0$, such that

$$h\,(x, 0) = f(x)$$

where $f(x)$ can be a constant value or a function of x. However, when using spectral analysis, which involves transforming time into frequency, the initial condition does not play a role in the solution. Thus, in this context, the initial condition is kept as

$$h\,(x, 0) = 0$$

Nevertheless, as the problems that we are dealing with are linear, in most cases, the initial condition can be added algebraically to the computed values during the post-processing.

Boundary Conditions

The boundary conditions elucidate how the groundwater system interacts with its surroundings for $t > 0$. They specify the inflow, outflow, or interaction of water within the system with the surrounding environment. The relevant boundary conditions for groundwater flow problems encompass Dirichlet, Neumann, and Robin conditions.

Dirichlet Boundary Condition: Prescribed Piezometric Head

The Dirichlet boundary condition, also known as the boundary condition of the *first kind* or the *essential* boundary condition, implies prescribing a piezometric head $h(x,t)$ on some points or surface areas on the boundary of a domain. In saturated groundwater flow problems, this boundary condition is typically given by

$$h\left(x_i,t\right)=H(t), \quad \text{or} \quad h\left(x_i,t\right)=h_0(t)$$

where $H(t)$ is some value of the piezometric head at $x=x_i$ that may be constant or time dependent.

For a 1D domain spanning in the range $0\le x\le L$ the Dirichlet boundary conditions might be given by

$$h\left(0,t\right)=H_1(t), \quad h\left(L,t\right)=H_2$$

where, in this case, $H_1(t)$ is a time-dependent head at $x=0$, and H_2 is a constant head at $x=L$.

Neumann Boundary Condition: Prescribed Water Flux or Water Discharge

9

The Neumann boundary condition, also known as the boundary condition of the *second kind* or the *natural* boundary condition, implies applying the value of the spatial derivative of the dependent variable (h, s or p) on some points or surface areas on the boundary of a domain.

In groundwater flow problems, the Neumann boundary condition implies specifying the value of water flux per unit area per unit time across a surface area on the boundary. It is applied with respect to an outward unit normal vector perpendicular to the surface **n**, such that

$$-(\mathbf{K}\cdot\nabla h)\cdot\mathbf{n}=q$$

where **K** (m s^{-1}) is the hydraulic conductivity vector and q (m s^{-1}) is some prescribed mass flux.

Consider the one-dimensional porous medium domain shown in Fig. 9.6. The Neumann boundary conditions on the end boundaries of this domain are

where K_x is the hydraulic conductivity of the domain in the

$$-K_x\left.\frac{\partial h}{\partial x}\right|_{x=0}=q_1(t)\cdot n_x, \quad -K_x\left.\frac{\partial h}{\partial x}\right|_{x=L}=q_2\cdot n_x$$

x-direction. In this case, $q_1(t)$ is a time-dependent mass flux at $x=0$, and q_2 is constant at $x=L$. In 0, n_x on the left-hand side of the domain is -1 because its direction is opposite to the direction of the x-axis, and its magnitude on the right-hand side is $+1$. If a specified water flux is set to enter the domain at $x=0$ and another to leave the domain at $x=L$, as that depicted in Fig. 9.6, then the Neumann boundary conditions are expressed as

$$K_x\left.\frac{\partial h}{\partial x}\right|_{x=0}=q_1(t), \quad -K_x\left.\frac{\partial h}{\partial x}\right|_{x=L}=q_2$$

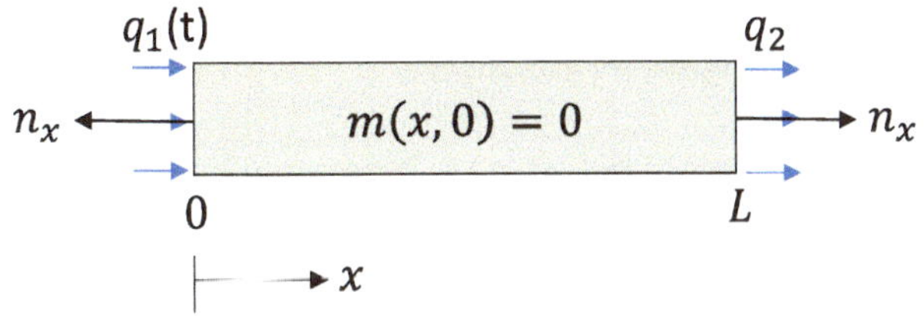

Fig. 9.6 Neumann boundary conditions

If the surface area at $x = 0$ is effectively impervious that allows no water flow, then $q = 0$, and the boundary condition on this location is given by

$$\left.\frac{\partial h}{\partial x}\right|_{x=0} = 0$$

Robin Boundary Condition: Semipervious Boundary

The Robin boundary condition, also known as the boundary condition of the *third kind*, corresponds to a linear combination of the value of the dependent variable and its derivative on the boundary. This refers to imposing a linear function combining a Dirichlet and a Neumann boundary condition on a specific point or area of the boundary.

In groundwater flow problems, this kind of boundary conditions occurs when there exists a sublayer with relatively small thickness, B', that is fully saturated and behaves as a semipervious membrane with flow *resistance* c_r (s). Aquitards are examples of such layers, see Fig. 9.5. This boundary condition is given by

$$q = \frac{h - h_0}{c_r}, \quad c_r = \frac{B'}{K'}$$

in which h_0 is the piezometric head in the semipervious membrane, assumed constant, and K' is its hydraulic conductivity. The reciprocal of c_r is referred to as the *leakance*.

For a 1D domain $[0, L]$, the Robin boundary condition at the two ends of the domain implies

$$-K_x \left.\frac{\partial h}{\partial x}\right|_{x=0} = \frac{h(0,t) - h_0(t)}{c_r} \cdot n_x,$$
$$-K_x \left.\frac{\partial h}{\partial x}\right|_{x=L} = \frac{h(L,t) - h_0(t)}{c_r} \cdot n_x$$

where h_0 is some piezometric head at the boundaries that can be constant or time dependent, and n_x can be $+1$ or -1, depending on the direction of leakage.

Homogeneous Boundary Conditions

The above given Dirichlet and Neumann boundary conditions are nonhomogeneous since they are not identically zero. Should these boundary conditions be

$$h(x_i, t) = 0, \quad \frac{\partial h}{\partial n} = 0$$

they are said to be homogeneous. Similarly, the above given Robin boundary condition is nonhomogeneous, but if $h_0 = 0$, i.e.:

$$-K\frac{\partial h}{\partial n} - \frac{h(x_i, t)}{c_r} = 0$$

it becomes homogeneous at $x = x_i$.

9.5 Consolidation Theory

9

Groundwater flow is not solely driven by hydraulic pressure gradients, it can also result from mechanical forces inducing groundwater movement due to mechanical pore pressure generation. This phenomenon is explained by the theory of consolidation, pioneered by Terzaghi, alongside Fillunger, see, for example, Lambe and Whitman (1969) and Berry and Reid (1987). Essentially, this theory postulates that external forces exerted by buildings or embankments initially compel the nearly incompressible groundwater to bear the entire load. This load generates *excess pore pressure*, driving water out of the loaded area. This process persists until all excess pore pressure dissipates. In highly permeable soils like sand and gravel, consolidation occurs almost instantly, while in low permeability soils like clay, this process can take years.

The consolidation theory entails that there is no storage of mass, rather, all excess pore pressure induced by the external load must dissipate over time. As excess pore pressure dissipates, the pores shrink, resulting in a change in volume. For the control volume depicted in Fig. 9.4, the continuity of water volume entails

$$\text{Inflow per time} - \text{outflow per time} = \text{rate of volume change}$$

which clearly indicates that the formulation of the governing equations for consolidation can closely be followed from the mass balance equation given in ► Sect. 9.3. Recall the mass balance Eq. (9.19), without a source term,

$$\nabla \cdot (\mathbf{K} \cdot \nabla h) = S_s \frac{\partial h}{\partial t}$$

where h is the piezometric head. The left-hand side term in this equation expresses the net inflow and outflow of mass. Terzaghi's consolidation theory assumes that the porous medium domain is fully saturated, implying that the volume of water squeezed out of the voids is equal to the change of volume of the control volume. Following this, the mass balance equation can be modified to express the volume rate change, such that

$$\nabla \cdot (\mathbf{K} \cdot \nabla h) = \frac{1}{V}\frac{\partial V}{\partial t}$$

where V (m^3) is the volume of the control volume.

Recall that $V = V_s + V_v$, where V_s is the volume of the solid phase and V_v is the volume of the voids (pores). The consolidation theory assumes that the water and the soil grains are incompressible, entailing that the volume change dV is in effect the change of the voids dV_v, leading to

$$\nabla \cdot (\mathbf{K} \cdot \nabla h) = \frac{1}{V}\frac{\partial V_v}{\partial t} \tag{9.23}$$

The original consolidation theory of Terzaghi was initially derived for one-dimensional domains, and as such, (9.23) may be expressed as

$$k_v \frac{\partial^2 h}{\partial z^2} = \frac{1}{V}\frac{\partial V_v}{\partial t} \tag{9.24}$$

where k_v (m s^{-1}) is the vertical permeability (in the z-axis).

If e denotes the void ratio and V_s the volume of the solid phase, then by definition $V_v = eV_s$. Substituting this relationship into (9.24), and noting that the solid volume V_s is constant, it gives

$$k_v \frac{\partial^2 h}{\partial z^2} = \frac{V_s}{V}\frac{\partial e}{\partial t}$$

And, knowing that $V_s/V = 1 + e$, we obtain

$$k_v \frac{\partial^2 h}{\partial z^2} = \frac{1}{1+e}\frac{\partial e}{\partial t} \tag{9.25}$$

Furthermore, this equation can be modified by expressing the total head in terms of its components, see (9.1),

$$k_v \frac{\partial^2 (h_{el} + h_p)}{\partial z^2} = \frac{1}{1+e}\frac{\partial e}{\partial t}$$

but as the elevation head h_{el} is constant, this equation is expressed in terms of the hydraulic head, as

$$k_v \frac{\partial^2 h_p}{\partial z^2} = \frac{1}{1+e}\frac{\partial e}{\partial t} \tag{9.26}$$

where

$$h_p = \frac{p}{\rho_w g} = \frac{u_h + u_e}{\gamma_w}$$

in which ρ_w (kg m^{-3}) is the water mass density and g (m s^{-2}) is the gravitational acceleration, γ_w (N m^{-3}) is the unit weight, and p is the pore pressure, which can be delineated into its components: $u_h = \gamma_w z$, the hydrostatic pore pressure and u_e, the excess pore pressure that arises from the consolidation process. In the equilibrium condition, u_h varies linearly with depth, so that $\partial^2 u_h/\partial z^2 = 0$, leading (9.26) to become

$$\frac{k_v(1+e)}{\gamma_w}\frac{\partial^2 u_e}{\partial z^2} = \frac{\partial e}{\partial t} \tag{9.27}$$

To complete this equation, we need a constitutive relationship relating the void ratio and the *effective stresses* in the soil mass. Using the chain rule, the rate of change of the void volume can be expressed as a function of the effective stress σ' as

$$\frac{\partial e}{\partial t} = \frac{\partial e}{\partial \sigma'}\frac{\partial \sigma'}{\partial t}$$

Terzaghi proposed a linear relationship between the effective stress and the void ratio, such that

9

$$a_v = -\frac{\partial e}{\partial \sigma'}$$

where a_v (m^2 N^{-1}) is referred to as the *coefficient of compressibility*. This entails

$$\frac{\partial e}{\partial t} = -a_v \frac{\partial \sigma'}{\partial t} \tag{9.28}$$

In a porous medium domain, the effective stress and the pore water pressure $u = u_e + u_h$ are related via the total stress, such that

$$\sigma = \sigma' + u_e + u_h$$

where u_h is the hydrostatic pore water pressure. Assuming that the total stress and the hydrostatic pressure are not affected by the consolidation process, the rate of change of the stresses in the soil mass is then

$$\frac{\partial \sigma'}{\partial t} + \frac{\partial u_e}{\partial t} = 0$$

This entails

$$\frac{\partial \sigma'}{\partial t} = -\frac{\partial u_e}{\partial t}$$

Thus, (9.28) can be expressed as

$$\frac{\partial e}{\partial t} = a_v \frac{\partial u_e}{\partial t} \tag{9.29}$$

Substituting this relationship into (9.27) leads to

$$\frac{k_v(1+e)}{\gamma_w a_v}\frac{\partial^2 u_e}{\partial z^2} = \frac{\partial u_e}{\partial t} \tag{9.30}$$

In soil engineering, this equation is expressed in a more compact form, such as

$$c_{\mathrm{v}} \frac{\partial^2 u_e}{\partial z^2} = \frac{\partial u_{\mathrm{e}}}{\partial t} \tag{9.31}$$

where

$$c_{\mathrm{v}} = \frac{k_v(1+e)}{\gamma_w a_{\mathrm{v}}} \tag{9.32}$$

which is referred to as the *coefficient of vertical consolidation* ($m^2\ s^{-1}$). In some literature (9.31) is written as

$$\frac{k_v}{\gamma_w m_{\mathrm{v}}} \frac{\partial^2 u_e}{\partial z^2} = \frac{\partial u_{\mathrm{e}}}{\partial t} \tag{9.33}$$

in which

$$m_{\mathrm{v}} = \frac{a_{\mathrm{v}}}{1+e} \tag{9.34}$$

which is referred to as the *coefficient of volume compressibility* ($m^2\ N^{-1}$).

Equation (9.31) is the governing one-dimensional consolidation equation, as derived by Terzaghi in 1932. Later, this equation has been extended to two-dimensional domains that reads

$$c_{\mathrm{h}} \frac{\partial^2 u_e}{\partial x^2} + c_{\mathrm{v}} \frac{\partial^2 u_e}{\partial z^2} = \frac{\partial u_{\mathrm{e}}}{\partial t} \tag{9.35}$$

where c_{h} is *the coefficient of horizontal consolidation* ($m^2\ s^{-1}$).

9.6 Electrokinetic Consolidation

Electrokinetic is a phenomenon that involves a relative motion between a liquid containing ions, such as an electrolyte solution, and a charged solid surface. Electro-osmosis is one of the most studied electrokinetic effects in soils. Applications of electro-osmosis to improve subsoil behavior have been conducted since the 40s of the last century.

In 1968, Esrig extended Terzaghi's consolidation theory to account for the influence of the electro-osmosis effect. The fundamental assumption in Esrig's development is that the groundwater flux $\mathbf{q}$ ($m\ s^{-1}$) in the presence of a hydraulic gradient and an electric field can be described as a superposition of both effects, that is:

$$\mathbf{q} = \mathbf{q}_h + \mathbf{j}_e \tag{9.36}$$

in which $\mathbf{q}_h$ is the hydraulically induced fluid flux and $\mathbf{j}_e$ is the electrically generated fluid flux.

In a one-dimensional domain, the electrically induced flux is given by *Ohm's law*, where

$$J_e = k_e \frac{\partial E}{\partial z} \tag{9.37}$$

in which E (V) is the *electric potential* and k_e (m^2 s^{-1} V^{-1}) is the *coefficient of the electro-osmotic permeability*.

On the other hand, the hydraulically induced fluid flux follows Darcy's law as

$$q_h = \frac{k_z}{\rho_w g} \frac{\partial u_e}{\partial z} \tag{9.38}$$

where Darcy's law is expressed in terms of the excess pore pressure u_e (Pa).

Following Esrig hypothesis, the continuity equation describing the consolidation process in a soil mass in the presence of an electro-osmosis effect is given via

$$\nabla \cdot \mathbf{q} = \frac{\partial h}{\partial t} \tag{9.39}$$

9

where h (m) is the piezometric head and $\mathbf{q}$ is given in (9.36). Thus, following the derivation of the continuity equation of the fluid flow due to hydraulic consolidation given in ▶ Sect. 9.5, the continuity equation for a one-dimensional domain influenced by the electro-osmosis effect can then be expressed as

$$k_e \frac{\partial^2 E}{\partial z^2} + \frac{k_z}{\rho_w g} \frac{\partial^2 u_e}{\partial z^2} = m_v \frac{\partial u_e}{\partial t} \tag{9.40}$$

Making use of (9.32), this equation can be written as

$$\frac{\partial^2 u_e}{\partial z^2} + \gamma_w \frac{k_e}{k_z} \frac{\partial^2 E}{\partial z^2} = \frac{1}{c_v} \frac{\partial u_e}{\partial t} \tag{9.41}$$

The conservation equation of the electric field in a soil mass is

$$\nabla \cdot \mathbf{J}_e = 0 \tag{9.42}$$

Thus, employing Ohm's law (9.37), the conservation equation for the electro-osmosis field in a one-dimensional domain reads

$$k_e \frac{\partial^2 E}{\partial z^2} = 0 \tag{9.43}$$

Combining these two conservation equations, the coupled continuity equations for hydraulic and electro-osmosis effects in a general multi-dimensional domain can be expressed as

$$\begin{aligned} k_h \nabla^2 u_e + k_e \gamma_w \nabla^2 E &= m_v \gamma_w \frac{\partial u_e}{\partial t} \\ k_e \nabla^2 E &= 0 \end{aligned} \tag{9.44}$$

where k_h represents the hydraulic conductivity and k_e represents the electro-osmosis conductivity. Note that, as the second equation in (9.44) is zero, the second term on the left-hand side of the first equation vanishes. This reduces the governing balance equations to become

$$\begin{aligned} k_h \nabla^2 u_e &= m_v \gamma_w \frac{\partial u_e}{\partial t} \\ k_e \nabla^2 E &= 0 \end{aligned} \tag{9.45}$$

Problems

1. What factors influence the velocity and direction of groundwater flow in a porous medium?
2. Explain the concept of hydraulic conductivity and its significance in understanding groundwater flow.
3. Discuss the importance of establishing appropriate initial and boundary conditions when modeling groundwater flow in engineering applications.
4. What is the fundamental principle behind Terzaghi's consolidation theory, and how does it explain the behavior of saturated soils under external loads?
5. Following Terzaghi's theory, explain the concept of excess pore water pressure and its role in the consolidation process.
6. What is electrokinetic consolidation theory, and how does it differ from the traditional consolidation theory, such as Terzaghi's theory?
7. Explain the principles behind electrokinetic consolidation and how electric fields can be used to enhance the consolidation process in clay soils.
8. Discuss the factors that influence the effectiveness of electrokinetic consolidation and how they differ from factors involved in the traditional consolidation theory.

References

Berry P, Reid D (1987) An introduction to soil mechanics. McGraw-Hill Book Company

MI Esrig 1968 Pore pressures, consolidation, and electrokinetics J Soil Mech Found Div 94 4 899 922

TW Lambe RV Whitman 1969 Soil mechanics Wiley New York

10 Analytical Groundwater Flow

R. Al Khoury and C. Kasbergen, *Analytical Subsurface Environment*,
https://doi.org/10.1007/978-3-031-93910-5_10

Studying heat transfer and contaminant transport in subsurface formations is closely associated with studying groundwater flow. The groundwater flow constitutes one of the main mechanisms that drive heat and contaminants away from their sources. Throughout the years, numerous mathematical models have been formulated to analyze the motion of groundwater in subsurface aquifers possessing diverse physical characteristics and subjected to various boundary conditions. The aquifers are typically represented by a porous medium domain often constituting two phases: a water phase and a solid phase. In devising analytical models, developers entice to idealize the geometry in order to make the solution feasible. The aquifers are commonly approximated as semi-infinite, homogeneous and isotropic, with the possibility of leakage into surrounding aquitards. They are treated as either confined or unconfined, depending on the geometry and the initial and boundary conditions.

In this chapter, we will address solutions to boundary value problems for both confined and leaky aquifers exhibiting varying geometry and boundary conditions. We will primary use $h = h_0 - s$ to indicate the change of piezometric head, where h_0 denotes the initial hydraulic head, and s represents the drawdown generated by the applied boundary conditions. Alternatively, we will also use $s = h_0 - h$ for describing the governing equations for groundwater flow.

10.1 Confined Aquifer

A confined aquifer is a fully saturated subsurface layer bounded by impermeable formations, known as *aquicludes*. The confinement exerts a pressure that is higher than the atmospheric pressure, such that when a well is installed in a confined aquifer, the water in the well arises above the top boundary of the aquifer. ◘ Figure 10.1 shows schematically a confined aquifer, together with aquicludes.

The governing equation of a transient groundwater flow in a confined aquifer is

$$\frac{\partial^2 h}{\partial x^2} - \frac{S}{T}\frac{\partial h}{\partial t} = 0 \tag{10.1}$$

where h (m) is the piezometric head; $S = S_s b$ is a dimensionless storage coefficient with S_s (m^{-1}) the specific storage and b (m) the thickness of the aquifer; and $T = Kb$ is the transmissivity coefficient (m^2 s^{-1}) with K the permeability (hydraulic conductivity) (m s^{-1}) of the aquifer.

This equation can be written in terms of the drawdown $s = h_0 - h$ to express the groundwater flow in a confined aquifer, such that

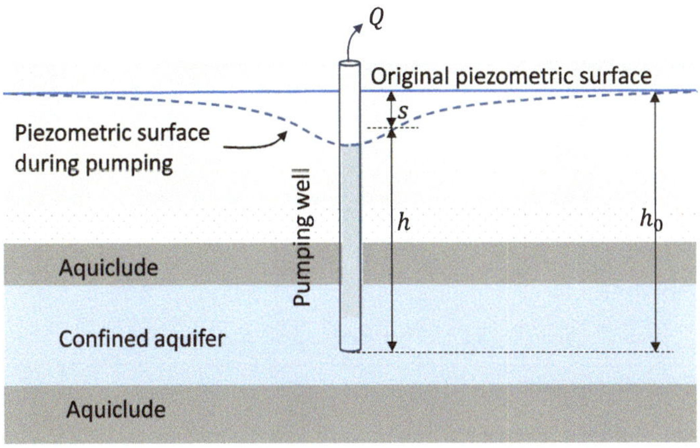

Fig. 10.1 Typical confined aquifer characteristics

$$\frac{\partial^2 s}{\partial x^2} - \frac{1}{\alpha}\frac{\partial s}{\partial t} = 0 \tag{10.2}$$

where $\alpha = T/S$ or equivalently $\alpha = K/S_s$.

In the following sub-sections, we will address confined aquifers featuring different geometries subjected to varying boundary conditions.

10.1.1 Constant Drawdown: Dirichlet Boundary Condition

In groundwater flow engineering, a constant drawdown refers to a condition where the hydraulic head, or water level, is maintained at a constant height in the well or at a specified point within the aquifer. The constant drawdown is often achieved by continuously pumping water from the well, at a rate calibrated to balance the natural inflow of groundwater with the volume of water being withdrawn. This technique is commonly used to assess the impact of pumping on nearby regions and wells, determine well efficiency, evaluate aquifer properties, and analyze the movement of contaminants in groundwater systems. By understanding how groundwater responds under constant drawdown conditions, engineers can make informed decisions regarding groundwater management and sustainable utilization of water resources. The Dirichlet boundary condition for this scenario implies prescribing a constant hydraulic head (or drawdown) at a fix point in the domain.

The boundary value problem of a constant drawdown process in a semi-infinite, one-dimensional domain can be expressed as

$$\frac{\partial^2 s}{\partial x^2} - \frac{1}{\alpha}\frac{\partial s}{\partial t} = 0$$

$$\begin{aligned} &\text{IC:} \quad s(x,0) = 0 \\ &\text{BC:} \quad s(0,t) = S_w \\ &\qquad\;\; s(\infty,t) = 0 \end{aligned} \tag{10.3}$$

with

$$\alpha = T/S = K/S_s, \quad T = Kb, \quad S = S_s b$$

General solution

Applying the Fourier transform to (10.3), gives

$$\frac{d^2\hat{s}}{dx^2} - \frac{i\omega}{\alpha}\hat{s} = 0$$

$$\begin{aligned} &\text{BC:} \quad \hat{s}(0,\omega) = \hat{S}_w \\ &\qquad\;\; \hat{s}(\infty,\omega) = 0 \end{aligned} \tag{10.4}$$

As the coefficient of this ordinary differential equation is constant, the trial solution can expressed as

10

$$\hat{s} = ce^{\kappa x}$$

Substituting this trial function into the differential equation in (10.4), yields

$$c\kappa^2 e^{\kappa x} - \frac{i\omega}{\alpha}ce^{\kappa x} = 0$$

Rearranging leads to

$$\kappa^2 - \frac{i\omega}{\alpha} = 0$$

This gives

$$\kappa = \mp\sqrt{\frac{i\omega}{\alpha}}$$

Thus, the general solution can be expressed as

$$\hat{s} = c_1 e^{-\kappa x} + c_2 e^{\kappa x} \tag{10.5}$$

Boundary conditions

As $x \to \infty$, $c_2 = 0$, otherwise the solution becomes unbounded at far distances, leading to

$$\hat{s} = c_1 e^{-\kappa x}$$

At $x = 0$, the first boundary condition in (10.4) implies

$$c_1 = \hat{S}_w$$

BVP solution

Following the above derivation, the solution of this BVP in the frequency domain would then read

$$\hat{s}(x, \omega_n) = \hat{S}_w e^{-\kappa x}, \quad \kappa = \sqrt{i\omega_n / \alpha} \tag{10.6}$$

Having solved for the frequency domain, the solution in the time domain can then be obtained using the inverse fast Fourier transform (IFFT) algorithm, such that

$$s(x, t) = \sum_n \hat{s}(x, \omega_n)\, e^{i\omega_n t}$$

Once the drawdown is computed, the piezometric head can be obtained by

$$h = h_0 - s$$

where h_0 is the initial head in the aquifer.

10.1.2 Constant Discharge: Neumann Boundary Condition

Considering the physical features of the previous case, we now address the Neumann boundary condition in which a constant flow rate applied at the boundary of the domain.

The relevant boundary value problem for this case is

$$\frac{\partial^2 s}{\partial x^2} - \frac{1}{\alpha}\frac{\partial s}{\partial t} = 0$$

$$\begin{aligned} \text{IC:} \quad & s(x, 0) = 0 \\ \text{BC:} \quad & -T \left.\frac{\partial s}{\partial x}\right|_{x=0} = q_w \\ & s(\infty, t) = 0 \end{aligned} \tag{10.7}$$

with

$$\alpha = T / S = K / S_s, \quad T = Kb, \quad S = S_s b$$

General solution

As given in the previous case, the general solution of the differential equation in (10.7) in the frequency domain is given by

$$\hat{s} = c_1 e^{-\kappa x}, \quad \kappa = \sqrt{\frac{i\omega}{\alpha}} \tag{10.8}$$

Boundary condition

Taking the derivative of (10.8) with respect to x, gives

$$\frac{d\hat{s}}{dx} = -\kappa c_1 e^{-\kappa x}$$

At $x = 0$, the first boundary condition in (10.7) implies

$$T\kappa c_1 = \hat{q}_w$$

This yields

$$c_1 = \frac{\hat{q}_w}{T\kappa}$$

BVP solution

Following this, the solution of this BVP problem for a frequency ω_n would then be

$$\hat{s}(x, \omega_n) = \frac{\hat{q}_w}{T\kappa} e^{-\kappa x}, \quad \kappa = \sqrt{i\omega_n / \alpha}$$

Having the solution in the frequency domain, the final solution in the time domain can be obtained using the inverse fast Fourier transform (IFFT) algorithm, such that

$$s(x, t) = \sum_n \hat{s}(x, \omega_n) e^{i\omega_n t}$$

Once solving for the drawdown, the piezometric head can be obtained via

$$h = h_0 - s$$

10.1.3 Cylindrical Drawdown Source or Sink: Known Radius of Influence

The drawdown is commonly a result of pumping groundwater from confined aquifers using cylindrical wells. Cylindrical wells are typically constructed by a cylindrical shaft that extends into the aquifer and allows the withdrawal of water. This process creates a pressure gradient in the aquifer that draws water towards the well, and eventually to the ground surface. The wells are of finite lengths, but for certain geometries and boundary conditions, it is reasonable to model them as infinite cylindrical sources or sinks. Groundwater flow resulting from pumping from an infinite cylindrical well can be conceptualized as a circular source generating groundwater flow within a planar 2D domain, exhibiting an axial (radial) symmetric effect.

The boundary value problem of an infinite cylindrical well with a known radius of influence can be expressed as

$$\frac{S_s}{K_r}\frac{\partial s}{\partial t} = \frac{\partial^2 s}{\partial r^2} + \frac{1}{r}\frac{\partial s}{\partial r}, \quad r > a$$

$$\begin{aligned} \text{IC}: \quad & s(r, 0) = 0 \\ \text{BC}: \quad & s(a, t) = S_w(t) \\ & s(R, t) = 0 \end{aligned} \tag{10.9}$$

where S_s (m^{-1}) is the specific (elastic) storage, K_r ($m\ s^{-1}$) is the radial hydraulic conductivity (permeability), a (m) is the radius of the well, and R (m) is the radius of influence. Note that S_s/K_r is equivalent to S/T (see the definitions in ▶ Sect. 10.1).

General solution

Applying the Fourier transform to (10.9) gives

$$\frac{d^2\hat{s}}{dr^2}+\frac{1}{r}\frac{d\hat{s}}{dr}-\frac{i\omega S_s}{K_r}\hat{s}=0$$
$$\text{BC}: \quad \hat{s}\,(R,\omega)=0 \qquad (10.10)$$
$$\hat{s}\,(a,\omega)=\hat{S}_w(\omega)$$

As described in ▶ Chap. 5, Sect. 5.2, the differential equation in (10.10) is a sort of the modified Bessel equation, with its general solution given by

$$\hat{s}=AK_0(\kappa r)+BI_0(\kappa r), \qquad \kappa=\sqrt{\frac{i\omega S_s}{K_r}} \tag{10.11}$$

in which I_0 and K_0 are the modified Bessel functions of zero order of the first and second kind, respectively. A and B are arbitrary constants, need to be determined from the boundary conditions.

Boundary conditions

At $r=a$, the second boundary condition in (10.10) implies

$$AK_0(\kappa a)+BI_0(\kappa a)=\hat{S}_w$$

At $r=R$, the first boundary condition implies

$$AK_0(\kappa R)+BI_0(\kappa R)=0$$

Putting these two equations in a matrix form yields

$$\begin{pmatrix} K_0(\kappa a) & I_0(\kappa a) \\ K_0(\kappa R) & I_0(\kappa R) \end{pmatrix}\begin{pmatrix} A \\ B \end{pmatrix}=\begin{pmatrix} \hat{S}_w \\ 0 \end{pmatrix}$$

and solving for A and B, we obtain

$$\begin{pmatrix} A \\ B \end{pmatrix}=\frac{1}{\Delta}\begin{pmatrix} I_0(\kappa R) & -I_0(\kappa a) \\ -K_0(\kappa R) & K_0(\kappa a) \end{pmatrix}\begin{pmatrix} \hat{S}_w \\ 0 \end{pmatrix}$$
$$\Delta=K_0(\kappa a)I_0(\kappa R)-I_0(\kappa a)K_0(\kappa R)$$

which leads to

$$A=\frac{1}{\Delta}I_0(\kappa R)\,\hat{S}_w, \qquad B=-\frac{1}{\Delta}K_0(\kappa R)\,\hat{S}_w \tag{10.12}$$

BVP solution

Substituting A and B in (10.12) into (10.11) yields

$$\hat{s}(r,\omega_n) = \frac{\hat{S}_w}{\Delta}\left[I_0(\kappa R)K_0(\kappa r) - K_0(\kappa R)I_0(\kappa r)\right]$$
$$\kappa = \sqrt{i\omega_n S_s / K_r} \qquad (10.13)$$
$$\Delta = K_0(\kappa a)I_0(\kappa R) - I_0(\kappa a)K_0(\kappa R)$$

The final solution in the time domain can then be obtained using the inverse fast Fourier transform (IFFT) algorithm, as

$$s(r,t) = \sum_n \hat{s}(r,\omega_n)\, e^{i\omega_n t}$$

10.1.4 Cylindrical Discharge Source or Sink: Known Radius of Influence

Considering the physical features of the previous case, we now address the problem under the Neumann boundary condition, in which a flow rate is applied at the well boundary.

The relevant boundary value problem of this case is given by

$$\frac{S_s}{K_r}\frac{\partial s}{\partial t} = \frac{\partial^2 s}{\partial r^2} + \frac{1}{r}\frac{\partial s}{\partial r}, \qquad r > a$$

$$\text{IC:} \quad s(r,0) = 0 \qquad (10.14)$$
$$\text{BC:} \quad -2\pi a T \left.\frac{\partial s}{\partial r}\right|_{r=a} = q_w(t)$$
$$s(R,t) = 0$$

where a is the radius of the well, and $T = K_r b$.

General solution

Applying the Fourier transform to (10.14) gives

$$\frac{d^2\hat{s}}{dr^2} + \frac{1}{r}\frac{d\hat{s}}{dr} - \frac{i\omega S_s}{K_r}\hat{s} = 0$$

$$\text{BC:} \quad -2\pi a T \left.\frac{d\hat{s}}{dr}\right|_{r=a} = \hat{q}_w(\omega) \qquad (10.15)$$
$$\hat{s}(R,\omega) = 0$$

As given above, the general solution of this equation is

$$\hat{s} = AK_0(\kappa r) + BI_0(\kappa r), \qquad \kappa = \sqrt{\frac{i\omega S_s}{K_r}} \qquad (10.16)$$

in which I_0 and K_0 are the modified Bessel functions of zero order of the first and second kind, respectively. A and B are arbitrary constants, which need to be determined from the boundary conditions.

Boundary conditions

Taking the derivative of (10.16) with respect to r gives

$$\frac{d\hat{s}}{dr} = -A\kappa K_1(\kappa r) + B\kappa I_1(\kappa r)$$

At $r = a$, the first boundary condition in (10.15) implies

$$-A\kappa K_1(\kappa a) + B\kappa I_1(\kappa a) = -\frac{\hat{q}_w}{2\pi a T}$$

where I_1 and K_1 are the modified Bessel function of first order of the first and second kind, respectively.

At $r = R$, the second boundary condition in (10.15) implies

$$AK_0(\kappa R) + BI_0(\kappa R) = 0$$

Putting these two equations in a matrix form yields

$$\begin{pmatrix} -\kappa K_1(\kappa a) & \kappa I_1(\kappa a) \\ K_0(\kappa R) & I_0(\kappa R) \end{pmatrix} \begin{pmatrix} A \\ B \end{pmatrix} = \begin{pmatrix} \frac{-\hat{q}_w}{2\pi a T} \\ 0 \end{pmatrix}$$

Solving this system of two equations for the constants vector, we obtain

$$\begin{pmatrix} A \\ B \end{pmatrix} = \frac{1}{\Delta} \begin{pmatrix} I_0(\kappa R) & -\kappa I_1(\kappa a) \\ -K_0(\kappa R) & -\kappa K_1(\kappa a) \end{pmatrix} \begin{pmatrix} \frac{-\hat{q}_w}{2\pi a T} \\ 0 \end{pmatrix}$$

$$\Delta = -\kappa K_1(\kappa a) I_0(\kappa R) - \kappa I_1(\kappa a) K_0(\kappa R)$$

from which

$$A = -\frac{1}{\Delta} I_0(\kappa R) \frac{\hat{q}_w}{2\pi a T}, \quad B = \frac{1}{\Delta} K_0(\kappa R) \frac{\hat{q}_w}{2\pi a T} \tag{10.17}$$

BVP solution

Substituting the constants A and B in (10.17) into (10.16), the solution of this boundary value problem in the frequency domain yields

$$\hat{s}(r, \omega_n) = \frac{1}{\Delta} \frac{\hat{q}_w}{2\pi a T} [-I_0(\kappa R) K_0(\kappa r) + K_0(\kappa R) I_0(\kappa r)]$$

$$\kappa = \sqrt{\frac{i\omega_n S_s}{K_r}} \tag{10.18}$$

$$\Delta = -\kappa K_1(\kappa a) I_0(\kappa R) - \kappa I_1(\kappa a) K_0(\kappa R)$$

Having this, the final solution in the time domain can then be reconstructed using the inverse fast Fourier transform (IFFT) algorithm, as

$$s(r, t) = \sum\nolimits_n \hat{s}(r, \omega_n) e^{i\omega_n t}$$

10.1.5 Cylindrical Discharge Source or Sink: Unknown Radius of Influence

The above two cases apply when the radius of influence, R, is known. However, if R is not known, a formulation with an infinite boundary at the radial direction must be adopted. In this case, the calculation will, by definition, determine the influence region at each time step.

The boundary value problem of this case is given by

$$\frac{S_s}{K_r}\frac{\partial s}{\partial t} = \frac{\partial^2 s}{\partial r^2} + \frac{1}{r}\frac{\partial s}{\partial r}, \qquad r > a$$

$$\begin{aligned} \text{IC}: &\quad s(r,0) = 0 \\ \text{BC}: &\quad s(r \to \infty, t) = 0 \\ &\quad -2\pi a T \left.\frac{\partial s}{\partial r}\right|_{r=a} = q_w(t) \end{aligned} \tag{10.19}$$

where a is the radius of the well.

General solution

10

Applying the Fourier transform to (10.19) gives

$$\begin{aligned} &\frac{d^2\hat{s}}{dr^2} + \frac{1}{r}\frac{d\hat{s}}{dr} - \frac{i\omega S_s}{K_r}\hat{s} = 0 \\ \text{BC}: &\quad \hat{s}(r \to \infty, \omega) = 0 \\ &\quad -2\pi a T \left.\frac{d\hat{s}}{dr}\right|_{r=a} = \hat{q}_w(\omega) \end{aligned} \tag{10.20}$$

As indicated earlier, the general solution of this differential equation is

$$\hat{s} = AK_0(\kappa r) + BI_0(\kappa r), \qquad \kappa = \sqrt{\frac{i\omega S_s}{K_r}} \tag{10.21}$$

Boundary conditions

As $r \to \infty$, I_0 becomes unbounded, and the first boundary condition in (10.20) would imply $B = 0$, leading the solution to become

$$\hat{s}(r,\omega) = AK_0(\kappa r) \tag{10.22}$$

Taking the derivative of (10.22) with respect to r, gives

$$\frac{d\hat{s}}{dr} = -A\kappa K_1(\kappa r)$$

where K_1 is the modified Bessel function of the second kind of order 1. At $r = a$, the second boundary condition in (10.20) implies

$$A = \frac{\hat{q}_w(\omega)}{2\pi a T \kappa K_1(\kappa a)}$$

BVP solution

Substituting A into (10.22) yields

$$\hat{s}\,(r,\omega_n) = \frac{\hat{q}_w(\omega_n)}{2\pi a T \kappa K_1(\kappa a)} K_0(\kappa r), \quad \kappa = \sqrt{i\omega_n S_s / K_r} \tag{10.23}$$

As usual, having solved the problem in the frequency domain, the solution in the time domain can be obtained using the inverse fast Fourier transform (IFFT) algorithm, as

$$s\,(r,t) = \sum_n \hat{s}\,(r,\omega_n)\, e^{i\omega_n t}$$

If needed, the piezometric head can be obtained simply by

$$h = h_0 - s$$

10.1.6 Multiple Cylindrical Well System

The use of multiple cylindrical wells in a region is a common set-up in groundwater engineering. The interaction between multiple wells can have a significant effect on the flow patterns within the aquifer. Factors such as the spacing between wells, the pumping rates of neighbouring wells, and the hydraulic properties of the aquifer, all contribute to the overall behaviour of the groundwater flow.

Groundwater flow models are utilized to simulate these interactions and provide insights into the behaviour of the system. Each well acts as a source or a sink. For Neumann problems, which involve discharge, the interaction between wells can simply be done using the superposition principle.

Considering a system of infinite cylindrical wells with known radius of influence of each individual well (▶ Sect. 10.1.4), the drawdown of well i in the region can readily be calculated as

$$\hat{s}_i(r,\omega_n) = \frac{\hat{q}_{wi}}{2\pi a T \Delta_i}\left[-I_0(\kappa R_i)K_0(\kappa r_i) + K_0(\kappa R_i)I_0(\kappa r_i)\right]$$

$$\kappa = \sqrt{\frac{i\omega_n S_s}{K_r}} \tag{10.24}$$

$$\Delta_i = -\kappa K_1(\kappa a_i) I_0(\kappa R_i) - \kappa I_1(\kappa a_i) K_0(\kappa R_i)$$

where r_i is the distance from well i to an observation point in the domain. This solution is repeated for all wells. Then, upon transforming the results to the time domain, the drawdown in the region can be determined simply via an algebraic sum of all wells, as

$$s\,(r,t) = \sum_i s_i(r,t) \tag{10.25}$$

Example 10.1: Multiple Pumping Wells

Consider a confined aquifer with a permeability K of 1×10^{-3} m s^{-1}, specific storage S_s of 1 m^{-1}, and thickness b of 15 m. Two fully penetrating pumping wells 0.3 m in diameter are operating in the aquifer. They are located as shown in Fig. 10.2. Well 1 has a pumping rate of 0.05 m^3 s^{-1}, and well 2 has a pumping rate of 0.1 m^3 s^{-1}. The pumping flow rates of the wells and the associated time distribution of the well operations are provided in Fig. 10.3. The radius of influence of both wells is 10 m.

Calculate the drawdown around the two wells after 20 and 100 min of operation.

Solution

As usual, we first transform the input pulses in Fig. 10.3 from the time domain to the frequency domain using the forward FFT. Next, we solve the system for every frequency using (10.24). Subsequently, we reconstruct the system back to the time domain using the inverse FFT. Later, we sum the effect of the individual wells at some grid points surrounding the two wells using (10.25).

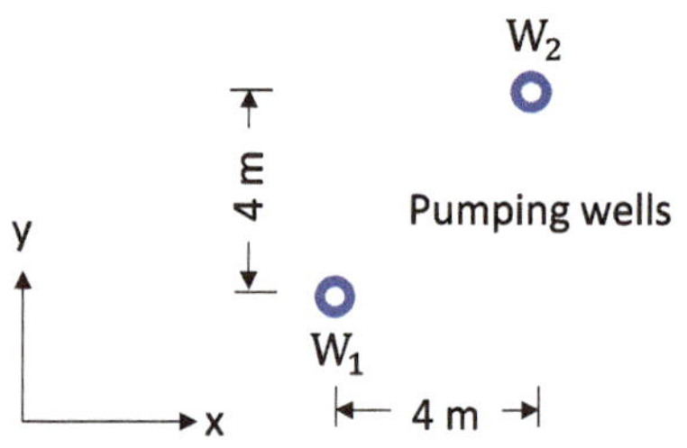

Fig. 10.2 Location of the pumping wells

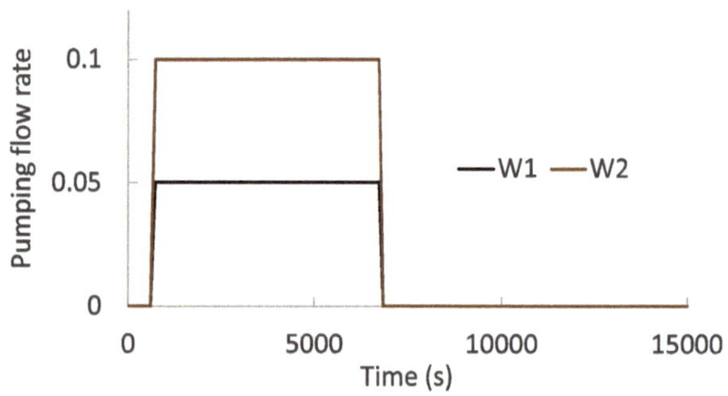

Fig. 10.3 Operating time of the two wells

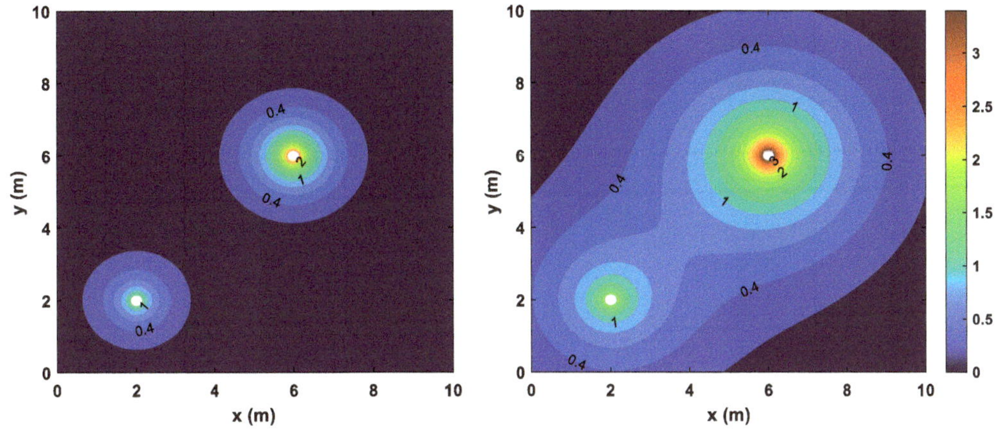

Fig. 10.4 Contour plots of the drawdown after 20 min (left) and 100 min (right)

Figure 10.4 demonstrates the contour plots of the water drawdown after 20 and 100 min of operation. The figure clearly shows that the drawdown is spatially exponential. It is rather large at the vicinity of the wells, beyond which the drawdown reduces significantly. In this case, the drawdown in the vicinity of well 2 is around 3.4 m, and just 2 m away from it, the drawdown drops to 0.7 m.

10.1.7 Line Discharge Source: Unknown Radius of Influence

In groundwater flow modeling, it is common to represent wells, even though cylindrical, as line sources or sinks. In this approach, the well is conceptualized as a line that extends vertically into the aquifer.

The boundary value problem governing groundwater flow due to pumping from a line well pump embedded in a semi-infinite domain is given by

$$\frac{S_s}{K_r}\frac{\partial s}{\partial t} = \frac{\partial^2 s}{\partial r^2} + \frac{1}{r}\frac{\partial s}{\partial r}$$

$$\begin{aligned} \text{IC}: &\quad s(r,0) = 0 \\ \text{BC}: &\quad s(r \to \infty, t) = 0 \\ &\quad T\frac{\partial s}{\partial r}\cdot 2\pi r\Big|_{r\to 0} = -q_w(t) \end{aligned} \tag{10.26}$$

General solution

Applying the Fourier transform to (10.26), we obtain

$$\frac{d^2\hat{s}}{dr^2}+\frac{1}{r}\frac{d\hat{s}}{dr}-\frac{i\omega S_s}{K_r}\hat{s}=0$$

$$\text{BC}: \quad \hat{s}\,(r\to\infty,\omega)=0 \tag{10.27}$$

$$T\frac{d\hat{s}}{\partial r}\cdot 2\pi r\Big|_{r\to 0}=-\hat{q}_w(\omega)$$

As given earlier, the solution of the ordinary differential equation in (10.27), considering the first boundary condition, is

$$\hat{s}\,(r,\omega)=AK_0(\kappa r) \tag{10.28}$$

Boundary condition

Taking the derivative of (10.28) with respect to r, the second boundary condition in (10.27) entails

$$2\pi TA\kappa rK_1(\kappa r)|_{r\to 0}=\hat{q}_w(\omega) \tag{10.29}$$

Knowing that

10

$$\lim_{r\to 0}\kappa r\,K_1(\kappa r)=1 \tag{10.30}$$

where this limiting case has been derived based on (see Abramowitz and Stegun, 1972, Ref. 0).

$$\lim_{x\to 0}K_\nu(x)=\frac{1/2\cdot\Gamma(\nu)}{(1/2\cdot x)^\nu}$$

Following this, (10.29) leads to

$$A=\frac{\hat{q}_w(\omega)}{2\pi T} \tag{10.31}$$

BVP solution

Substituting (10.31) into (10.28), the solution of this boundary value problem for a given frequency ω_n results in

$$\hat{s}\,(r,\omega_n)=\frac{\hat{q}_w(\omega_n)}{2\pi T}K_0(\kappa r),\quad \kappa=\sqrt{i\omega_n S_s/K_r} \tag{10.32}$$

Having done that, the final solution in the time domain can be obtained using the inverse fast Fourier transform (IFFT) algorithm, as

$$s\,(r,t)=\sum_n\hat{s}(r,\omega_n)e^{i\omega_n t}$$

For multiple line sources, the drawdown is simply the algebraic sum of each individual well, such that

$$s\,(r,t)=\sum_i s_i(r,t) \tag{10.33}$$

where the subscript i stands for a well, and r_i is the distance from well i to an observation point in the domain.

10.1.8 Line Discharge Source: Known Radius of Influence

The boundary value problem governing groundwater flow resulting from pumping water from a line well embedded in a semi-infinite domain is given by

$$\frac{S_s}{K_r}\frac{\partial s}{\partial t} = \frac{\partial^2 s}{\partial r^2} + \frac{1}{r}\frac{\partial s}{\partial r}, \quad a < r < R$$

$$\begin{aligned} \text{IC}: &\quad s(r,0) = 0 \\ \text{BC}: &\quad T\frac{\partial s}{\partial r}\cdot 2\pi r\bigg|_{r\to 0} = -q_w(t) \\ &\quad s(R,t) = 0 \end{aligned} \tag{10.34}$$

where R is the radius of influence.

General solution

The general solution of the differential equation in (10.34) in the frequency domain is given in ▶ Sect. 10.1.4 as

$$\hat{s} = AK_0(\kappa r) + BI_0(\kappa r), \quad \kappa = \sqrt{\frac{i\omega S_s}{K_r}} \tag{10.35}$$

Boundary conditions

Taking the derivative of (10.35) with respect to r, yields

$$\frac{d\hat{s}}{dr} = -A\kappa K_1(\kappa r) + B\kappa I_1(\kappa r)$$

At $r = 0$, $I_1(\kappa r) = 0$, leading to

$$-A\kappa K_1(\kappa r) 2\pi r T|_{r\to 0} = -\hat{q}_w \tag{10.36}$$

Following the procedure outlined in ▶ Sect. 10.1.7, the limiting case of (10.36) as $r \to 0$ gives

$$A = \frac{\hat{q}_w}{2\pi T} \tag{10.37}$$

At $r = R$, the second boundary condition in (10.34) implies

$$AK_0(\kappa R) + BI_0(\kappa R) = 0$$

Giving

$$B = -\frac{\hat{q}_w}{2\pi T}\frac{K_0(\kappa R)}{I_0(\kappa R)} \tag{10.38}$$

BVP solution

Substituting (10.37) and (10.38) into (10.35), the BVP solution for a given frequency ω_n is therefore

$$\hat{s}_n \equiv \hat{s}(r, \omega_n) = \frac{\hat{q}_w}{2\pi T}\left[K_0(\kappa r) - \frac{K_0(\kappa R)}{I_0(\kappa R)} I_0(\kappa r)\right], \quad \kappa = \sqrt{\frac{i\omega_n S_s}{K_r}} \tag{10.39}$$

The final solution in the time domain can then be obtained using the inverse fast Fourier transform (IFFT) algorithm, as

$$s(r, t) = \sum_n \hat{s}_n \, e^{i\omega_n t}$$

10.1.9 Infinite Domain with a Submerged Water Pump

Submerged pumping refers to the process of extracting or pumping water from a well or a body of water via a submerged source. In this set-up, the pump is fully immersed, and the water is drawn into the pump through an intake system and discharged to a desired location. The pump acts as a point source or sink.

10

The boundary value problem governing groundwater flow due to a submerged pump in an infinite domain $-\infty \leq r \leq \infty$ is conveniently given in the spherical coordinate system, such that

$$\frac{S_s}{K_r}\frac{\partial s}{\partial t} = \frac{\partial^2 s}{\partial r^2} + \frac{2}{r}\frac{\partial s}{\partial r}$$

$$\begin{aligned} \text{IC:} \quad & s(r, 0) = 0 \\ \text{BC:} \quad & -4\pi r^2 T \left.\frac{\partial s}{\partial r}\right|_{r=0} = q_w(t) \\ & s(\infty, t) = 0 \end{aligned} \tag{10.40}$$

General solution

Applying the Fourier transform to (10.40) gives

$$\frac{d^2\hat{s}}{dr^2} + \frac{2}{r}\frac{d\hat{s}}{dr} - \frac{i\omega S_s}{K_r}\hat{s} = 0$$

$$\begin{aligned} \text{BC:} \quad & r^2 \left.\frac{d\hat{s}}{dr}\right|_{r=0} = -\frac{\hat{q}_w(\omega)}{4\pi T} \\ & \hat{s}(\infty, \omega) = 0 \end{aligned} \tag{10.41}$$

The ordinary differential equation in (10.41) can be written as

$$\frac{1}{r}\frac{d^2}{dr^2}(r\hat{s}) - \kappa^2\hat{s} = 0, \quad \kappa^2 = \frac{i\omega S_s}{K_r}$$

Re-arranging this equation gives

$$\frac{d^2}{dr^2}(r\hat{s}) - \kappa^2 r\hat{s} = 0$$

If we let $\widehat{u} = r\widehat{s}$, this equation takes the form

$$\frac{d^2\hat{u}}{dr^2} - \kappa^2\hat{u} = 0$$

The general solution of this differential equation is obviously

$$\hat{u} = Ae^{-\kappa r} + Be^{\kappa r}$$

Substituting back $\widehat{u} = r\widehat{s}$, the general solution becomes

$$\hat{s} = \frac{1}{r}(Ae^{-\kappa r} + Be^{\kappa r}), \quad \kappa = \sqrt{\frac{i\omega S_s}{K_r}} \tag{10.42}$$

Boundary conditions

The second boundary condition in (10.41) entails $B = 0$, otherwise, the solution becomes unbounded. Thus, the general solution (10.42) reduces to

$$\hat{s} = \frac{1}{r}Ae^{-\kappa r} \tag{10.43}$$

Taking the derivative of this solution with respect to r, substituting it into the first boundary condition in (10.41) and carrying out the limit as $r \to 0$ it leads to

$$A = \frac{\hat{q}_w}{4\pi T}$$

BVP solution

Substituting A into (10.43), the solution of this boundary value problem for a given frequency ω_n becomes

$$\hat{s}_n = \frac{\hat{q}_w}{4\pi Tr}e^{-\kappa r}, \quad \kappa = \sqrt{\frac{i\omega_n S_s}{K_r}} \tag{10.44}$$

Conducting the inverse fast Fourier transform (IFFT) algorithm, the final solution in the time domain is thus

$$s\,(r,t) = \sum_n \hat{s}_n e^{i\omega_n t}$$

Having determined the drawdown, the piezometric head can be obtained via

$$h\,(r,t) = h_0 - s\,(r,t)$$

10.2 Leaky Aquifer

Leaky aquifers are a type of underground formations that exhibit seeping of water at their boundaries with other soil layers. Unlike confined aquifers, which are surrounded by effectively impermeable layers of rock or clay that prevent water from seeping, leaky aquifers embedded within layers that are partially permeable, allowing water to seep in and out, as schematically shown in Fig. 10.5. The exchange of water between underground formations can influence both the quantity and quality of water in the aquifer and the surface water bodies. Therefore, it is essential to study their behavior under different boundary conditions.

10.2.1 1D Leaky Confined Aquifer: Neumann Problem

The boundary value problem of a leaky aquifer under the influence of a Neumann boundary condition is given by

$$\frac{S}{T}\frac{\partial h}{\partial t} - \frac{\partial^2 h}{\partial x^2} - \frac{1}{B^2}(h_0 - h) = 0$$

$$\begin{aligned} \text{IC}: &\quad h(x,0) = 0 \\ \text{BC}: &\quad h(x \to \infty, t) = 0 \\ &\quad T\left.\frac{\partial h}{\partial x}\right|_{x=0} = q_w(t) \end{aligned} \tag{10.45}$$

where:

$S = S_s b$ with S_s the specific storage (m^{-1}) and b the thickness of the aquifer (m).

$T = Kb$ is the transitivity ($m^2\ s^{-1}$) with K the permeability of the aquifer ($m\ s^{-1}$).

$B = \sqrt{Kbb_{aqt}/K_{aqt}}$ is the leakage factor with b_{aqt} is the thickness of the aquitard and K_{aqt} is the permeability of the aquitard, see Fig. 10.5.

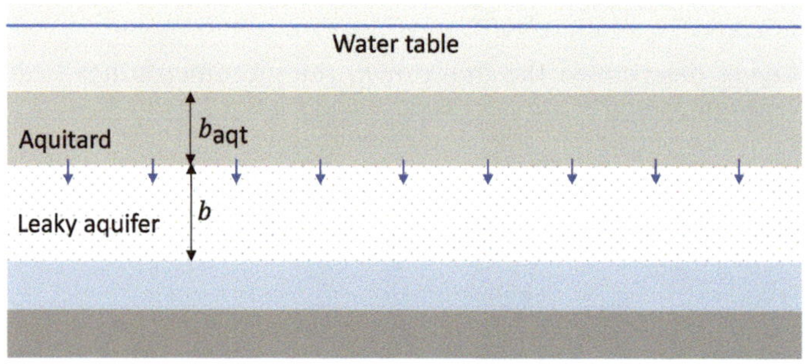

Fig. 10.5 Leaky aquifer

In (10.45) the leakage term $h_0 - h$ is in effect the discharge (drawdown), normally expressed as s. In the previous section we formulated the problems explicitly with s, but as this formulation is commonly available in literature, we here address both formulations.

General solution

Applying the Fourier transform to (10.45) gives

$$\frac{d^2\hat{h}}{dx^2} - \left(\frac{i\omega S}{T} + \frac{1}{B^2}\right)\hat{h} = -\frac{h_0}{B^2}$$

$$\text{BC}: \quad \hat{h}(x \to \infty, \omega) = 0 \qquad (10.46)$$

$$T\frac{d\hat{h}}{dx}\bigg|_{x=0} = \hat{q}_w(\omega)$$

Apparently, this ordinary differential equation is nonhomogeneous, and its general solution requires solving for the complementary and particular equations, such that

$$\hat{h} = \hat{h}_c + \hat{h}_p$$

where $\widehat{h_c}$ is the complementary function representing the general solution of the associated homogeneous differential equation of (10.46), and $\widehat{h_p}$ is the particular solution of the nonhomogeneous equation, arising from the presence of h_0/B^2.

The relevant homogeneous equation of (10.46) is

$$\frac{d^2\hat{h}_c}{dx^2} - \kappa^2\hat{h}_c = 0, \quad \kappa^2 = \frac{i\omega S}{T} + \frac{1}{B^2} \qquad (10.47)$$

The solution of this differential equation can readily be obtained via

$$\hat{h}_c = c_1 e^{-\kappa x} + c_2 e^{\kappa x}$$

On the other hand, the particular equation of (10.46) is

$$\frac{d^2\hat{h}_p}{dx^2} - \kappa^2\hat{h}_p = -\frac{h_0}{B^2} \qquad (10.48)$$

As the right-hand side is constant, the trial solution can be any arbitrary constant, such that

$$\hat{h}_p = A$$

Substituting this trial solution into (10.48) gives

$$A = \frac{h_0}{B^2\kappa^2} = \hat{h}_p$$

Collecting both solutions, the general solution of the governing differential equation in (10.46) is then

$$\begin{aligned}\hat{h} &= \hat{h}_c + \hat{h}_p \\ &= c_1 e^{-\kappa x} + c_2 e^{\kappa x} + \frac{h_0}{B^2 \kappa^2}\end{aligned} \tag{10.49}$$

Boundary conditions

The first boundary condition in (10.46) entails that the piezometric head must be bounded at infinity, leading $c_2 = 0$ in (10.49). Following this, the solution becomes

$$\hat{h} = c_1 e^{-\kappa x} + \frac{h_0}{B^2 \kappa^2} \tag{10.50}$$

Taking the derivative of $\hat{h}$ with respect to x yields

$$\frac{d\hat{h}}{dx} = -\kappa c_1 e^{-\kappa x}$$

At $x = 0$, the second boundary condition implies

$$c_1 = -\frac{\hat{q}_w(\omega)}{T\kappa}$$

BVP solution

10

Substituting c_1 into (10.50), we obtain the solution of this boundary value problem in the frequency domain, as

$$\hat{h}(x, \omega_n) = -\frac{\hat{q}_w(\omega_n)}{T\kappa_n} e^{-\kappa_n x} + \frac{h_0}{B^2 \kappa_n^2}, \quad \kappa_n = \sqrt{\frac{i\omega_n S}{T} + \frac{1}{B^2}} \tag{10.51}$$

Having solved the problem in the frequency domain, the solution in the time domain can be reconstructed using the inverse fast Fourier transform (IFFT) algorithm, such that

$$h(x, t) = \sum_n \hat{h}(x, \omega_n) e^{i\omega_n t}$$

10.2.2 1D Leaky Confined Aquifer: Dirichlet Problem

Here, we address the previous case, but with a Dirichlet boundary condition. The governing boundary value problem of this case is thus

$$\begin{aligned}&\frac{S}{T}\frac{\partial h}{\partial t} - \frac{\partial^2 h}{\partial x^2} - \frac{1}{B^2}(h_0 - h) = 0 \\ &\text{IC}: \quad h(x, 0) = 0 \\ &\text{BC}: \quad h(x \to \infty, t) = 0 \\ &\qquad\quad h(0, t) = H_w(t)\end{aligned} \tag{10.52}$$

General solution

As explained in the previous case, the general solution of this nonhomogeneous differential equation in the frequency domain is given by

$$\hat{h} = c_1 e^{-\kappa x} + c_2 e^{\kappa x} + \frac{h_0}{B^2 \kappa^2}$$

Boundary conditions

At far distances, $\widehat{h}$ must be bounded, implying $c_2 = 0$, and the solution reduces to

$$\hat{h} = c_1 e^{-\kappa x} + \frac{h_0}{B^2 \kappa^2} \tag{10.53}$$

At $x = 0$, the second boundary condition implies

$$c_1 = \hat{H}_w(\omega) - \frac{h_0}{B^2 \kappa^2}$$

BVP solution

Following this, the solution of this boundary value problem for a frequency ω_n is then

$$\hat{h}_n = \hat{H}_w(\omega)\, e^{-\kappa_n x} - \frac{h_0}{B^2 \kappa_n^2}(e^{-\kappa_n x} - 1), \quad \kappa_n = \sqrt{\frac{i\omega_n S}{T} + \frac{1}{B^2}} \tag{10.54}$$

Having done that, the solution in the time domain can be reconstructed via the inverse fast Fourier transform (IFFT) algorithm, given by

$$h(x, t) = \sum_n \hat{h}_n e^{i\omega_n t}$$

Special case: steady state

As $t \to \infty$, i.e. $\omega \to 0$, the system reaches the steady state. In this case

$$\kappa = \frac{1}{B}$$

Hence, the solution becomes

$$h = H_w e^{-x/B} - h_0 \left(e^{-x/B} - 1 \right) \tag{10.55}$$

which can be written in terms of the drawdown $s = h_0 - h$, leading to

$$s = S_w e^{-x/B} \tag{10.56}$$

where $S_w = h_0 - H_w$.

10.2.3 Alternative Formulation: 1D Leaky Confined Aquifer

The boundary value problem of the leaky aquifer presented in the previous sections can also be described by the drawdown $s = h_0 - h$, where h_0 is assumed constant.

Neumann case

Taking the derivatives of $h = h_0 - s$ with respect to t and x, and substituting into (10.52), the relevant boundary value problem in terms of s can be expressed, for the Neumann case, as

$$\frac{S}{T}\frac{\partial s}{\partial t} - \frac{\partial^2 s}{\partial x^2} + \frac{s}{B^2} = 0$$

$$\begin{aligned} \text{IC}: \quad & s\,(x,0) = 0 \\ \text{BC}: \quad & s\,(x \to \infty, t) = 0 \\ & -T\left.\frac{\partial s}{\partial x}\right|_{x=0} = q \end{aligned} \tag{10.57}$$

where, unlike the previous formulation, the governing differential equation is homogeneous.

Applying the Fourier transform to (10.57) yields

$$\frac{\partial^2 \hat{s}}{\partial x^2} - \kappa^2 \hat{s} = 0, \qquad \kappa^2 = \frac{i\omega S}{T} + \frac{1}{B^2}$$

$$\begin{aligned} \text{BC}: \quad & \hat{s}\,(x \to \infty, \omega) = 0 \\ & -T\left.\frac{\partial^2 \hat{s}}{\partial x^2}\right|_{x=0} = \hat{q}_w(\omega) \end{aligned} \tag{10.58}$$

The general solution of this differential equation is given by

$$\hat{s} = c_1 e^{-\kappa x} + c_2 e^{\kappa x}$$

As $x \to \infty$, the first boundary condition in (10.58) implies $c_2 = 0$, leading to

$$\hat{s} = c_1 e^{-\kappa x} \tag{10.59}$$

Taking the derivative of (10.59) with respect to x and applying the second boundary condition at $x = 0$, gives

$$c_1 = \frac{\hat{q}_w(\omega)}{T\kappa} \tag{10.60}$$

Substituting (10.60) into (10.59) yields the solution of this boundary value problem for a given frequency ω_n as

$$\hat{s}_n = \frac{\hat{q}_w(\omega_n)}{T\kappa_n} e^{-\kappa_n x}, \quad \kappa_n = \sqrt{\frac{i\omega_n S}{T} + \frac{1}{B^2}} \tag{10.61}$$

Dirichlet case

The boundary value problem for the Dirichlet case is

$$\frac{S}{T}\frac{\partial s}{\partial t} - \frac{\partial^2 s}{\partial x^2} + \frac{s}{B^2} = 0$$

$$\begin{aligned} \text{IC}: &\quad s(x,0) = 0 \\ \text{BC}: &\quad s(x \to \infty, t) = 0 \\ &\quad s(x = 0, t) = S_w(t) \end{aligned} \tag{10.62}$$

Applying the Fourier transform to (10.62) gives

$$\frac{\partial^2 \hat{s}}{\partial x^2} - \kappa^2 \hat{s} = 0, \qquad \kappa^2 = \frac{i\omega S}{T} + \frac{1}{B^2}$$

$$\begin{aligned} \text{BC}: &\quad \hat{s}(x \to \infty, \omega) = 0 \\ &\quad \hat{s}(0, \omega) = \hat{S}_w(\omega) \end{aligned} \tag{10.63}$$

The general solution of this differential equation is

$$\hat{s} = c_1 e^{-\kappa x} + c_2 e^{\kappa x}$$

As $x \to \infty$, the first boundary condition implies $c_2 = 0$, leading to

$$\hat{s} = c_1 e^{-\kappa x}$$

The second boundary condition in (10.63) implies

$$c_1 = \hat{S}_w \tag{10.64}$$

Following this, the solution of this BVP for a given frequency ω_n is then

$$\hat{s}_n = \hat{S}_w e^{-\kappa_n x}, \quad \kappa_n = \sqrt{\frac{i\omega_n S}{T} + \frac{1}{B^2}} \tag{10.65}$$

Having solved the two problems in the frequency domain, the solution in the time domain can be obtained via the IFFT algorithm, such that

$$s(x,t) = \sum_n \hat{s}_n e^{i\omega_n t}$$

Special case: steady state

As $t \to \infty$, i.e. $\omega \to 0$, the system reaches the steady state. In this case

$$\kappa = \frac{1}{B}$$

And the solution becomes

$$s = S_w e^{-x/B}$$

which is identical to (10.56) that was obtained by solving a nonhomogeneous equation.

10.2.4 Leaky Confined Aquifer with a Cylindrical Well

The boundary value problem of a leaky aquifer encompassing a cylindrical well going through the whole thickness of the aquifer can be expressed as

$$
\begin{aligned}
&\frac{S}{T}\frac{\partial h}{\partial t} - \frac{\partial^2 h}{\partial r^2} - \frac{1}{r}\frac{\partial h}{\partial r} - \frac{1}{B^2}(h_0 - h) = 0 \\
&\text{IC}: \quad h(r,0) = 0 \\
&\text{BC}: \quad h(r \to \infty, t) = 0 \\
&\qquad -\left.\frac{\partial h}{\partial r}\right|_{r=a} = \frac{q_w(t)}{2\pi a T}
\end{aligned}
\tag{10.66}
$$

where it is obvious that the system is axial-symmetric, as it can be deduced from Fig. 10.6.

General solution

Applying the Fourier transform to (10.66) yields

$$
\begin{aligned}
&\frac{d^2\hat{h}}{dr^2} + \frac{1}{r}\frac{d\hat{h}}{dr} - \kappa^2\hat{h} = -\frac{h_0}{B^2}, \quad \kappa^2 = \frac{i\omega S}{T} + \frac{1}{B^2} \\
&\text{BC}: \quad \hat{h}(r \to \infty, t) = 0 \\
&\qquad -\left.\frac{d\hat{h}}{dr}\right|_{r=a} = \frac{\hat{q}_w(\omega)}{2\pi a T}
\end{aligned}
\tag{10.67}
$$

This ordinary differential equation is nonhomogeneous, and the general solution can be attained by finding the complementary $\widehat{h_c}$ and particular $\widehat{h_p}$ solutions, that leads to

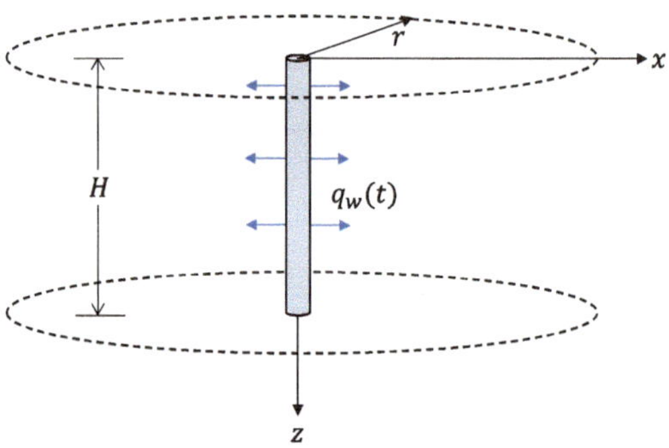

Fig. 10.6 Leaky aquifer with cylindrical well go through the aquifer thickness

$$\hat{h} = \hat{h}_c + \hat{h}_p$$

The associated homogeneous differential equation of (10.67) is

$$\frac{d^2\hat{h}_c}{dr^2} + \frac{1}{r}\frac{d\hat{h}_c}{dr} - \kappa^2\hat{h}_c = 0$$

which is the parametric form of the modified Bessel differential equation of order zero. The solution of this equation is $\widehat{h}_c = c_1 I_0(\kappa r) + c_2 K_0(\kappa r)$, but as $I_0(\kappa r)$ is unbounded at far distances, the first boundary condition in (10.67) implies $c_1 = 0$, and the solution becomes

$$\hat{h}_c(r,\omega) = c_2 K_0(\kappa r) \tag{10.68}$$

where K_0 is the modified Bessel function of the second kind of order zero.

The associated nonhomogeneous differential equation of (10.67) is

$$\frac{d^2\hat{h}_p}{dr^2} + \frac{1}{r}\frac{d\hat{h}_p}{dr} - \kappa^2\hat{h}_p = -\frac{h_0}{B^2} \tag{10.69}$$

As the right-hand side of this equation is constant, the nonhomogeneous trial solution can be any constant, such that

$$\hat{h}_p = A$$

Substituting this trial solution into (10.69) and solving for A, leads to

$$A = \frac{h_0}{B^2\kappa^2}$$

Thus, the general solution becomes

$$\hat{h}\,(r,\omega) = c_2 K_0(\kappa r) + \frac{h_0}{B^2\kappa^2} \tag{10.70}$$

Boundary condition

Taking the derivative of (10.70) with respect to r, and solving for $r = a$, the second boundary condition in (10.67) implies

$$c_2 = \frac{\hat{q}_w(\omega)}{2\pi a T\kappa\, K_1(\kappa a)}$$

Solution of BVP

Having determined the constants, the solution of the BVP for a given frequency ω_n is then

$$\hat{h}\,(r,\omega_n) = \frac{h_0}{B^2\kappa^2} + \frac{\hat{q}_w(\omega_n)}{2\pi a T\kappa\, K_1(\kappa_n a)} K_0(\kappa r), \quad \kappa = \sqrt{\frac{i\omega_n S}{T} + \frac{1}{B^2}} \tag{10.71}$$

Having solved the problem in the frequency domain, we can then reconstruct the solution back to the time domain using the IFFT algorithm, such that

$$h(r,t) = \sum_n \hat{h}(x,\omega_n)\, e^{i\omega_n t}$$

Special case 1: line well

As $a \to 0$, the cylindrical well reaches a line well, entailing

$$\lim_{a\to 0} [aK_1(\kappa a)] \approx \frac{1}{\kappa}$$

Applying this limiting relationship on (10.71), reveals the solution for a line source in the frequency domain, as

$$\hat{h}_n = \frac{h_0}{B^2\kappa^2} + \frac{\hat{q}_w(\omega_n)}{2\pi T} K_0(\kappa r), \quad \kappa = \sqrt{\frac{i\omega_n S}{T} + \frac{1}{B^2}} \quad (10.72)$$

Special case 2: steady state

As $t \to \infty$, i.e. $\omega \to 0$, the system reaches the steady state. This entails

$$\kappa = \frac{1}{B}, \quad \hat{q}_w(\omega_n) = q_w$$

leading, the solution of the cylindrical well in the steady state to be given by

$$h(r) = h_0 + \frac{Bq_w}{2\pi aT\, K_1(a/B)} K_0(r/B)$$

and for the line well, it is given by

$$h(r) = h_0 + \frac{q_w}{2\pi T} K_0(r/B)$$

Having determined the hydraulic head, we can readily determine the drawdown, as

$$s(r) = \frac{q_w}{2\pi T} K_0(r/B)$$

10.2.5 Leaky Confined Aquifer with a Finite Cylindrical Well

In the previous case we addressed a boundary value problem for a cylindrical well that extends through the whole thickness of the layer. Here, we extend this problem to the case where the well is shorter than the aquifer thickness, making the prescribed water flux a function of z and time. We also examine leakage from the upper boundary, as depicted in ◘ Fig. 10.7.

Using the drawdown formulation, the boundary value problem of this case can be expressed as

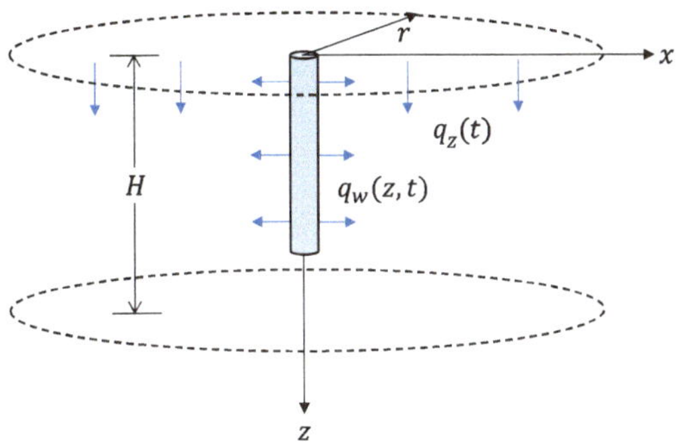

Fig. 10.7 Finite cylindrical well in a leaky confined aquifer

$$
\begin{aligned}
&\frac{1}{\alpha}\frac{\partial s}{\partial t} - \frac{\partial^2 s}{\partial r^2} - \frac{1}{r}\frac{\partial s}{\partial r} - \frac{\partial^2 s}{\partial z^2} = 0 \\
&\text{IC}: \quad s(r, z, 0) = 0 \\
&\text{BC}: \quad s(r \to \infty, z, t) = 0 \\
&\qquad -R\frac{\partial s}{\partial z}(r, 0, t) = q_z(t) \\
&\qquad \frac{\partial s}{\partial z}(r, H, t) = 0 \\
&\qquad -2\pi a T \left.\frac{\partial s}{\partial r}\right|_{r=a} = q_w(z, t), \qquad 0 \le z \le L
\end{aligned}
\tag{10.73}
$$

where H is the thickness of the aquifer that is larger than the length of the well, L, and $\alpha = K/S_s = T/S$. R ($\text{m}^2\ \text{s}^{-1}$) is some leakage coefficient, defining the leakage flow rate at the boundary between the aquifer and the aquitard per meter. This coefficient might be considered as the hydraulic diffusivity of the aquitard $K'b'/S'$, where K' is the permeablity of the aquitard, b' is its thickness and S' is its storage.

This boundary value problem involves two nonhomogeneous boundary conditions, and in order to solve it using the eigenfunction expansion, it is necessary to split it into two sub-problems, such that

$$
s(r, z, t) = s_1(z, t) + s_2(r, z, t) \tag{10.74}
$$

Inserting (10.74) into the partial differential equation in (10.73) gives

$$
\frac{1}{\alpha}\frac{\partial s_1}{\partial t} + \frac{1}{\alpha}\frac{\partial s_2}{\partial t} - \frac{\partial^2 s_1}{\partial z^2} - \frac{\partial^2 s_2}{\partial r^2} - \frac{1}{r}\frac{\partial s_2}{\partial r} - \frac{\partial^2 s_2}{\partial z^2} = 0 \tag{10.75}
$$

This equation can be split into two sub-equations, which can be expressed with their relevant boundary conditions, as

Sub-problems 1

$$\frac{\partial^2 s_1}{\partial z^2} - \frac{1}{\alpha}\frac{\partial s_1}{\partial t} = 0$$

$$\begin{aligned} \text{IC}: \quad & s_1(z,0) = 0 \\ \text{BC}: \quad & -R\frac{\partial s_1}{\partial z}(0,t) = q_z(t) \\ & \frac{\partial s_1}{\partial z}(H,t) = 0 \end{aligned} \tag{10.76}$$

Sub-problem 2

$$\frac{\partial^2 s_2}{\partial r^2} + \frac{1}{r}\frac{\partial s_2}{\partial r} + \frac{\partial^2 s_2}{\partial z^2} - \frac{1}{\alpha}\frac{\partial s_2}{\partial t} = 0$$

$$\begin{aligned} \text{IC}: \quad & s_2(r,z,0) = 0 \\ \text{BC}: \quad & s_2(r \to \infty, z, t) = 0 \\ & \frac{\partial s_2}{\partial z}(r,0,t) = 0 \\ & \frac{\partial s_2}{\partial z}(r,H,t) = 0 \\ & -2\pi a T \left.\frac{\partial s_2}{\partial r}\right|_{r=a} = q_w(z,t) \end{aligned} \tag{10.77}$$

10

Sub-problem 1

Applying the Fourier transform to (10.76) yields

$$\frac{d^2\hat{s}_1}{dz^2} - \frac{i\omega}{\alpha}\hat{s}_1 = 0$$

$$\begin{aligned} \text{BC:} \quad & \frac{d\hat{s}_1}{dz}(0,\omega) = -\frac{\hat{q}_z(\omega)}{R} \\ & \frac{d\hat{s}_1}{dz}(H,\omega) = 0 \end{aligned} \tag{10.78}$$

The general solution of the differential equation in (10.78) is

$$\hat{s}_1 = a_1 e^{-\kappa_1 z} + a_2 e^{-\kappa_2 z}, \quad \kappa_1 = \sqrt{\frac{i\omega}{\alpha}}, \quad \kappa_2 = -\sqrt{\frac{i\omega}{\alpha}} \tag{10.79}$$

Taking the derivative of $\hat{s}_1$ with respect to z gives

$$\frac{d\hat{s}_1}{dz} = -\kappa_1 a_1 e^{-\kappa_1 z} - \kappa_2 a_2 e^{-\kappa_2 z}$$

At $z = 0$, the first boundary condition in (10.78) implies

$$-\kappa_1 a_1 - \kappa_2 a_2 = -\frac{\hat{q}_z}{R}$$

At $z = H$, the second boundary condition in (10.78) implies

$$-\kappa_1 a_1 e^{-\kappa_1 H} - \kappa_2 a_2 e^{-\kappa_2 H} = 0$$

Solving these two equations simultaneously yields

$$a_1 = \frac{\hat{q}_z e^{-\kappa_2 H}}{\kappa_1 (e^{-\kappa_2 H} - e^{-\kappa_1 H}) R}, \quad a_2 = -\frac{\hat{q}_z e^{-\kappa_1 H}}{\kappa_2 (e^{-\kappa_2 H} - e^{-\kappa_1 H}) R}$$

Thus, the solution of the BVP in (10.78) is

$$\hat{s}_1(z,\omega) = \frac{\hat{q}_z e^{-\kappa_2 H}}{\kappa_1 (e^{-\kappa_2 H} - e^{-\kappa_1 H}) R} e^{-\kappa_1 z} - \frac{\hat{q}_z e^{-\kappa_1 H}}{\kappa_2 (e^{-\kappa_2 H} - e^{-\kappa_1 H}) R} e^{-\kappa_2 z} \tag{10.80}$$

Sub-problem 2

Applying the Fourier transform to (10.77) results in

$$\frac{\partial^2 \hat{s}_2}{\partial r^2} + \frac{1}{r}\frac{\partial \hat{s}_2}{\partial r} + \frac{\partial^2 \hat{s}_2}{\partial z^2} - \frac{i\omega}{\alpha}\hat{s}_2 = 0$$

$$\begin{aligned} \text{BC}: \quad & \hat{s}_2(r \to \infty, z, \omega) = 0 \\ & \frac{\partial \hat{s}_2}{\partial z}(r, 0, \omega) = 0 \\ & \frac{\partial \hat{s}_2}{\partial z}(r, H, \omega) = 0 \\ & -2\pi a T \left.\frac{\partial \hat{s}_2}{\partial r}\right|_{r=a} = \hat{q}_w(z,\omega), \quad 0 \le z \le L \end{aligned} \tag{10.81}$$

Solution of the partial differential equation in (10.81) can be carried out using the method of separation of variables. Assume

$$\hat{s}_2(r,z) = R(r)Z(z) = RZ \tag{10.82}$$

Substituting (10.82) into the PDE in (10.81), and equating the result to a separation constant, say λ, which can be zero, positive or negative, gives

$$Z'' + \lambda Z = 0 \quad \text{and} \quad R'' + \frac{1}{r}R' - \left(\frac{i\omega}{\alpha} + \lambda\right)R = 0$$

Applying the boundary conditions on these two equations, we can readily verify that for $\lambda = -\beta^2 < 0$ it only leads to the trivial solution $\hat{s}_2 = 0$. For $\lambda = 0$, and $\lambda = \beta^2 > 0$ the solution becomes non-trivial. Thus, the two DEs become

$$\frac{d^2 Z}{dz^2} + \beta^2 Z = 0 \tag{10.83}$$

$$\frac{d^2 R}{dr^2} + \frac{1}{r}\frac{dR}{dr} - \left(\frac{i\omega}{\alpha} + \beta^2\right)R = 0 \tag{10.84}$$

The solution of (10.83) is

$$Z = c_1 \cos(\beta z) + c_2 \sin(\beta z)$$

where c_1 and c_2 are constants to be determined from boundary conditions.

The solution of (10.84) is

$$R(r) = c_3 I_0(\kappa_r r) + c_4 K_0(\kappa_r r), \quad \kappa_r = \sqrt{\frac{i\omega}{\alpha} + \beta^2}$$

in which I_0 and K_0 are the modified Bessel functions of zero order of the first and second kind, respectively, and c_3 and c_4 are arbitrary constants, need to be determined from the boundary conditions. Since I_0 is unbounded at far distances, the first boundary condition in (10.81) implies $c_3 = 0$, and the solution reduces to

$$R(r) = c_4 K_0(\kappa_r r)$$

Combining both solutions for $R(r)$ and $Z(z)$ gives

$$\hat{s}_2 = R(r)Z(z) = K_0(\kappa_r r)[c_1 \cos(\beta z) + c_2 \sin(\beta z)] \quad (10.85)$$

where c_4 is dissolved within c_1 and c_2. Taking the derivative of (10.85) with respect to z yields

$$\frac{\partial \hat{s}_2}{\partial z} = K_0(\kappa_r r)[-\beta c_1 \sin(\beta z) + \beta c_2 \cos(\beta z)]$$

At $z = 0$, the second boundary condition in (10.81), valid to all r, implies $c_2 = 0$, leading to

$$\hat{s}_2 = cK_0(\kappa_r r) \cos(\beta z) \quad (10.86)$$

where we replaced c_1 with c. Taking the derivative of (10.86) with respect to z yields

$$\frac{\partial \hat{s}_2}{\partial z} = -\beta c K_0(\kappa_r r) \sin(\beta z)$$

At $z = H$, the third boundary condition in (10.81) implies

$$-\beta c K_0(\kappa_r r) \sin(\beta H) = 0$$

which entails

$$\sin(\beta H) = 0$$

that is true only if

$$\beta h = m\pi \Rightarrow \beta = \frac{m\pi}{H}, \quad m = 1, 2, \ldots$$

Since the problem is linear, the superposition principle implies that the solution becomes

$$\hat{s}_2(r,z,\omega) = \sum_m c_m K_0(\kappa_r r)\cos(\beta_m z)$$

with

$$\kappa_r = \sqrt{\kappa^2 + \beta_m^2}$$
$$\kappa^2 = \frac{i\omega}{\alpha}$$
$$\beta_m = \frac{m\pi}{H}, \quad m = 0, 1, 2, \ldots \tag{10.87}$$

Note that we assigned m ≥ 0, since, as indicated above, $\lambda = 0$ is also an eigenvalue.

Taking the derivative of (10.87) with respect to r yields

$$\frac{\partial \hat{s}_2}{\partial r} = -\sum_m c_m \cos\left(\frac{m\pi z}{H}\right) \kappa_r K_1(\kappa_r r)$$

Then, at $r = a$, the fourth boundary condition in (10.81) implies

$$2\pi a T \sum_m c_m \cos\left(\frac{m\pi z}{H}\right) \kappa_r K_1(\kappa_r a) = \hat{q}_w(z,\omega_n)$$

This is a cosine Fourier series with its coefficients c_m expressed as

$$c_0 = \frac{1}{H}\int_0^L \frac{\hat{q}_w(z,\omega_n)}{2\pi a T \kappa_r K_1(\kappa_r a)} dz$$
$$= \frac{1}{2\pi a T H \kappa_r K_1(\kappa_r a)} \int_0^L \hat{q}_w(z,\omega_n)\, dz, \qquad m = 0$$

$$c_m = \frac{2}{H}\int_0^L \frac{\hat{q}_w(z,\omega_n)}{2\pi a T \kappa_r K_1(\kappa_r a)} \cos\left(\frac{m\pi z}{H}\right) dz$$
$$= \frac{1}{\pi a T H \kappa_r K_1(\kappa_r a)} \int_0^L \hat{q}_w(z,\omega_n) \cos\left(\frac{m\pi z}{h}\right) dz, \qquad m = 1, 2, \ldots \tag{10.88}$$

Following this, the solution of $\widehat{s}_2$ can then be expressed as

$$\hat{s}_2(r,z,\omega) = \sum_m c_m K_0(\kappa_r r)\cos(\beta_m z), \quad m = 0, 1, 2, \ldots \tag{10.89}$$

with c_m given in (10.88).

Special cases

In the special case where $\widehat{q}_w$ is constant, the Fourier coefficients become

$$c_0 = \frac{\hat{q}_w(\omega_n)L}{2\pi a T H \kappa_r K_1(\kappa_r a)}$$

$$c_m = \frac{\hat{q}_w(\omega_n)}{\pi a T H \kappa_r K_1(\kappa_r a)} \int_0^L \cos\left(\frac{m\pi z}{H}\right) dz \qquad (10.90)$$

$$= \frac{\hat{q}_w(\omega_n)}{m\pi^2 a T \kappa_r K_1(\kappa_r a)} \sin\left(\frac{m\pi L}{H}\right)$$

Moreover, for the special case where $\widehat{q}_w$ is constant and $L = H$, the solution becomes

$$\hat{s}_2(r, z, \omega) = \frac{\hat{q}_w(\omega_n)}{2\pi a T \kappa_r K_1(\kappa_r a)} K_0(\kappa_r r)$$

BVP solution

Recall (10.74), which is, in the frequency domain, expressed as

$$\hat{s}\,(r, z, \omega_n) = \hat{s}_1(z, \omega_n) + \hat{s}_2(r, z, \omega_n) \qquad (10.91)$$

Substituting the solutions of $\widehat{s}_1$ (10.80) and $\widehat{s}_2$ (10.89) into (10.91) yields

$$\hat{s}\,(r, z, \omega_n) = -\frac{\hat{q}_z e^{-\kappa_2 H}}{\kappa_1 \left(e^{-\kappa_2 H} + e^{-\kappa_1 H}\right) R} e^{-\kappa_1 z} + \frac{\hat{q}_z e^{-\kappa_1 H}}{\kappa_2 \left(e^{-\kappa_2 H} + e^{-\kappa_1 H}\right) R} e^{-\kappa_2 z} + \sum_m c_m K_0(\kappa_r r) \cos(\beta_m z), \qquad m = 0, 1, 2, \ldots \qquad (10.92)$$

with c_m given in (10.88) and

$$\kappa_1 = \sqrt{\frac{i\omega}{\alpha}}, \quad \kappa_2 = -\kappa_1, \quad \kappa_r = \sqrt{\frac{i\omega}{\alpha} + \beta_m^2}, \quad \beta_m = \frac{m\pi}{H}$$

Employing the inverse fast Fourier transform (IFFT) algorithm gives the solution in the time domain, such that

$$s\,(r, z, t) = \sum_n \hat{s}\,(r, z, \omega_n)\, e^{i\omega_n t}$$

EXAMPLE 10.2: Leaky Aquifer

Consider a leaky confined aquifer with a hydraulic conductivity K of 1×10^{-2} m s^{-1}, specific storage S_s of 1 m^{-1}, and thickness b of 10 m. A pumping well 5 m in length and 0.2 m in diameter is installed in the aquifer. The water leakage occurs at a rate of 0.1 m^2 s^{-1} per meter, and the pumping is operated at a rate 1 m^3 s^{-1}. The time distributions of the pumping and leaking flow rates are shown in Fig. 10.8. The leakage coefficient R is assumed 1 m^2 s^{-1} per meter. Note that these relatively unrealistic parameters were chosen to highlight the capability of the model to capture the interplay between pumping and leaking.

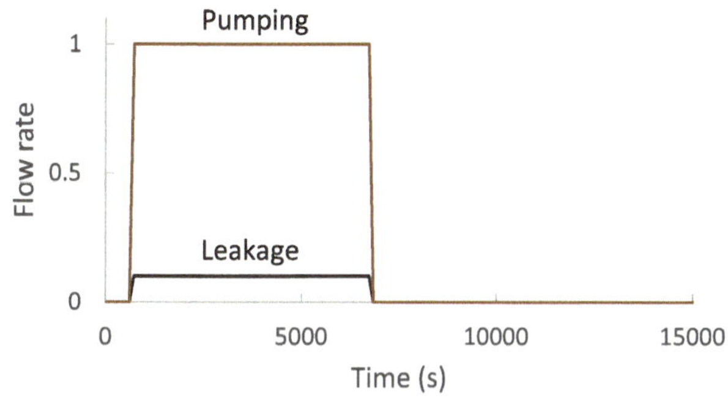

Fig. 10.8 Pumping and leaking flow rates

Calculate the drawdown around the well after 20 and 100 min of operation. And discuss how the leakage influences the response of the aquifer to pumping.

Solution

As usual, we begin by transforming the input signals from the time domain to the frequency domain using the forward FFT. Here we used 4096 samples with 60 s sampling rate. Next, we employ (10.92) to calculate the drawdown for every frequency. After that, we use the IFFT to reconstruct the solution back to the time domain.

Figure 10.9 shows the drawdown in the aquifer after 20 and 100 min of operation. The figure shows that the area close to the pumping well exhibits drawdown, reading around 2.7 m at the vicinity of the well and diminishing nearly 3 m away from it. The zero contour line designates the boundary between the pumping-influence region (positive values) and the leakage-influence region (negative values). The figure displays a clear coupling between the pumping and the leakage

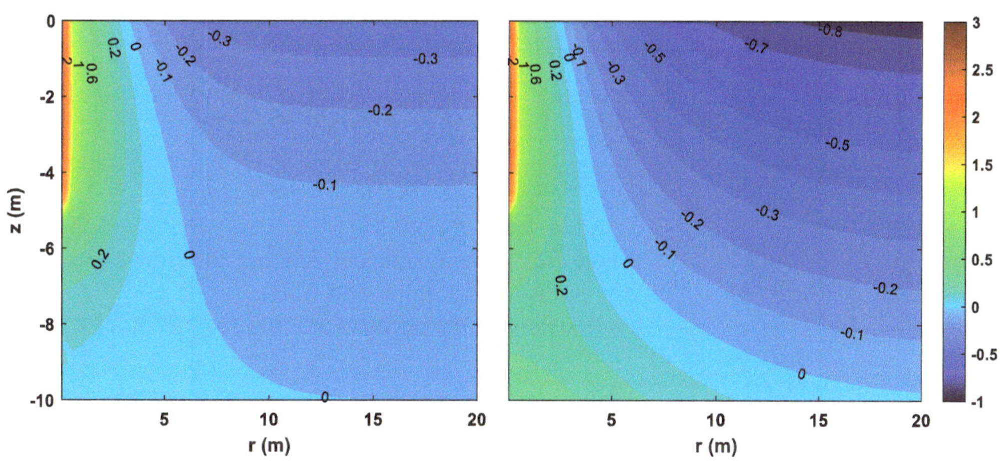

Fig. 10.9 Contour plots of the drawdown after 20 min (left) and 100 min (right)

processes. With elapsed time there is an increase in the region affected by the pumping, but at the same time there is an increase of the hydraulic head (negative s) on the region away from the well.

Problems

1. Solve the case for the constant drawdown (▶ Sect. 10.1.1) for a domain with a radius of influence.
   Hint: set the second boundary condition in (10.3) to $s(L, t) = 0$.
2. Solve the case for the constant discharge (▶ Sect. 10.1.2) for a domain with a radius of influence.
   Hint: set the second boundary condition in (10.7) to $s(L, t) = 0$.
3. Implement the line discharge source given in ▶ Sect. 10.1.7 with the well-known This solution (see Ref. 0)

$$s\left(r,t\right) = \frac{q}{4\pi T}\mathrm{Ei}\left(-\frac{Sr^2}{4tT}\right)$$

   where Ei is the exponential integral. Use $q = 10$ units and $S = T = 1$ units.

10

4. Compare the solution in ▶ Sect. 10.2.4, with $h_0 = 0$, to the solution given by Hantush and Jacob, (1955) (see Ref. 0). Compare also the steady state solutions.
5. Solve the case of ▶ Sect. 10.2.5 using the piezometric head (h) formulation.
6. Solve this boundary value problem:

$$\frac{1}{\alpha}\frac{\partial s}{\partial t} - \frac{\partial^2 s}{\partial r^2} - \frac{1}{r}\frac{\partial s}{\partial r} - \frac{\partial^2 s}{\partial z^2} = 0$$

$$\begin{aligned}
\mathrm{IC}: \quad & s(r, z, 0) = 0 \\
\mathrm{BC}: \quad & s(r \to \infty, z, t) = 0 \\
& -R\frac{\partial s}{\partial z}(r, 0, t) = q_z(t) \\
& \frac{\partial s}{\partial z}(r, H, t) = 0 \\
& -2\pi a T\left.\frac{\partial s}{\partial r}\right|_{r=a} = b\left[s - S_w(t)\right], \qquad 0 \le z \le L
\end{aligned}$$

   where b is some coefficient relating the drawdown in the domain with that in the well, $S_w(t)$.
   Hint: delineate the problem as: $s\left(r, z, t\right) = s_1\left(r, z, t\right) + s_2\left(r, z, t\right)$.

References

Abramowitz, Stegun (1972) Handbook of mathematical functions. Dover Publications, New York

Hantush MS, Jacob CE (1955) Non-steady radial flow in an infinite leaky aquifer. Eos Trans AGU 36(1):95–100

Sethi R, Molfetta D (2019) Groundwater engineering. Springer Nature

Further Reading

Berry P, Reid D (1987) An introduction to soil mechanics. McGraw-Hill Book Company.

Sethi R, Di Molfetta A (2019) Groundwater engineering, a technical approach to hydrogeology, contaminant transport and groundwater remediation. Springer

Hydraulic Consolidation

R. Al Khoury and C. Kasbergen, *Analytical Subsurface Environment*,
https://doi.org/10.1007/978-3-031-93910-5_11

Hydraulic consolidation is an important process in soil engineering and mechanics, wherein water is expelled from a saturated or a partially saturated soil mass due to an increase in external pressure. This pressure is commonly induced by the placement of additional soil layers, the construction of heavy structures, or the application of other external loads. As the water is expelled, the soil particles move closer to one another, resulting in a reduction of the soil's volume. This reduction, known as soil consolidation, enhances the overall stability, bearing capacity, and strength of the soil mass. The process is particularly important in geo-engineering projects, where controlling the settlement and ensuring the stability of foundational soil layers is essential. Hydraulic consolidation plays a crucial role in large-scale infrastructure developments such as embankments, foundations, roadways, and dam construction. By mitigating potential soil deformation and instability, this process helps to ensure the long-term durability and safety of such structures.

In this chapter, we will explore various models that address hydraulic consolidation under different geometries, configurations, and boundary conditions. These models are designed to simulate real-world scenarios, providing insight into the consolidation behavior of soil layers under a range of loading and drainage conditions.

11.1 Terzaghi 1D Consolidation

Terzaghi's one-dimensional consolidation theory is a widely used model for simulating soil consolidation under vertical loads. The theory proposes that during consolidation, water is expelled from the soil pores due to the application of vertical loads, resulting in an increase in the effective stress and a decrease in the soil volume. This process continues until the rate of water expulsion decreases, and the soil reaches a stable, fully consolidated state. Terzaghi's formulation assumes that the generated excess pore pressure is equal to the applied load at time $t = 0$, and that the whole soil layer initially exhibits the same excess pore pressure. With time, the excess pore pressure dissipates via the seepage surfaces. If the seepage surface is at the upper boundary of the layer, $z = 0$, the initial value problem for this case can be expressed as

$$\frac{\partial^2 u_{ei}}{\partial z^2} - \frac{1}{c_v}\frac{\partial u_{ei}}{\partial t} = 0$$

$$\begin{aligned} \text{IC:} \quad & u_{ei}(z,0) = P \\ \text{BC:} \quad & u_{ei}(0,t) = 0 \\ & \left.\frac{\partial u_{ei}}{\partial z}\right|_{z=H} = 0 \end{aligned} \tag{11.1}$$

where the subscript *ei* refers to the initial excess pore pressure, and $c_v\left(m^2\ s^{-1}\right)$ is the coefficient of vertical consolidation.

This formulation has two main shortcomings:

1. The external load is assumed to be continuous and instantly applied, which leads to thc assumption that the pore water carries the entire load immediately after load application. This assumption oversimplifies the process by neglecting the time-dependent nature of pore pressure propagation, which is governed by the soil's permeability and compressibility. Consequently, Terzaghi's assumption may not accurately reflect the actual mechanics, particularly in thicker layers or soils with low permeability.
2. Even if this assumption is reasonable for uniform, one-dimensional domains, it is not valid for two and three-dimensional domains, where the excess pore pressure dissipation occurs not only via the seepage surfaces, but also to directions away from the area of the applied load. Therefore, Terzaghi's model is not suitable for cases with radial or multi-dimensional seepage, which can significantly influence the rate of consolidation.

To address these shortcomings, we rely on the physical fact that the domain is initially with zero excess pore pressure. When an external load is applied, excess pore pressure starts to rise, with the pore water at the ground surface, $z = 0$, being the first to bear the load. This formulation accounts for the construction period, as typically observed in practice. Since groundwater movement is dependent on the pressure gradient and permeability, the commence of consolidation in deeper layers and regions away from the applied load experiences a time delay.

11.2 1D Consolidation: Single Drainage

We begin by modeling a one-dimensional consolidation of a single drainage (seepage) soil layer, resting on an impermeable boundary. Figure 11.1 shows schematically a soil layer bounded by a top seepage surface and a bottom impermeable layer.

The boundary value problem of this case can be expressed as

$$\frac{\partial^2 u_e}{\partial z^2} - \frac{1}{c_v}\frac{\partial u_e}{\partial t} = 0$$

$$\begin{aligned} \text{IC:}\quad & u_e(z,0) = 0 \\ \text{BC:}\quad & u_e(0,t) = P(t) \\ & \left.\frac{\partial u_e}{\partial z}\right|_{z=H} = 0 \end{aligned} \tag{11.2}$$

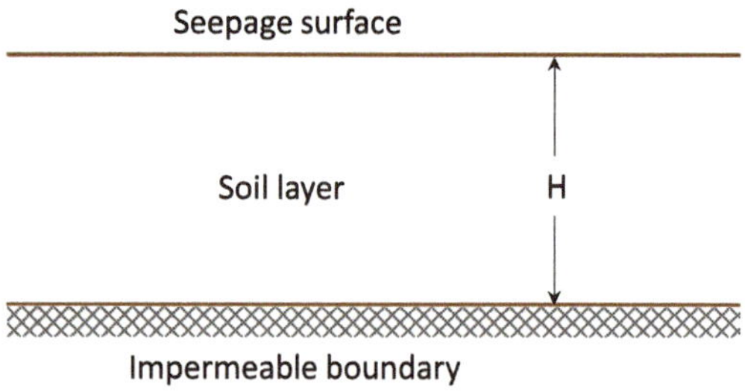

Fig. 11.1 A soil layer with top seepage surface and bottom impermeable layer

General Solution

Applying the Fourier transform to (11.2) gives

$$\frac{d^2\hat{u}_e}{dz^2} - \frac{i\omega}{c_v}\hat{u}_e = 0$$

$$\text{BC:} \quad \hat{u}_e(0,\omega) = \hat{P}(\omega) \qquad \left.\frac{d\hat{u}_e}{dz}\right|_{z=H} = 0 \tag{11.3}$$

This differential equation is one of the standard forms, and the solution can be readily expressed as

$$\hat{u}_e = a_1 e^{-\kappa z} + a_2 e^{\kappa z}, \quad \kappa = \sqrt{i\omega/c_v} \tag{11.4}$$

Boundary Conditions

At $z = 0$, the first boundary condition in (11.3) implies

$$a_1 + a_2 = \hat{P} \tag{11.5}$$

Taking the derivative of (11.4) with respect to z gives

$$\frac{d\hat{u}_e}{dz} = -a_1 \kappa e^{-\kappa z} + a_2 \kappa e^{\kappa z}$$

At $z = H$, the second boundary condition in (11.3) implies

$$-a_1 e^{-\kappa H} + a_2 e^{\kappa H} = 0 \tag{11.6}$$

Putting (11.5) and (11.6) in a matrix form and solving for the constants yields

$$a_1 = \frac{\hat{P}}{e^{\kappa H} + e^{-\kappa H}} e^{\kappa H}, \quad a_2 = \frac{\hat{P}}{e^{\kappa H} + e^{-\kappa H}} e^{-\kappa H}$$

BVP Solution

Having determined the constants, the solution of this BVP in the frequency domain would then read

$$\hat{u}_e(z,\omega_n) = \frac{\hat{P}\,(\omega_n)}{e^{\kappa H} + e^{-\kappa H}}\left[e^{-\kappa(z-H)} + e^{\kappa(z-H)}\right]$$

with

$$\kappa = \sqrt{i\omega/c_v} \tag{11.7}$$

Average Consolidation

In geo-engineering, it is common practice to determine the average value of the excess pore pressure. This can be done via

$$\hat{u}_{av} = \frac{1}{H}\int_0^H \hat{u}_e dz \tag{11.8}$$

Applying this integral to (11.7), it can readily be shown that the average value for this case is

$$\hat{u}_{av} = \frac{\hat{P}}{H\kappa}\left(\frac{e^{2\kappa H} - 1}{e^{2\kappa H} + 1}\right) \tag{11.9}$$

Or

$$\hat{u}_{av} = \frac{\hat{P}}{H\kappa}\left(\frac{1 - e^{-2\kappa H}}{1 + e^{-2\kappa H}}\right) \tag{11.10}$$

Having solved the problem in the frequency domain, the solution in the time domain can be obtained using the inverse fast Fourier transform (IFFT) algorithm, as

$$u_e(z,t) = \sum_n \hat{u}_e(z,\omega_n)\, e^{i\omega_n t}$$

Degree of Consolidation

Terzaghi has introduced the concept of *degree of consolidation*, which is effectively the degree of dissipation of excess pore pressure. It is defined as

$$U_v = 1 - \frac{u_e}{u_i}$$

where u_i is the initial excess pore pressure, taken equals to the applied pressure P.

Example 11.1: Terzaghi 1D Consolidation

Consider a single-drain soil layer, 5m thick with infinite lateral boundaries. A uniform embankment exerting 10 kPa pressure has been placed on the upper surface of the soil layer. Figure 11.2 shows the geometry of the domain along with the soil parameters. A load resembling an embankment is applied for nearly 1400 h, after which it was removed, a time when the consolidation process effectively begins, see Fig. 11.3. (A relatively long time is implemented in this example to ensure that the entire soil layer undergoes an excess pore pressure equal to the applied pressure, approximating the initial condition in Terzaghi's formulation.) Calculate the excess pore pressure in the domain and discuss the results. Also, plot the degree of consolidation versus time at $z = 0, 1$ and 5 m.

Solution

In solving this problem, we begin by transforming the applied load signal shown in 0 from the time domain to the frequency domain using the forward FFT. Here we used 4096 FFT samples with a sampling rate of 7 h (25,200 s). Having the frequency spectrum and the associated magnitudes, we then solve the system for every frequency using (11.7). Finally, we employ the inverse FFT to restore the solution to the time domain.

11

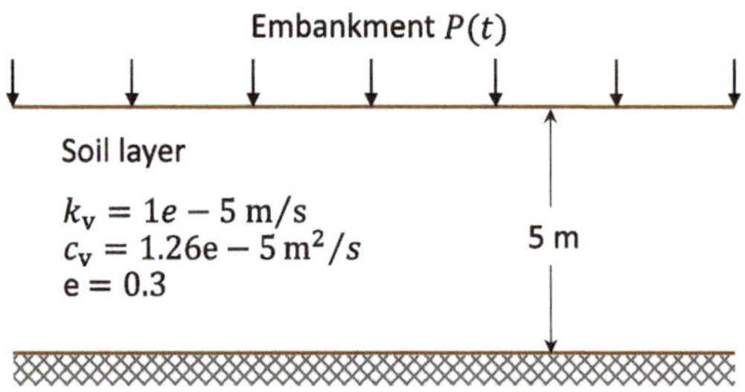

Fig. 11.2 Geometry, boundary conditions, and material parameters of Example 12.1

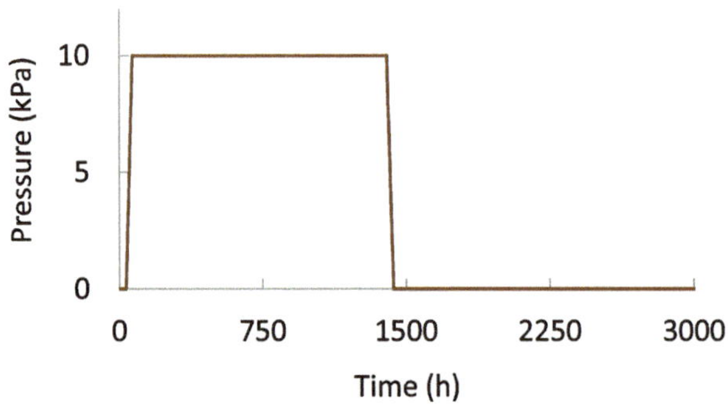

Fig. 11.3 Applied load representing the placement of the embankment

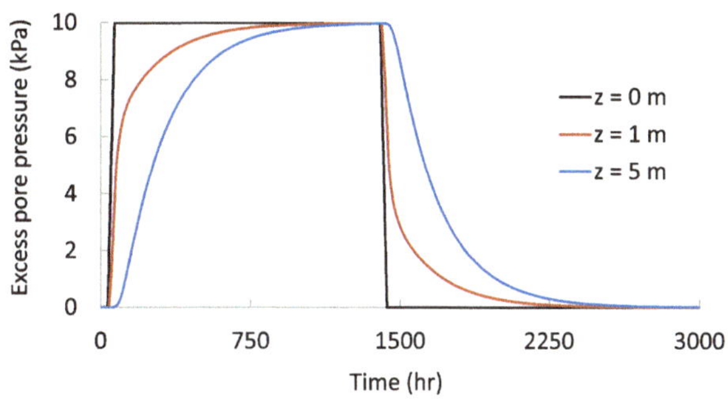

Fig. 11.4 Calculated excess pore pressure at different depths

Figure 11.4 shows the calculated excess pore pressure at various depths within the soil layer. Two distinct processes can be observed in this figure: construction and consolidation. During construction, the pore water at the seepage surface, $z = 0$ m, carries the applied pressure immediately. Upon removing the load, the accumulated excess pore pressure during construction instantly seeps away from the upper boundary. Deeper into the layer, the buildup of excess pore pressure takes time, with the bulk of the buildup time occurring at the bottom of the soil layer at $z = 5$ m. When the external pressure is totally released, the consolidation process begins, with points closer to the seepage surface exhibiting faster dissipation of excess pore pressure than those located deeper within the soil layer.

In engineering practice, the consolidation of a soil layer is presented by a semi-log diagram of the degree of consolidation $U_v = 1 - u_e/P$ versus the time factor $T_v = c_v t/H^2$, where t is time and H is the thickness of the layer. Figure 11.5 shows the degree of consolidation at different depths within the layer versus the time factor. Apparently, the excess pore pressure at the seepage surface dissipates immediately after the construction period stops.

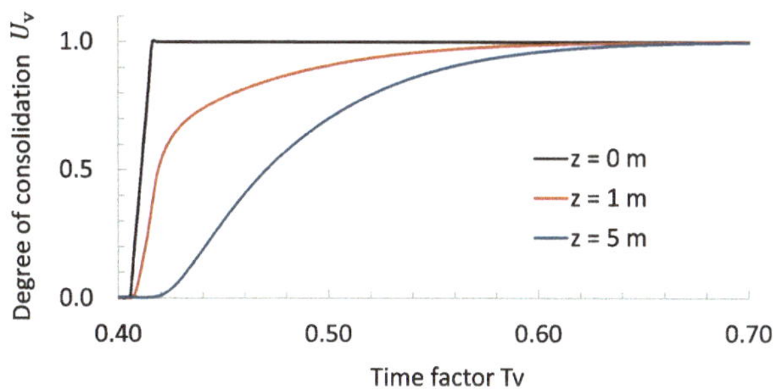

Fig. 11.5 Degree of consolidation at different depths

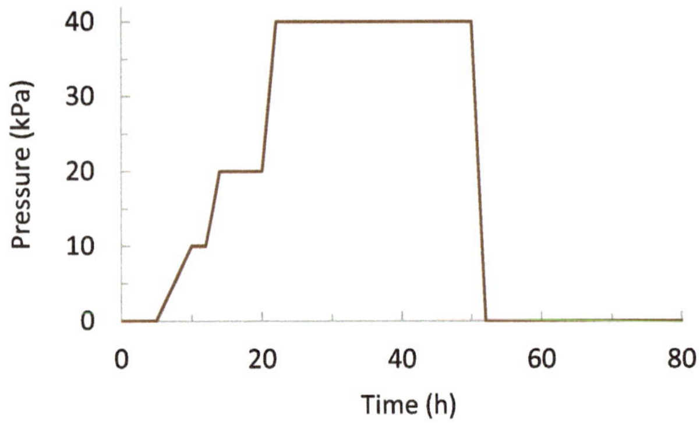

Fig. 11.6 Staged construction of an external pressure

Example 11.2: Staged Construction

Illustrate the effect of staged construction on the buildup of excess pore pressure and consolidation at different points in a soil layer, highlighting on the limitations of Terzaghi's formulation. Then, using the geometry and material given in Fig. 11.2, calculate the buildup and dissipation of excess pore pressure resulting from a surcharge applied in stages, as shown in Fig. 11.6.

Solution

In practice, embankments and buildings are often constructed in stages. This means that, at any given time, points at different depths in the soil layers exhibit varying excess pore pressures. The consolidation process at these points effectively begins after the buildup of excess pore pressure is complete during construction. This mounting of excess pore pressure during construction and the commence of the consolidation process at different points in time cannot be modeled using classical Terzaghi's formulation. The assumption that all points in the soil layer experience the same excess pore pressure at time $t = 0$ does not hold for staged construction. Instead, the formulation provided in this section is required.

Following the solution procedure outlined in the previous example, this problem is addressed using (11.7) and subsequently conducting the inverse FFT algorithm to transform the solution from the frequency domain to the time domain. Figure 11.7 shows the computed excess pore pressure at various depths within the soil layer. Consistent with physical observations, the figure reveals that points at different depths in the soil layer exhibit variations in the magnitude of excess pore pressure and the onset times of the consolidation process.

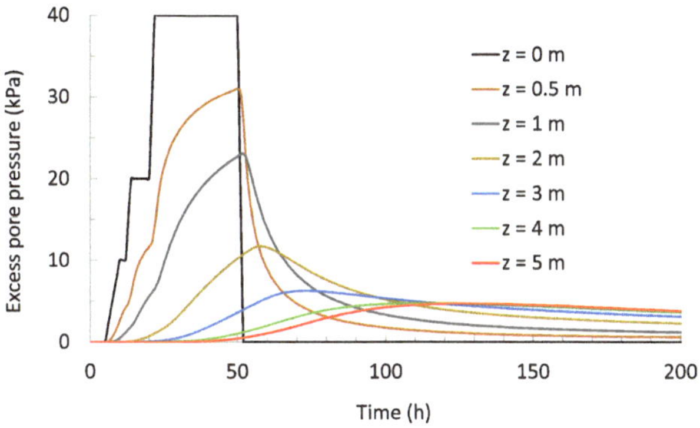

Fig. 11.7 Excess pore pressure due to staged construction

11.3 1D Consolidation: Double Drainage

In many subsurface formations, soil layers may rest on a fully permeable layer allowing water to seep out through both the top and bottom surfaces. Figure 11.8 schematically depicts a soil layer undergoing one-dimensional consolidation with double drainage surfaces.

The boundary value problem of this case is

$$\frac{\partial^2 u_e}{\partial z^2} - \frac{1}{c_v}\frac{\partial u_e}{\partial t} = 0$$

$$\begin{aligned} \text{IC:} \quad & u_e(z,0) = 0 \\ \text{BC:} \quad & u_e(0,t) = P_1(t) \\ & u_e(H,t) = P_2(t) \end{aligned} \qquad (11.11)$$

where P_1 is the pressure exerted by a building or an embankment, and P_2 is some pressure exerted on the bottom boundary of the soil layer, most probably equals to P_1.

General Solution

As for the previous case, the general solution of the differential equation in (11.11), in the frequency domain, is given by

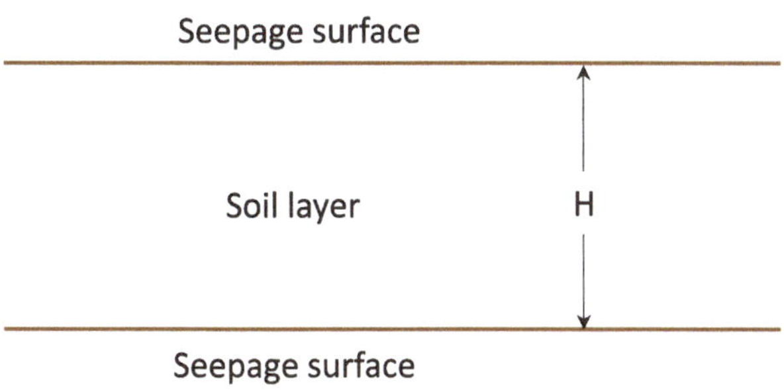

Fig. 11.8 Double drainage soil layer

$$\hat{u}_e = a_1 e^{-\kappa z} + a_2 e^{\kappa z}, \quad \kappa = \sqrt{i\omega / c_v} \tag{11.12}$$

Boundary Conditions

At $z = 0$, the first boundary condition in (11.11), in the frequency domain, implies

$$a_1 + a_2 = \hat{P}_1$$

At $z = H$, the second boundary condition implies

$$a_1 e^{-\kappa H} + a_2 e^{\kappa H} = \hat{P}_2$$

Solving these equations simultaneously, we get

$$a_1 = \frac{e^{\kappa H}\hat{P}_1 - \hat{P}_2}{e^{\kappa H} - e^{-\kappa H}}, \quad a_2 = \frac{-e^{-\kappa H}\hat{P}_1 + \hat{P}_2}{e^{\kappa H} - e^{-\kappa H}} \tag{11.13}$$

BVP Solution

Substituting (11.13) into (11.12), the solution of this BVP in the frequency domain would read

$$\hat{u}_e(z, \omega_n) = \frac{1}{e^{\kappa H} - e^{-\kappa H}} \begin{bmatrix} \left(e^{\kappa H}\hat{P}_1 - \hat{P}_2\right) e^{-\kappa z} \\ + \left(\hat{P}_2 - e^{-\kappa H}\hat{P}_1\right) e^{\kappa z} \end{bmatrix} \tag{11.14}$$

11

Average Consolidation

As for the previous case, the average consolidation can be calculated via

$$\hat{u}_{av} = \frac{1}{H} \int_0^H \hat{u}_e dz \tag{11.15}$$

Applying this integral to (11.14), it can readily be shown that the average consolidation is

$$\hat{u}_{av} = \frac{\left(\hat{P}_1 + \hat{P}_2\right)\left(e^{\kappa H} - 1\right)}{H\kappa \left(e^{\kappa H} + 1\right)} \tag{11.16}$$

Having solved the boundary value problem in the frequency domain, the solution in the time domain can then be obtained using the inverse fast Fourier transform (IFFT) algorithm, as

$$u_e(z, t) = \sum_n \hat{u}_e(z, \omega_n)\, e^{i\omega_n t}.$$

11.4 Axial Symmetric Consolidation: Circular Tanks

Circular tanks are used all over the world for various purposes, such as drinking water storage, sedimentation, rainwater storage, aquaculture, and oil refineries, among others. During both construction and usage stages, these tanks generate excess pore pressure due to the continuous filling and emptying of fluids. The tank load generates an axial symmetric excess pore pressure in soil that dissipates through the seepage surfaces and in directions away from the tank.

The relevant boundary value problem of this case can be expressed as

$$c_r \frac{\partial^2 u_e}{\partial r^2} + \frac{c_r}{r}\frac{\partial u_e}{\partial r} + c_z \frac{\partial^2 u_e}{\partial z^2} - \frac{\partial u_e}{\partial t} = 0$$

$$\begin{aligned} &\text{IC}: \quad u_e(r, z, 0) = 0 \\ &\text{BC}: \quad u_e(r, 0, t) = U_0(r, t), \quad 0 \le r \le a \\ &\qquad\quad u_e(\ell, z, t) = 0, \qquad\qquad \ell \gg a \end{aligned} \tag{11.17}$$

where a is the radius of the tank, and ℓ is some far distance, beyond which the consolidation is no longer significant. The ground surface acts here as a seepage surface, see Fig. 11.9.

General Solution

Applying the Fourier transform to (11.17), the boundary value problem in the frequency domain becomes

$$c_r \frac{\partial^2 \hat{u}_e}{\partial r^2} + \frac{c_r}{r}\frac{\partial \hat{u}_e}{\partial r} + c_z \frac{\partial^2 \hat{u}_e}{\partial z^2} - i\omega \hat{u}_e = 0$$

$$\begin{aligned} &\text{BC}: \quad \hat{u}_e(r, 0, \omega) = \hat{U}_0(r, \omega), \quad 0 \le r \le a \\ &\qquad\quad \hat{u}_e(\ell, z, \omega) = 0, \qquad\qquad \ell \gg a \end{aligned} \tag{11.18}$$

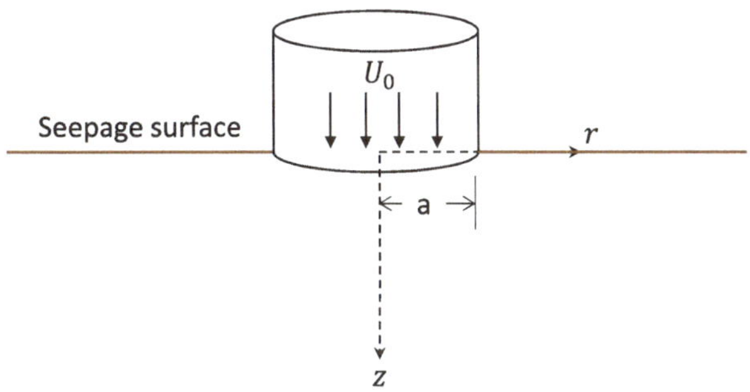

Fig. 11.9 A circular tank on a half space

Using the separation of variables, we first assume

$$\hat{u}_e = R(r)Z(z) \tag{11.19}$$

Taking the first and second derivatives of this equation with respect to r and z, then, substituting the results into (11.18), and dividing by $R(r)Z(z)$, gives

$$\frac{c_r}{R(r)}\frac{\partial^2 R(r)}{\partial r^2} + \frac{c_r}{R(r)}\frac{1}{r}\frac{\partial R(r)}{\partial r} + \frac{c_z}{Z(z)}\frac{\partial^2 Z(z)}{\partial z^2} - i\omega = 0 \tag{11.20}$$

This equation can be divided into two separate equations by equating them to some constant, say $-\kappa_z^2$, yielding

$$\begin{aligned} &\frac{d^2 R(r)}{\partial r^2} + \frac{1}{r}\frac{dR(r)}{\partial r} + \left(\frac{\kappa_z^2 - i\omega}{c_r}\right)R(r) = 0 \\ &\frac{d^2 Z(z)}{\partial z^2} - \frac{\kappa_z^2}{c_z}Z(z) = 0 \end{aligned} \tag{11.21}$$

The solution of $Z(z)$ in (11.21), considering that z goes to infinity, can be expressed as

$$Z(z) = a_1 e^{-\kappa_z z/\sqrt{c_z}}$$

where a_1 is an arbitrary constant that needs to be determined from the boundary conditions.

The differential equation of $R(r)$ in (11.21) is a sort of a Bessel equation, with its general solution given by

$$R(r) = a_2 J_0(\kappa_r r) + a_3 Y_0(\kappa_r r), \quad \kappa_r = \sqrt{\frac{\kappa_z^2 - i\omega}{c_r}}$$

in which J_0 and Y_0 are the Bessel functions of zero order of the first and second kind, respectively, and a_2 and a_3 are arbitrary constants. As Y_0 is unbounded at $r = 0$, a_3 must be zero, yielding

$$R(r) = a_2 J_0(\kappa_r r)$$

Substituting the solutions of $Z(z)$ and $R(r)$ into (11.19) gives

$$\hat{u}_e = A e^{-\kappa_z z/\sqrt{c_z}} J_0(\kappa_r r) \tag{11.22}$$

where $A = a_1 a_2$.

Boundary Conditions

The second boundary condition in (11.18) implies that at some far distance from the applied load, ℓ, the consolidation becomes insignificant. This applies to all z, leading to

$$J_0(\kappa_r \ell) = 0$$

This condition can be satisfied at the infinitely many positive roots μ_m of J_0 function, implying

$$\mu_m = \kappa_r \ell, \quad \Rightarrow \quad \kappa_r = \frac{\mu_m}{\ell}, \quad m = 1, 2, \ldots$$

which represent the eigenvalues of the system, and

$$J_0(\kappa_r r) = J_0\left(\frac{\mu_m}{\ell} r\right)$$

are their corresponding eigenfunctions.

Following the principle of superposition, the solution of this eigenfunction problem becomes

$$\hat{u}_e(r, z, \omega) = \sum_m A_m e^{-\kappa_z z/\sqrt{c_z}} J_0(\kappa_r r) \tag{11.23}$$

At $z = 0$, the first boundary condition in (11.18) implies

$$\sum_m A_m J_0(\kappa_r r) = \hat{U}_0(r), \quad 0 \le r \le a$$

This is a Bessel–Fourier series with its coefficients determined by (see Sect. 3.4 in ▶ Chap. 3)

$$A_m = \frac{2}{\ell^2 J_1^2(\mu_m)} \int_0^a r \hat{U}_0(r) J_0(\kappa_r r) dr \tag{11.24}$$

BVP Solution

The solution of the BVP (11.18) for a given frequency ω_n is thus

$$\hat{u}_e(r, z, \omega_n) = \sum_m A_m e^{-\kappa_z z/\sqrt{c_z}} J_0(\kappa_r r)$$

with

$$A_m = \frac{2}{\ell^2 J_1^2(\mu_m)} \int_0^a r\, \hat{U}_0(r) J_0(\kappa_r r)\, dr \tag{11.25}$$

$$\kappa_r = \frac{\mu_m}{\ell}$$

$$\kappa_z = \sqrt{c_r \kappa_r^2 + i\omega_n}$$

Usually $\hat{U}_0$ varies with time (frequency), but constant over $r \le a$. In this case the Fourier–Bessel coefficients can readily be expressed as

$$A_m = \frac{2a\, \hat{U}_0 J_1(\mu_m a/l)}{\mu_m l J_1^2(\mu_m)}$$

Having the solution in the frequency domain, the solution in the time domain can be obtained using the inverse fast Fourier transform (IFFT) algorithm, as

$$u_e(r, z, t) = \sum_n \hat{u}_e(r, z, \omega_n)\, e^{i\omega_n t}$$

▪ Example 11.3: Consolidation Due to a Circular Tank

A circular tank, 10 m in diameter, is placed on a fully saturated soil mass. ◘ Figure 11.10 shows schematically the soil mass and its material parameters. The figure also shows that the tank is loaded in two stages. It is required to calculate the excess pore pressure at three points along the axis of symmetry for $z = 0, 1$ and 2 m. Also, show the contour plots at different time intervals.

▪ Solution

Upon transforming the applied load signal from the time domain to the frequency domain, the excess pore pressure anywhere in the domain can be calculated using (11.25). Subsequently, by performing the inverse FFT algorithm, the solution is restored to the time domain.

◘ Figure 11.11 shows the calculated excess pore pressure at $r = 0$ and $z = 0, 1$ and 2 m. The figure demonstrates that the point directly underneath the tank experiences an immediate rise in excess pore pressure, which dissipates instantly upon

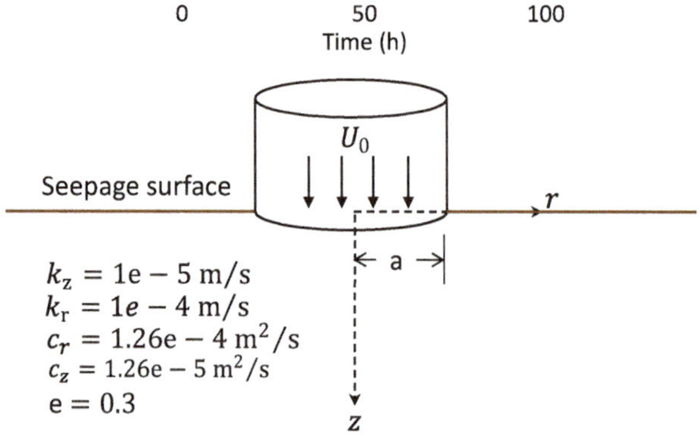

◘ **Fig. 11.10** Geometry, material, and boundary conditions

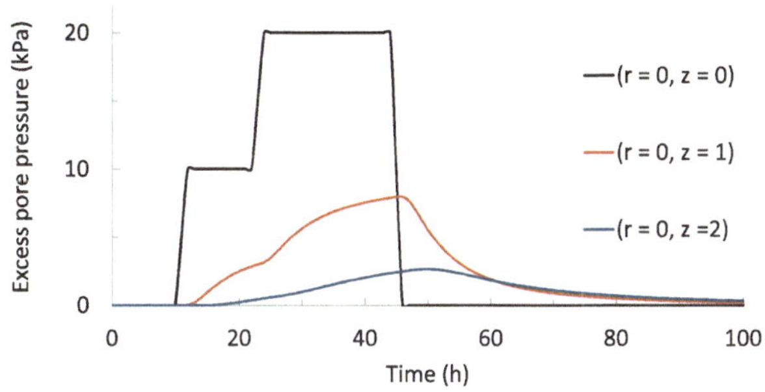

Fig. 11.11 Calculated excess pore pressure at three depths

the release of the external pressure. Deeper in the soil layer, the excess pore pressure begins to rise after some time. At a depth of $z = 1$ m, the soil senses the construction load after around 14 h, whereas at $z = 2$ m, it starts after 17 h.

Figure 11.12 shows the contour plots after 12, 20, 34, and 40 h of applying the load. The plots depict the buildup of the

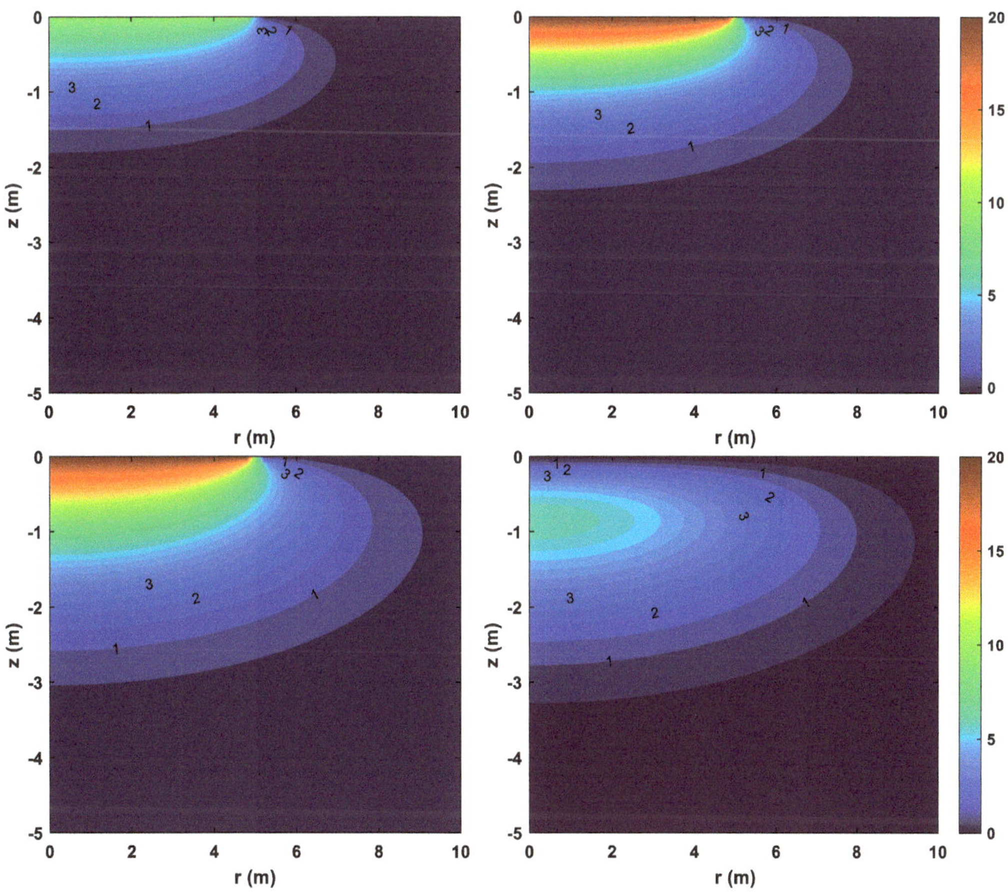

Fig. 11.12 Contour plots of excess pore pressure after 12 h (up left), 20 h (up right), 34 h (bottom left) and 40 h (bottom right) from the beginning of loading

excess pore pressure following the staged construction shown in Fig. 11.10. A remarkable behavior of the model can be recognized from the bottom right plot, where it shows the dissipation of the excess pore pressure after removing the load.

11.5 2D Consolidation: A Plane Strain Condition

Two-dimensional consolidation typically arises from the construction of roads and earth embankments, as illustrated schematically in Fig. 11.13. Excess pore pressure generated within these structures generally dissipates under a two-dimensional plane strain condition. Consequently, the boundary value problem for this geometry can be expressed as

$$c_x \frac{\partial^2 u_e}{\partial x^2} + c_z \frac{\partial^2 u_e}{\partial z^2} - \frac{\partial u_e}{\partial t} = 0$$

$$\begin{aligned} &\text{IC :} \quad u_e(x, z, 0) = 0 \\ &\text{BC:} \quad u_e(x, 0, t) = U_0(x, t), \quad -a \le x \le a \\ &\qquad\;\; u_e(\ell, z, t) = 0, \qquad \ell >> a \end{aligned} \tag{11.26}$$

where a is half the width of the embankment, and ℓ is some far distance where the consolidation becomes insignificant.

General Solution

Applying the Fourier transform to (11.26) leads to

$$c_x \frac{\partial^2 \hat{u}_e}{\partial x^2} + c_z \frac{\partial^2 \hat{u}_e}{\partial z^2} - i\omega \hat{u}_e = 0$$

$$\begin{aligned} &\text{BC:} \quad \hat{u}_e(x, 0, \omega) = \hat{U}_0(x, \omega), \quad 0 \le x \le a \\ &\qquad\;\; \hat{u}_e(\ell, z, \omega) = 0, \qquad \ell >> a \end{aligned} \tag{11.27}$$

where here we considered only half of the domain due to its symmetry.

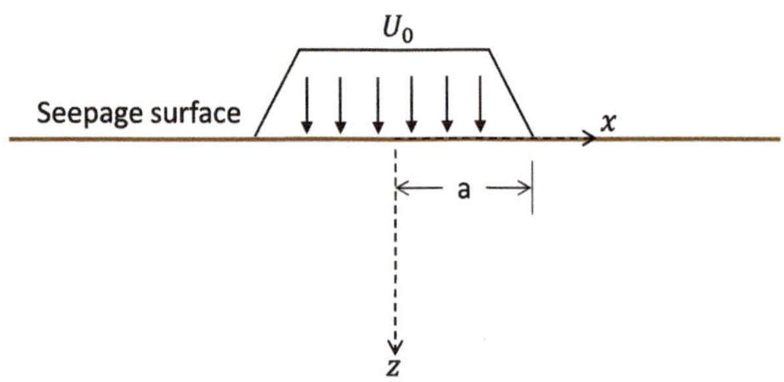

Fig. 11.13 Two-dimensional plane strain geometry

Using the separation of variables, we first assume that

$$\hat{u}_e = X(x)Z(z) \tag{11.28}$$

Taking the second derivatives of this function with respect to x and z, substituting the derivatives into (11.27), dividing by $X(x)Z(z)$, and equating the resulting two equations to some constant, say $-\kappa_z^2$, we obtain

$$\begin{aligned} &\frac{d^2X(x)}{\partial x^2} + \kappa_x^2 X(x) = 0, \quad \kappa_x^2 = \frac{\kappa_z^2 - i\omega}{c_x} \\ &\frac{d^2Z(z)}{\partial z^2} - \frac{\kappa_z^2}{c_z} Z(z) = 0 \end{aligned} \tag{11.29}$$

The general solution of $Z(z)$ in (11.29), considering that z goes to infinity, is

$$Z(z) = a_1 e^{-\kappa_z z/\sqrt{c_z}} \tag{11.30}$$

where a_1 is an arbitrary constant that needs to be determined from the boundary conditions.

The general solution of $X(x)$ in (11.29) is

$$X(x) = a_2 \cos(\kappa_x x) + a_3 \sin(\kappa_x x)$$

As the problem is symmetric around the z-axis, the solution can be reduced to

$$X(x) = a_2 \cos(\kappa_x x) \tag{11.31}$$

Combining the solutions given in (11.30) and (11.31), the general solution can then be expressed as

$$\hat{u}_e = A \cos(\kappa_x x)\, e^{-\kappa_z z/\sqrt{c_z}} \tag{11.32}$$

where $A = a_1 a_2$.

Boundary Conditions

The second boundary condition in (11.27), valid to all z, implies

$$\cos(\kappa_x \ell) = 0$$

This condition can be satisfied only if

$$\kappa_x \equiv \kappa_{xm} = \frac{1}{l}\left(\frac{\pi}{2} + m\pi\right)$$

Accordingly, following (11.29), we obtain

$$\kappa_z \equiv \kappa_{zm} = \sqrt{c_x \kappa_{xm}^2 + i\omega}$$

Thus, the solution becomes

$$\hat{u}_e = \sum_m A_m \cos(\kappa_{xm} x) e^{-\kappa_{zm} z / \sqrt{c_z}} \tag{11.33}$$

At $z = 0$, the first boundary condition in (11.27) implies

$$\hat{U}_0(x) = \sum_m A_m \cos(\kappa_{xm} x), \quad m = 0, 1, 2, \ldots$$

This is a cosine Fourier series with its coefficient given by

$$A_m = \frac{2}{\ell} \int_0^a \hat{U}_0(x) \cos(\kappa_{xm} x) dx, \quad m = 0, 1, 2, \ldots \tag{11.34}$$

BVP Solution

The solution of the BVP (11.27) for a given frequency ω_n is thus

$$\hat{u}_e(x, z, \omega_n) = \sum_m A_m \cos(\kappa_{xm} x) e^{-\kappa_{zm} z / \sqrt{c_z}}, \quad m = 0, 1, 2, \ldots$$

with

$$\begin{aligned} A_m &= \frac{2}{\ell} \int_0^a \hat{U}_0(x) \cos(\kappa_{xm} x) dx \\ \kappa_{xm} &= \frac{1}{\ell} \left(\frac{\pi}{2} + m\pi \right) \\ \kappa_{zm} &= \sqrt{c_x \kappa_{xm}^2 + i\omega_n} \end{aligned} \tag{11.35}$$

Having solved the problem in the frequency domain, the solution in the time domain can be obtained using the inverse fast Fourier transform (IFFT) algorithm, as

$$u_e(x, z, t) = \sum_n \hat{u}_e(x, z, \omega_n) e^{i\omega_n t}$$

11.6 2D Consolidation: Aquifer Bounded by an Impermeable Layer

In this section, we address again the plane strain consolidation, but for an aquifer bounded from the bottom side by an impermeable layer, see Fig. 11.14. The relevant boundary value problem of this case can be expressed as

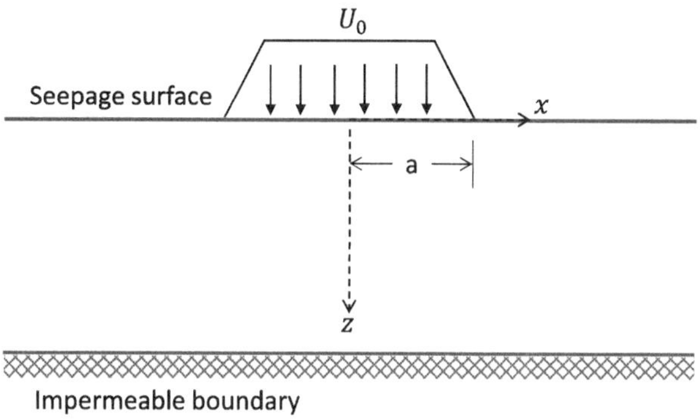

Fig. 11.14 Two-dimensional plane strain bounded aquifer

$$
\begin{aligned}
& c_x \frac{\partial^2 u_e}{\partial x^2} + c_z \frac{\partial^2 u_e}{\partial z^2} - \frac{\partial u_e}{\partial t} = 0 \\
& \text{IC}: \quad u_e(x, z, 0) = 0 \\
& \text{BC}: \quad u_e(x, 0, t) = U_0(x, t), \quad 0 \le x \le a \\
& \qquad \left. \frac{\partial u_e}{\partial z} \right|_{z=H} = 0 \\
& \qquad u_e(\ell, z, t) = 0, \qquad \ell \gg a
\end{aligned}
\tag{11.36}
$$

where a is half the width of the embankment, and ℓ is some far distance where the consolidation vanishes.

General Solution

Applying the Fourier transform to (11.36), this boundary value problem becomes

$$
\begin{aligned}
& c_x \frac{\partial^2 \hat{u}_e}{\partial x^2} + c_z \frac{\partial^2 \hat{u}_e}{\partial z^2} - i\omega \hat{u}_e = 0 \\
& \text{BC}: \quad \hat{u}_e(x, 0, \omega) = \hat{U}_0(x, \omega), \quad 0 \le x \le a \\
& \qquad \left. \frac{\partial \hat{u}_e}{\partial z} \right|_{z=H} = 0 \\
& \qquad \hat{u}_e(\ell, z, \omega) = 0, \qquad \ell \gg a
\end{aligned}
\tag{11.37}
$$

Using the separation of variables, we first assume

$$\hat{u}_e = X(x)Z(z) \tag{11.38}$$

Then, taking the second derivative of this function with respect to x and z, substituting the resulting derivatives into (11.37), dividing by $X(x)Z(z)$, and equating the result to some constant, say $-\kappa_z^2$, we obtain

$$\frac{d^2X(x)}{\partial x^2} + \kappa_x^2 X(x) = 0, \quad \kappa_x^2 = \frac{\kappa_z^2 - i\omega}{c_x}$$
$$\frac{d^2Z(z)}{\partial z^2} - \frac{\kappa_z^2}{c_z} Z(z) = 0 \tag{11.39}$$

The general solution of $Z(z)$ in (11.39) is

$$Z(z) = a_1 e^{-\kappa_z z/\sqrt{c_z}} + a_2 e^{\kappa_z z/\sqrt{c_z}} \tag{11.40}$$

where a_1 and a_2 are arbitrary constants that need to be determined from the boundary conditions.

The general solution of $X(x)$ in (11.39) is given by

$$X(x) = a_3 \cos(\kappa_x x) + a_4 \sin(\kappa_x x)$$

Since the problem is symmetric around the z-axis, the general solution can be reduced to

$$X(x) = a_3 \cos(\kappa_x x) \tag{11.41}$$

Combining both solutions in (11.40) and (11.41), the general solution of the differential equation in (11.37) can then be expressed as

$$\hat{u}_e = \cos(\kappa_x x)\left(A_1 e^{-\kappa_z z/\sqrt{c_z}} + A_2 e^{\kappa_z z/\sqrt{c_z}}\right) \tag{11.42}$$

where $A_1 = a_1 a_3$ and $A_2 = a_2 a_3$.

Boundary Conditions

Taking the derivative of (11.42) with respect to z yields

$$\frac{\partial \hat{u}_e}{\partial z} = \cos(\kappa_x x)\left[-A_1\left(\kappa_z/\sqrt{c_z}\right) e^{-\kappa_z z/\sqrt{c_z}} + A_2\left(\kappa_z/\sqrt{c_z}\right) e^{\kappa_z z/\sqrt{c_z}}\right]$$

At $z = H$, the second boundary condition in (11.37), valid for all x, implies

$$-A_1\left(\kappa_z/\sqrt{c_z}\right) e^{-\kappa_z H/\sqrt{c_z}} + A_2\left(\kappa_z/\sqrt{c_z}\right) e^{\kappa_z H/\sqrt{c_z}} = 0$$

This results in

$$A_2 = e^{-2\kappa_z H/\sqrt{c_z}} A_1$$

Thus

$$\hat{u}_e = A_1 \cos(\kappa_x x)\left[e^{-\kappa_z z/\sqrt{c_z}} + e^{-2\kappa_z H/\sqrt{c_z}} e^{\kappa_z z/\sqrt{c_z}}\right]$$

The third boundary condition in (11.37), implies, for all z,

$$\cos(\kappa_x \ell) = 0$$

This entails

$$\kappa_x \equiv \kappa_{xm} = \frac{1}{\ell}\left(\frac{\pi}{2} + m\pi\right)$$

from which, and according to (11.39), we obtain

$$\kappa_z \equiv \kappa_{zm} = \sqrt{c_x \kappa_{xm}^2 + i\omega}$$

Thus, the solution so far is

$$\hat{u}_e = \sum_m A_m \cos(\kappa_{xm} x) e^{-\kappa_{zm} z/\sqrt{c_z}} \left(1 + e^{-2\kappa_{zm}(H-z)/\sqrt{c_z}}\right), \quad m = 0, 1, 2, \ldots \tag{11.43}$$

where we replaced A_1 by A_m.

At $z = 0$, the first boundary condition in (11.37) implies

$$\hat{U}_0(x) = \sum_m A_m \cos(\kappa_{xm} x)\left(1 + e^{-2\kappa_{zm} H/\sqrt{c_z}}\right)$$

This is a typical Fourier cosine series with its coefficients given by

$$A_m = \frac{2}{\ell\left(1 + e^{-2\kappa_{zm} H/\sqrt{c_z}}\right)} \int_0^a \hat{U}_0(x)\cos(\kappa_{xm} x)\, dx, \quad m = 0, 1, 2, \ldots \tag{11.44}$$

BVP Solution

The solution of the BVP (11.37) for a given frequency ω_n is thus

$$\hat{u}_e(x, z, \omega_n) = \sum_m A_m \cos(\kappa_{xm} x) e^{-\kappa_{zm} z/\sqrt{c_z}} \left(1 + e^{-2\kappa_{zm}(H-z)/\sqrt{c_z}}\right), \quad m = 0, 1, 2, \ldots$$

with

$$A_m = \frac{2}{\ell\left(1 + e^{-2\kappa_{zm} H/\sqrt{c_z}}\right)} \int_0^a \hat{U}_0(x)\cos(\kappa_{xm} x) dx$$

$$\kappa_{xm} = \frac{1}{\ell}\left(\frac{\pi}{2} + m\pi\right)$$

$$\kappa_{zm} = \sqrt{c_x \kappa_{xm}^2 + i\omega_n} \tag{11.45}$$

Having solved the problem in the frequency domain, the solution in the time domain can then be obtained via the IFFT, as

$$u_e(x, z, t) = \sum_n \hat{u}_e(x, z, \omega_n) e^{i\omega_n t}.$$

Example 11.4: Consolidation Due to an Embankment

An embankment, 10 m wide, is placed on a 15 m deep fully saturated aquifer, bounded on the top by a seepage surface and on the bottom by an impermeable layer. Figure 11.15 shows schematically the aquifer and its material parameters. The figure also shows the embankment load signal during construction. Calculate the excess pore pressure in the aquifer and make contour plots showing the development of the excess pore pressure during construction and its dissipation afterwards.

Solution

The excess pore pressure for this bounded aquifer can be calculated in the frequency domain using (11.45). The frequency content and their associated magnitudes are determined by transforming the applied load signal shown in Fig. 11.15 from the time domain to the frequency domain using the FFT algorithm. Then, by solving the system in the frequency domain, the solution in the time domain can be reconstructed using the inverse FFT algorithm.

Figure 11.16 illustrates contour plots depicting the evolution of excess pore pressure during both the construction and post-construction phases of the embankment. The top two plots depict the progression of excess pore pressure at 20 and 80 h into the construction period. These plots highlight the immediate development of excess pore pressure directly

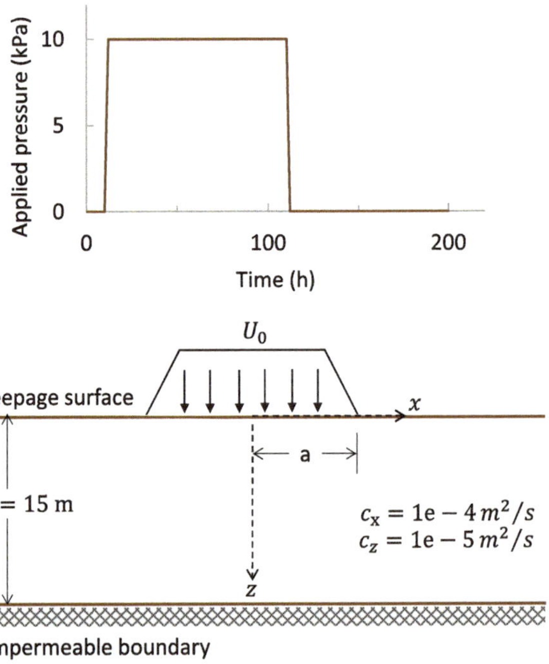

Fig. 11.15 Geometry of the aquifer and the applied embankment load

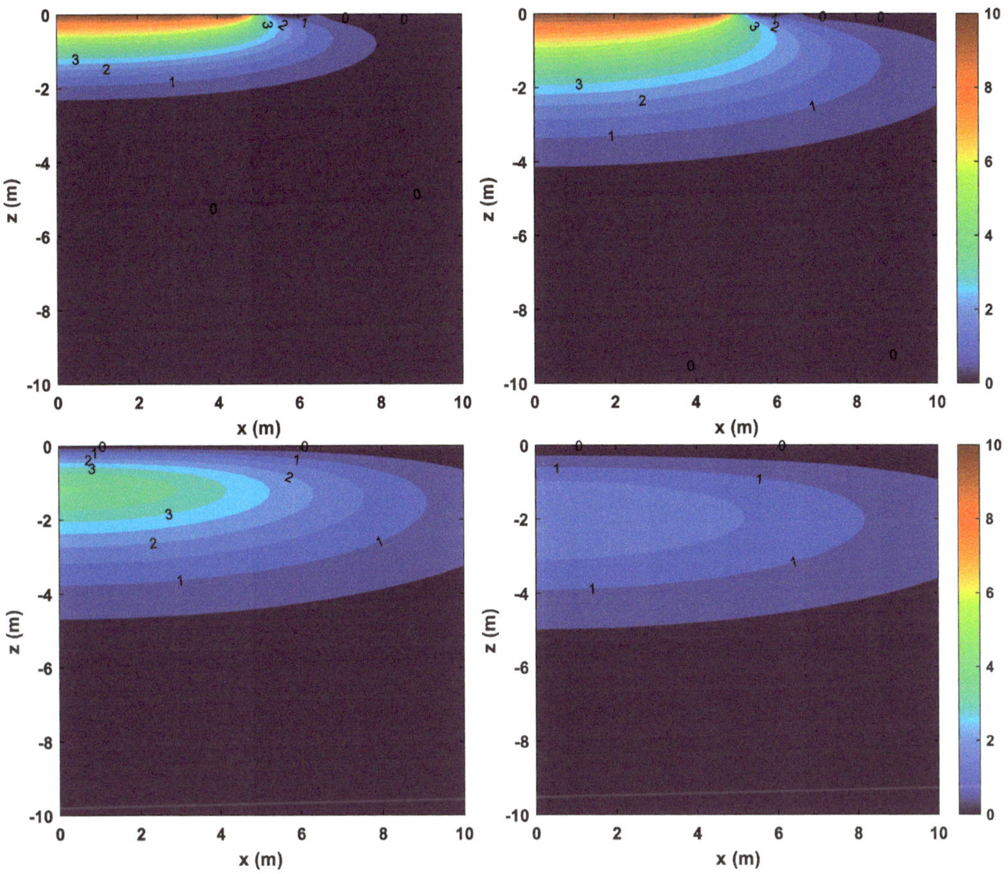

Fig. 11.16 Contour plots of excess pore pressure after 20 h (up left), 80 h (up right), 110 h (bottom left) and 130 h (bottom right) from the beginning of loading

beneath the applied load, contrasting with the absence of excess pore pressure on the seepage surface away from the load. Over time, the excess pore pressure diffuses into the aquifer, with greater dispersion observed in the x-direction due to its higher coefficient of consolidation. The bottom plots demonstrate the dissipation of excess pore pressure 10 and 30 h after the completion of the construction phase. These plots illustrate the gradual dissipation of excess pore pressure both through the seepage surface and across the entire domain.

11.7 Consolidation with Finite Radial Drainage: With Radius of Influence

In engineering practice, consolidation rates are often accelerated by installing vertical cylindrical drainage pipes or sand columns in the ground. The drainage pipe is often designed to have a radius of influence, r_e, beyond which the drain is ineffective, as shown schematically in Fig. 11.17. The seep-

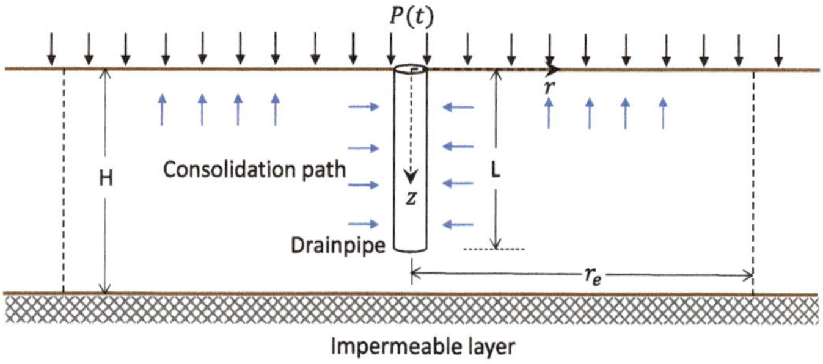

Fig. 11.17 Consolidation via seepage surface and a finite drainage pipe with a radius of influence

age in this system occurs in both vertical and radial directions, resulting in an effectively axially symmetric flow through the drainage pipe.

The boundary value problem of a soil layer resting on an impermeable layer and allowing seepage from its upper surface and via a finite drainage pipe with length L smaller than the layer thickness H can be expressed as

$$
\begin{aligned}
& c_r \frac{\partial^2 u_e}{\partial r^2} + \frac{c_r}{r}\frac{\partial u_e}{\partial r} + c_z \frac{\partial^2 u_e}{\partial z^2} - \frac{\partial u_e}{\partial t} = 0 \\
& \text{IC}: \quad u_e(r, z, 0) = 0 \\
& \text{BC}: \quad u_e(r, 0, t) = P(t) \\
& \qquad \left.\frac{\partial u_e}{\partial z}\right|_{z=H} = 0 \\
& \qquad 2\pi a K_s \left.\frac{\partial u_e}{\partial r}\right|_{r=a} = K_d[u_e - u_d(z)], \quad 0 \le z \le L \\
& \qquad u_e(r_e, z, t) = 0
\end{aligned}
\tag{11.46}
$$

where $P(t)$ is a prescribed surcharge pressure, a is the radius of the drain pipe, and r_e is its radius of influence and u_d is some excess pore pressure generated in the drain, which can be close to atmospheric. Notice that we have employed the Robin boundary condition along the drain boundary since, in this case, the dissipation rate of excess pore pressure depends not only on the consolidation parameters of the soil mass, such as permeability, but also on the discharge capacity of the drain. In this context, K_s represents a measure of the hydraulic conductivity of the drainage pipe, which depends on the perforation of the pipe material, or it may refer to the permeability of the smeared soil at the interface between the sand drain and the soil mass. Meanwhile, K_d denotes the permeability of the sand column. Alternative constitutive models may also be employed for this purpose.

General Solution

The boundary value problem in (11.46) has two nonhomogeneous boundaries on both r and z, and to facilitate the solution, we divide it into two functions, such that

$$u_e(r, z, t) = u_1(z, t) + u_2(r, z, t) \tag{11.47}$$

where u_1 is a function of z and time, and u_2 is a function of r, z, and time. Substituting (11.47) into (11.46) gives

$$c_z \frac{\partial^2 u_1}{\partial z^2} - \frac{\partial u_1}{\partial t} + c_z \frac{\partial^2 u_2}{\partial z^2} + c_r \frac{\partial^2 u_2}{\partial r^2} + \frac{c_r}{r} \frac{\partial u_2}{\partial r} - \frac{\partial u_2}{\partial t} = 0 \tag{11.48}$$

This equation can be split into two equations, expressed with their relevant boundary conditions as follows.

Sub-problem 1

$$\begin{aligned} &\frac{\partial^2 u_1}{\partial z^2} - \frac{1}{\alpha} \frac{\partial u_1}{\partial t} = 0 \\ &\text{IC}: \quad u_1(z, 0) = 0 \\ &\text{BC}: \quad u_1(0, t) = P(t) \\ &\qquad \left. \frac{\partial u_1}{\partial z} \right|_{z=H} = 0 \end{aligned} \tag{11.49}$$

Sub-problem 2

$$\begin{aligned} &\frac{\partial^2 u_2}{\partial z^2} + \frac{\partial^2 u_2}{\partial r^2} + \frac{1}{r} \frac{\partial u_2}{\partial r} - \frac{1}{\alpha} \frac{\partial u_2}{\partial t} = 0 \\ &\text{IC}: \quad u_2(r, z, 0) = 0 \\ &\text{BC}: \quad u_2(r, 0, t) = 0 \\ &\qquad \left. \frac{\partial u_2}{\partial z} \right|_{z=H} = 0 \\ &\qquad 2\pi a \left. \frac{\partial u_2}{\partial r} \right|_{r=a} = b[u_1(z) + u_2(a, z) - u_d(z)] \\ &\qquad u_2(r_e, z, t) = 0 \end{aligned} \tag{11.50}$$

where we assume $\alpha = c$ with $c_r = c_z = c$, and b is some drainage property that can be specified as K_d/K_s.

Solving Sub-problem 1

The solution of the BVP (11.49) in the frequency domain is given in ▶ Sect. 11.2, that lead to

$$\hat{u}_1(z, \omega_n) = \frac{\hat{P}(\omega_n)}{e^{\kappa_z H} + e^{-\kappa_z H}} \left[e^{-\kappa_z (z-H)} + e^{\kappa_z (z-H)} \right], \quad \kappa_z = \sqrt{i\omega / c_z} \tag{11.51}$$

Solving Sub-problem 2

Applying the Fourier transform to (11.50) results in

$$\frac{\partial^2 \hat{u}_2}{\partial r^2} + \frac{1}{r}\frac{\partial \hat{u}_2}{\partial r} + \frac{\partial^2 \hat{u}_2}{\partial z^2} - \frac{i\omega}{\alpha}\hat{u}_2 = 0$$

$$\begin{aligned} \text{BC}:\ & \hat{u}_2(r,0,\omega) = 0 \\ & \left.\frac{\partial \hat{u}_2}{\partial z}\right|_{z=H} = 0 \\ & 2\pi a \left.\frac{\partial \hat{u}_2}{\partial r}\right|_{r=a} = b\left[\hat{u}_1(z) + \hat{u}_2(a,z) - \hat{u}_d(z)\right] \\ & \hat{u}_2(r_e, z, \omega) = 0 \end{aligned} \tag{11.52}$$

Solution of the partial differential equation in (11.52) can be carried out using the method of separation of variables. Assume

$$\hat{u}_2(r,z) = R(r)Z(z) = RZ \tag{11.53}$$

Substituting (11.53) into the PDE in (11.52), and equating the result to a separation constant, say η, which can be zero, positive or negative, gives

11

$$Z'' + \eta Z = 0 \quad \text{and} \quad R'' + \frac{1}{r}R\prime - \left(\frac{i\omega}{\alpha} + \eta\right)R = 0 \tag{11.54}$$

Depending on η, the three possible general solutions of $Z(z)$ in (11.54) are given by

$$\begin{aligned} Z(z) &= c_1 + c_2 z, & \eta &= 0 \\ Z(z) &= c_1\cosh(\beta z) + c_2\sinh(\beta z), & \eta &= -\beta^2 < 0 \\ Z(z) &= c_1\cos(\beta z) + c_2\sin(\beta z), & \eta &= \beta^2 > 0 \end{aligned} \tag{11.55}$$

where c_1 and c_2 are arbitrary constants. To utilize the eigenfunction expansion, and considering that the drainage pipe represents only a portion of the layer depth, we seek a solution that is periodic in z-direction. The second solution in (11.55) is evidently nonperiodic and can therefore be discarded. The first solution is periodic if $c_2 = 0$, with $c_1 \neq 0$. Consequently, $\eta = 0$ is an eigenvalue. Finally, the third solution, with $\eta = \beta^2$, can be periodic, rendering it suitable for solving the problem.

The second equation in (11.54) is a sort of Bessel equation, with its solution given by

$$R(r) = c_3 I_0(\kappa_r r) + c_4 K_0(\kappa_r r), \quad \kappa_r = \sqrt{\frac{i\omega}{\alpha} + \beta^2} \tag{11.56}$$

in which I_0 and K_0 are the modified Bessel functions of zero order of the first and second kind, respectively, and c_3 and c_4 are arbitrary constants need to be determined from the boundary conditions.

Collecting $Z(z)$ and $R(r)$ solutions, the general solution of $\hat{u}_2$ is thus

$$\hat{u}_2(r, z, \omega) = [a_1 I_0(\kappa_r r) + a_2 K_0(\kappa_r r)][c_1 \cos(\beta z) + c_2 \sin(\beta z)]$$

Boundary Conditions

At $r = r_e$, the fourth boundary condition in (11.52), valid for all z, implies

$$a_1 I_0(\kappa_r r_e) + a_2 K_0(\kappa_r r_e) = 0$$

Leading to

$$a_1 = -\frac{K_0(\kappa_r r_e)}{I_0(\kappa_r r_e)} a_2$$

Thus

$$\hat{u}_2 = \left[-\frac{K_0(\kappa_r r_e)}{I_0(\kappa_r r_e)} I_0(\kappa_r r) + K_0(\kappa_r r)\right]\left[A \cos(\beta z) + B \sin(\beta z)\right] \tag{11.57}$$

where $A = a_2 c_1$ and $B = a_2 c_2$.

At $z = 0$, the first boundary condition in (11.52), valid for all r, implies $A = 0$, leading to

$$\hat{u}_2 = \left[-\frac{K_0(\kappa_r r_e)}{I_0(\kappa_r r_e)} I_0(\kappa_r r) + K_0(\kappa_r r)\right] B \sin(\beta z) \tag{11.58}$$

Taking the derivative of (11.58) with respect to z yields

$$\frac{\partial \hat{u}_2}{\partial z} = \left[-\frac{K_0(\kappa_r r_e)}{I_0(\kappa_r r_e)} I_0(\kappa_r r) + K_0(\kappa_r r)\right] \beta B \cos(\beta z)$$

At $z = H$, the second boundary condition in (11.52), valid for all r, implies

$$\cos(\beta H) = 0, \quad \Rightarrow \beta \equiv \beta_m = \frac{1}{H}\left(\frac{\pi}{2} + m\pi\right)$$

Since the problem is linear, any linear combination of solutions is also a solution, leading the solution of $\hat{u}_2$ to become

$$\hat{u}_2 = \sum_m \left[-\frac{K_0(\kappa_r r_e)}{I_0(\kappa_r r_e)} I_0(\kappa_r r) + K_0(\kappa_r r)\right] B_m \sin(\beta_m z)$$

with (11.59)

$$\kappa_r = \sqrt{\frac{i\omega}{\alpha} + \beta_m^2}$$

Taking the derivative of (11.59) with respect to r, for an m, gives

$$\frac{\partial \hat{u}_{2m}}{\partial r} = \left[-\frac{K_0(\kappa_r r_e)}{I_0(\kappa_r r_e)} \kappa_r I_1(\kappa_r r) - \kappa_r K_1(\kappa_r r) \right] B_m \sin(\beta_m z)$$

At $r = a$, the third boundary condition in (11.52) implies

$$\sum_m B_m \sin(\beta_m z) \left[\begin{array}{l} 2\pi a \left\{ -\frac{K_0(\kappa_r r_e)}{I_0(\kappa_r r_e)} \kappa_r I_1(\kappa_r a) - \kappa_r K_1(\kappa_r a) \right\} \\ - b \left\{ -\frac{K_0(\kappa_r r_e)}{I_0(\kappa_r r_e)} I_0(\kappa_r a) + K_0(\kappa_r a) \right\} \end{array} \right]$$
$$= b\left[\hat{u}_1(z) - \hat{u}_d(z)\right]$$

This equation can be written as

$$\sum_m B_m \sin(\beta_m z) = \frac{b\hat{u}_1(z) - \hat{u}_d(z)}{G_m} \tag{11.60}$$

with

$$G_m = 2\pi a \kappa_r \left[-\frac{K_0(\kappa_r r_s)}{I_0(\kappa_r r_s)} I_1(\kappa_r a) - K_1(\kappa_r a) \right]$$
$$- b\left[-\frac{K_0(\kappa_r r_s)}{I_0(\kappa_r r_s)} I_0(\kappa_r a) + K_0(\kappa_r a) \right]$$

The equation in (11.60) is a typical Fourier sine series with its coefficients given by

$$B_m = \frac{2}{H} \int_0^L \frac{b\left[\hat{u}_1(z) - \hat{u}_d(z)\right]}{G_m} \sin(\beta_m z)\, dz$$
$$= \frac{2b}{G_m H} \int_0^L [\hat{u}_1(z) - \hat{u}_d(z)] \sin(\beta_m z)\, dz, \quad m = 0, 1, 2, \ldots \tag{11.61}$$

Note that we stated $m \geq 0$ because $\sin(\beta_m z)$ is not necessarily zero for $m = 0$, and, as indicated above, zero is an eigenvalue.

BVP Solution

Combining (11.51) and (11.59), the solution of the BVP (11.46) in the frequency domain yields

$$\hat{u}_e(r, z, \omega_n) = \hat{u}_1(z, \omega_n) + \hat{u}_2(r, z, \omega_n)$$

with

$$\hat{u}_1(z, \omega_n) = \frac{\hat{P}(\omega_n)}{e^{\kappa_z H} + e^{-\kappa_z H}} \left[e^{-\kappa_z (z-H)} + e^{\kappa_z (z-H)} \right]$$
$$\hat{u}_2(r, z, \omega_n) = \sum_m \left[-\frac{K_0(\kappa_r r_e)}{I_0(\kappa_r r_e)} I_0(\kappa_r r) + K_0(\kappa_r r) \right] B_m \sin(\beta_m z), \quad m = 0, 1, 2, \ldots \tag{11.62}$$

where

$$\kappa_r = \sqrt{\kappa_z^2 + \beta_m^2}, \quad \beta_m = \frac{1}{H}\left(\frac{\pi}{2} + m\pi\right)$$
$$\kappa_z = \sqrt{\frac{i\omega_n}{\alpha}}$$
$$B_m = \frac{2b}{G_m H}\int_0^L \left[\hat{u}_1(z) - \hat{u}_d(z)\right] \sin(\beta_m z)\, dz \tag{11.63}$$
$$G_m = 2\pi a\left[-\frac{K_0(\kappa_r r_e)}{I_0(\kappa_r r_e)}\kappa_r I_1(\kappa_r a) - \kappa_r K_1(\kappa_r a)\right] - b\left[-\frac{K_0(\kappa_r r_e)}{I_0(\kappa_r r_e)}I_0(\kappa_r a) + K_0(\kappa_r a)\right]$$

Having solved the problem in the frequency domain, the solution in the time domain can readily be obtained using the inverse fast Fourier transform (IFFT) algorithm, as

$$u_e(r,z,t) = \sum_n \hat{u}_e(r,z,\omega_n)e^{i\omega_n t}$$

Special case 1: $L = H$and u_d is constant

For the special case where the length of the drainpipe is equal to the depth of the soil layer, i.e. $L = H$, and that the excess pore pressure at the drainpipe u_d, is constant, the Fourier coefficients become

$$B_m = \frac{2b}{G_m H}\int_0^H \left[\hat{u}_1(z) - \hat{u}_d\right] \sin(\beta_m z)\, dz = \frac{2b}{G_m H}\left(\frac{\beta_m P(\omega)}{\beta_m^2 + \kappa_z^2} - \frac{\hat{u}_d}{\beta_m}\right)$$

Special case 1: $L = H$and u_d is a function of z

For the special case where the length of the drain pipe is equal to the depth of the soil layer, i.e. $L = H$, and that the excess pore pressure at the drain pipe is a function of z, such as $u_d = U_0 z$, where U_0 is some multiplier, the Fourier coefficients become

$$B_m = \frac{2b}{G_m H}\int_0^H \left[\hat{u}_1(z) - \hat{U}_0\, z\right] \sin(\beta_m z)\, dz = \frac{2b}{G_m H}\left(\frac{\beta_m P(\omega)}{\beta_m^2 + \kappa_z^2} - \frac{\hat{U}_0 (-1)^m}{\beta_m^2}\right)$$

Example 11.5: Consolidation with Radial Drainage

Consider a 10 m deep fully saturated soil layer resting on an impermeable layer. An embankment exerting a pressure of 10 kPa has been constructed at the ground surface over a period of 480 h (20 days). To accelerate the consolidation process, sand columns with a radius of $a = 0.1$ m have been installed. Each sand column is designed to have an influence radius equals five time its circumference, i.e. $r_e = 10\pi a = 3.14$ m. Figure 11.18 shows the top view of the sand columns and their region of influence, and Fig. 11.19 depicts the profile of one of the sand columns and the surrounding soil mass, along with the relevant geometrical and material parameters. This figure also shows the construction time of the embankment. The task at hand is to compute the excess pore pressure distribution due to a single sand column considering these two scenarios:

1. The excess pore pressure in the sand column, u_d, is maintained at atmospheric pressure, i.e. zero at all times. Show the spatial profile of the excess pore pressure at the end of the construction time at soil points 2 m below the ground surface. Also, show their evolution with time.
2. The excess pore pressure in the sand column, u_d, is made −4 kPa with a time distribution similar to the applied surcharge load. Make contour plots.

Solution

The solution of this problem can readily be solved by implementing (11.62) in a computer code. The process begins by transforming the input time signals into the frequency domain using the forward FFT algorithm. Next, the excess pore pressure is calculated for every frequency using the solution given in this equation and its associated coefficients and parameters. After that, the time response is reconstructed using the inverse FFT algorithm.

For the case where $u_d = 0$ kPa, Fig. 11.20 depicts the spatial distribution of excess pore pressure at the end of the 480 h construction period, computed at soil points located 2 m below the ground surface. The profile spans from the outer boundary of the sand drain to a distance 3 m away from it. Apparently, there is a noticeable seepage at points closer to the sand drain as compared to those located further away. Note that even though the drain is designed to have an area of

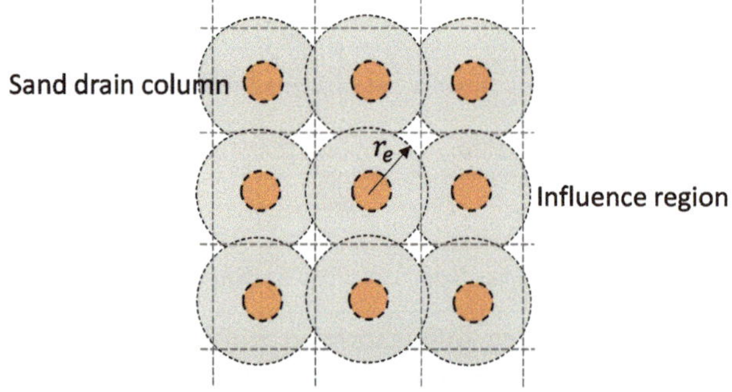

Fig. 11.18 Top view of sand drain columns

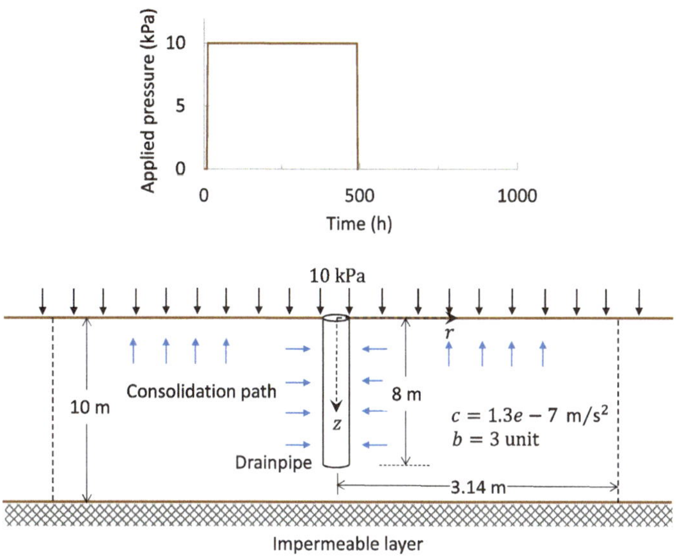

Fig. 11.19 Side view of one sand drain column with the time signal of constructing the surcharge

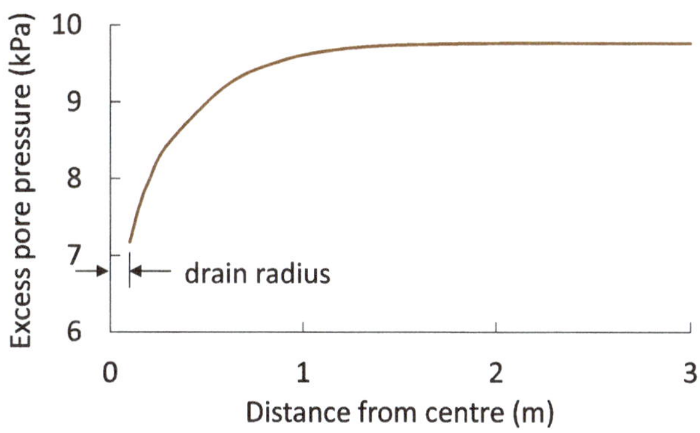

Fig. 11.20 Excess pore pressure versus distance

influence, the observed excess pore pressure near r_e is due to the surcharge loading at the surface.

Figure 11.21 presents the temporal distribution of excess pore pressure at various soil points along the radial direction. As shown in the figure, the rate of excess pore pressure dissipation varies between points located closer to the sand drain and those farther away. It is evident that there is a dynamic interplay between the buildup and dissipation of excess pore pressure during both the application and removal of the surcharge.

In the case where $u_d = -4$ kPa, Fig. 11.22 shows the contours of excess pore pressure distribution after 100 and 400 h from the start of the construction phase. The figure highlights the noticeable influence of the drainage column, even during the construction of the embankment. Figure 11.23 depicts

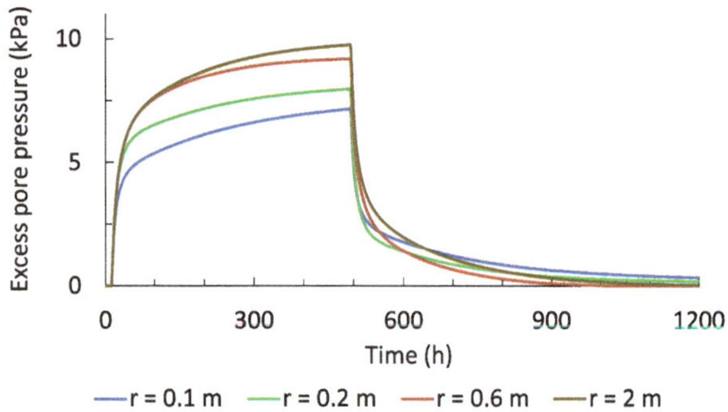

Fig. 11.21 Excess pore pressure versus time

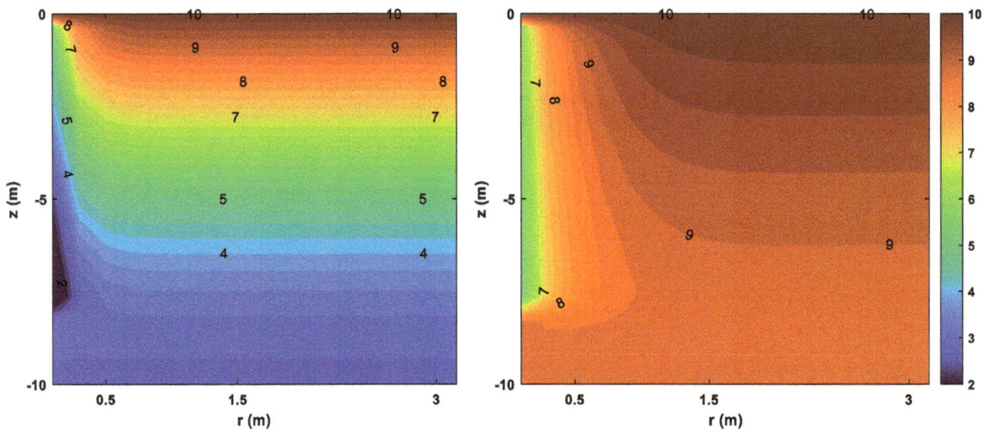

Fig. 11.22 Excess pore pressure after 100 h (left) and 400 h (right) from beginning the construction phase

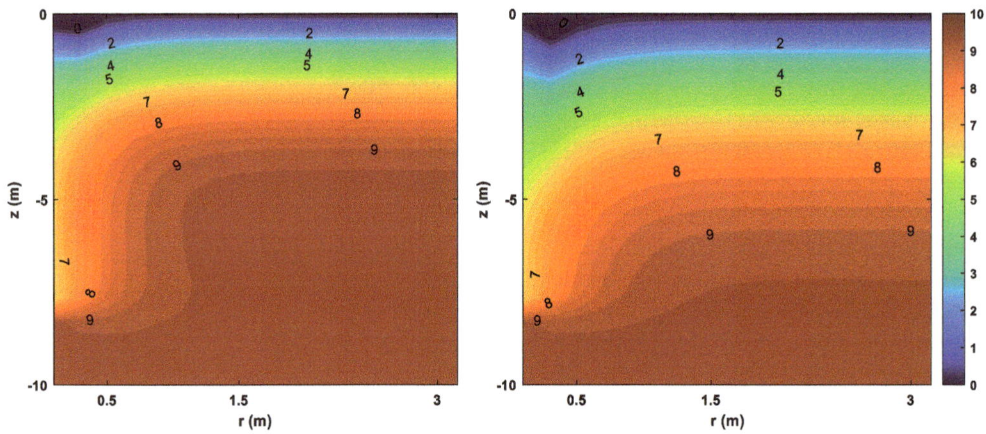

Fig. 11.23 Excess pore pressure after 12 h (left) and 20 h (right) from completing the construction phase

the dissipation of excess pore pressure following the end of the construction phase, clearly showing how the pore pressure dissipates toward the upper seepage surface and thorough the drainage pipe.

11.8 Consolidation with Finite Radial Drainage: No Radius of Influence

In this section, we extend the domain of the previous case to infinity in the radial direction, effectively eliminating the concept of a finite radius of influence, see Fig. 11.24.

The boundary value problem of an infinite soil layer resting on an impermeable layer and allowing seepage from its upper surface and through a finite drainage pipe installed along its central axis can be expressed as

$$c_r \frac{\partial^2 u_e}{\partial r^2} + \frac{c_r}{r}\frac{\partial u_e}{\partial r} + c_z \frac{\partial^2 u_e}{\partial z^2} - \frac{\partial u_e}{\partial t} = 0$$

$$\begin{aligned}
\text{IC}: \quad & u_e(r,z,0) = 0 \\
\text{BC}: \quad & u_e(r,0,t) = P(t) \\
& \left.\frac{\partial u_e}{\partial z}\right|_{z=H} = 0 \\
& 2\pi a K_s \left.\frac{\partial u_e}{\partial r}\right|_{r=a} = K_d[(u_e - u_d(z)], \quad 0 \le z \le L \\
& u_e(\infty,z,t) = 0
\end{aligned} \tag{11.64}$$

where $P(t)$ is the prescribed embankment pressure, a is the radius of the drain pipe, and u_d is a measure of the excess pore pressure in the drain. The drain pipe has a length $L \le H$, where H represents the thickness of the soil layer.

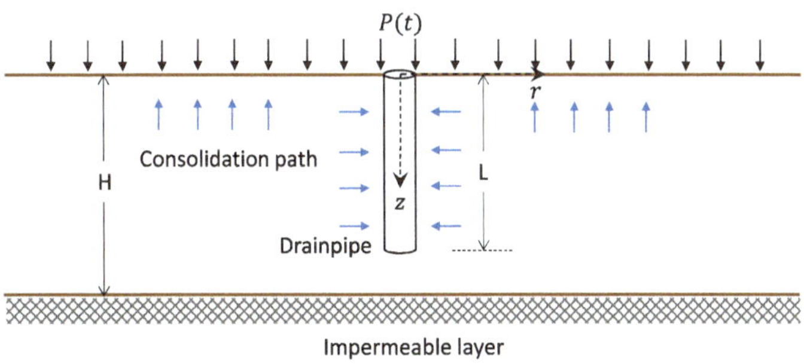

Fig. 11.24 Consolidation with a finite drain pipe in an infinite domain, $r_e \to \infty$

General Solution

The boundary value problem in (11.64) involves two nonhomogeneous boundary conditions in both r and z. Therefore, it is best addressed by dividing it into two sub-problems such that

$$u_e(r,z,t) = u_1(z,t) + u_2(r,z,t) \tag{11.65}$$

where u_1 is a function of z and time, and u_2 is a function of r, z, and time. Substituting (11.65) into (11.64) gives

$$c_z\frac{\partial^2 u_1}{\partial z^2} - \frac{\partial u_1}{\partial t} + c_z\frac{\partial^2 u_2}{\partial z^2} + c_r\frac{\partial^2 u_2}{\partial r^2} + \frac{c_r}{r}\frac{\partial u_2}{\partial r} - \frac{\partial u_2}{\partial t} = 0 \tag{11.66}$$

This equation can be split into two equations, expressed with their relevant boundary conditions as follows.

Sub-problem 1

$$\begin{aligned}
&\frac{\partial^2 u_1}{\partial z^2} - \frac{1}{\alpha}\frac{\partial u_1}{\partial t} = 0\\
&\text{IC}: \quad u_1(z,0) = 0\\
&\text{BC}: \quad u_1(0,t) = P(t)\\
&\qquad\quad \left.\frac{\partial u_1}{\partial z}\right|_{z=H} = 0
\end{aligned} \tag{11.67}$$

Sub-problem 2

$$\begin{aligned}
&\frac{\partial^2 u_2}{\partial z^2} + \frac{\partial^2 u_2}{\partial r^2} + \frac{1}{r}\frac{\partial u_2}{\partial r} - \frac{1}{\alpha}\frac{\partial u_2}{\partial t} = 0\\
&\text{IC}: \quad u_2(r,z,0) = 0\\
&\text{BC}: \quad u_2(r,0,t) = 0\\
&\qquad\quad \left.\frac{\partial u_2}{\partial z}\right|_{z=H} = 0\\
&\qquad\quad 2\pi a\left.\frac{\partial u_2}{\partial r}\right|_{r=a} = b\left[u_1(z) + u_2(a,z) - u_d(z)\right]\\
&\qquad\quad u_2(\infty,z,t) = 0
\end{aligned} \tag{11.68}$$

where we assume $\alpha = c$ with $c_r = c_z = c$, and b represents a drainage coefficient that can be specified as K_d/K_s.

Solving Sub-problem 1

The solution of the BVP (11.67) in the frequency domain is given in ▶ Sect. 11.2, that lead to

$$\hat{u}_1(z,\omega_n) = \frac{\hat{P}(\omega_n)}{e^{\kappa_z H} + e^{-\kappa_z H}}\left[e^{-\kappa_z(z-H)} + e^{\kappa_z(z-H)}\right], \quad \kappa_z = \sqrt{i\omega/c_z} \tag{11.69}$$

Solving Sub-problem 2

Applying the Fourier transform to (11.68) yields

$$\frac{\partial^2 \hat{u}_2}{\partial r^2} + \frac{1}{r}\frac{\partial \hat{u}_2}{\partial r} + \frac{\partial^2 \hat{u}_2}{\partial z^2} - \frac{i\omega}{\alpha}\hat{u}_2 = 0$$

$$\begin{aligned} \text{BC}: \ & \hat{u}_2(r,0,\omega) = 0 \\ & \left.\frac{\partial \hat{u}_2}{\partial z}\right|_{z=H} = 0 \\ & 2\pi a \left.\frac{\partial \hat{u}_2}{\partial r}\right|_{r=a} = b\left[\hat{u}_1(z) + \hat{u}_2(a,z) - \hat{u}_d(z)\right] \\ & \hat{u}_2(\infty, z, \omega) = 0 \end{aligned} \tag{11.70}$$

Following the solution procedure outlined in the previous case, and considering the fourth boundary condition, the general solution of partial differential equation leads to

$$\hat{u}_2(r,z,\omega) = K_0(\kappa_r r)A\cos(\beta z) + B\sin(\beta z) \tag{11.71}$$

where we eliminated I_0 because it is unbounded at far distances.

Boundary Conditions

At $z = 0$, the first boundary condition in (11.70), valid for all r, implies $A = 0$, leading the solution to become

$$\hat{u}_2 = B\sin(\beta z)K_0(\kappa_r r)$$

Taking the derivative of this equation with respect to z yields

$$\frac{\partial \hat{u}_2}{\partial z} = \beta B\cos(\beta z)K_0(\kappa_r r)$$

At $z = H$, the second boundary condition in (11.70), valid for all r, implies

$$\cos(\beta H) = 0, \quad \Rightarrow \beta \equiv \beta_m = \frac{1}{H}\left(\frac{\pi}{2} + m\pi\right)$$

Since the problem is linear, any linear combination of solutions is also a solution, yielding the solution of $\hat{u}_2$ to become

$$\begin{aligned} & \hat{u}_2 = \sum_m K_0(\kappa_r r)B_m\sin(\beta_m z), \quad m = 0,1,2,\ldots \\ & \text{with} \\ & \kappa_r = \sqrt{\frac{i\omega}{\alpha} + \beta_m^2} \end{aligned} \tag{11.72}$$

Taking the derivative of (11.72) with respect to r, for an m, gives

$$\frac{d\hat{u}_{2m}}{dr} = -\kappa_r K_1(\kappa_r r)B_m\sin(\beta_m z)$$

At $r = a$, the third boundary condition in (11.70) entails

$$\sum_m [-2\pi a\kappa_r K_1(\kappa_r a) - bK_0(\kappa_r a)]B_m \sin(\beta_m z)$$
$$= b\left[\hat{u}_1(z) - \hat{u}_d(z)\right]$$

This equation can be written as

$$\sum_m B_m \sin(\beta_m z) = \frac{-b\left[\hat{u}_1(z) - \hat{u}_d(z)\right]}{2\pi a\kappa_r K_1(\kappa_r a) + bK_0(\kappa_r a)} \tag{11.73}$$

This is a typical Fourier sine series with its coefficients given by

$$\begin{aligned} B_m &= \frac{2}{H}\int_0^L \frac{-b\left[\hat{u}_1(z) - \hat{u}_d(z)\right]}{2\pi a\kappa_r K_1(\kappa_r a) + bK_0(\kappa_r a)} \sin(\beta_m z)\, dz \\ &= \frac{-2b}{[2\pi a\kappa_r K_1(\kappa_r a) + bK_0(\kappa_r a)]H} \\ &\quad \int_0^L \left[\hat{u}_1(z) - \hat{u}_d(z)\right] \sin(\beta_m z)\, dz \end{aligned} \tag{11.74}$$

BVP Solution

Combining (11.69) and (11.72), the solution of the BVP (11.64) in the frequency domain yields

$\hat{u}_e(r, z, \omega_n) = \hat{u}_1(z, \omega_n) + \hat{u}_2(r, z, \omega_n)$

with

$$\begin{aligned} \hat{u}_1(z, \omega_n) &= \frac{\hat{P}(\omega_n)}{e^{\kappa_z H} + e^{-\kappa_z H}}\left[e^{-\kappa_z(z-H)} + e^{\kappa_z(z-H)}\right] \\ \hat{u}_2(r, z, \omega_n) &= \sum_m B_m \sin(\beta_m z) K_0(\kappa_r r), \quad m = 0, 1, 2, \ldots \end{aligned} \tag{11.75}$$

where

$$\begin{aligned} \kappa_r &= \sqrt{\kappa_z^2 + \beta_m^2}, \quad \beta_m = \frac{1}{H}\left(\frac{\pi}{2} + m\pi\right) \\ \kappa_z &= \sqrt{\frac{i\omega_n}{\alpha}} \\ B_m &= \frac{-2b}{[2\pi a\kappa_r K_1(\kappa_r a) + bK_0(\kappa_r a)]H} \\ &\quad \int_0^L \left[\hat{u}_1(z) - \hat{u}_d(z)\right] \sin(\beta_m z)\, dz \end{aligned} \tag{11.76}$$

Having solved the problem in the frequency domain, the solution in the time domain can readily be obtained using the inverse fast Fourier transform (IFFT) algorithm, as

$$u_e(r,z,t) = \sum_n \hat{u}_e(r,z,\omega_n)\, e^{i\omega_n t}$$

11.9 Consolidation in a Confined Aquifer

In the previous two cases, we examined boundary value problems related to soil consolidation involving a cylindrical drainage pipe and a soil layer with a seepage surface on the top boundary. Now, we turn to the consolidation in an aquifer of thickness H, which is bounded by impervious layers on both the upper and lower surfaces. This scenario resembles a confined aquifer located deeper in the ground, where consolidation occurs solely due to radial seepage through a sand column with length $L \leq H$ and radius a, positioned along the central axis of the aquifer, refer to Fig. 11.25.

The boundary value problem of this case can be expressed as

$$\frac{\partial^2 u_e}{\partial r^2} + \frac{1}{r}\frac{\partial u_e}{\partial r} + \frac{\partial^2 u_e}{\partial z^2} - \frac{1}{\alpha}\frac{\partial u_e}{\partial t} = 0$$

$$\begin{aligned} \text{IC}: &\quad u_e(r,z,0) = 0 \\ \text{BC}: &\quad u_e(\infty,z,t) = 0 \\ &\quad \left.\frac{\partial u_e}{\partial z}\right|_{z=0} = 0 \\ &\quad \left.\frac{\partial u_e}{\partial z}\right|_{z=H} = 0 \\ &\quad 2\pi a \left.\frac{\partial u_e}{\partial r}\right|_{r=a} = b[u_e(a,z) - u_d(z,t)] \end{aligned} \tag{11.77}$$

where $\alpha = c$ with $c_r = c_z = c$, and b is a measure of the drainage property that can be specified as K_d/K_s.

General Solution

Applying the Fourier transform to (11.77) yields

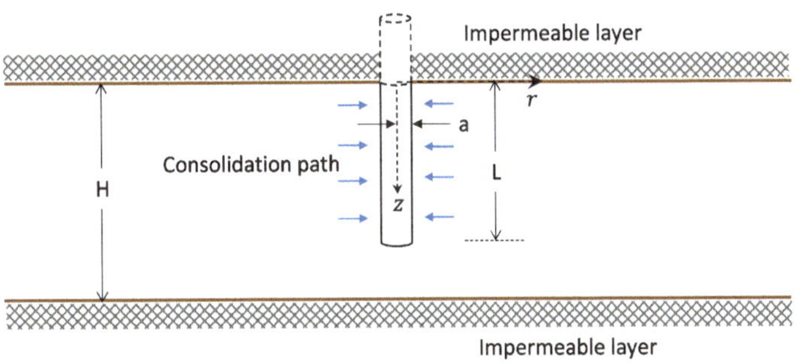

Fig. 11.25 Consolidation in a confined aquifer

$$\frac{\partial^2 \hat{u}_e}{\partial r^2} + \frac{1}{r}\frac{\partial \hat{u}_e}{\partial r} + \frac{\partial^2 \hat{u}_e}{\partial z^2} - \frac{i\omega}{\alpha}\hat{u}_e = 0$$

$$\begin{aligned} \text{BC:}\quad & \hat{u}_e(\infty, z, \omega) = 0 \\ & \left.\frac{\partial \hat{u}_e}{\partial z}\right|_{z=0} = 0 \\ & \left.\frac{\partial \hat{u}_e}{\partial z}\right|_{z=H} = 0 \\ & 2\pi a \left.\frac{\partial \hat{u}_e}{\partial r}\right|_{r=a} = b\left[\hat{u}_e(a,z) - \hat{u}_d(z)\right] \end{aligned} \tag{11.78}$$

This problem involves a single nonhomogeneous boundary condition at a and can therefore be solved using the method of separation of variables. Following the solution procedure outlined in ▶ Sect. 11.8, the general solution of the differential equation in (11.78) is given by

$$\begin{aligned} & \hat{u}_e = K_0(\kappa_r r)[A\cos(\beta z) + B\sin(\beta z)] \\ & \text{with} \\ & \kappa_r = \sqrt{\kappa_z^2 + \beta^2}, \quad \kappa_z^2 = \frac{i\omega}{\alpha} \end{aligned} \tag{11.79}$$

11

Boundary Conditions

Taking the derivative of (11.79) with respect to z yields

$$\frac{\partial \hat{u}_e}{\partial z} = K_0(\kappa_r r)[-\beta A \sin(\beta z) + \beta B \cos(\beta z)]$$

At $z = 0$, the second boundary condition in (11.78), valid for all r, implies $B = 0$, leading to

$$\hat{u}_e = A K_0(\kappa_r r)\cos(\beta z)$$

Taking the derivative of this equation with respect to z gives

$$\frac{\partial \hat{u}_e}{\partial z} = -\beta A K_0(\kappa_r r)\sin(\beta z)$$

At $z = H$, the third boundary condition in (11.78) implies

$$-\beta A K_0(\kappa_r r)\sin(\beta H) = 0$$

This entails

$$\sin(\beta H) = 0$$

which is true only if

$$\beta H = m\pi \Rightarrow \beta \equiv \beta_m = \frac{m\pi}{H}, \quad m = 1, 2, \cdots$$

Thus, since the problem is linear, the superposition principle implies

$$\hat{u}_e(r,z,\omega) = \sum_m A_m K_0(\kappa_r r)\cos(\beta_m z)$$

with (11.80)

$$\kappa_r = \sqrt{\frac{i\omega}{\alpha} + \beta_m^2}, \quad \beta_m = \frac{m\pi}{H}, \quad m = 0, 1, 2, \ldots$$

Note that we here stated $m \geq 0$, since, as indicated in ▶ Sect. 11.7, $\lambda = 0$ is also an eigenvalue.

Taking the derivative of (11.80) with respect to r gives

$$\frac{\partial \hat{u}_e}{\partial r} = -\sum_m A_m \cos\left(\frac{m\pi z}{H}\right)\kappa_r K_1(\kappa_r r)$$

Then, at $r = a$, the fourth boundary condition in (11.78) implies

$$-\sum_m A_m \kappa_r K_1(\kappa_r a)\cos(\beta_m z) = \frac{b}{2\pi a}\left[\sum_m A_m K_0(\kappa_r a)\cos(\beta_m z) - \hat{u}_d(z)\right]$$

Or

$$\sum_m A_m\left[\kappa_r K_1(\kappa_r a) + \frac{b}{2\pi a}K_0(\kappa_r a)\right]\cos(\beta_m z) = \frac{b}{2\pi a}\hat{u}_d(z) \tag{11.81}$$

This is a Fourier cosine series with its coefficients A_m expressed as

$$A_0 = \frac{1}{H}\int_0^L \frac{b\,\hat{u}_d(z)}{2\pi a[\kappa_r K_1(\kappa_r a) + \frac{b}{2\pi a}K_0(\kappa_r a)]}dz$$

$$= \frac{b}{2\pi a[\kappa_r K_1(\kappa_r a) + \frac{b}{2\pi a}K_0(\kappa_r a)]H}\int_0^L \hat{u}_d(z)\,dz, \qquad m = 0$$

$$A_m = \frac{2}{H}\int_0^L \frac{b\,\hat{u}_d(z)}{2\pi a[\kappa_r K_1(\kappa_r a) + \frac{b}{2\pi a}K_0(\kappa_r a)]}\cos\left(\frac{m\pi z}{H}\right)dz$$

$$= \frac{b}{\pi a[\kappa_r K_1(\kappa_r a) + \frac{b}{2\pi a}K_0(\kappa_r a)]H}\int_0^L \hat{u}_d(z)\cos\left(\frac{m\pi z}{H}\right)dz, \qquad m = 1, 2, \ldots \tag{11.82}$$

BVP Solution

The solution of this boundary value problem in the frequency domain can then be expressed as

$$\hat{u}_e(r,z,\omega) = \sum_m A_m K_0(\kappa_r r)\cos(\beta_m z), \quad m = 0, 1, 2, \ldots \tag{11.83}$$

with A_m given in (11.82), and

$$\kappa_r = \sqrt{\frac{i\omega}{\alpha} + \beta_m^2}, \quad \beta_m = \frac{m\pi}{H}$$

Upon employing the inverse fast Fourier transform (IFFT) algorithm, the solution in the time domain can be reconstructed via

$$u_e(r,z,t) = \sum_n \hat{u}_e(r,z,\omega_n)e^{i\omega_n t}.$$

Special Cases

In the special case where $\hat{u}_d$ is constant (not a function of z), the Fourier coefficients become

$$A_0 = \frac{bL\hat{u}_d}{2\pi a\left[\kappa_r K_1(\kappa_r a) + \frac{b}{2\pi a}K_0(\kappa_r a)\right]H}, \quad m = 0$$

$$A_m = \frac{b\hat{u}_d}{m\pi^2 a\left[\kappa_r K_1(\kappa_r a) + \frac{b}{2\pi a}K_0(\kappa_r a)\right]}\sin\left(\frac{m\pi L}{H}\right), \quad m = 1, 2, \ldots \tag{11.84}$$

Moreover, for the special case where $\hat{u}_d$ is constant, and $L = H$, $A_m = 0$, and the solution becomes

$$\hat{u}_e(r,z,\omega) = \frac{b\hat{u}_d}{2\pi a\left[\kappa_r K_1(\kappa_r a) + \frac{b}{2\pi a}K_0(\kappa_r a)\right]}K_0(\kappa_r r) \tag{11.85}$$

■ **Example 11.6: Consolidation with Radial Drainage**

Consider a fully saturated soil layer, 10 m deep, confined by impermeable layers at both the top and bottom boundaries. Initially, the pore pressure in the aquifer is 10 kPa. To consolidate the aquifer, a sand column with a radius $a = 0.1$ m has been installed, extending 9 m from the surface along the central axis of the aquifer. The sand drain is designed to be fully permeable, ensuring the pressure inside remains atmospheric at all time. It is operated for nearly 20 days, see Fig. 11.26. Figure 11.27 shows the profile of the sand column and the surrounding soil mass, along with the relevant geometrical and material parameters.

The task at hand is to compute the excess pore pressure distribution due to the sand column. Make contour plots and show the time variation of the pore pressure at 1 m below the surface.

Solution

As explained in discussing the Fourier method in ▶ Chap. 3, spectral analysis, which involves transforming signals from the time domain to the frequency domain, does not account for the initial conditions at $t = 0$. Yet, since this problem is linear, the initial condition can be added algebraically at the post-processing stage. In this example, the domain initially has a pore pressure of 10 kPa. Thus, to maintain atmospheric pressure at $t = 0$, we simply apply a pressure of -10 kPa at the drain pipe. In this example, the drain pipe is operated for 20 days, and hence the time distribution of the drain is as shown in Fig. 11.27.

Having accounted for the initial condition, we proceed by transforming the applied pore pressure signal from the time domain to the frequency domain using the forward FFT. In this case, 4096 FFT samples are employed, with a time step of 3600 s. Once the frequency spectrum is determined, the solution at each frequency is obtained via (11.83) and (11.84). Finally, the solution can be reconstructed to the time domain using the inverse FFT algorithm.

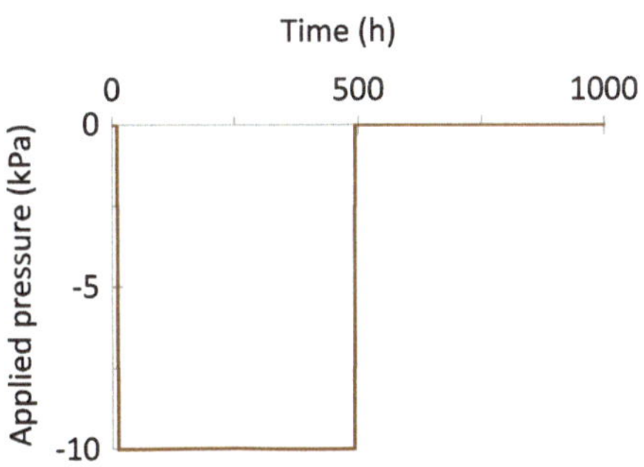

Fig. 11.26 Geometry and material parameters

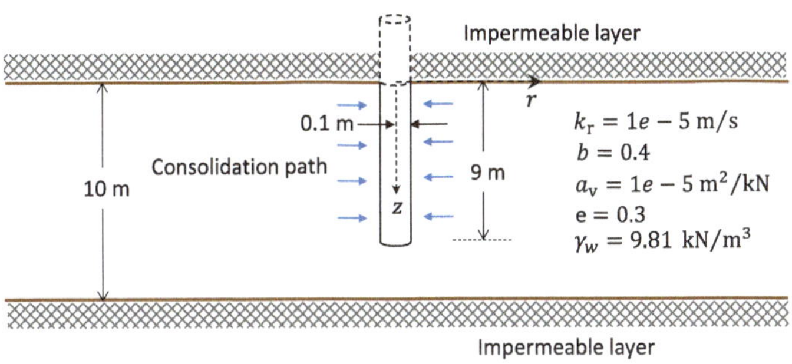

Fig. 11.27 Applied pore pressure at the drain pipe

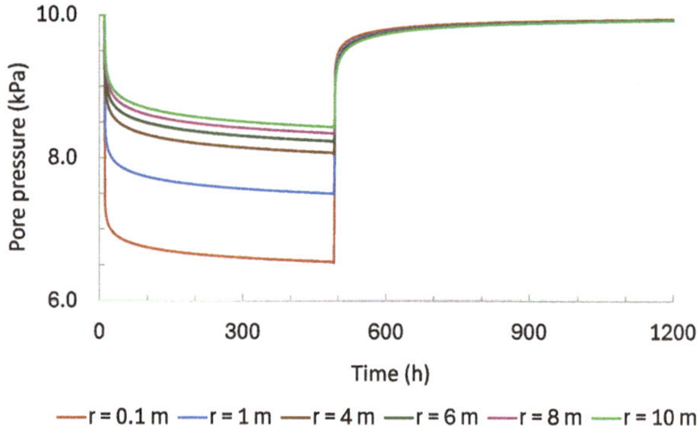

Fig. 11.28 Time variation of pore pressure in the radial direction, at $z = 1$ m below the surface

Figure 11.28 presents the computed pore pressure distribution in the radial direction at a depth of 1 m below the surface. The figure illustrates that near the drain pipe, the pore pressure decreases by approximately 3.5 kPa over a period of 20 days. At a greater distance, specifically 10 m from the drain, the pore pressure reduction is around 1.5 kPa. Figure 11.29 displays contours of the computed pore pressure in the surrounding soil mass. The Figure highlights how the presence of the drain accelerates pore pressure dissipation, contributing to the overall consolidation of the soil domain.

Upon reviewing this example, it is clear that the Fourier method, combined with spectral analysis, is a powerful tool capable of addressing complex cases such as the one we have just examined. This approach allows for efficient analysis and solution of intricate problems, demonstrating its effectiveness in various applications.

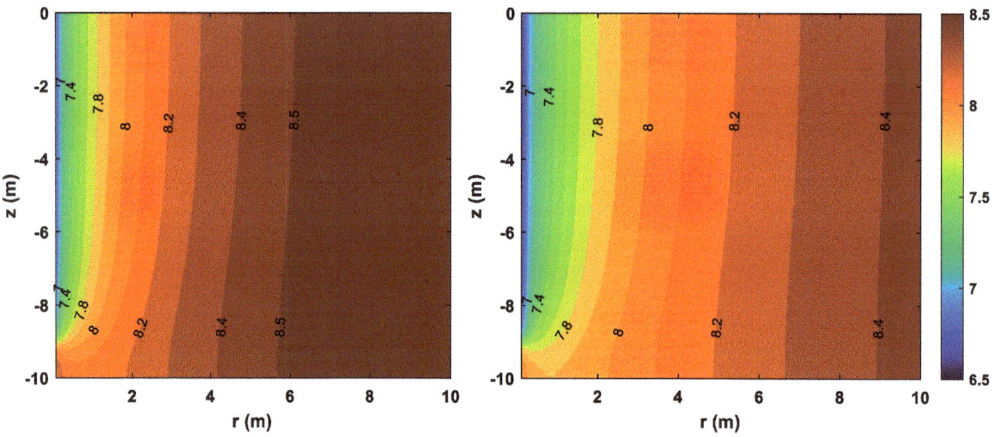

Fig. 11.29 Computed pore pressure after 4 days (left) and 20 days (right)

Problems

1. Show that the solution of Terzaghi's initial value problem

$$\frac{\partial^2 u_e}{\partial z^2} - \frac{1}{c_v}\frac{\partial u_e}{\partial t} = 0$$

$$\begin{aligned}
\text{IC:}\quad & u_e(z,0) = P, \quad 0 \leq z \leq H \\
\text{BC:}\quad & u_e(0,t) = 0 \\
& \left.\frac{\partial u_e}{\partial z}\right|_{z=H} = 0
\end{aligned}$$

is given by

$$u_e = P\sum_m \frac{2}{\beta_m} \sin\left[\beta_m\left(1 - \frac{z}{H}\right)\right] e^{-\beta_m^2 T_v}, \quad m = 0, 1, 2, \ldots$$

with

$$\beta_m = \left(\frac{\pi}{2} + m\pi\right)$$

$$T_v = \frac{c_v t}{H^2}$$

Hint: see Berry and Reid (1987).

2. Having a closer look at Terzaghi's formulation, explain why it fails to accurately capture the physics of the problem.
3. An oil tank with a radius a is to be placed on a saturated soil layer with a thickness H. When fully filled, the tank is expected to exert a pressure $P(t)$ on the surface of the soil layer. The soil layer rests on an impermeable bedrock and has a permeable surface that allows seepage. Formulate the boundary value problem for the consolidation process and solve it.
Hint: see the case given in ▶ Sect. 11.4.
4. In solving the plane strain case in ▶ Sect. 11.5 (also similarly in ▶ Sect. 11.6), we noted that, due to the symmetry of the problem, the sine term in the x-direction is eliminated, reducing the solution to $X(x) = a_2\cos(\kappa_x x)$. Prove that this statement is physically valid.
Hint: solve the boundary value problem (11.26) by imposing a no-flux condition at the axis of symmetry:

$$\left.\frac{\partial u_e}{\partial x}\right|_{x=0} = 0$$

5. The difference between the consolidation cases in Sects. 11.8 and 11.9 lies in the embankment pressure P applied at the surface. Using the solution given in (11.75) and (11.76), show that by setting $P = 0$, the solution reduces to that given in (11.83) and (11.84).

References

Berry PL, Reid D (1987) An introduction to soil mechanics. McGraw-Hill Book Company, UK

Further Reading

Lambe TW, Whitman RV (1979) Soil mechanics. John Wiley & Sons, Singapore

Verruijt A (2018) An introduction to soil mechanics. Springer Nature, Switzerland

Electro-osmosis Consolidation

R. Al Khoury and C. Kasbergen, *Analytical Subsurface Environment*,
https://doi.org/10.1007/978-3-031-93910-5_12

Electro-osmosis consolidation is a process by which an electric field is applied to a soil layer, causing the water to move towards pre-installed electrodes. The electro-osmosis effect is induced by an electric field generated by applying a voltage potential across anodes with positive charges and cathodes with negative charges. The movement of water reduces the water pressure in the pores and allows the soil particles to settle tightly together. In geotechnical engineering, electro-osmosis is mainly used to accelerate hydraulic consolidation, thus reducing the time required for the soil to reach its desired densification. To further expedite consolidation, the electrodes are installed in the ground in rows with different geometrical configurations that can facilitate the dissipation of pore water via seepage drains.

In this chapter, we will present various models that address the applications of electro-osmosis consolidation under different geometries, configurations, and boundary conditions. We study scenarios for electro-osmosis effects in one-dimensional (1D), axial symmetric, and two-dimensional (2D) configurations.

12

12.1 Electro-osmosis Consolidation

Electro-osmosis consolidation in a one-dimensional configuration involves the installation of anode and cathode electrodes in rows with lengths that are greater than the spacing between them, as shown in Fig. 12.1. In this configuration, pore water moves from the positive anode row to the negative cathode row in a nearly one-dimensional manner.

The coupled relationship between the hydraulically induced consolidation and the electro-osmosis consolidation for a 1D configuration is given by

$$\begin{aligned}
&\frac{K_z}{\rho_w g}\frac{\partial^2 u_e}{\partial z^2}+K_e\frac{\partial^2 E}{\partial z^2}=m_v\frac{\partial u_e}{\partial t}\\
&K_e\frac{\partial^2 E}{\partial z^2}=C_p\frac{\partial E}{\partial t}
\end{aligned} \tag{12.1}$$

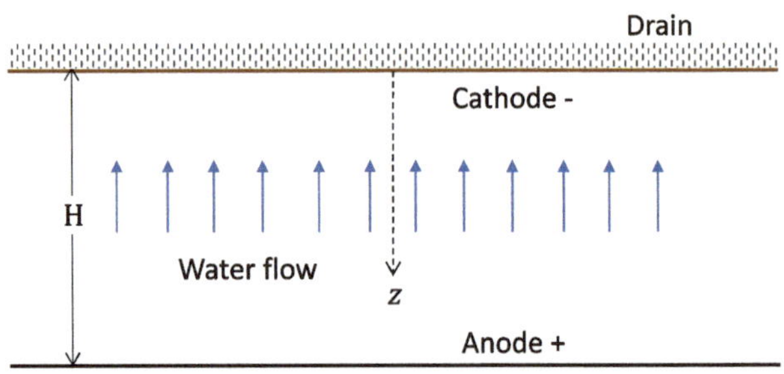

Fig. 12.1 One-dimensional electro-osmosis configuration

in which K_e (m^2 s^{-1} V^{-1}) is the coefficient of the electro-osmotic permeability, m_v (m^2 N^{-1}) is the coefficient of volume compressibility, K_z (m s^{-1}) is the soil permeability, ρ_w (kg m^{-3}) is the water mass density, g (m s^{-2}) is the gravitational acceleration, and C_p (V^{-1}) is the electric capacitance of soil.

Since the electric capacitance C_p of soil is negligible, the electric field is reaching the steady state condition soon after operating the system, leading the relevant boundary value problem to become

$$
\begin{aligned}
&\frac{K_z}{\gamma_w}\frac{\partial^2 u_e}{\partial z^2} + K_e\frac{d^2 E}{dz^2} = m_v\frac{\partial u_e}{\partial t}\\
&\frac{d^2 E}{dz^2} = 0\\
&\text{IC}: \quad u_e(z,0) = 0\\
&\text{BC}: \quad u_e(0,t) = 0\\
&\qquad \left.\left(-\frac{K_z}{\gamma_w}\frac{\partial u_e}{\partial z} - K_e\frac{\partial E}{\partial z}\right)\right|_{z=H} = 0\\
&\qquad E(0) = 0\\
&\qquad E(H) = E_0
\end{aligned}
\tag{12.2}
$$

in which E_0 (V) is a prescribed electric potential at the anode, and $\gamma_w = \rho_w g$ (N m^{-3}). Note that the second boundary condition is formally expressed as

$$
\left.\left(-\frac{K_z}{\gamma_w}\frac{\partial u_e}{\partial z}\cdot n_x - K_e\frac{\partial E}{\partial z}\cdot n_x\right)\right|_{z=H} = 0
$$

where n_x is the outward unit normal vector. In this case $n_x = 1$, since the outward normal vector to the Anode axis aligns with the direction of the z-axis, see 0.

BVP Solution

Applying the Fourier transform to (12.2) results in

$$
\begin{aligned}
&\frac{K_z}{\gamma_w}\frac{d^2 \hat{u}_e}{dz^2} + K_e\frac{d^2 \hat{E}}{dz^2} = i\omega m_v \hat{u}_e\\
&\frac{d^2 \hat{E}}{dz^2} = 0\\
&\text{BC}: \quad \hat{u}_e(0,t) = 0\\
&\qquad \left.\left(\frac{K_z}{\gamma_w}\frac{d\hat{u}_e}{dz} + K_e\frac{d\hat{E}}{dz}\right)\right|_{z=H} = 0\\
&\qquad \hat{E}(0) = 0\\
&\qquad \hat{E}(H) = \hat{E}_0
\end{aligned}
\tag{12.3}
$$

where it is worth noting that while E is always assumed to be in a steady state condition, its boundary may vary as a function of time (or frequency). Refer to the Note below for further details.

The general solution of the second differential equation in (12.3) is

$$\hat{E} = c_1 z + c_2$$

The third boundary condition in (12.3) implies $c_2 = 0$, and the fourth boundary condition implies $c_1 = \hat{E}_0/H$, leading the solution of the electric field equation to become

$$\hat{E} = \frac{z}{H}\hat{E}_0 \tag{12.4}$$

The general solution of the first differential equation in (12.3) is

$$\hat{u}_e = a_1 e^{-\kappa z} + a_2 e^{\kappa z}, \quad \kappa = \sqrt{\frac{m_v \gamma_w i\omega}{K_z}}$$

At $z = 0$, the first boundary condition in (12.3) implies $a_2 = -a_1$, leading to

$$\hat{u}_e = a_1(e^{-\kappa z} - e^{\kappa z}) \tag{12.5}$$

Taking the derivative of $\widehat{u}_e$ in (12.5) and $\widehat{E}$ in (12.4) with respect to z gives

12

$$\frac{d\hat{u}_e}{dz} = a_1\kappa\left(-e^{-\kappa z} - e^{\kappa z}\right), \quad \text{and } \hat{E} = \frac{\hat{E}_0}{H}$$

Following these, at $z = H$, the second boundary condition in (12.3) implies

$$a_1\kappa\left(-e^{-\kappa H} - e^{\kappa H}\right) = -\frac{\gamma_w K_e}{K_z}\frac{\hat{E}_0}{H}$$

This yields

$$a_1 = \frac{K_e \gamma_w}{K_z \kappa\left(e^{-\kappa H} + e^{\kappa H}\right)}\frac{\hat{E}_0}{H} \tag{12.6}$$

Substituting (12.6) into (12.5) gives the solution of the boundary value problem for a given frequency ω_n, as

$$\hat{u}_e(z, \omega_n) = \frac{K_e \gamma_w \hat{E}_0}{K_z \kappa H\left(e^{-\kappa H} + e^{\kappa H}\right)}\left(e^{-\kappa z} - e^{\kappa z}\right) \tag{12.7}$$

with

$$\kappa = \sqrt{\frac{m_v \gamma_w i\omega_n}{K_z}}$$

Having solved the problem in the frequency domain, the solution in the time domain can then be obtained using the inverse FFT, such that

$$\begin{aligned} u_e(z,t) &= \sum\nolimits_n \hat{u}_e(z,\omega_n)\, e^{i\omega_n t} \\ E(z,t) &= \sum\nolimits_n \hat{E}(z,\omega_n)\, e^{i\omega_n t} \end{aligned}$$

Special Case: Steady State

At the steady state when $t \to \infty$, the frequency approaches zero, $\omega \to 0$, and the solution (12.7) becomes

$$u_e(z) = -\frac{K_e \gamma_w E_0}{K_z H} z \tag{12.8}$$

Note

If we consider the time-dependent boundary value problem of the electric field:

$$K_e \frac{\partial^2 E}{\partial z^2} = C_p \frac{\partial E}{\partial t}$$

BC :

$$\begin{aligned} E(0) &= 0 \\ E(H) &= E_0 \end{aligned}$$

the solution in the frequency domain becomes

$$\hat{E}(z,\omega) = \frac{\hat{E}_0(z,\omega)}{e^{\eta H} - e^{-\eta H}} \left(e^{\eta z} - e^{-\eta z}\right), \quad \eta = \sqrt{\frac{i\omega C_p}{K_e}}$$

Given that C_p of the soil is negligible, taking the limit of $\widehat{E}(z,\omega)$ as $\eta \to 0$ results in

$$\hat{E} = \frac{z}{H} \hat{E}_0$$

which is identical to (12.4).

This note applies to all the cases discussed in this chapter.

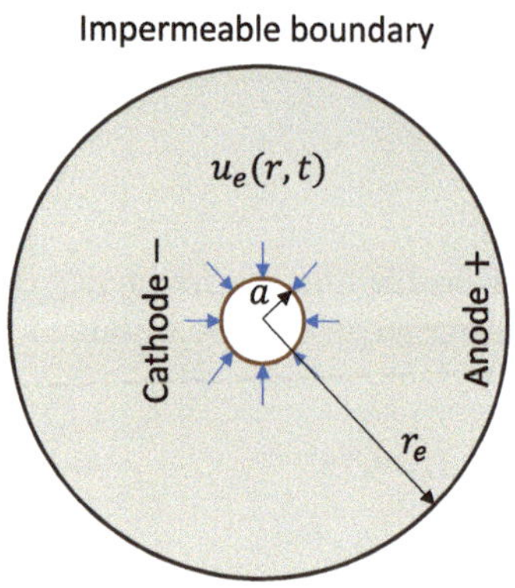

Fig. 12.2 Axial symmetric electro-osmosis configuration

12.2 Axial Symmetric Electro-osmosis Consolidation

Consider a soil mass encompassing at its center a row of cathode electrodes forming a circle with radius a. Away from the cathode electrodes circle, a row of anode electrodes is installed forming a cylinder with radius r_e, as illustrated in Fig. 12.2. By applying an electrical potential, the pore water in the soil moves from the positively charged outer surface to the negatively charged center in an axial-symmetric manner. The cathode electrodes here act as a drain well, allowing the pore water to be discharged.

Neglecting the electrical capacitance of the soil mass, the relevant boundary value problem of this case is

$$K_r\left(\frac{1}{r}\frac{\partial u_e}{\partial r}+\frac{\partial^2 u_e}{\partial r^2}\right)+\gamma_w K_e\left(\frac{1}{r}\frac{dE}{dr}+\frac{d^2E}{dr^2}\right)=m_v\gamma_w\frac{\partial u_e}{\partial t}$$
$$K_e\left(\frac{1}{r}\frac{dE}{dr}+\frac{d^2E}{dr^2}\right)=0$$

$$\begin{aligned}\text{IC}:&\quad u_e(r,0)=0\\ \text{BC}:&\quad u_e(a,t)=0\\ &\quad\left(-K_r\frac{\partial u_e}{\partial r}-\gamma_w K_e\frac{dE}{dr}\right)\bigg|_{r=r_e}=0\\ &\quad E(a)=0\\ &\quad E(r_e)=E_0(t)\end{aligned} \tag{12.9}$$

Considering the second differential equation in (12.9), it is obvious that the electro-osmosis term on the first differential equation must vanish, leading this BVP to be expressed as

$$K_r\left(\frac{1}{r}\frac{\partial u_e}{\partial r}+\frac{\partial^2 u_e}{\partial r^2}\right)=m_v\gamma_w\frac{\partial u_e}{\partial t}$$
$$\frac{1}{r}\frac{dE}{dr}+\frac{d^2E}{dr^2}=0$$

$$\begin{aligned}\text{IC}:\quad & u_e(r,0)=0\\ \text{BC}:\quad & u_e(a,t)=0\\ & \left(-K_r\frac{\partial u_e}{\partial r}-\gamma_w K_e\frac{dE}{dr}\right)\Big|_{r=r_e}=0\\ & E(a)=0\\ & E(r_e)=E_0(t)\end{aligned} \tag{12.10}$$

BVP Solution

The general solution of the second differential equation in (12.10) is

$$E=c_1+c_2\ln(r) \tag{12.11}$$

At $r=a$, the third boundary condition implies

$$c_1+c_2\ln(a)=0$$

At $r=r_e$, the fourth boundary condition implies

$$c_1+c_2\ln(r_e)=E_0$$

Solving these two equations simultaneously yields

$$c_1=-\frac{E_0}{\ln(r_e/a)}\ln(a),\quad c_2=\frac{E_0}{\ln(r_e/a)}$$

Considering that E_0 may be a function of time (refer to the note in ▶ Sect. 12.1), then upon substituting these constants into (12.11), gives the solution in the frequency domain of the boundary value problem of the electric field, as

$$\hat{E}=\frac{\hat{E}_0}{\ln(r_e/a)}[\ln(r)-\ln(a)] \tag{12.12}$$

Applying the Fourier transform to the first differential equation in (12.10) gives

$$\frac{1}{r}\frac{d\hat{u}_e}{dr}+\frac{d^2\hat{u}_e}{dr^2}-\kappa^2\hat{u}_e=0,\quad \kappa^2=\frac{m_v\gamma_w}{K_r}i\omega \tag{12.13}$$

which is a kind of the modified Bessel equation, with its general solution given by

$$\hat{u}_e=c_3I_0(\kappa r)+c_4K_0(\kappa r) \tag{12.14}$$

in which I_0 and K_0 are the modified Bessel functions of zero order of the first and second kind, respectively, and c_3 and c_4 are arbitrary constants.

At $r = a$, the first boundary condition in (12.10), in the frequency domain, implies

$$c_3 I_0(\kappa a) + c_4 K_0(\kappa a) = 0$$

This gives

$$c_4 = -\frac{I_0(\kappa a)}{K_0(\kappa a)} c_3$$

Thus

$$\hat{u}_e = \left[I_0(\kappa r) - \frac{I_0(\kappa a)}{K_0(\kappa a)} K_0(\kappa r) \right] c_3 \tag{12.15}$$

Taking the derivative of $\widehat{u}_e$ with respect to r yields

$$\frac{d\hat{u}_e}{dr} = \kappa \left[I_1(\kappa r) + \frac{I_0(\kappa a)}{K_0(\kappa a)} K_1(\kappa r) \right] c_3$$

Also, taking the derivative of $\widehat{E}$ in (12.12) with respect to r gives

$$\frac{d\hat{E}}{dr} = \frac{\hat{E}_0}{\ln\left(r_e/a\right) r}$$

12

Then, at $r = r_e$, the second boundary condition in (12.10) implies

$$-K_r c_3 \kappa \left[I_1(\kappa r_e) + \frac{I_0(\kappa a)}{K_0(\kappa a)} K_1(\kappa r_e) \right] - \gamma_w K_e \frac{\hat{E}_0}{\ln(r_e/a) r_e} = 0$$

This gives

$$c_3 = -\frac{\gamma_w K_e}{K_r \kappa \left[I_1(\kappa r_e) + \frac{I_0(\kappa a)}{K_0(\kappa a)} K_1(\kappa r_e) \right]} \frac{\hat{E}_0}{\ln(r_e/a) r_e}$$

where it is obvious that this constant is a function of the electric field.

Thus, upon substituting c_3 into (12.15) we obtain the solution of this boundary value problem in the frequency domain, as

$$\hat{u}_e(r, \omega_n) = -\frac{\gamma_w K_e \hat{E}_0(\omega_n)}{K_r \kappa \ln\left(r_e/a\right) r_e} \cdot \frac{I_0(\kappa r) - \frac{I_0(\kappa a)}{K_0(\kappa a)} K_0(\kappa r)}{I_1(\kappa r_e) + \frac{I_0(\kappa a)}{K_0(\kappa a)} K_1(\kappa r_e)} \tag{12.16}$$

where

$$\kappa = \sqrt{\frac{m_v \gamma_w}{K_r} i\omega_n}$$

Together with

$$\hat{E}(r, \omega_n) = \frac{\hat{E}_0(\omega_n)}{\ln(r_e/a)}[\ln(r) - \ln(a)] \tag{12.17}$$

Having solved the problem in the frequency domain, then, upon applying the inverse FFT to (12.16) we obtain the solution in the time domain, such that

$$u_e(r, t) = \sum_n \hat{u}_e(r, \omega_n) e^{i\omega_n t}$$

$$E(r, t) = \sum_n \hat{E}(r, \omega_n) e^{i\omega_n t}$$

12.3 Plane Strain Electro-osmosis Consolidation

This section aims to expand upon the models presented in the previous two sections, incorporating not only consolidation resulting from electro-osmosis but also consolidation induced by a surcharge applied to the ground surface. We consider a soil mass under a plane strain condition that undergoes a surcharge applied at its surface boundary. To speed up the consolidation process, rows of anode and cathode electrodes are arranged in parallel, as depicted in Fig. 12.3. Sand drains are installed in such a way that, upon application of the surcharge and electric field, the pore water will dissipate through the surface boundary and via the drains surrounding the cathode electrode row. In this discussion, we consider a scenario where the electrodes have a finite length $L_e \leq H$, resulting in a two-dimensional electric field.

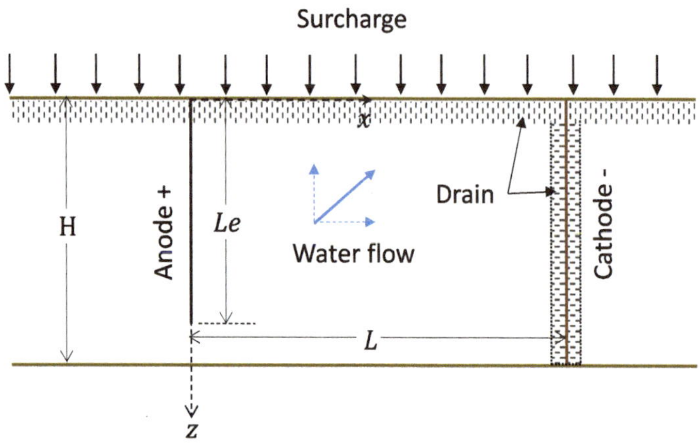

Fig. 12.3 Electro-osmosis with a finite length electrodes

The relevant boundary value problem of this configuration can be

$$K_x \frac{\partial^2 u_e}{\partial x^2} + K_z \frac{\partial^2 u_e}{\partial z^2} + \gamma_w K_e \left(\frac{d^2 E}{dx^2} + \frac{d^2 E}{dz^2} \right) = m_v \gamma_w \frac{\partial u_e}{\partial t}$$

$$\frac{d^2 E}{dx^2} + \frac{d^2 E}{dz^2} = 0$$

$$\begin{aligned}
\text{IC}: \quad & u_e(x, z, 0) = 0 \\
\text{BC}: \quad & u_e(x, 0, t) = P(t) \\
& \left. \frac{\partial u_e}{\partial z} \right|_{z=H} = 0 \\
& u_e(L, z, t) = 0 \\
& \left. \left(-K_x \frac{\partial u_e}{\partial x} - \gamma_w K_e \frac{dE}{dx} \right) \right|_{x=0} = 0, \quad 0 < z < L_e \\
& E(0, z, t) = E_0(z), \quad 0 < z < L_e \\
& E(L, z, t) = 0 \\
& \left. \frac{dE}{dz} \right|_{z=0} = 0 \\
& \left. \frac{dE}{dz} \right|_{z=H} = 0
\end{aligned} \tag{12.18}$$

BVP Solution

The solution of this problem can be attained by dividing it into two sub-problems, such that

$$u_e(x, z, t) = u_1(x, z, t) + u_2(x, z, t) \tag{12.19}$$

This division leads to solving for two sub-problems, expressed with their relevant initial and boundary conditions as

- **Sub-problems 1**

$$K_x\frac{\partial^2 u_1}{\partial x^2}+K_z\frac{\partial^2 u_1}{\partial z^2}+\gamma_w K_e\left(\frac{d^2E}{dx^2}+\frac{d^2E}{dz^2}\right)=m_v\gamma_w\frac{\partial u_1}{\partial t}$$

$$\frac{d^2E}{dx^2}+\frac{d^2E}{dz^2}=0$$

$$\begin{aligned}
\text{IC}:\quad & u_1(x,z,0)=0\\
\text{BC}:\quad & u_1(L,z,t)=0\\
& \left.\left(K_x\frac{\partial u_1}{\partial x}+\gamma_w K_e\frac{dE}{dx}\right)\right|_{x=0}=0,\quad 0<z<L_e\\
& u_1(x,0,t)=0\\
& \left.\frac{\partial u_1}{\partial z}\right|_{z=H}=0\\
& E(0,t)=E_0(t),\quad 0<z<L_e\\
& E(L,t)=0\\
& \left.\frac{dE}{dz}\right|_{z=0}=0\\
& \left.\frac{dE}{dz}\right|_{z=H}=0
\end{aligned}\tag{12.20}$$

- **Sub-problem 2**

$$K_x\frac{\partial^2 u_2}{\partial x^2}+K_z\frac{\partial^2 u_2}{\partial z^2}=m_v\gamma_w\frac{\partial u_2}{\partial t}$$

$$\begin{aligned}
\text{IC}:\quad & u_2(x,z,0)=0\\
\text{BC}:\quad & u_2(x,0,t)=P(t)\\
& \left.\frac{\partial u_2}{\partial z}\right|_{z=H}=0\\
& \left.\frac{\partial u_2}{\partial x}\right|_{x=0}=0\\
& u_2(L,z,t)=0
\end{aligned}\tag{12.21}$$

Solving Sub-problem 1

This problem involves a two-dimensional coupling between the electric field and the mass flow field. We begin here with solving the electric field boundary value problem.

Solution of $E(x,z)$

The BVP of the electric field is

$$\frac{d^2E}{dx^2} + \frac{d^2E}{dz^2} = 0$$

$$\begin{aligned} \text{BC}: \; & E(0,t) = E_0(t), \quad 0 < z < L_e \\ & E(L,t) = 0 \\ & \left.\frac{dE}{dz}\right|_{z=0} = 0 \\ & \left.\frac{dE}{dz}\right|_{z=H} = 0 \end{aligned} \tag{12.22}$$

This is a typical form of the Laplace equation. To solve this BVP we seek a solution that decays in the x-direction and satisfies an eigenfunction expansion in the z-direction.

Assuming $E(x,z) = X(x)Z(z)$ and applying the method of separation of variables, we readily obtain

$$Z'' + \lambda Z = 0, \quad X'' - \lambda X = 0 \tag{12.23}$$

where λ is a separation constant that can be 0, $-\alpha^2 < 0$, and $\alpha^2 > 0$. To determine λ, we need to solve an eigenfunction expansion problem, which here is manifested by the $Z(z)$ boundary value problem:

$$Z'' + \lambda Z = 0, \quad Z'(0) = 0 \quad \text{and} \quad Z'(H) = 0 \tag{12.24}$$

For $\lambda = 0$, the solution of the this differential equation is

$$Z = c_1 + c_2 z$$

The boundary conditions in (12.24) imply $c_2 = 0$, leading the solution to become

$$Z = c_1, \quad c_1 \neq 0$$

to be nontrivial.

For $\lambda = -\alpha^2 < 0$, the solution will not lead to solving eigenfunction expansion, giving the trivial outcome $Z = 0$.

For $\lambda = \alpha^2 > 0$, the general solution is given by

$$Z = c_1 \cos(\alpha z) + c_2 \sin(\alpha z)$$

At $z = 0$, the first boundary condition in (12.24) implies $c_2 = 0$, resulting in

$$Z = c_1 \cos(\alpha z)$$

At $z = H$, the second boundary condition in (12.24) implies

$$-\alpha c_1 \sin(\alpha H) = 0, \quad \Rightarrow \quad \alpha = \frac{m\pi}{H}$$

Thus, the solution of $Z(z)$ is

$$Z = c_1 \cos\left(\frac{m\pi}{H} z\right), \quad m = 0, 1, 2, \ldots$$

Having solved the $Z(z)$ differential equation, we now move to solving $X(x)$ boundary value problem:

$$X'' - \lambda X = 0, \quad E(0) = E_0 \quad \text{and} \quad E(L) = 0 \tag{12.25}$$

which is subject to a single homogeneous boundary condition at $x = L$. The general solution for this differential equation is given by

$$X = c_3 \cosh(\alpha x) + c_4 \sinh(\alpha x), \quad \alpha = \frac{m\pi}{H}$$

At $x = L$, the second boundary condition in (12.25) implies

$$c_3 \cosh(\alpha L) + c_4 \sinh(\alpha L) = 0, \quad \Rightarrow \quad c_3 = -\tanh(\alpha L)\, c_4$$

This leads to

$$X = c_4[-\tanh(\alpha L)\cosh(\alpha x) + \sinh(\alpha x)]$$

Combining the two solutions we obtain

$$E(x, z) = A\cos(\alpha z)[-\tanh(\alpha L)\cosh(\alpha x) + \sinh(\alpha x)] \tag{12.26}$$

where $A = c_1 c_4$.

Since the problem is linear, the superposition principle yields

$$E(x, z) = -\sum_{m=0}^{\infty} A_m \tanh(\alpha_m L)\cos(\alpha_m z)\left[\cosh(\alpha_m x) - \frac{\sinh(\alpha_m x)}{\tanh(\alpha_m L)}\right]$$

with

$$\alpha_m = \frac{m\pi}{H} \tag{12.27}$$

This formulation, although mathematically correct, suffers from being identically zero for $m = 0$. To address this issue, we adopt this alternative formulation:

$$E(x, z) = \sum_{m=0}^{\infty} \overline{A}_m \cos(\alpha_m z)\left[\cosh(\alpha_m x) - \frac{\sinh(\alpha_m x)}{\tanh(\alpha_m L)}\right] \tag{12.28}$$

with

$$\overline{A}_m = -A_m \tanh(\alpha_m L)$$

Following this, at $x = 0$, the first boundary condition in (12.25) implies

$$\sum_{m=0}^{\infty} \overline{A}_m \cos(\alpha_m z) = E_0(z)$$

which is a half-range expansion of E_0 in Fourier cosine series, with its coefficients given by

$$\begin{aligned}\overline{A}_0 &= \frac{1}{H}\int_0^{L_e} E_0(z)\,dz \\ \overline{A}_m &= \frac{2}{H}\int_0^{L_e} E_0(z)\cos(\alpha_m z)\,dz\end{aligned} \tag{12.29}$$

As indicated in ▶ Sect. 12.1, even though the differential equation for the electric field, (12.22), represents a steady state condition, the applied electric potential E_0 can still vary with time. Consequently, the solution (12.28), together with (12.29), can be expressed as

$$\hat{E}(x,z) = \sum_{m=0}^{\infty} \overline{A}_m \cos(\alpha_m z)\left[\cosh(\alpha_m x) - \frac{\sinh(\alpha_m x)}{\tanh(\alpha_m L)}\right]$$

with

$$\begin{aligned}\overline{A}_0 &= \frac{1}{H}\int_0^{L_e} \hat{E}_0(z)\,dz \\ \overline{A}_m &= \frac{2}{H}\int_0^{L_e} \hat{E}_0(z)\cos(\alpha_m z)\,dz\end{aligned} \tag{12.30}$$

12

For $m = 0$, $\widehat{E}(x,z)$ can be determined via

$$\begin{aligned}\hat{E}(x,z)\Big|_{m=0} &= \overline{A}_0 \lim_{\alpha_m \to 0}\left[\cosh(\alpha_m x) - \frac{\sinh(\alpha_m x)}{\tanh(\alpha_m L)}\right] \\ &= \overline{A}_0\left(1 - \frac{x}{L}\right)\end{aligned} \tag{12.31}$$

For a constant $\widehat{E}_0$ (i.e. not a function of z), the first coefficient in (12.30) results in

$$\overline{A}_0 = \frac{L_e}{H}\hat{E}_0$$

This leads (12.31) to become

$$\hat{E}(x,z)\Big|_{m=0} = \frac{L_e}{H}\left(1 - \frac{x}{L}\right)E_0$$

Solution of $u_1(x,z)$

Transforming the pore water pressure boundary value problem in (12.20) into the frequency domain results in

$$K_x \frac{\partial^2 \hat{u}_1}{\partial x^2} + K_z \frac{\partial^2 \hat{u}_1}{\partial z^2} - m_v \gamma_w i\omega \hat{u}_1 = 0$$

$$\begin{aligned} \text{BC}: \quad & \hat{u}_1(L,t) = 0 \\ & \left(K_x \frac{\partial \hat{u}_1}{\partial x} + \gamma_w K_e \frac{d\hat{E}}{dx} \right) \bigg|_{x=0} = 0 \\ & \hat{u}_1(x,0,t) = 0 \\ & \frac{\partial \hat{u}_1}{\partial z} \bigg|_{z=H} = 0 \end{aligned} \tag{12.32}$$

where we eliminated the electric field part because it is identically zero.

Following the derivation procedure outlined in the previous section, and keeping in mind that our goal is to obtain a general solution leading to eigenfunction expansion, the general solution of the partial differential equation in (12.32) leads to

$$\hat{u}_1(x,z) = \left[A\cos(\kappa_z z) + B\sin(\kappa_z z) \right] \left[C\cosh(\kappa_x x) + D\sinh(\kappa_x x) \right]$$

with

$$\kappa_x = \sqrt{\frac{\kappa^2 + m_v \gamma_w i\omega}{K_x}}, \qquad \kappa_z = \sqrt{\frac{\kappa^2}{K_z}}$$

in which κ represents the separation constant, to be determined from the eigenfunction expansion of the system.

At $z = 0$, the third boundary condition in (12.32) implies $A = 0$, leading to

$$\hat{u}_1(x,z) = B\sin(\kappa_z z)[C\cosh(\kappa_x x) + D\sinh(\kappa_x x)]$$

Taking the derivative of this equation with respect to z yields

$$\frac{\partial \hat{u}_1}{\partial z} = B\kappa_z \cos(\kappa_z z)[C\cosh(\kappa_x x) + D\sinh(\kappa_x x)] \tag{12.33}$$

At $z = H$, the fourth boundary condition in (12.32) implies

$$\cos(\kappa_z H) = 0, \quad \Rightarrow \quad \kappa_z \equiv \kappa_{zn} = \frac{1}{H}\left(\frac{\pi}{2} + n\pi\right)$$

from which

$$\kappa = \sqrt{\kappa_{zn}^2 K_z}, \quad \text{and} \quad \kappa_{xn} = \sqrt{\frac{\kappa_{zn}^2 K_z + m_v \gamma_w i\omega}{K_x}}$$

Accordingly, the superposition principle entails

$$\hat{u}_1(x,z) = \sum\nolimits_n B_n \sin(\kappa_{zn} z)[C\cosh(\kappa_{xn}x) + D\sinh(\kappa_{xn}x)], \quad \kappa_{zn} = \frac{1}{H}\left(\frac{\pi}{2} + n\pi\right)$$

At $x = L$, the first boundary condition in (12.32) implies

$$C = -\tanh(\kappa_{xn}L)D$$

Thus

$$\hat{u}_1(x,z) = \sum_n B_n \sin(\kappa_{zn}z)[-\tanh(\kappa_{xn}L)\cosh(\kappa_{xn}x) + \sinh(\kappa_{xn}x)] \tag{12.34}$$

where the constant D is dissolved in B.

The solution in (12.34) can be conveniently written as

$$\hat{u}_1(x,z) = \sum_n \overline{B}_n \sin(\kappa_{zn}z)\left[\cosh(\kappa_{xn}x) - \frac{\sinh(\kappa_{xn}x)}{\tanh(\kappa_{xn}L)}\right]$$

with (12.35)

$$\overline{B}_n = -B_n \tanh(\kappa_{xn}L)$$

Taking the derivative of (12.35) with respect to x gives

$$\frac{\partial \hat{u}_1}{\partial x} = \sum_n \overline{B}_n \sin(\kappa_{zn}z)\kappa_{xn}\left[\sinh(\kappa_{xn}x) - \frac{\cosh(\kappa_{xn}x)}{\tanh(\kappa_{xn}L)}\right]$$

Taking the derivative of the electric field (12.30) with respect to x gives

$$\frac{\partial \hat{E}}{\partial x} = \sum_{m=0}^{\infty} \overline{A}_m \cos(\alpha_m z)\,\alpha_m\left[\sinh(\alpha_m x) - \frac{\cosh(\alpha_m x)}{\tanh(\alpha_m L)}\right]$$

Then, at $x = 0$, the second boundary condition in (12.32) implies

$$\sum_n \overline{B}_n \sin(\kappa_{zn}z) = -\frac{\gamma_w K_e}{\kappa_{xn}K_x}\sum_{m=0}^{\infty}\frac{\tanh(\kappa_{xn}L)}{\tanh(\alpha_m L)}\overline{A}_m \cos(\alpha_m z)\,\alpha_m,$$

$$n = 0, 1, 2, \ldots$$

where $\overline{A}_m$ is given in (12.30). This is a Fourier sine series with its coefficients given by

$$\overline{B}_n = \frac{2G_m}{H}\int_0^{L_e} \cos(\alpha_m z)\sin(\kappa_{zn}z)\,dz$$

with

$$G_m = -\frac{\gamma_w K_e}{\kappa_{xn}K_x}\sum_{m=0}^{\infty}\frac{\tanh(\kappa_{xn}L)}{\tanh(\alpha_m L)}\overline{A}_m\alpha_m, \quad n = 0, 1, 2, \ldots$$

Note that in the case where $\widehat{E}_0$ is constat, G_0 can be obtained by taking the limit as $m \to 0$, yielding

$$G_0 = -\frac{\gamma_w K_e E_0 L_e}{\kappa_{xn} K_x L H} \tanh\left(\kappa_{xn} L\right)$$

Solving Sub-problem 2

Applying the Fourier transform to (12.21), leads, together with the relevant boundary conditions, to

$$\begin{aligned}
&K_x \frac{\partial^2 \hat{u}_2}{\partial x^2} + K_z \frac{\partial^2 \hat{u}_2}{\partial z^2} - m_v \gamma_w i\omega \hat{u}_2 = 0 \\
&\text{BC}: \quad \hat{u}_2(x, 0, \omega) = \hat{P}(\omega) \\
&\qquad \left.\frac{\partial \hat{u}_2}{\partial z}\right|_{z=H} = 0 \\
&\qquad \left.\frac{\partial \hat{u}_2}{\partial x}\right|_{x=0} = 0 \\
&\qquad \hat{u}_2(L, z, \omega) = 0
\end{aligned} \tag{12.36}$$

Using the separation of variables method, we first assume

$$\hat{u}_2(x, z) = X(x)Z(z)$$

Substituting this identity into (12.36), dividing by XZ, and equating the result by a constant, say $-\eta^2$, yields

$$\frac{K_x}{X}\frac{d^2X}{dx^2} - m_v \gamma_w i\omega = -\frac{K_z}{Z}\frac{d^2Z}{dz^2} = -\eta^2$$

The X-equation gives

$$K_x \frac{d^2X}{dx^2} + \mu_x^2 X = 0, \quad \mu_x = \sqrt{\frac{\eta^2 - m_v \gamma_w i\omega}{K_x}}$$

The general solution of this differential equation is

$$X(x) = A\cos\left(\mu_x x\right) + B\sin\left(\mu_x x\right) \tag{12.37}$$

The Z-equation gives

$$\frac{d^2Z}{dz^2} - \mu_z^2 Z = 0, \quad \mu_z^2 = \frac{\eta^2}{K_z}$$

The general solution of this differential equation is

$$Z(z) = C\cosh\left(\mu_z z\right) + D\sinh\left(\mu_z z\right)$$

Combining the two solutions, the general solution of $\widehat{u}_2(x, z)$ becomes

$$\hat{u}_2(x, z) = \left[A\cos\left(\mu_x x\right) + B\sin\left(\mu_x x\right)\right]\left[C\cosh\left(\mu_z z\right) + D\sinh\left(\mu_z z\right)\right] \tag{12.38}$$

Taking the derivative of this equation with respect to x yields

$$\frac{\partial \hat{u}_2}{\partial x} = [-\mu_x A \sin(\mu_x x) + \mu_x B \cos(\mu_x x)]\left[C \cosh(\mu_z z) + D \sinh(\mu_z z)\right]$$

At $x = 0$, the third boundary condition in (12.36) implies

$$\mu_x B\left[C \cosh(\mu_z z) + D \sinh(\mu_z z)\right] = 0, \quad \Rightarrow \quad B = 0$$

This leads the solution to reduce to

$$\hat{u}_2(x, z) = A \cos(\mu_x x)\left[C \cosh(\mu_z z) + D \sinh(\mu_z z)\right]$$

Taking the derivative of this outcome with respect to z, gives

$$\frac{\partial \hat{u}_2}{\partial z} = A \cos(\mu_x x)\left[C\mu_z \sinh(\mu_z z) + D\mu_z \cosh(\mu_z z)\right]$$

At $z = H$, the second boundary condition in (12.36), valid for all x, entails

$$A \cos(\mu_x x)\left[C\mu_z \sinh(\mu_z H) + D\mu_z \cosh(\mu_z H)\right] = 0$$

or

$$D = -\tanh(\mu_z H)\, C$$

This reduces the solution to become

$$\hat{u}_2(x, z) = C \cos(\mu_x x)\left[\cosh(\mu_z z) - \tanh(\mu_z H) \sinh(\mu_z z)\right]$$

where the constant A has been dissolved in C.

At $x = L$, the fourth boundary condition in (12.36), valid for all z, implies

12

$$\cos(\mu_x L) = 0, \quad \Rightarrow \quad \mu_x \equiv \mu_{xk} = \frac{1}{L}\left(\frac{\pi}{2} + k\pi\right)$$

Since the problem is linear, any linear combinations of solutions is also a solution, leading to

$$\hat{u}_2(x, z, \omega) = \sum_k C_k \cos(\mu_{xk} x)\left[\cosh(\mu_{zk} z) - \tanh(\mu_{zk} H) \sinh(\mu_{zk} z)\right]$$

where

$$\mu_z \equiv \mu_{zk} = \sqrt{\frac{\mu_{xk}^2 K_x + m_v \gamma_w i\omega}{K_z}}$$

At $z = 0$, the first boundary condition in (12.36) entails

$$\sum_k C_k \cos(\mu_{xk} x) = \hat{P}(\omega)$$

This is a typical cosine Fourier series with its coefficients given by

$$C_k = \frac{2}{L} \int_0^L \hat{P}(\omega) \cos(\mu_{xk} x)\, dx, \quad k = 0, 1, 2, \ldots$$

BVP Solution

Collecting the solutions of the two sub-problems, gives

$$\hat{u}_e(x,z,\omega_s) = \hat{u}_1(x,z,\omega_s) + \hat{u}_2(x,z,\omega_s) \tag{12.39}$$

where

$$\hat{u}_1(x,z) = \sum_n \overline{B}_n \sin(\kappa_{zn}z)\left[\cosh(\kappa_{xn}x) - \frac{\sinh(\kappa_{xn}x)}{\tanh(\kappa_{xn}L)}\right], \quad n = 0,1,2,\ldots$$

with

$$\overline{B}_n = \frac{2G_m}{H}\int_0^{L_e} \cos(\alpha_m z)\sin(\kappa_{zn}z)\,dz$$

$$G_m = -\frac{\gamma_w K_e}{\kappa_{xn}K_x}\sum_{m=0}^{\infty}\frac{\tanh(\kappa_{xn}L)}{\tanh(\alpha_m L)}\overline{A}_m\,\alpha_m$$

$$\kappa_{zn} = \frac{1}{H}\left(\frac{\pi}{2}+n\pi\right),\ \kappa = \sqrt{\kappa_{zn}^2 K_z},\ \text{and}\ \kappa_{xn} = \sqrt{\frac{\kappa_{zn}^2 K_z + m_v\gamma_w i\omega}{K_x}}$$

$$\alpha_m = \frac{m\pi}{H}$$

and

$$\hat{u}_2(x,z,\omega_s) = \sum_k C_k \cos(\mu_{xk}x)\left[\cosh(\mu_{zk}z) - \tanh(\mu_{zk}H)\sinh(\mu_{zk}z)\right]$$

with

$$C_k = \frac{2}{L}\int_0^L \hat{P}(\omega)\cos(\mu_{xk}x)\,dx, \quad k = 0,1,2,\ldots$$

and

$$\mu_{xk} = \frac{1}{L}\left(\frac{\pi}{2}+k\pi\right), \quad \mu_{zk} = \sqrt{\frac{\mu_{xk}^2 K_x + m_v\gamma_w i\omega}{K_z}}$$

together with

$$\hat{E}(x,z) = \sum_{m=0}^{\infty}\overline{A}_m\cos(\alpha_m z)\left[\cosh(\alpha_m x) - \frac{\sinh(\alpha_m x)}{\tanh(\alpha_m L)}\right]$$

with

$$\overline{A}_0 = \frac{1}{H}\int_0^{L_e}\hat{E}_0(z)\,dz, \qquad m = 0$$

$$\overline{A}_m = \frac{2}{H}\int_0^{L_e}\hat{E}_0(z)\cos(\alpha_m z)\,dz, \qquad m = 1,2,\ldots$$

and

$$\alpha_m = \frac{m\pi}{H}$$

Applying the inverse FFT to (12.39) and its associated solutions restores the solution to the time domain, such that

$$u_e(x,z,t) = \sum_s \hat{u}_e(x,z,\omega_s)\, e^{i\omega_s t}$$
$$E(x,z,t) = \sum_s \hat{E}\,(x,z,\omega_s)\, e^{i\omega_s t}$$

Special case

In the special case where the length of the electrodes spans the entire thickness of the soil layer, i.e. $L_e = H$ (see Fig. 12.4), the solution simplifies as follows.

Electric BVP

For a constant $\widehat{E}_0$, i.e. independent of z, the Fourier coefficients become

$$\overline{A}_0 = \hat{E}_0(\omega)$$
$$\overline{A}_m = 0, \qquad m \geq 1$$

If the applied electric field is a function of z, say $\widehat{E}_0 z$, the Fourier coefficients become

$$\overline{A}_0 = \frac{1}{2}\hat{E}_0(\omega)H$$
$$\overline{A}_m = \frac{2\hat{E}_0(\omega)}{\alpha_m^2 H}[\alpha_m H \sin(\alpha_m H) + \cos(\alpha_m H) - 1]$$
$$= \frac{2\hat{E}_0(\omega)}{\alpha_m^2 H}[\alpha_m H \sin(m\pi) + \cos(m\pi) - 1]$$
$$= \frac{2\hat{E}_0(\omega)}{\alpha_m^2 H}[\,(-1)^m - 1\,], \qquad m \geq 1$$

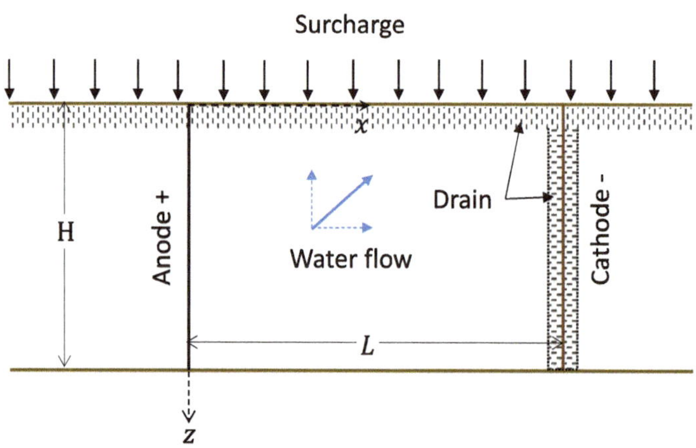

Fig. 12.4 Electrodes configuration with length $L_e = H$

$\hat{u}_1(x, z)$ BVP

Setting

$$a_1 = \alpha_m - \kappa_{zn}$$
$$a_2 = \alpha_m + \kappa_{zn}$$

with

$$\alpha_m = \frac{m\pi}{H}, \quad \kappa_{zn} = \frac{1}{H}\left(n\pi + \frac{1}{2}\pi\right)$$

Then, the Fourier coefficients of $\widehat{u}_1(x, z, \omega)$ become

$$\begin{aligned}\overline{B}_n &= \frac{2G_m}{H}\int_0^H \cos(\alpha_m z)\sin(\kappa_{zn} z)\,dz \\ &= \frac{2G_m}{H}\left[\frac{\cos(a_1 H)\,a_2 - \cos(a_2 H)\,a_1 - 2\kappa_{zn}}{2a_1 a_2}\right]\end{aligned}$$

And knowing that

$$\begin{aligned}\cos(a_1 H) &= \cos\left[(m - \left(n + \frac{1}{2}\right)\pi\right] \\ &= 0 \\ \cos(a_2 H) &= \cos\left[(m + \left(n + \frac{1}{2}\right)\pi\right] \\ &= 0\end{aligned}$$

we arrive at

$$\begin{aligned}\overline{B}_n &= \frac{2G_m}{H}\left(\frac{-2\kappa_{zn}}{2a_1 a_2}\right) \\ &= -\frac{2G_m}{H}\left(\frac{\kappa_{zn}}{\alpha_m^2 - \kappa_{zn}^2}\right)\end{aligned}$$

■ EXAMPLE 12.1: Hydraulic and Electro-Osmosis Consolidation

Consider a 10 m deep, fully saturated soil layer resting on an impermeable layer. Over a period of approximately 50 days, a surcharge exerting a pressure of maximum 100 kPa was applied at the ground surface, after which it was removed. Following this, to accelerate the consolidation process, rows of anode and cathode electrodes, 5 m apart, were installed. The anode was operated to exert an electric field of maximum 10 V. The time signals of both the surcharge application and the electric potential are shown in ▫ Fig. 12.5. ▫ Figure 12.6 depicts the soil layer with the electrodes configuration, along with the relevant geometrical and material parameters.

Compute the excess pore pressure distribution resulting from both hydraulic and electro-osmosis consolidation during the surcharge application, the activation of the electric field, and following the removal of the surcharge and electrodes.

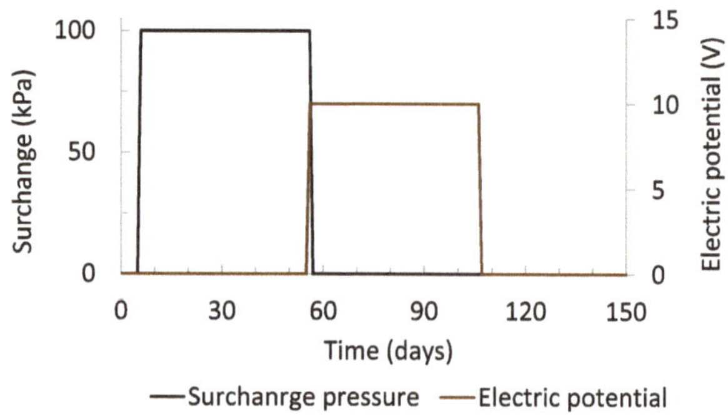

Fig. 12.5 Time signals of surcharge pressure and electric potential

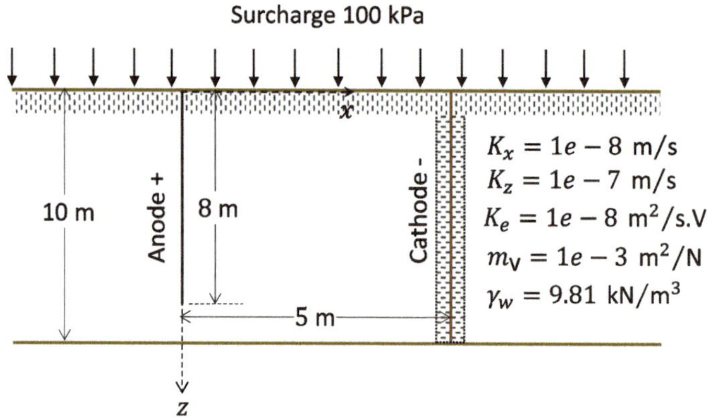

Fig. 12.6 Electrodes configuration with material parameters

12

Solution

As usual, the solution of this problem begins by transforming the input signals provided in Fig. 12.5 into the frequency domain using the forward FFT algorithm. The solution of the electro-osmosis consolidation in the frequency domain can then be obtained using (12.39) and its associated equations. Following this, the final solution can be reconstructed into the time domain using the inverse FFT algorithm.

Figure 12.7 shows the distribution of electric potential across the domain, highlighting the influence of the insulated top and bottom boundaries on the electric field flow. It also depicts the cathode at the right-hand side of the domain, maintained at zero potential.

Figure 12.8 depicts the excess pore pressure distribution approximately 6 and 50 days after the application of the surcharge load at the ground surface. The figure effectively demonstrates the temporal evolution of positive excess pore pressure in the domain, while showing that zero pressure is consistently maintained along the drainage surfaces.

▢ Figure 12.9 illustrates the distribution of excess pore pressure resulting from the electro-osmosis effect at approximately 10 and 25 days after the removal of the surcharge. The positive pore pressure generated by the surcharge decreases significantly within 10 days, primarily due to the free drainage boundaries and the influence of electro-osmosis. Over time, this reduction becomes more pronounced as negative pore pressure begins to develop. Even more reduction of pore pressure continues with time, as demonstrated in ▢ Fig. 12.10, where the pore pressure distributions, 35 and 50 days after the surcharge removal, are presented. The figure highlights the progressive dissipation of positive pore pressure and the further development of negative pore pressure, indicating the ongoing effectiveness of electro-osmosis and drainage in consolidating the soil.

▢ Figure 12.11 illustrates the distribution of excess pore pressure approximately 5 and 15 days after the electrodes were

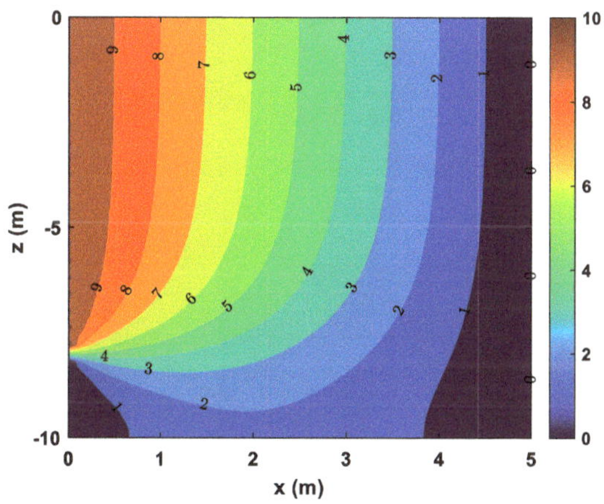

▢ **Fig. 12.7** Electric field distribution in the domain during activation

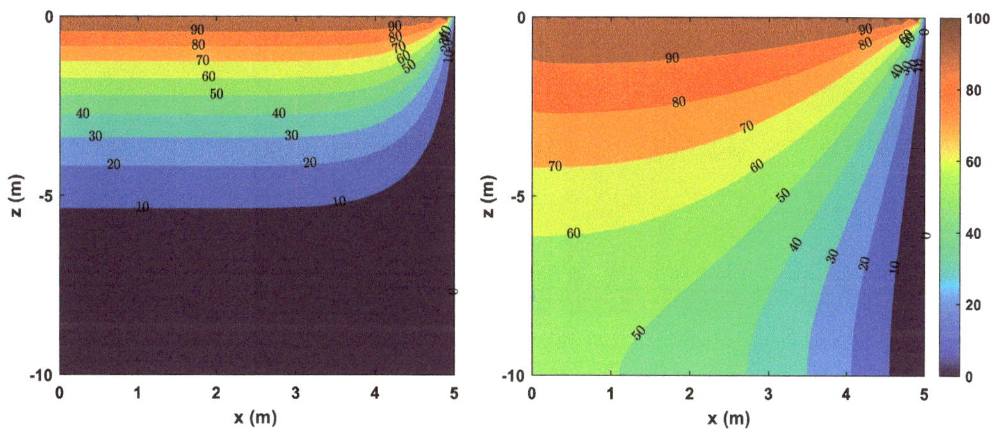

▢ **Fig. 12.8** Excess pore pressure after 6 days (left) and 50 days (right) following the application of the surcharge

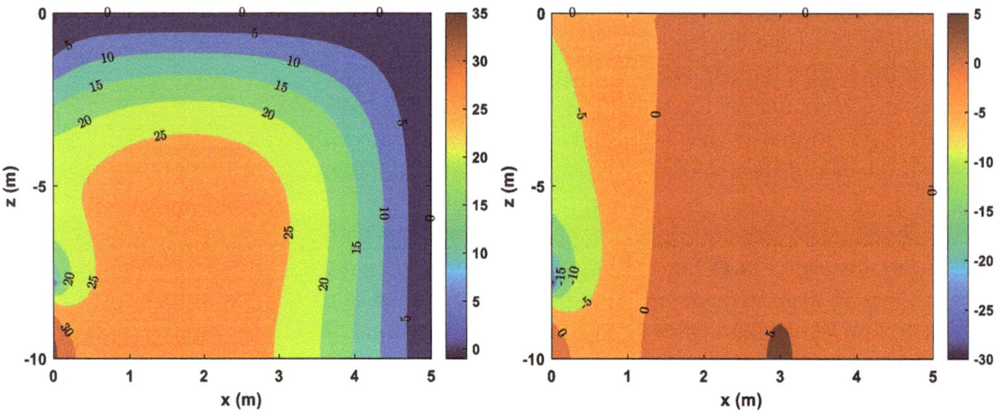

Fig. 12.9 Evolution of negative pore pressure after 10 days (left) and 25 days (right) of electrode activation following the removal of the surcharge

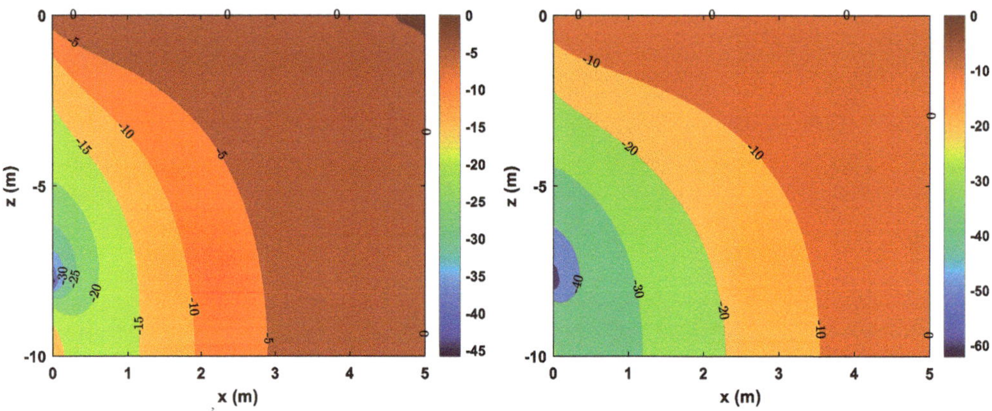

Fig. 12.10 Evolution of negative pore pressure after 35 days (left) and 50 days (right) of electrode activation following the removal of the surcharge

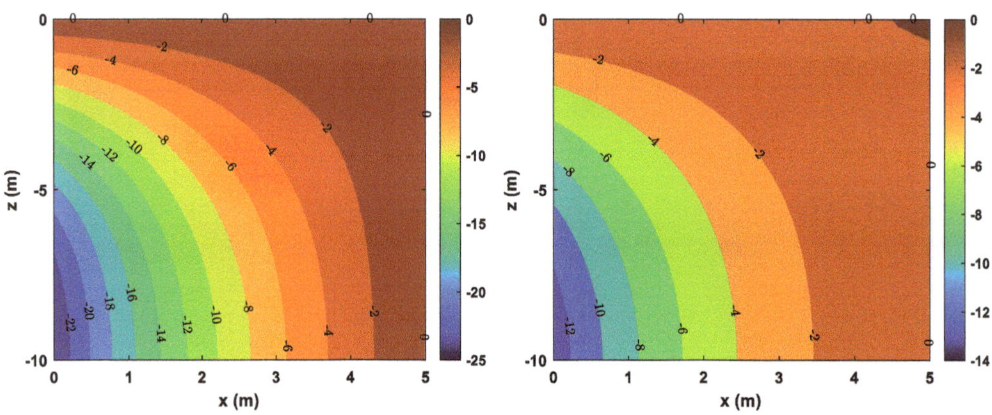

Fig. 12.11 Dissipation of excess pore pressure 5 days (left) and 15 days (right) after switching off the electrodes

deactivated. The figure effectively captures the temporal dynamics of pore pressure behavior, emphasizing the gradual recovery of pressure towards the initial condition.

12.4 Plane Strain Electro-osmosis Consolidation: Electrically Grounded Surface

This section addresses the plane strain model presented in the previous section, but this time by assuming that the ground surface is grounded, i.e. the electric intensity is zero at $z = 0$.

The boundary value problem of this configuration is

$$K_x \frac{\partial^2 u_e}{\partial x^2} + K_z \frac{\partial^2 u_e}{\partial z^2} + \gamma_w K_e \left(\frac{d^2 E}{dx^2} + \frac{d^2 E}{dz^2} \right) = m_v \gamma_w \frac{\partial u_e}{\partial t}$$

$$\frac{d^2 E}{dx^2} + \frac{d^2 E}{dz^2} = 0$$

$$\begin{aligned}
\text{IC}: \quad & u_e(x, z, 0) = 0 \\
\text{BC}: \quad & u_e(x, 0, t) = P(t) \\
& \left. \frac{\partial u_e}{\partial z} \right|_{z=H} = 0 \\
& u_e(L, z, t) = 0 \\
& \left. \left(-K_x \frac{\partial u_e}{\partial x} - \gamma_w K_e \frac{dE}{dx} \right) \right|_{x=0} = 0, \quad 0 < z < L_e \\
& E(0, z, t) = E_0(z), \quad 0 < z < L_e \\
& E(L, z, t) = 0 \\
& E(x, 0, t) = 0 \\
& \left. \frac{dE}{dz} \right|_{z=H} = 0
\end{aligned} \tag{12.40}$$

▪ BVP Solution

The solution of this problem can be attained by dividing it into two sub-problems, such that

$$u_e(x, z, t) = u_1(x, z, t) + u_2(x, z, t) \tag{12.41}$$

This division leads to solving for two sub-problems, expressed with their relevant initial and boundary conditions as

- **Sub-problems 1**

$$K_x\frac{\partial^2 u_1}{\partial x^2}+K_z\frac{\partial^2 u_1}{\partial z^2}+\gamma_w K_e\left(\frac{d^2E}{dx^2}+\frac{d^2E}{dz^2}\right)=m_v\gamma_w\frac{\partial u_1}{\partial t}$$
$$\frac{d^2E}{dx^2}+\frac{d^2E}{dz^2}=0$$

$$\begin{aligned}
\text{IC}:\quad & u_1(x,z,0)=0\\
\text{BC}:\quad & u_1(L,z,t)=0\\
& \left.\left(K_x\frac{\partial u_1}{\partial x}+\gamma_w K_e\frac{dE}{dx}\right)\right|_{x=0}=0,\quad 0<z<L_e\\
& u_1(x,0,t)=0\\
& \left.\frac{\partial u_1}{\partial z}\right|_{z=H}=0\\
& E(0,z,t)=E_0(t),\quad 0<z<L_e\\
& E(L,z,t)=0\\
& E(x,0,t)=0\\
& \left.\frac{dE}{dz}\right|_{z=H}=0
\end{aligned}\tag{12.42}$$

- **Sub-problem 2**

$$K_x\frac{\partial^2 u_2}{\partial x^2}+K_z\frac{\partial^2 u_2}{\partial z^2}=m_v\gamma_w\frac{\partial u_2}{\partial t}$$

$$\begin{aligned}
\text{IC}:\quad & u_2(x,z,0)=0\\
\text{BC}:\quad & u_2(x,0,t)=P(t)\\
& \left.\frac{\partial u_2}{\partial z}\right|_{z=H}=0\\
& \left.\frac{\partial u_2}{\partial x}\right|_{x=0}=0\\
& u_2(L,z,t)=0
\end{aligned}\tag{12.43}$$

Solving Sub-problem 1

This problem involves a two-dimensional coupling between the electric field and the mass flow field. We begin here by solving for the electric field.

Solution of $E(x,z)$

The BVP of the electric field is

$$\frac{d^2E}{dx^2}+\frac{d^2E}{dz^2}=0$$

$$\begin{aligned}\text{BC}:\quad & E(0,z,t)=E_0(t), \quad 0<z<L_e \\ & E(L,z,t)=0 \\ & E(x,0,t)=0 \\ & \left.\frac{dE}{dz}\right|_{z=H}=0\end{aligned} \tag{12.44}$$

Using the separation of variables, we first assume $E(x,z)=X(x)Z(z)$. Then upon applying the method of separation of variables, we obtain

$$Z''+\lambda Z=0, \quad Z(0)=0, \quad Z'(H)=0 \tag{12.45}$$

and

$$X''-\lambda X=0, \quad X(0)=E_0, \quad X(L)=0 \tag{12.46}$$

where λ is a separation constant that can be 0, $-\alpha^2<0$, and $\alpha^2>0$.

In the previous section we proved that in solving (12.45) $\lambda=0$ is an eigenvalue, and $\lambda=\alpha^2>0$ leads to solving an eigenfunction expansion with the general solution given by

$$Z=c_1\cos(\alpha z)+c_2\sin(\alpha z)$$

At $z=0$, the first boundary condition in (12.45) implies $c_1=0$, resulting in

$$Z=c_2\sin(\alpha z)$$

At $z=H$, the second boundary condition in (12.45) implies

$$\alpha c_2\cos(\alpha H)=0, \quad \Rightarrow \quad \alpha\equiv\alpha_m=\frac{1}{H}(\frac{\pi}{2}+m\pi)$$

Thus the solution of $Z(z)$ is

$$Z=c_2\sin(\alpha_m z), \quad m=0,1,2,\ldots$$

The general solution for the differential equation in (12.46) is

$$X=c_3\cosh(\alpha_m x)+c_4\sinh(\alpha_m x), \quad \alpha_m=\frac{1}{H}(\frac{\pi}{2}+m\pi)$$

At $x=L$, the second boundary condition in (12.46) implies

$$c_3\cosh(\alpha_m L)+c_4\sinh(\alpha_m L)=0, \quad \Rightarrow \quad c_3=-\tanh(\alpha_m L)\,c_4$$

This leads to

$$X=c_4[-\tanh(\alpha_m L)\cosh(\alpha_m x)+\sinh(\alpha_m x)]$$

Combining the two solutions we obtain

$$E(x,z) = A \sin(\alpha_m z)[-\tanh(\alpha_m L)\cosh(\alpha_m x) + \sinh(\alpha_m x)] \tag{12.47}$$

where $A = c_1 c_4$.

Since the problem is linear, the superposition principle yields

$$E(x,z) = \sum_{m=0}^{\infty} A_m \sin(\alpha_m z)[-\tanh(\alpha_m L)\cosh(\alpha_m x) + \sinh(\alpha_m x)]$$

with

$$\alpha_m = \frac{1}{H}\left(\frac{\pi}{2} + m\pi\right) \tag{12.48}$$

Finally, at $x = 0$, the first boundary condition in (12.46) implies

$$\sum_{m=0}^{\infty} A_m \sin(\alpha_m z)[-\tanh(\alpha_m L)] = E_0(z)$$

which is a half-range expansion of E_0 in Fourier sine series, with its coefficients given by

$$A_m = \frac{-2}{H\tanh(\alpha_m L)} \int_0^{L_e} E_0(z)\sin(\alpha_m z)\,dz, \quad m = 0, 1, 2, \ldots \tag{12.49}$$

12

Even though the differential equation in (12.44) is in a steady state condition, the applied electrical field, E_0, may still vary with time. Accordingly, the solution (12.48) along with (12.49) can be written as

$$\hat{E}(x,z,\omega) = \sum_{m=0}^{\infty} A_m \sin(\alpha_m z)[-\tanh(\alpha_m L)\cosh(\alpha_m x) + \sinh(\alpha_m x)]$$

with

$$A_m = \frac{-2}{H\tanh(\alpha_m L)} \int_0^{L_e} \hat{E}_0(z,\omega)\sin(\alpha_m z)\,dz, \quad m = 0, 1, 2, \ldots \tag{12.50}$$

Solution of $u_1(x, z)$

Transforming the boundary value problem in (12.42) involving the pore water pressure into the frequency domain results in

$$
\begin{aligned}
&K_x \frac{\partial^2 \hat{u}_1}{\partial x^2} + K_z \frac{\partial^2 \hat{u}_1}{\partial z^2} - m_v \gamma_w i\omega \hat{u}_1 = 0 \\
&\text{BC}: \quad \hat{u}_1(L, t) = 0 \\
&\qquad \left.\left(K_x \frac{\partial \hat{u}_1}{\partial x} + \gamma_w K_e \frac{d\hat{E}}{dx} \right)\right|_{x=0} = 0 \\
&\qquad \hat{u}_1(x, 0, t) = 0 \\
&\qquad \left.\frac{\partial \hat{u}_1}{\partial z}\right|_{z=H} = 0
\end{aligned}
\tag{12.51}
$$

where we eliminated the electric field part because it is identically zero.

Following the derivation procedure outlined in the previous section, and keeping in mind that our goal to obtain a general solution leading to eigenfunction expansion, the general solution of the partial differential equation in (12.51) leads to

$$
\hat{u}_1(x, z) = \left[A \cos(\kappa_z z) + B \sin(\kappa_z z)\right]\left[C \cosh(\kappa_x x) + D \sinh(\kappa_x x)\right]
$$

with

$$
\kappa_x = \sqrt{\frac{\kappa^2 + m_v \gamma_w i\omega}{K_x}}, \qquad \kappa_z = \sqrt{\frac{\kappa^2}{K_z}}
$$

in which κ represents the separation constant, to be determined from the eigenfunction expansion of the system.

At $z = 0$, the third boundary condition in (12.51) implies $A = 0$, leading to

$$
\hat{u}_1(x, z) = B \sin(\kappa_z z)[C \cosh(\kappa_x x) + D \sinh(\kappa_x x)]
$$

Taking the derivative of this equation with respect to z yields

$$
\frac{\partial \hat{u}_1}{\partial z} = B\kappa_z \cos(\kappa_z z)[C \cosh(\kappa_x x) + D \sinh(\kappa_x x)] \tag{12.52}
$$

At $z = H$, the fourth boundary condition in (12.51) implies

$$
\cos(\kappa_z H) = 0, \quad \Rightarrow \quad \kappa_z \equiv \kappa_{zn} = \frac{1}{H}\left(\frac{\pi}{2} + n\pi\right)
$$

from which

$$
\kappa = \sqrt{\kappa_{zn}^2 K_z}, \quad \text{and} \quad \kappa_{xn} = \sqrt{\frac{\kappa_{zn}^2 K_z + m_v \gamma_w i\omega}{K_x}}
$$

Accordingly, the superposition principle entails

$$\hat{u}_1(x,z) = \sum_n B_n \sin(\kappa_{zn} z)[C \cosh(\kappa_{xn} x) + D \sinh(\kappa_{xn} x)],$$

$$\kappa_{zn} = \frac{1}{H}(\frac{\pi}{2} + n\pi)$$

At $x = L$, the first boundary condition in (12.51) implies

$$C = -\tanh(\kappa_{xn} L) D$$

Thus

$$\hat{u}_1(x,z) = \sum_n B_n \sin(\kappa_{zn} z)[-\tanh(\kappa_{xn} L)\cosh(\kappa_{xn} x) + \sinh(\kappa_{xn} x)] \tag{12.53}$$

where the constant D is dissolved in B.

Taking the derivative of (12.53) with respect to x gives

$$\frac{\partial \hat{u}_1}{\partial x} = \sum_n B_n \sin(\kappa_{zn} z)\kappa_{xn}[-\tanh(\kappa_{xn} L)\sinh(\kappa_{xn} x) + \cosh(\kappa_{xn} x)]$$

Taking the derivative of the electric field (12.50) with respect to x gives

$$\frac{\partial \hat{E}}{\partial x} = \sum_{m=0}^{\infty} A_m \sin(\alpha_m z)\,\alpha_m[-\tanh(\alpha_m L)\,\sinh(\alpha_m x) + \cosh(\alpha_m x)]$$

12

Then, at $x = 0$, the second boundary condition in (12.51) implies

$$\sum_n B_n \sin(\kappa_{zn} z) = \frac{\gamma_w K_e}{\kappa_{xn} K_x} \sum_{m=0}^{\infty} A_m \sin(\alpha_m z)\,\alpha_m, \quad n = 0, 1, 2, \ldots$$

This is a Fourier sine series with its coefficients are given by

$$B_n = \frac{2G_m}{H} \int_0^{L_e} \sin(\alpha_m z)\sin(\kappa_{zn} z)\,dz$$

with

$$G_m = \frac{\gamma_w K_e}{\kappa_{xn} K_x} \sum_{m=0}^{\infty} A_m \alpha_m$$

Solving Sub-problem 2

Applying Fourier transform to (12.43), leads, together with the relevant boundary conditions, to

$$
\begin{aligned}
& K_x \frac{\partial^2 \hat{u}_2}{\partial x^2} + K_z \frac{\partial^2 \hat{u}_2}{\partial z^2} - m_v \gamma_w i\omega \hat{u}_2 = 0 \\
& \text{BC}: \quad \hat{u}_2(x, 0, \omega) = \hat{P}(\omega) \\
& \qquad \left. \frac{\partial \hat{u}_2}{\partial z} \right|_{z=H} = 0 \\
& \qquad \left. \frac{\partial \hat{u}_2}{\partial x} \right|_{x=0} = 0 \\
& \qquad \hat{u}_2(L, z, \omega) = 0
\end{aligned}
\tag{12.54}
$$

We addressed the solution of this boundary value problem in the previous section in detail. The solution resulted in

$$
\cos(\mu_x L) = 0, \quad \Rightarrow \quad \mu_x \equiv \mu_{xk} = \frac{1}{L}\left(\frac{\pi}{2} + k\pi\right)
$$

Since the problem is linear, any linear combinations of solutions is also a solution, leading to

$$
\hat{u}_2(x, z, \omega) = \sum_k C_k \cos(\mu_{xk} x) \left[\cosh(\mu_{zk} z) - \tanh(\mu_{zk} H) \sinh(\mu_{zk} z)\right]
$$

where

$$
C_k = \frac{2}{L} \int_0^L \hat{P}(\omega) \cos(\mu_{xk} x) dx, \quad k = 0, 1, 2, \ldots
$$

with

$$
\mu_{zk} = \sqrt{\frac{\mu_{xk}^2 K_x + m_v \gamma_w i\omega}{K_z}}, \quad \mu_{xk} = \frac{1}{L}\left(\frac{\pi}{2} + k\pi\right)
$$

BVP Solution

Collecting the solutions of the two sub-problems, gives

$$
\hat{u}_e(x, z, \omega_s) = \hat{u}_1(x, z, \omega_s) + \hat{u}_2(x, z, \omega_s) \tag{12.55}
$$

where

$$\hat{u}_1(x,z,\omega_s) = \sum_n B_n \sin(\kappa_{zn}z)[-\tanh(\kappa_{xn}L)\cosh(\kappa_{xn}x) + \sinh(\kappa_{xn}x)]$$

with

$$B_n = \frac{2G_m}{H}\int_0^{L_e} \sin(\alpha_m z)\sin(\kappa_{zn}z)\,dz$$

$$G_m = \frac{\gamma_w K_e}{\kappa_{xn}K_x}\sum_{m=0}^{\infty} A_m\alpha_m$$

$$\kappa_{zn} = \frac{1}{H}\left(\frac{\pi}{2}+n\pi\right),\ \kappa = \sqrt{\kappa_{zn}^2 K_z},\ \text{and}\ \kappa_{xn} = \sqrt{\frac{\kappa_{zn}^2 K_z + m_v\gamma_w i\omega}{K_x}}$$

$$\alpha_m = \frac{1}{H}\left(\frac{\pi}{2}+m\pi\right)$$

and

$$\hat{u}_2(x,z,\omega_s) = \sum_k C_k \cos(\mu_{xk}x)\left[\cosh(\mu_{zk}z) - \tanh(\mu_{zk}H)\sinh(\mu_{zk}z)\right]$$

with

$$C_k = \frac{2}{L}\int_0^{L} \hat{P}(\omega)\cos(\mu_{xk}x)dx, \quad k = 0,1,2,\ldots$$

12

and

$$\mu_{xk} = \frac{1}{L}\left(\frac{\pi}{2}+k\pi\right), \quad \mu_{zk} = \sqrt{\frac{\mu_{xk}^2 K_x + m_v\gamma_w i\omega}{K_z}}$$

together with

$$\hat{E}(x,z,\omega) = \sum_{m=0}^{\infty} A_m \sin(\alpha_m z)[-\tanh(\alpha_m L)\cosh(\alpha_m x) + \sinh(\alpha_m x)]$$

with

$$A_m = \frac{-2}{H\tanh(\alpha_m L)}\int_0^{L_e} \hat{E}_0(z,\omega)\sin(\alpha_m z)\,dz, \quad m = 0,1,2,\ldots$$

and

$$\alpha_m = \frac{1}{H}\left(\frac{\pi}{2}+m\pi\right)$$

Applying the inverse FFT to (12.39) and its associated solutions restores the solution to the time domain, such that

$$u_e(x,z,t) = \sum_s \hat{u}_e(x,z,\omega_s)\,e^{i\omega_s t}$$

$$E(x,z,t) = \sum_s \hat{E}(x,z,\omega_s)\,e^{i\omega_s t}$$

Special case.

In the special case where the length of the electrodes spans the entire thickness of the soil layer, i.e. $L_e = H$, the solution simplifies as follows.

Electric BVP

For a constant $\widehat{E}_0$, i.e. independent of z, the Fourier coefficients become

$$A_m = -\frac{2\hat{E}_0}{H \tanh(\alpha_m L)} \int_0^H \sin(\alpha_m z)\, dz, \quad \alpha_m = \frac{1}{H}\left[\left(m + \frac{1}{2}\right)\pi\right]$$

$$= -\frac{2\hat{E}_0}{H \tanh(\alpha_m L)\, \alpha_m} [\cos(\alpha_m H) - 1]$$

$$= \frac{2\hat{E}_0}{H \tanh(\alpha_m L)\, \alpha_m}$$

If the applied electric field is a function of z, say $\widehat{E}_0 z$, the Fourier coefficients become

$$A_m = -\frac{2\hat{E}_0}{H \tanh(\alpha_m L)} \int_0^H z \sin(\alpha_m z)\, dz, \quad \alpha_m = \frac{1}{H}\left[\left(m + \frac{1}{2}\right)\pi\right]$$

$$= \frac{2\hat{E}_0}{H \tanh(\alpha_m L)\, \alpha_m^2} [H\alpha_m \cos(\alpha_m H) - \sin(\alpha_m H)]$$

$$= \frac{2\hat{E}_0}{H \tanh(\alpha_m L)\, \alpha_m^2} (-1)^m$$

$\widehat{\boldsymbol{u}}_1(\boldsymbol{x}, \boldsymbol{z})$ BVP
Setting

$$a_1 = \alpha_m - \kappa_{zn}$$
$$a_2 = \alpha_m + \kappa_{zn}$$

with

$$\alpha_m = \frac{\pi}{H}\left(m + \frac{1}{2}\right), \quad \kappa_{zn} = \frac{\pi}{H}\left(n + \frac{1}{2}\right)$$

Then, the Fourier coefficients of $\widehat{u}_1(x, z, \omega)$ become

$$B_n = \frac{2G_m}{H} \int_0^H \sin(\alpha_m z) \sin(\kappa_{zn} z)\, dz$$

$$= \frac{2G_m}{H}\left[\frac{\sin(Ha_1)\alpha_m + \sin(Ha_1)\kappa_{zn} - \sin(Ha_2)\alpha_m + \sin(Ha_2)\kappa_{zn}}{2a_1 a_2}\right]$$

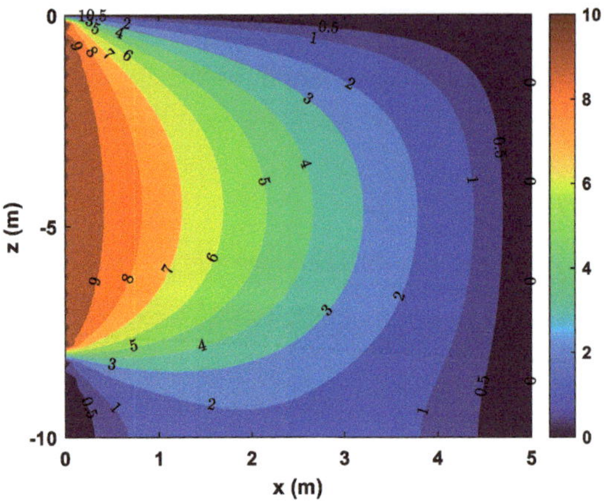

Fig. 12.12 Electric field distribution in the domain during activation

For $m = n$, $a_1 = 0$, then

$$\begin{aligned} B_n &= \frac{2G_m}{H} \lim_{\substack{a_1 \to 0 \\ (\alpha_m = \kappa_{zn})}} \left[\frac{\sin(Ha_1)\alpha_m + \sin(Ha_1)\kappa_{zn} - \sin(Ha_2)\alpha_m + \sin(Ha_2)\kappa_{zn}}{2a_1 a_2} \right] \\ &= \frac{2G_m}{H} \frac{H}{2} \\ &= G_m \end{aligned}$$

For $m \neq n$

$$B_n = 0$$

12

▪ EXAMPLE 12.2: Hydraulic and Electro-Osmosis Consolidation: Grounded Surface

Repeat Example 12.1 for a domain with its upper surface is electrically grounded.

▪ Solution

Following the usual solution procedure, the electric field and the pore pressure distributions can be calculated in the frequency domain using (12.55), followed by reconstructing the time domain using the IFFT algorithm.

Figure 12.12 shows the electric potential distribution across the domain. It demonstrates the influence of the insulated bottom boundary on the flow of the electric field, and highlights the impact of the grounded top boundary and the cathode, both maintained at zero potential. Compare to Fig. 12.7, this figure clearly shows the effect of the boundary conditions on the electric field distribution.

Figures 12.13 and 12.14 depict the excess pore pressure distributions at approximately 10, 25, 35, and 50 days after the surcharge removal. In comparison to Figs. 12.9 and 12.10, a notable difference between the two electric field configura-

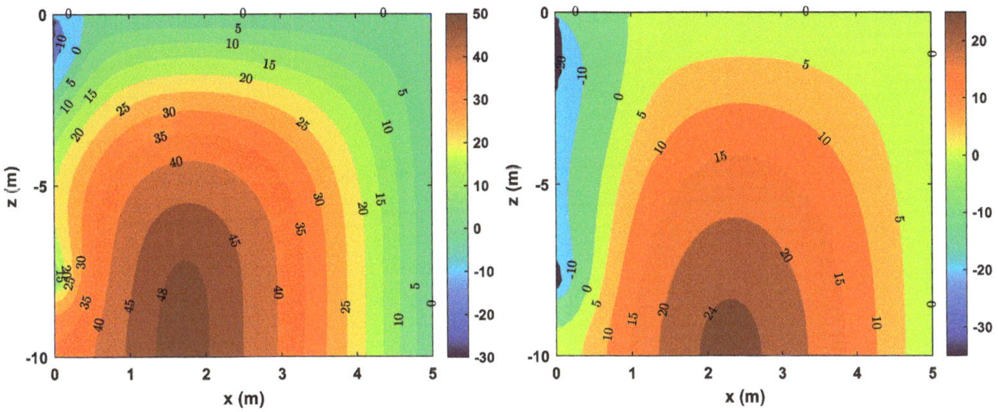

Fig. 12.13 Evolution of negative pore pressure after 10 days (left) and 25 days (right) of electrode activation following the removal of the surcharge

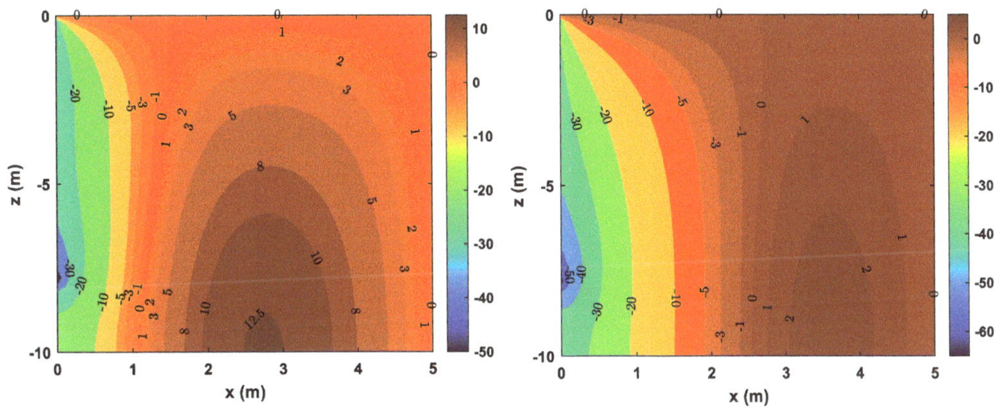

Fig. 12.14 Evolution of negative pore pressure after 35 days (left) and 50 days (right) of electrode activation following the removal of the surcharge

tions is evident. A similar observation can be deduced after the deactivation of the electrodes in Fig. 12.15, as compared to Fig. 12.11.

EXAMPLE 12.3: Hydraulic and Electro-Osmosis Consolidation: $L_e = H$

Repeat Example 12.6 for a domain with its electrodes go through the whole thickness of the soil layer, i.e. with $L_e = H$.

Solution

This problem can be addressed using the solution given in the Special case of ► Sect. 12.4.

Figure 12.16 demonstrates the electric field distribution in the domain. The figure highlights the length of the anode and demonstrates how the grounded surface and cathode maintain zero electric potential at their respective boundaries.

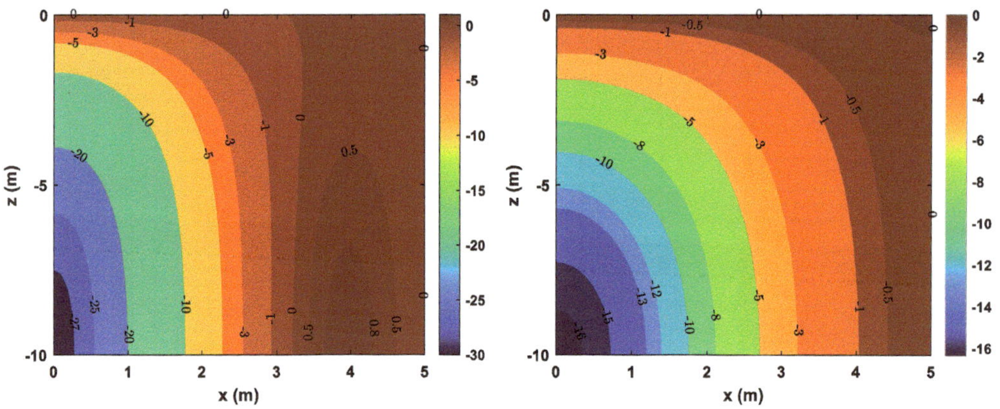

Fig. 12.15 Dissipation of the negative pore pressure at 5 days (left) and 15 days (right) after switching off the electrodes

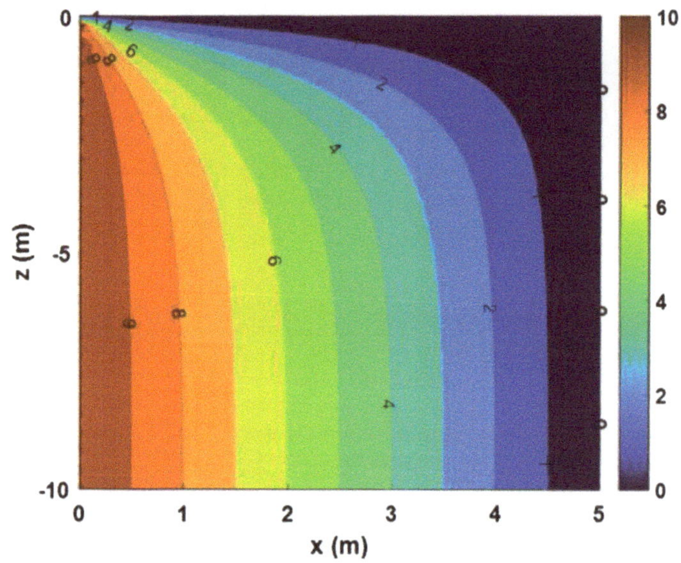

Fig. 12.16 Electric field distribution

Figure 12.17 depicts the evolution of pore pressure resulting from the electro-osmosis field, approximately 35 and 50 days after the anode activation. Figure 12.18 demonstrates the pore pressure distribution at 5 and 15 days following the deactivation of the electric potential. These figures illustrate the dynamic interplay between hydraulic consolidation, reflected in the remaining positive pore pressure, and the electro-osmosis potential, indicated by the negative pore pressure.

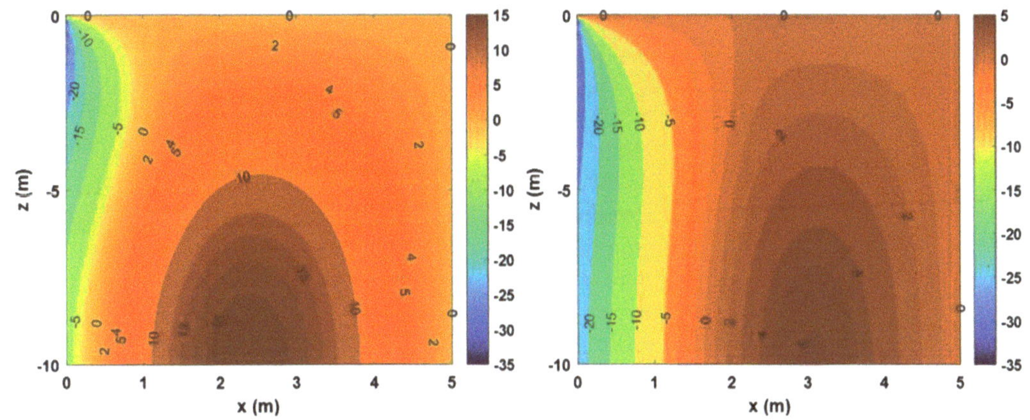

Fig. 12.17 Pore pressure distribution after 35 days (left) and 50 days (right) of electrode activation following the removal of the surcharge

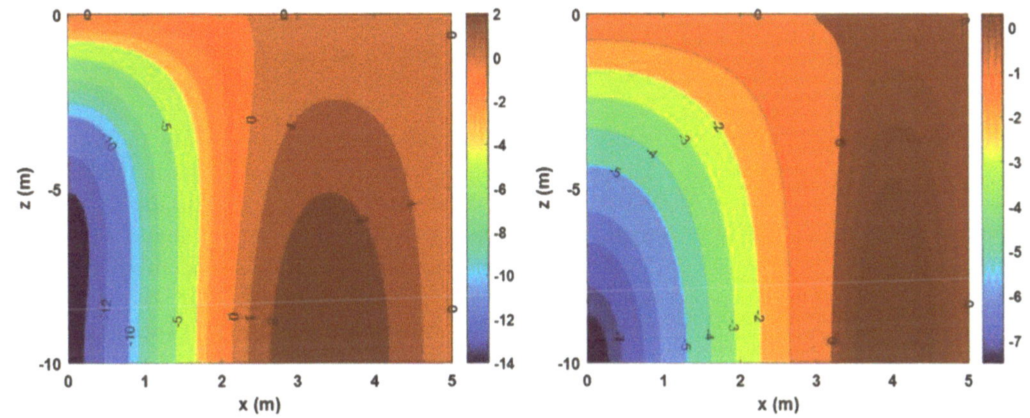

Fig. 12.18 Dissipation of the pore pressure at 5 days (left) and 15 days (right) after switching off the electrodes

Problems

1. Esrig (1968) solved the one-dimensional electro-osmosis consolidation, presented in ▶ Sect. 12.1. The Esrig's solution is given by

$$u_e(z,t) = -\frac{K_e\gamma_w E}{K_z} + \frac{2K_e\gamma_w E_0}{K_z\pi^2}\sum_{n=0}^{\infty}\frac{(-1)^n}{m^2}\sin\left(\frac{m\pi z}{H}\right)e^{-m^2\pi^2 T_v}$$

where

$$E = \frac{z}{H}E_0,\quad T_v = \frac{c_v t}{H^2},\quad c_v = \frac{K_z}{m_v\gamma_w},\quad m = n + \frac{1}{2}$$

Determine the excess pore pressure in a 1 m deep domain, bounded by a row of cathodes at its surface and a row of anodes at the bottom. The applied electric potential at the anodes is $E_0 = 10$ V, induced for 6000 s. Given that $K_e = 2\text{e-}9\ \text{m}^2/\text{V.s}$, $K_z = 3\text{e-}5$ m/s, $m_v = 0.1\text{e-}3\ \text{m}^2/\text{N}$,

$\gamma_w = 9810\,\text{N/m}^3$, calculate the excess pore pressure along the depth of the domain at 6000 s, using:
a. The Esrig model, and
b. The solution given in (12.7).
Hint: The excess pore pressure at $z = 0.5$ m after 6000 s of operating the electrodes is −0.89 Pa.

2. Continue with the Esrig formulation and assign large time (something like $e10$) to get the steady state solution. Compare the result with (12.8). Discuss the complexity of the two solution methods.
Hint: The excess pore pressure at $z = 0.5$ m after 10×10 seconds of operating the electrodes is −3.27 Pa.
3. Solve the boundary value problem of the plane strain electro-osmosis consolidation by replacing the specified Neumann boundary condition at $x = 0$, by a Robin boundary condition of the form:

$$-\frac{K_x}{\gamma_w}\frac{\partial u_e}{\partial x}\bigg|_{x=0} = \alpha u_e(0, z) - \beta E\,(0, z)$$

where α and β are coefficients representing the coupling between the electric field and the pore pressure field.
4. Based on the mathematical formulation of the axial symmetric case given in ► Sect. 12.2, re-formulate the problem for a Robin boundary condition at the anode row and solve the resulting boundary value problem.
5. Show that the derivations given in the two special cases in Sects. 12.3 and 12.4 are accurate.

12

Reference

Esrig MI (1968) Pore pressures, consolidation, and electrokinetics. J Soil Mech Found Div 94(4):899–922

Contaminant Transport

Contaminant Transport

R. Al Khoury and C. Kasbergen, *Analytical Subsurface Environment*,
https://doi.org/10.1007/978-3-031-93910-5_13

Contaminant transport refers to the dispersion and spread of pollutants or harmful substances within a given medium, such as air, water, or soil. The contaminants can arise from various sources, including industrial activities and agricultural runoff. Studying mass transport in ground layers with a focus on contaminant transport provides valuable insights into the movement and fate of pollutants, enabling engineers and scientists to mitigate their impacts and work towards a cleaner and healthier environment.

Formulating the contaminant transport equation is, in principle, similar to formulating the groundwater flow equation. The main difference, however, is that in groundwater flow, the transported quantity is the mass of the fluid phase itself, while in the contaminant transport, the transported quantity is the mass of the contaminant (chemical species) carried by the hosting fluid. In many cases, in contaminant transport we deal with several interacting species. In groundwater flow, the intensive quantity is the fluid mass density, whereas in the contaminant transport, it is the concentration of the species.

The goal of this chapter is to formulate the contaminant transport equation applicable to various contamination scenarios occurring in the ground subsurface. Emphasis will be given to linear problems that are solvable using the Fourier method.

13.1 Subsurface Contamination

Globally, contamination of subsurface soil and groundwater represents a challenging and increasingly pressing issue affecting water supplies. Enormous quantities of diverse pollutants, predominantly anthropogenic organic and inorganic chemicals, are released onto the ground surface. These pollutants subsequently infiltrate through the layers of soil, eventually reaching and affecting groundwater aquifers. Once at the aquifer, the contaminants will be transported along the groundwater flow, spreading to broader areas, damaging the delicate subsurface ecosystem, and degrading the quality of groundwater. Contaminant transport can also reach surface waters and agriculture irrigation systems, causing harm to the food chain and the biodiversity of rivers.

Contaminants are substances or particles in various forms, such as gases, liquids, or solids. In this book we focus on the transport of dissolved chemical species within groundwater. In literature, terms like contaminants, chemical species, solvent, and solute are frequently used interchangeably to refer to substances transported by groundwater.

The concentration of a contaminant is a measure of its mass per unit volume of the fluid phase, typically expressed as

$$c_\alpha^i = \frac{m_\alpha^i}{V_\alpha}$$

in which c_α^i is the mass concentration of species i dissolved in fluid phase α, m_α^i is the mass of the contaminant in a control volume V_α of the fluid phase. The unit of mass concentration is Kgm^{-3}, which is, obviously, equivalent to mass density. For simplicity of notation, in this book, we will be using c instead of c_α^i to represent species i dissolved in fluid phase α, unless necessary.

Modelling contaminant transport falls within the realm of mass conservation, particularly, conservation of species. Different mechanisms and modes govern the transport of species in porous media, including advection, molecular diffusion, dispersion, and chemical or biochemical reactions. In this chapter, our focus is on formulating the governing equations describing the conservation of species that will serve as the foundation for analysing a variety of contaminant transport phenomena. In subsequent chapters, we will indulge into solving a range of practical contaminant transport problems, considering both one-dimensional (1D) and two-dimensional (2D) cases that are of practical interest.

13.2 Conservation of Species

Consider the control volume, $dxdydz$, shown in Fig. 13.1. A fluid phase α occupies the control volume, and it carries a chemical species c_α^i that might accumulate, leave, adsorb, or chemically react with other species and/or with the phase itself. The conservation of mass of the species in the control volume can thus be stated as

$$\text{Accumulated mass} = \text{Net mass entering} + \text{Net production} \tag{13.1}$$

Net mass entering

Assume that the rate of mass concentration entering the control volume in the x-direction is J_{x1} and that leaving is $J_{x2} = J_{x1} + (\partial J_{x1}/\partial x)\partial x$. Hence, the net mass flux in the control volume in the x-direction is

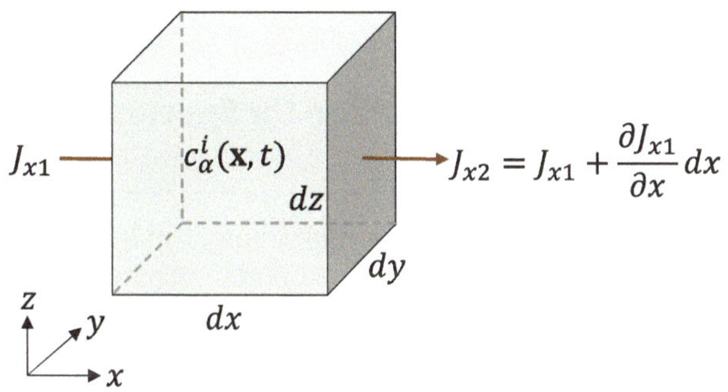

Fig. 13.1 Control volume of a contaminant transport

$$(J_{x1} - J_{x2})\, dydz = -\frac{\partial J_{x1}}{\partial x} dxdydz. \quad (13.2)$$

Similarly, the net mass flux crossing the other two directions are

$$\begin{aligned}(J_{y1} - J_{y2})\, dxdz &= -\frac{\partial J_{y1}}{\partial y} dxdydz \\ (J_{z1} - J_{z2})\, dxdy &= -\frac{\partial J_{z1}}{\partial z} dxdydz.\end{aligned} \quad (13.3)$$

Summing (13.2) and (13.3) the net mass flux in the control volume is hence

$$-\left(\frac{\partial J_{x1}}{\partial x} + \frac{\partial J_{y1}}{\partial y} + \frac{\partial J_{z1}}{\partial z}\right) dV, \quad \text{or} \quad -\nabla \cdot \mathbf{J} dV \quad (13.4)$$

where $dV = dxdydz$.

Accumulated mass

The accumulated mass is given by

$$\frac{\partial c_\alpha^i}{\partial t} dV \quad (13.5)$$

which is the rate of increase of the species in the control volume.

Net production

13

The net production is given by

$$\dot{m} dV \quad (13.6)$$

which is the rate of production that might result from chemical reactions or physical processes.

Collecting the three conservation modes, (13.4)–(13.6), and inserting into (13.1), the mass balance equation can then be expressed as

$$\frac{\partial c_\alpha^i}{\partial t} = -\nabla \cdot \mathbf{J} + \dot{m}. \quad (13.7)$$

13.3 Mass Balance Equation for Species in a Porous Domain

Equation (13.7) describes the conservation of mass of species i in a control volume, which is a fixed point in space at a microscopic level. To formulate the macroscopic mass balance equation for a fully saturated porous medium domain, volume averaging over a representative elementary volume (REV) can be applied, that, for a single fluid phase and one species, lead to

$$\underbrace{\partial(\varphi c)/\partial t}_{\text{accumulation}} = \underbrace{-\nabla\cdot(\varphi \mathbf{J})}_{\text{net mass flux}} + \underbrace{\varphi \dot{m}}_{\text{production rate}} \tag{13.8}$$

where φ is the porosity of the medium. This balance equation distinguishes three conservation modes: mass accumulation, mass flux, and net production.

Mass accumulation

In a fully saturated porous medium domain, mass accumulation can take place through a process called *adsorption*. Adsorption refers to the physical or chemical trapping of the solute (species or contaminant) in the porous medium. This phenomenon occurs when the solute molecules adhere to the surface of the fluid or solid phases in the porous medium. Adsorption can significantly impact the transport and fate of contaminants in porous media and plays a crucial role in their behaviour.

Some of the species' mass is adsorbed in the liquid phase that occupies the pores, while another portion is adsorbed by the solid phase of the porous medium matrix. These two adsorption mechanisms are interdependent, such that an increase in the concentration in the liquid phase corresponds to an increase in the solid phase, and vice versa. This relationship is described by what is known as *adsorption isotherm*, which is a function that relates the amount of adsorbed species in the solid phase to that in the liquid phase, under isothermal and chemical equilibrium conditions. Unlike adsorbed concentration in the liquid, expressed in units of kg m^{-3}, the adsorbed concentration in the solid phase is expressed in kg kg^{-1}.

The term on the left-hand side of (13.8) represents the adsorbed (dissolved) species in the liquid phase. In the solid phase, the adsorbed quantity is commonly given by

$$\rho s \tag{13.9}$$

in which ρ (kg m^{-3}) is the mass density of the aquifer material, and s is normally described by a linear isotherm relation of the form

$$s = K_d c \tag{13.10}$$

where K_d ($m^3 kg^{-1}$) is a *distribution coefficient*, also known as the *partition coefficient*, defining the affinity of the species for adsorption by the solid relative to the fluid.

Net mass flux

The net mass flux $\mathbf{J}$ represents the overall rate of species inflow and outflow within a control volume. It is primarily composed of three main fluxes: *advection*, *diffusion*, and *dispersion*. These fluxes can be expressed as

$$\mathbf{J} = \mathbf{J}_{\text{adv}} + \mathbf{J}_{\text{dif}} + \mathbf{J}_{\text{dis}} \tag{13.11}$$

The advective mass flux, $\mathbf{J}_{\text{adv}}$, occurs as a result of motion of the fluid carrying the contaminant mass in the domain. It is typically expressed as the product of the fluid velocity and the concentration of the contaminant, such that

$$\mathbf{J}_{\text{adv}} = \mathbf{v}c \tag{13.12}$$

where $\mathbf{v}$ (ms^{-1}) is the fluid velocity, mainly driven by pressure (hydraulic) gradients and governed by Darcy's law:

$$\mathbf{v} = -\mathbf{K}\nabla h \tag{13.13}$$

in which $\mathbf{K}$ (ms^{-1}) is the *hydraulic conductivity*, a tensor that can have different magnitudes in different directions; and ∇h (m m^{-1}) is the *hydraulic gradient*.

The diffusive mass flux $\mathbf{J}_{\text{dif}}$ in (13.11) occurs as a result of random movement of species molecules in the host domain, governed by the *Brownian motion*. Diffusion, also known as *molecular diffusion*, is expressed by Fick's law, entailing

$$\mathbf{J}_{\text{dif}} = -\mathbf{D}_{\text{dif}}\nabla c \tag{13.14}$$

where $\mathbf{D}_{\text{dif}}$ is the *molecular diffusion coefficient* (m^{2}s^{-1}). This coefficient is typically in the order of 10^{-9} m^{2}s^{-1} and might be a function of temperature and tortuosity.

The diffusion flux can be dominant in low-permeability porous media, such as clay layers; where there is no groundwater flow; or a flow with velocity less than 0.1m d^{-1}. However, studies have shown that contamination due to diffusion in typical porous media exhibit only a few decimeters of pollution transport per year. Compared to dispersion and advection transport, the amount of contamination caused by diffusion might be, in many cases, negligible.

13

The dispersive mass flux $\mathbf{J}_{\text{dis}}$ in (13.11) occurs as a result of disparity of fluid velocity in different locations of the domain. It arises from the heterogeneity of the medium, flow channelling, and other factors. In subsurface aquifers, pores' geometry and cross-sections vary considerably from one location to another, causing the fluid crossing these pores to exhibit different velocities. This variation in fluid velocity at the pores gives rise to the spread of species on a wider area, that, in many cases, can be significant and must be taken into consideration. ◘ Figure 13.2 shows schematically two scenarios highlighting the effect of dispersion on the species distribution in the domain, as compared to a case where the dispersion is not taken into consideration.

This kind of mass flux is also known as *mechanical dispersion*, which is, unlike diffusion, a function of fluid velocity. Dispersion is larger in the direction of fluid flow. It is distinguished between a *longitudinal dispersion*, which is in the direction of flow, and a *transverse dispersion* that is perpendicular to the flow direction. Nevertheless, despite arising from

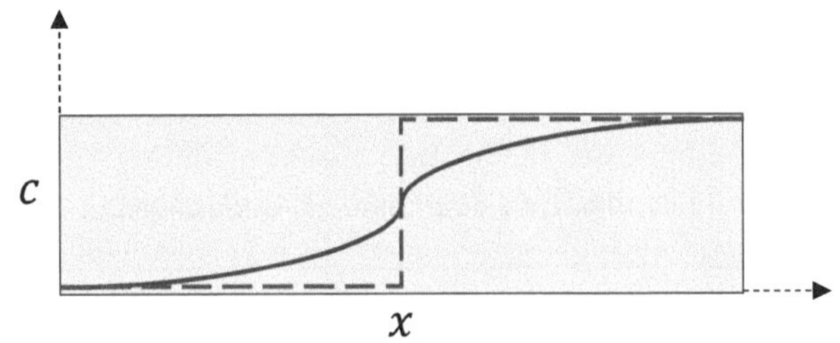

Fig. 13.2 Mass flux fronts with no dispersion (dotted) and with dispersion (blue line)

disparity in the fluid velocity at the pore scale, dispersion at the macro-scale behaves like diffusion and, thus, can be expressed using Fick's law. Accordingly, the longitudinal and transverse dispersions can be expressed as

$$\mathbf{J}_{\text{dis}} = -\mathbf{D}_{\text{dis}} \frac{\partial c}{\partial \mathbf{x}} \tag{13.15}$$

where $\mathbf{D}_{\text{dis}}$ is the *dispersion coefficients* (m^2s^{-1}) vector. The dispersion coefficient is usually expressed as a function of the average pore fluid velocity, such that if we align the coordinate axes so that groundwater velocity is in the positive x-direction, it can be defined as

$$D_{\text{dis},x} = a_x v_x \tag{13.16}$$

where a_x is the *dispersivity* in the x-direction and v_x is the corresponding average *pore velocity*.

Despite this distinguishing between the diffusion flux and the dispersion flux, they occur simultaneously and are effectively inseparable. The deviation of fluid flux at the macroscopic level from the pure advection flux is thus due to both molecular diffusion and mechanical dispersion. This close relation between the two fluxes is referred to as the *hydrodynamic dispersion*, given by

$$\mathbf{D} \equiv \mathbf{D}_{\text{dif}} + \mathbf{D}_{\text{dis}} \tag{13.17}$$

Production rate

The production rate term $\dot{m}$ in (13.8) describes the chemical and/or biological processes that act as sources or sinks of the species in the porous medium domain. Contaminants such as heavy metals and organic compounds can react with aquifer materials and groundwater, resulting in an increase or decay of the species' mass. The reactions, whether chemical, biological, or a combination of physio-biochemical are usually called *chemical reactions*. Within the scope of this book, $\dot{m}$ will give an approximation to the real chemical processes, which naturally involve complicated details that cannot be incorporated in analytical models. The interest here is in chemical reactions that occur while the species is being advected, diffused, or dispersed via the fluid, regardless of the thermo-biochemical-mechanical conditions of the hosting porous medium.

The chemical reactions are usually simulated using *zero-order production* and *first-order decay*. In a porous medium domain constituting a water phase and a solid (matrix) phase, the zero-order production in these two phases is given by

$$\varphi\gamma_w, \quad \rho\gamma_s \tag{13.18}$$

in which γ_w (kg m^{-3}s^{-1}) and γ_s (s^{-1}) are rate constants of production of the species in the water and solid matrix, respectively. The first-order decay of the water and solid phases, on the other hand, is given by

$$\mu_w\varphi c, \quad \mu_s\rho s \tag{13.19}$$

in which μ_w (s^{-1}) and μ_s (s^{-1}) are rate constants of species decay in the water phase and solid matrix, respectively.

13.4 Contaminant Transport Equation

The mass conservation Eq. (13.8) is valid for all kinds of macroscopic field problems in porous media. Tailoring it to the contaminant transport necessitates including the constitutive and physical relationships given in the previous section. Including solid adsorption (13.9), advection flux (13.12), diffusion flux (13.14), dispersion flux (13.15), together with the rate of production and decay chemical reactions, (13.18) and (13.19), the mass conservation equation of a species with concentration c flowing with water in a porous medium domain can be expressed as

13

$$\underbrace{\frac{\partial(\varphi c + \rho s)}{\partial t}}_{\text{accumulation}} + \underbrace{\mathbf{v}\cdot\nabla(\varphi c)}_{\text{advection}} = \underbrace{\mathbf{D}\nabla\cdot\nabla(\varphi c)}_{\text{hydrodynamic disp.}} + \underbrace{\mu_w\varphi c + \mu_s\rho s}_{\text{decay rate}} + \underbrace{\varphi\gamma_w + \rho\gamma_s}_{\text{production rate}} \tag{13.20}$$

where the hydrodynamic dispersion is given by

$$\underbrace{\mathbf{D}\nabla\cdot\nabla(\varphi c)}_{\text{hydrodynamic dispersion}} \equiv \underbrace{\mathbf{D}_{\text{dif}}\nabla\cdot\nabla(\varphi c)}_{\text{diffusion flux}} + \underbrace{\mathbf{D}_{\text{dis}}\nabla\cdot\nabla(\varphi c)}_{\text{dispersion flux}}. \tag{13.21}$$

Substituting the constitutive relation for solid absorption in (13.10), and by introducing a *retardation factor* and effective rate coefficients, (13.20) can be rewritten as

$$\frac{\partial(\varphi R c)}{\partial t} + \mathbf{v}\cdot\nabla(\varphi c) = \mathbf{D}\nabla\cdot\nabla(\varphi c) + \mu\varphi c + \varphi\gamma \tag{13.22}$$

where

$$R = 1 + \rho K_d/\varphi, \quad \mu = \mu_w + \mu_s \rho K_d/\varphi, \quad \gamma = \gamma_w + \gamma_s \rho/\varphi \tag{13.23}$$

in which R is a coefficient that expresses the partitioning of species in a unit volume of the porous medium, such that for every unit of mass in the liquid phase, the solid phase adsorbs $\rho K_d/\varphi$ mass units of species. For a case with no solid adsorption, $R = 1$.

Assuming constant porosity, the transport balance equation reduces to

$$\underbrace{R\frac{\partial c}{\partial t}}_{\text{accumulation}} + \underbrace{\mathbf{v} \cdot \nabla c}_{\text{advection}} = \underbrace{\mathbf{D}\nabla \cdot \nabla c}_{\text{hyd. dispersion}} + \underbrace{\mu c}_{\substack{\text{first-order}\\ \text{decay rate}}} + \underbrace{\gamma}_{\substack{\text{zero-order}\\ \text{production rate}}} . \tag{13.24}$$

This equation is commonly known as the *advective-dispersive-reactive transport* (ADR). It has been utilized for modelling field problems in a wide range of engineering, applied physics, and applied chemistry applications. The transport of ammonium and nitrate in subsurface aquifers, transport of radioactive materials, and intrusion of saltwater in coastal areas are only a few examples for its utilization.

The ADR Eq. (13.24) is generic and encompasses different mechanisms of contaminant transport in porous media domains. These mechanisms often occur simultaneously, as for the ADR case. However, in some cases, mechanisms of different nature can be more dominant relative to others, depending on the chemistry of the species and the environment where it resides. The following cases are identified in literature:

(a) Contaminant transport with no production or decay.
(b) Contaminant transport with zero-order production only.
(c) Contaminant transport with first-order decay only.
(d) Contaminant transport with simultaneous zero-order production and first-order decay (ADR).

In subsequent chapters we will delve into solving the governing equations of these cases using the Fourier method.

13.5 Initial and Boundary Conditions

The state variable of a contaminant transport problem is the species concentration, commonly denoted as c. The initial condition implies specifying the concentration value at time $t = 0$, while the boundary conditions imply prescribing certain values, potentially including zero, at the boundaries of the domain.

In general, the initial condition for a transport problem can be specified as

$$c(\mathbf{x},0) = f(\mathbf{x})$$

where $f(\mathbf{x})$ may represent a constant value or a function of $\mathbf{x}$. However, when employing spectral analysis, which involves transforming the time domain into the frequency domain, the initial condition does not play a role in the solution. Consequently, in this book, the initial condition for contaminant transport problems is consistently assumed to be

$$c(\mathbf{x},0) = 0$$

Nevertheless, because the problems addressed are linear, initial conditions can often be incorporated algebraically into the results during post-processing.

The relevant boundary conditions for contaminant transport problems are the Dirichlet, Neumann, and Robin.

Dirichlet boundary condition: Prescribed concentration

The Dirichlet boundary condition (also known as the boundary condition of the *first kind* or the *essential* boundary condition) implies prescribing a species concentration at certain spatial points or surface areas on the boundary of a domain. In contaminant transport problems, it might be constant, time-dependent, or a combination of both. For a one-dimensional domain spanning in the range $0 \leq x \leq L$ the Dirichlet boundary conditions may take the form

$$c\,(0,t) = C_1(t), \quad c\,(L,t) = C_2$$

where, in this case, $C_1(t)$ (kg m^{-3}) is a prescribed time-dependent concentration at $x = 0$, and C_2 is a prescribed constant concentration at $x = L$.

Neumann boundary condition: Applied mass flux

The Neumann boundary condition (also known as the boundary condition of the *second kind* or the *natural* boundary condition) implies applying the value of spatial derivative of the species concentration on a surface area on the boundary of a domain.

In contaminant transport problems, it implies specifying the value of contaminant flux per unit area per unit time across a surface area on the boundary. It is applied with respect to an outward unit normal vector perpendicular to the surface (n_x in Fig. 13.3). Using Fix's law, the Neumann boundary condition on the end boundaries of a 1D domain can be expressed as

$$-DA_x \frac{\partial c}{\partial x}\bigg|_{x=0} n_x = J_1(t), \quad -DA_x \frac{\partial c}{\partial x}\bigg|_{x=L} n_x = J_2$$

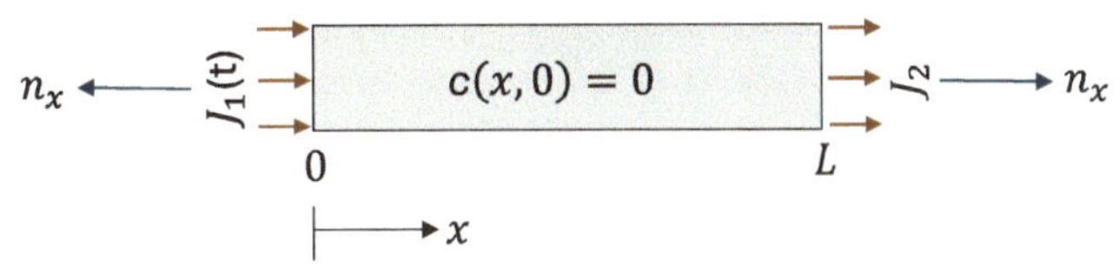

Fig. 13.3 Neumann mass flux

where D is the *hydrodynamic dispersion coefficient* (m^2s^{-1}) and A_x (m^2) is the cross-sectional area of the domain in the x-direction. In this case, $J_1(t)$ (kg m^2s^{-1}) is a time-dependent flux at $x = 0$, and J_2 is a constant flux at $x = L$. In Fig. 13.3, n_x on the left-hand side of the rod is -1, since its direction is opposite to that of the x-axis, and its magnitude on the right-hand side is $+1$. If a specified concentration is set to enter the domain at $x = 0$ and another to leave the domain at $x = L$, the Neumann boundary conditions are then expressed as

$$DA_x \frac{\partial c}{\partial x}\bigg|_{x=0} = J_1(t), \quad -DA_x \frac{\partial c}{\partial x}\bigg|_{x=L} = J_2.$$

If the surface area at $x = 0$ is effectively isolated, then $J_1 = 0$, and the boundary condition on this boundary is given by

$$\frac{\partial c}{\partial x}\bigg|_{x=0} = 0$$

Robin boundary condition

The Robin boundary condition (also known as the boundary condition of the *third kind*) corresponds to a linear combination of the value of the contaminant concentration and its derivative on the boundary. This corresponds to imposing a linear function combining a Dirichlet and a Neumann boundary condition on a specific point or area on the boundary.

In a diffusive-advective transport problem, the Robin boundary conditions, for a one-dimensional domain $[0, L]$, may typically be given by

$$-D\frac{\partial c}{\partial x}\bigg|_{x=0} n_x = v_x \left[C_\infty(t) - c(0,t)\right],$$
$$-D\frac{\partial c}{\partial x}\bigg|_{x=L} n_x = v_x \left[C_\infty - c(L,t)\right]$$

where v_x ($m\,s^{-1}$) is the fluid velocity and C_∞ (kg m^{-3}) is some concentration existing at the surrounding environment. They can be constant or time-dependent.

Sometimes, this boundary condition is written as

$$-D\left.\frac{\partial c}{\partial x}\right|_{x=0} n_x + \alpha_1 c(0,t) = h_1(t),$$
$$-D\left.\frac{\partial c}{\partial x}\right|_{x=L} n_x + \alpha_2 c(L,t) = h_2(t)$$

where α_1 and α_2 (m s^{-1}) are coefficients that represent the proportionality between the flux and the concentration at the boundaries $x = 0$ and $x = L$, respectively; and $h_1(t)$ and $h_2(t)$ (kg m^{-2}s^{-1}) are corresponding boundary values that can be constant or time-dependent.

Homogeneous boundary conditions

The above given Dirichlet and Neumann boundary conditions are nonhomogeneous since they are not identically zero. Should these boundary conditions be

$$c(0,t) = 0, \quad \left.\frac{\partial c}{\partial x}\right|_{x=L} = 0$$

they are said to be homogeneous. Similarly, the above given Robin boundary condition is nonhomogeneous, but if $C_\infty = 0$, i.e.

$$-D\left.\frac{\partial u}{\partial x}\right|_{x=L} + v_x c(L,t) = 0$$

it becomes homogeneous at $x = L$.

13

Problems

1. What are the key factors and parameters that influence contaminant transport in porous media?
2. How can mathematical models be used to simulate and predict contaminant transport in porous media?
3. How would you express the ADR transport equation for a nonporous medium?
4. How can the properties of a porous medium, such as porosity and permeability, be incorporated into contaminant transport models?

5. What are the challenges and uncertainties associated with modelling contaminant transport in porous media analytically?

Further Reading

Van Genuchten MTh (1982) Analytical solutions of one dimensional convective dispersive solute transport equations. ResearchGate (freely available in internet)

Rausch R, Schafer W, Therrien R, Wagner C (2005) Solute transport modelling: an introduction to models and solution strategies. Gebruder Borntraeger Verlagsbuchhandlung, Berlin-Stuttgart

1D Contaminant Transport

R. Al Khoury and C. Kasbergen, *Analytical Subsurface Environment*,
https://doi.org/10.1007/978-3-031-93910-5_14

In the preceding chapter we established the mass balance equation for contaminant transport in porous media. In this chapter, our attention focuses towards addressing boundary value problems that govern contaminant transport in one-dimensional porous medium domains. We solve these boundary value problems of varying complexity using the Fourier method. While the geometries studied may seem simplistic, these models hold practical value, particularly during the preliminary stages of investigation. They offer reliable estimations of the qualitative implications arising from various parameters and boundary conditions on the transport process. Furthermore, the presented models provide systematic methodology for solving the governing equations analytically and give insight into whether more intricate models can be addressed.

14.1 Contaminant Transport Equation

The contaminant transport equation mathematically describes and predicts how pollutants move through porous media. This equation is derived based on the principle of mass conservation, fluid mechanics, and chemical reactions. It takes into account factors such as advection (bulk flow of the fluid), diffusion (molecular movement of the contaminant), and dispersion (spreading of the contaminant due to velocity variations in the pores). Solving this equation provides valuable insights into how contaminants spread, react, and interact with the surrounding medium. This understanding is essential for assessing potential risks and designing effective pollution mitigation and remediation strategies.

14

In this chapter, we present analytical solutions for contaminant transport problems occurring in both finite and infinite domains. Specifically, we focus on solving the one-dimensional Advection–Dispersion-Reaction (ADR) equation with varying complexity. Recall from ▶ Chap. 12, this equation is given by

$$\underbrace{R\frac{\partial c}{\partial t}}_{\text{accumulation}} + \underbrace{v\frac{\partial c}{\partial x}}_{\text{advection}} = \underbrace{D\frac{\partial^2 c}{\partial x^2}}_{\text{hyd. dispersion}} + \underbrace{\mu c}_{\substack{\text{first - order}\\ \text{decay rate}}} + \underbrace{\dot{m}}_{\substack{\text{zero - order}\\ \text{production rate}}} \tag{14.1}$$

in which c (x,t) (kg m^{-3}) is the species concentration, D (m^{2}s^{-1}) is the hydrodynamic dispersion coefficient, v (ms^{-1}) is the fluid velocity, and R is given by

$$R = 1 + \rho K_d / \varphi$$

with ρ (kg m^{-3}) the bulk density of the porous medium, K_d (m^{3}kg^{-1}) the distribution coefficient, also known as the *partition coefficient*, and φ the porosity. For a non-adsorption case, $R = 1$. Note that in here we used $\dot{m}$ (kg m^{-3}s^{-1}) instead of γ

(see ▶ Chap. 13) to clearly indicate that the species production is in reality a mass production rate.

14.2 Mass Transport with No Production or Decay

Mass transport with no production or decay refers to a scenario where a chemical species is transported through a medium without being generated or consumed. In this case, the concentration of the species remains constant throughout the transport process and there are no additional sources or sinks affecting its quantity.

Understanding mass transport with no production or decay is important for assessing the behavior and distribution of the species within the medium, as well as for predicting its movement over time. Mathematically, this scenario implies the use of the Advection–Dispersion-Reaction (ADR) Eq. (14.1), with the last two terms on the right-hand side excluded. We proceed to solve this problem for various geometries and boundary conditions, aiming to understand the impact of different factors on the transport of the contaminant within the porous medium.

14.2.1 Semi-infinite 1D Domain: Dirichlet Problem

The boundary value problem for a pollutant transport in a semi-infinite domain subjected to a Dirichlet boundary condition is

$$
\begin{aligned}
&D\frac{\partial^2 c}{\partial x^2} - v\frac{\partial c}{\partial x} - R\frac{\partial c}{\partial t} = 0 \\
&\text{IC:}\ \ c(x,0) = 0 \\
&\text{BC:}\ \ c(0,t) = C_0(t) \\
&\qquad\ \ c(\infty,t) = 0
\end{aligned}
\tag{14.2}
$$

in which $C_0(t)$ (kg m^{-1}) per unit area is some species concentration at boundary $x = 0$ that may be constant or a function of time.

Applying the Fourier transform to (14.2) gives

$$
\begin{aligned}
&\frac{d^2\hat{c}}{dx^2} - \frac{v}{D}\frac{d\hat{c}}{dx} - \frac{i\omega}{\alpha}\hat{c} = 0 \\
&\text{BC:}\ \ \hat{c}(0,\omega) = \hat{C}_0(\omega) \\
&\qquad\ \ \hat{c}(\infty,\omega) = 0
\end{aligned}
\tag{14.3}
$$

in which $\alpha = D/R$ (m^{2}s^{-1}).

General solution

Since the coefficients in (14.3) are constant, the trial solution can be

$$\hat{c} = Ae^{\kappa x} \tag{14.4}$$

in which A and κ are both constants that need to be determined.

Taking the first and second derivatives of (14.4) and substituting into the differential equation in (14.3) leads to

$$\kappa^2 - \frac{v}{D}\kappa - \frac{i\omega}{\alpha} = 0$$

Using the quadratic formula to find the roots of this algebraic equation, yields

$$\begin{aligned} \kappa_1 &= \frac{1}{2D}\left(v - \sqrt{v^2 + 4\frac{i\omega}{\alpha}D^2}\right) \\ \kappa_2 &= \frac{1}{2D}\left(v + \sqrt{v^2 + 4\frac{i\omega}{\alpha}D^2}\right) \end{aligned} \tag{14.5}$$

which represent the two eigenvalues of the system. These eigenvalues correspond to two eigenfunctions, leading the general solution to be expressed as

$$\hat{c} = a_1 e^{\kappa_1 x} + a_2 e^{\kappa_2 x} \tag{14.6}$$

Boundary conditions

14

As κ_2 in (14.5) gives always positive values, the second boundary condition in (14.3) entails $a_2 = 0$ in (14.6), otherwise the solution becomes unbounded at far distances. Thus, the solution reduces to

$$\hat{c} = a_1 e^{\kappa_1 x}$$

At $x = 0$, the first boundary condition implies

$$a_1 = \hat{C}_0$$

BVP solution

Hence, for a given frequency ω_n, the solution of (14.3) becomes

$$\hat{c}\,(x, \omega_n) = \hat{C}_0\, e^{\kappa_1 x} \tag{14.7}$$

Having obtained the solution in the frequency domain, the contaminant distribution in the time domain can then be calculated using the inverse fast Fourier transform (IFFT) algorithm, such that

$$c(x,t) = \sum_n \hat{c}(x,\omega_n)\, e^{i\omega_n t}$$

■ **Example 14.1: 1D domain with no species production**

A fully saturate one-dimensional porous medium domain have a bulk density 1600 kg/m^3 and porosity 0.3. The groundwater flow velocity in the domain is 1.0e−6 m/s. At its boundary, $x = 0$, the domain is exposed for nearly 20 h to a 0.5 kg/m per unit area concentration of a chemical species, with a time pulse given in ◘ Fig. 14.1. The chemical species is characterized by a distribution coefficient of 0.8 m^3/kg and a hydrodynamic dispersion coefficient of 1.5 m^2/s. Compute the concentrations at 0.0, 1.0, 5.0 and 10.0 m away from the source.

■ **Solution**

Solving this problem using the spectral analysis requires:

(a) Transforming the input concentration pulse from the time domain to the frequency domain using the forward fast Fourier transform (FFT). The transformed signal of the input pulse given in ◘ Fig. 14.1 is shown in ◘ Fig. 14.2. This figure depicts the frequency spectrum of the signal and highlights the significant range of frequencies.
(b) Solving the problem using (14.7) for all involved frequencies.
(c) Transforming the computed concentrations back to the time domain using the inverse fast Fourier transform (IFFT).

The outcome of these three steps is presented in ◘ Fig. 14.3. The figure shows the variation of the species concentrations with time and distance from the source. Apparently, even with no generation, the species exhibits decay with distance due to dissipation and dispersion.

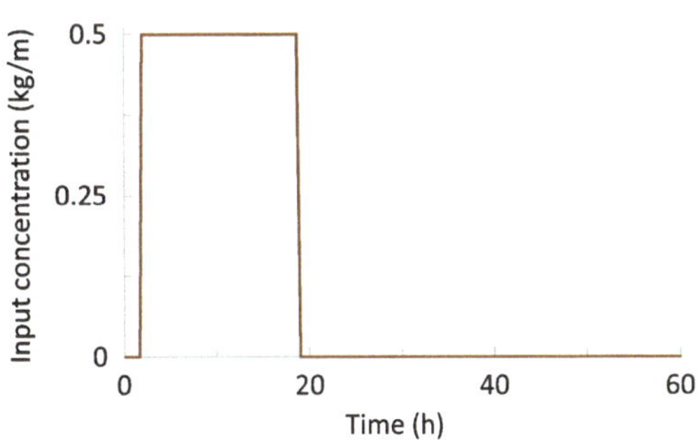

◘ **Fig. 14.1** Input concentration pulse

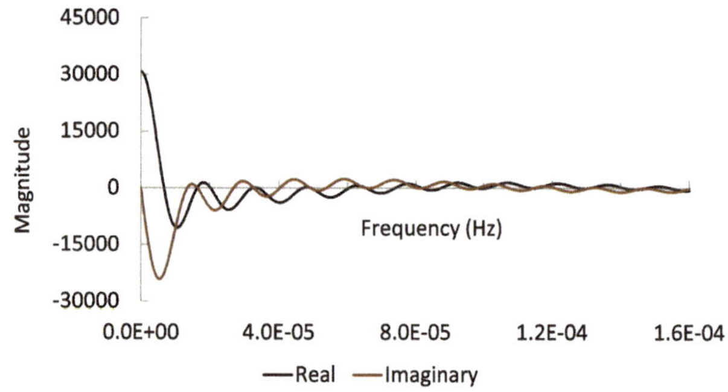

Fig. 14.2 Frequency spectrum of input concentration pulse

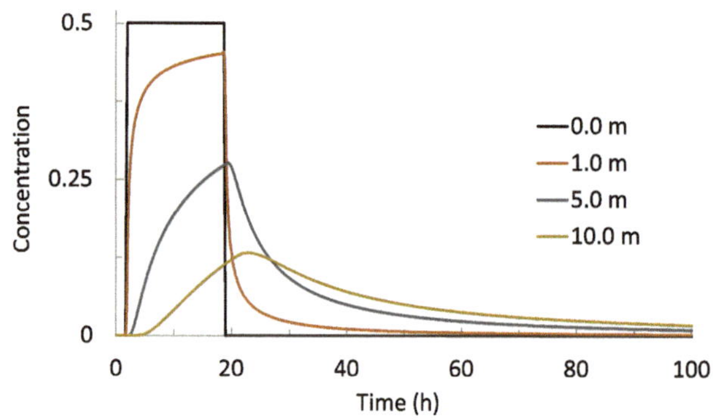

Fig. 14.3 Computed species concentrations

14.2.2 Finite 1D Domain: Dirichlet–Neumann

The boundary value problem of a contaminant transport in a finite 1D domain with a Dirichlet condition on one boundary and insulated on the other boundary is given by

$$D\frac{\partial^2 c}{\partial x^2} - v\frac{\partial c}{\partial x} - R\frac{\partial c}{\partial t} = 0$$

$$\begin{aligned} &\text{IC:} \quad c(x,0) = 0 \\ &\text{BC:} \quad c(0,t) = C_0(t) \\ &\qquad \left.\frac{\partial c(x,t)}{\partial x}\right|_{x=L} = 0 \end{aligned} \tag{14.8}$$

where L is the length of the domain.

General solution

Applying the Fourier transform to (14.8) gives

$$\frac{d^2\hat{c}}{dx^2} - \frac{v}{D}\frac{d\hat{c}}{dx} - \frac{i\omega}{\alpha}\hat{c} = 0$$

$$\text{BC: } \hat{c}\,(0,\omega) = \hat{C}_0(\omega) \qquad (14.9)$$

$$\left.\frac{d\hat{c}\,(x,\omega)}{dx}\right|_{x=L} = 0$$

As illustrated in the previous case, the general solution of the differential equation in (14.9) is

$$\hat{c} = a_1 e^{\kappa_1 x} + a_2 e^{\kappa_2 x}$$

with

$$\kappa_1 = \frac{1}{2D}\left(v - \sqrt{v^2 + 4\frac{i\omega}{\alpha}D^2}\right), \quad \kappa_2 = \frac{1}{2D}\left(v + \sqrt{v^2 + 4\frac{i\omega}{\alpha}D^2}\right) \qquad (14.10)$$

Boundary conditions

The first boundary condition in (14.9) at $x = 0$ implies

$$a_1 + a_2 = \hat{C}_0$$

Taking the derivative of $\hat{c}$ with respect to x and applying the boundary condition at $x = L$ leads to

$$a_1\kappa_1 e^{\kappa_1 L} + a_2\kappa_2 e^{\kappa_2 L} = 0$$

Putting these two equations in a matrix format gives

$$\begin{pmatrix} 1 & 1 \\ \kappa_1 e^{\kappa_1 L} & \kappa_2 e^{\kappa_2 L} \end{pmatrix}\begin{pmatrix} a_1 \\ a_2 \end{pmatrix} = \begin{pmatrix} \hat{C}_0 \\ 0 \end{pmatrix}$$

Solving for the constants vector $\boldsymbol{a} \equiv \{a_1 a_2\}$ yields

$$\begin{pmatrix} a_1 \\ a_2 \end{pmatrix} = \frac{1}{\kappa_2 e^{\kappa_2 L} - \kappa_1 e^{\kappa_1 L}}\begin{pmatrix} \kappa_2 e^{\kappa_2 L} & -1 \\ -\kappa_1 e^{\kappa_1 L} & 1 \end{pmatrix}\begin{pmatrix} \hat{C}_0 \\ 0 \end{pmatrix}$$

from which we obtain

$$a_1 = \frac{\hat{C}_0}{1 - \frac{\kappa_1}{\kappa_2} e^{(\kappa_1 - \kappa_2)L}}, \qquad a_2 = \frac{\hat{C}_0}{1 - \frac{\kappa_2}{\kappa_1} e^{(\kappa_2 - \kappa_1)L}} \qquad (14.11)$$

BVP solution

Substituting (14.11) into (14.10), the solution of this boundary value problem for a given frequency ω_n becomes

$$\hat{c}\left(x, \omega_n\right) = \frac{\hat{C}_0}{1 - \frac{\kappa_1}{\kappa_2} e^{(\kappa_1 - \kappa_2)L}} e^{\kappa_1 x} + \frac{\hat{C}_0}{1 - \frac{\kappa_2}{\kappa_1} e^{(\kappa_2 - \kappa_1)L}} e^{\kappa_2 x} \quad (14.12)$$

Having solved the problem in the frequency domain, the solution in the time domain can then be obtained by the use of the inverse fast Fourier transform (IFFT) algorithm, such that

$$c\left(x, t\right) = \sum_n \hat{c}\left(x, \omega_n\right) e^{i\omega_n t}$$

14.2.3 Finite 1D Domain: Double Dirichlet Problem

The boundary value problem of a contaminant transport in a finite 1D domain with Dirichlet boundary conditions at both ends can be expressed as

$$D\frac{\partial^2 c}{\partial x^2} - v\frac{\partial c}{\partial x} - R\frac{\partial c}{\partial t} = 0$$

$$\begin{aligned} &\text{IC:} \quad c\left(x, 0\right) = 0 \\ &\text{BC:} \quad c\left(0, t\right) = C_0(t) \\ &\qquad\;\; c\left(L, t\right) = C_L(t) \end{aligned} \quad (14.13)$$

where L is the length of the domain.

General solution

As illustrated earlier, the general solution of the differential equation in (14.13), in the frequency domain, leads to

$$\hat{c} = a_1 e^{\kappa_1 x} + a_2 e^{\kappa_2 x}$$

with

$$\kappa_1 = \frac{1}{2D}\left(v - \sqrt{v^2 + 4\frac{i\omega}{\alpha}D^2}\right), \quad \kappa_2 = \frac{1}{2D}\left(v + \sqrt{v^2 + 4\frac{i\omega}{\alpha}D^2}\right) \quad (14.14)$$

Boundary conditions

Applying the boundary conditions, implies
At $x = 0$

$$a_1 + a_2 = \hat{C}_0$$

At $x = L$

$$a_1 e^{\kappa_1 L} + a_2 e^{\kappa_2 L} = \hat{C}_L$$

14

Putting these two equations in a matrix format and solving for the constants, results in

$$a_1 = \frac{e^{\kappa_2 L}\hat{C}_0 - \hat{C}_L}{e^{\kappa_2 L} - e^{\kappa_1 L}}, \qquad a_2 = \frac{-e^{\kappa_1 L}\hat{C}_0 + \hat{C}_L}{e^{\kappa_2 L} - e^{\kappa_1 L}} \tag{14.15}$$

BVP solution

Substituting (14.15) into (14.14), the solution of this boundary value problem for a given frequency ω_n becomes

$$\hat{c}(x, \omega_n) = \frac{e^{\kappa_2 L}\hat{C}_0 - \hat{C}_L}{e^{\kappa_2 L} - e^{\kappa_1 L}} e^{\kappa_1 x} + \frac{-e^{\kappa_1 L}\hat{C}_0 + \hat{C}_L}{e^{\kappa_2 L} - e^{\kappa_1 L}} e^{\kappa_2 x} \tag{14.16}$$

Having done this, the solution in the time domain can then be obtained using the inverse fast Fourier transform (IFFT) algorithm, as

$$c(x, t) = \sum\nolimits_n \hat{c}(x, \omega_n)\, e^{i\omega_n t}$$

14.2.4 Finite 1D Domain: Robin Problem

The boundary value problem of a pollutant transport in a finite 1D domain with Robin boundary condition at one end, and insulated at the other end is given by

$$D\frac{\partial^2 c}{\partial x^2} - v\frac{\partial c}{\partial x} - R\frac{\partial c}{\partial t} = 0$$

$$\begin{aligned} &\text{IC:} \quad c(x, 0) = 0 \\ &\text{BC:} \quad \left(-D\frac{\partial c}{\partial x} + vc\right)\bigg|_{x=0} = vC_\infty(t) \\ &\qquad\quad \frac{\partial c}{\partial x}\bigg|_{x=L} = 0 \end{aligned} \tag{14.17}$$

in which C_∞ (kg m^{-3}) is a contamination source at $x = 0$.

General solution

As explained above, the general solution of the differential equation in (14.17) is

$$\hat{c} = a_1 e^{\kappa_1 x} + a_2 e^{\kappa_2 x}$$

with

$$\kappa_1 = \frac{1}{2D}\left(v - \sqrt{v^2 + 4\frac{i\omega}{\alpha}D^2}\right), \quad \kappa_2 = \frac{1}{2D}\left(v + \sqrt{v^2 + 4\frac{i\omega}{\alpha}D^2}\right) \tag{14.18}$$

Boundary conditions

Taking the derivative of (14.18) with respect to x, gives

$$\frac{\partial \hat{c}}{\partial x} = a_1 \kappa_1 e^{\kappa_1 x} + a_2 \kappa_2 e^{\kappa_2 x}$$

Applying the boundary conditions implies:
At $x = 0$

$$a_1(-D\kappa_1 + v) + a_2(-D\kappa_2 + v) = v\hat{C}_\infty$$

At $x = L$

$$a_1 \kappa_1 e^{\kappa_1 L} + a_2 \kappa_2 e^{\kappa_2 L} = 0$$

Solving these two equations simultaneously yields

$$\begin{aligned} a_1 &= \frac{\kappa_2 v \hat{C}_\infty}{(-D\kappa_1 + v)\,\kappa_2 - (-D\kappa_2 + v)\,\kappa_1 e^{(\kappa_1 - \kappa_2)L}} \\ a_2 &= \frac{-\kappa_1 v \hat{C}_\infty}{(-D\kappa_1 + v)\,\kappa_2 e^{(\kappa_2 - \kappa_1)L} - (-D\kappa_2 + v)\,\kappa_1} \end{aligned} \tag{14.19}$$

BVP solution

Substituting (14.19) into (14.18), the solution of this boundary value problem for a given frequency ω_n leads to

$$\begin{aligned} \hat{c}\,(x, \omega_n) = &\frac{\kappa_2 v \hat{C}_\infty}{(-D\kappa_1 + v)\,\kappa_2 - (-D\kappa_2 + v)\,\kappa_1 e^{(\kappa_1 - \kappa_2)L}} e^{\kappa_1 x} \\ &+ \frac{-\kappa_1 v \hat{C}_\infty}{(-D\kappa_1 + v)\,\kappa_2 e^{(\kappa_2 - \kappa_1)L} - (-D\kappa_2 + v)\,\kappa_1} e^{\kappa_2 x} \end{aligned} \tag{14.20}$$

14

Using the IFFT algorithm, the solution of the boundary value problem (14.17) in the time domain can then be obtained via

$$c\,(x, t) = \sum\nolimits_n \hat{c}\,(x, \omega_n)\, e^{i\omega_n t}$$

Example 14.2: Finite 1D domain

A 10 m fully saturated porous medium domain with groundwater flow velocity 1.0e−3 m/s is exposed at its boundary, $x = 0$, to a Robin boundary condition generated by a chemical species with a maximum $C_\infty(t) = 0.5$ kg/m per unit area. The contamination from this species lasted for nearly 20 h, with the time distribution shown in Fig. 14.1. The chemical species has a hydrodynamic dispersion coefficient $D = 1\text{e}{-3}$ m^2/s and $R = 10$. Compute the concentrations at 0.0, 1.0, 3.0 and 6.0 m away from the source.

Solution

Following the solution procedure outlined in Example 14.1, this problem can be solved in the frequency domain using (14.20). Then, by transforming the outcome back to the time domain through the IFFT algorithm, the time domain distribution of the species across the domain can be readily determined.

Figure 14.4 shows the fate of the species with time for different distances from the source, and Fig. 14.5 depicts the distribution of the species in the domain at different times.

14.3 Mass Transport with Zero-Order Production

Mass transport with zero-order production refers to a scenario where a chemical species is transported through a medium while being generated at a constant rate. In such a case, the transport process is influenced by advection, dispersion, and other transport mechanisms, while the production of the species remains constant throughout the system. Understanding mass transport with zero-order production is essential for assessing the impact of continuous generation on the distribution and concentration of species in underground formations.

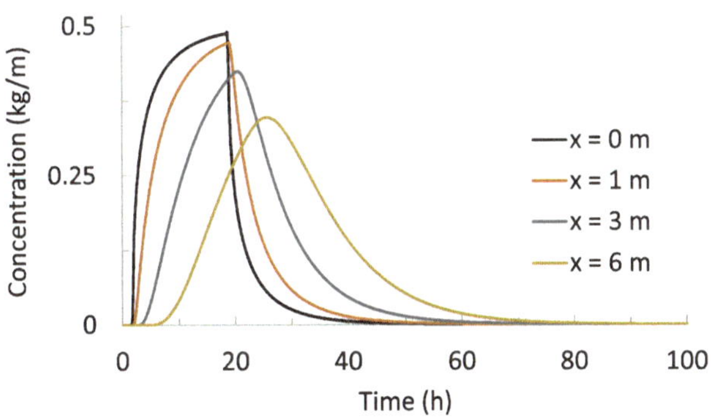

Fig. 14.4 Time variation of the species concentration with distance

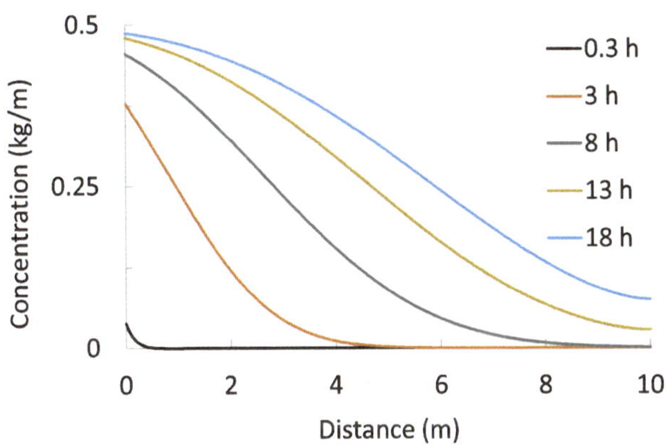

Fig. 14.5 Spatial variation of the species concentration at different times

Mathematically, this scenario implies that all terms in the Advection–Dispersion-Reaction (ADR) Eq. (14.1) are taken into account, except for the second term on the right-hand side. We proceed to solve this problem for various geometries and boundary conditions, but first we start by discussing the steady state condition.

14.3.1 Finite 1D Domain: Steady State with Dirichlet–Neumann Boundary Conditions

We here consider a boundary value problem representing a steady state contaminant transport in a finite 1D domain, with one boundary undergoes Dirichlet condition and another Neumann condition, representing a non-pervious boundary. This BVP is expressed as

$$\frac{d^2c}{dx^2} - \frac{v}{D}\frac{dc}{dx} + \frac{\dot{m}}{D} = 0, \qquad 0 \leq x \leq L$$

$$\text{BC:} \quad c\,(0) = C_0 \qquad\qquad (14.21)$$

$$\left.\frac{dc}{dx}\right|_{x=L} = 0$$

where $\dot{m}$ ($\text{kg m}^{-3}\text{s}^{-1}$) is the zero-order production rate of the species.

General solution

The differential equation in (14.21) is nonhomogeneous due to the presence of the production term $\dot{m}$. The solution can be approached by separately addressing the homogeneous part of the equation and its corresponding particular (nonhomogeneous) part, such that

$$c = c_h + c_p \qquad (14.22)$$

with c_h representing the homogeneous solution and c_p representing the particular solution.

The corresponding homogenous part of the differential equation in (14.21) is

$$\frac{d^2c_h}{dx^2} - \frac{v}{D}\frac{dc_h}{dx} = 0 \qquad (14.23)$$

As the coefficient of this equation is constant, the trial solution can be expressed as

$$c_h = ae^{\kappa x}$$

Substituting this trial function into (14.23) and rearranging, leads to

$$\kappa\left(\kappa-\frac{v}{D}\right)=0$$

This gives

$$\kappa_1=0 \quad \text{and} \quad \kappa_2=\frac{v}{D}$$

which represent the two eigenvalues of the system. The corresponding eigenfunctions form the general solution of the homogeneous equation, such that

$$c_h=a_1+a_2e^{vx/D} \tag{14.24}$$

The relevant particular part of the differential equation in (14.21) is

$$\frac{d^2c_p}{dx^2}-\frac{v}{D}\frac{dc_p}{dx}+\frac{\dot{m}}{D}=0 \tag{14.25}$$

As the particular term, $\dot{m}/D$, is constant, the method of undetermined coefficients assumes a trial function that can be any constant, say A. However, one of the solutions of the homogeneous equation is constant, a_1 in (14.24). Thus we ought to modify the trial function by multiplying it by x, giving

$$c_p=Ax$$

Taking the first and second derivatives of this trial function with respect x and substituting them into (14.25) leads to

$$-\frac{v}{D}A+\frac{\dot{m}}{D}=0$$

This readily gives

$$A=\frac{\dot{m}}{v}$$

Thus, the general solution of the particular part is

$$c_p=\frac{\dot{m}}{v}x \tag{14.26}$$

Substituting the homogeneous solution (14.24) and the particular solution (14.26) into (14.22), the general solution of the problem becomes

$$c\left(x\right)=a_1+a_2e^{vx/D}+\frac{\dot{m}}{v}x \tag{14.27}$$

Boundary conditions

At $x=0$, the first boundary condition in (14.21) implies

$$a_1+a_2=C_0 \quad \Rightarrow \quad a_1=C_0-a_2$$

Having this relationship, the solution (14.27) can then be modified to read

$$c = C_0 - a_2(1 - e^{vx/D}) + \frac{\dot{m}}{v}x \tag{14.28}$$

Taking the derivation of this equation with respect to x, gives

$$\frac{dc}{dx} = \frac{v}{D}a_2 e^{vx/D} + \frac{\dot{m}}{v}$$

At $x = L$, the second boundary condition in (14.21) implies

$$\frac{v}{D}a_2 e^{vL/D} + \frac{\dot{m}}{v} = 0$$

This gives

$$a_2 = -\frac{D}{v^2}\dot{m}e^{-vL/D} \tag{14.29}$$

BVP solution

Substituting (14.29) into (14.28), the solution of the boundary value problem (14.21) becomes

$$c(x) = C_0 + \frac{D\dot{m}}{v^2}\left(e^{-vL/D} - e^{v(x-L)/D}\right) + \frac{\dot{m}}{v}x \tag{14.30}$$

Special case: semi-infinite domain

For a semi-infinite domain, as L approaches ∞, the second term on the right-hand side of (14.30) vanishes, simplifying the solution to:

$$c(x) = C_0 + \frac{\dot{m}}{v}x$$

14

14.3.2 Finite 1D Domain: Transient Dirichlet–Neumann Problem

The boundary value problem of a contaminant transport with zero-order production in a finite 1D domain, subjected to a Dirichlet boundary condition on one side and a Neumann boundary condition, representing a non-pervious domain, on the other side is given by

$$\frac{1}{\alpha}\frac{\partial c}{\partial t} - \frac{\partial^2 c}{\partial x^2} + \frac{v}{D}\frac{\partial c}{\partial x} + \frac{\dot{m}}{D} = 0, \qquad 0 \le x \le L$$

$$\begin{aligned} \text{IC:} \quad & c(x,0) = 0 \\ \text{BC:} \quad & c(0,t) = C_0 \\ & \left.\frac{\partial c}{\partial x}\right|_{x=L} = 0 \end{aligned} \tag{14.31}$$

where $\alpha = D/R$.

General solution

Applying the Fourier transform to the BVP (14.31) gives

$$\frac{i\omega}{\alpha}\hat{c} - \frac{d^2\hat{c}}{dx^2} + \frac{v}{D}\frac{d\hat{c}}{dx} + \frac{\dot{m}}{D} = 0$$

$$\text{BC:} \quad \hat{c}\,(0,\omega) = \hat{C}_0(\omega) \qquad (14.32)$$

$$\left.\frac{d\hat{c}}{dx}\right|_{x=L} = 0$$

As for the previous case, this boundary value problem is nonhomogeneous and be solved via partitioning it into a homogeneous solution and a particular solution, such that

$$\hat{c} = \hat{c}_h + \hat{c}_p \qquad (14.33)$$

The relevant homogenous part of the ODE in (14.32) is

$$\frac{i\omega}{\alpha}\hat{c}_h - \frac{d^2\hat{c}_h}{dx^2} + \frac{v}{D}\frac{d\hat{c}_h}{dx} = 0 \qquad (14.34)$$

As the coefficients are constant, the trial solution can be expressed as

$$\hat{c}_h = ae^{\kappa x}$$

Substituting this trial solution into (14.34) gives

$$\kappa^2 - \frac{v}{D}\kappa - \frac{i\omega}{\alpha} = 0$$

Solving this algebraic equation using the quadratic formula, we obtain

$$\kappa_1 = \frac{\alpha v - \sqrt{\alpha^2 v^2 + 4i\alpha\omega D^2}}{2\alpha D}, \qquad \kappa_2 = \frac{\alpha v + \sqrt{\alpha^2 v^2 + 4i\alpha\omega D^2}}{2\alpha D}$$

Thus, the general solution of the homogeneous part is given by

$$\hat{c}_h = a_1 e^{\kappa_1 x} + a_2 e^{\kappa_2 x} \qquad (14.35)$$

The relevant particular part of the ODE in (14.32) is

$$\frac{i\omega}{\alpha}\hat{c}_p - \frac{d^2\hat{c}_p}{dx^2} + \frac{v}{D}\frac{d\hat{c}_p}{dx} + \frac{\dot{m}}{D} = 0 \qquad (14.36)$$

As the nonhomogeneous (reaction) term is constant, the trial solution can be any constant, say

$$\hat{c}_p = A$$

Substituting this trial solution into (14.36) gives

$$\frac{i\omega}{\alpha}A + \frac{\dot{m}}{D} = 0 \quad \Rightarrow \quad A = -\frac{\dot{m}\alpha}{i\omega D}$$

Hence, the general solution of the particular part is given by

$$\hat{c}_p = -\frac{\dot{m}\alpha}{i\omega D} \tag{14.37}$$

Substituting the homogeneous solution (14.35) and the particular solution (14.37) into (14.33) gives the general solution of the differential equation in (14.32), in the form

$$\hat{c} = a_1 e^{\kappa_1 x} + a_2 e^{\kappa_2 x} - \frac{\dot{m}\alpha}{i\omega D} \tag{14.38}$$

Boundary Conditions
At $x = 0$, the first boundary condition in (14.32) implies

$$a_1 + a_2 - \frac{\dot{m}\alpha}{i\omega D} = \hat{C}_0$$

At $x = L$, the second boundary condition implies

$$a_1 \kappa_1 e^{\kappa_1 L} + a_2 \kappa_2 e^{\kappa_2 L} = 0$$

Solving these two equations simultaneously yields

$$a_1 = \frac{1}{\kappa_2 e^{\kappa_2 L} - \kappa_1 e^{\kappa_1 L}} \left[\kappa_2 \left(\hat{C}_0 + \frac{\dot{m}\alpha}{i\omega D} \right) e^{\kappa_2 L} \right]$$

$$a_2 = \frac{1}{\kappa_2 e^{\kappa_2 L} - \kappa_1 e^{\kappa_1 L}} \left[-\kappa_1 \left(\hat{C}_0 + \frac{\dot{m}\alpha}{i\omega D} \right) e^{\kappa_1 L} \right]$$

Or

$$a_1 = \frac{\hat{C}_0 + \dfrac{\dot{m}\alpha}{i\omega D}}{1 - \dfrac{\kappa_1}{\kappa_2} e^{(\kappa_1 - \kappa_2)L}}, \qquad a_2 = \frac{\hat{C}_0 + \dfrac{\dot{m}\alpha}{i\omega D}}{1 - \dfrac{\kappa_2}{\kappa_1} e^{(\kappa_2 - \kappa_1)L}} \tag{14.39}$$

14

BVP Solution

Substituting (14.39) into (14.38) gives the solution of the boundary value problem (14.32) in the frequency domain, as

$$\hat{c}\,(x, \omega_n) = \frac{\hat{C}_0 + \dfrac{\dot{m}\alpha}{i\omega D}}{1 - \dfrac{\kappa_1}{\kappa_2} e^{(\kappa_1 - \kappa_2)L}} e^{\kappa_1 x} - \frac{\hat{C}_0 + \dfrac{\dot{m}\alpha}{i\omega D}}{\dfrac{\kappa_2}{\kappa_1} e^{(\kappa_2 - \kappa_1)L} - 1} e^{\kappa_2 x} - \frac{\dot{m}\alpha}{i\omega_n D} \tag{14.40}$$

in which

$$\alpha = \frac{D}{R}$$

$$\kappa_1 = \frac{\alpha v - \sqrt{\alpha^2 v^2 + 4i\alpha\omega D^2}}{2\alpha D}, \qquad \kappa_2 = \frac{\alpha v + \sqrt{\alpha^2 v^2 + 4i\alpha\omega D^2}}{2\alpha D}$$

Having solved the problem in the frequency domain, the contaminant distribution in the time domain can then be obtained using the inverse fast Fourier transform (IFFT) algorithm, such that

$$c\,(x,t) = \sum\nolimits_n \hat{c}\,(x,\omega_n)\, e^{i\omega_n t}$$

Special case: semi-infinite domain

As L approaches ∞ the second term on the right-hand side of (14.40) becomes unbounded and can be eliminated. Also, the denominator of the first term becomes 1, leading the solution to be

$$\hat{c}\,(x,\omega_n) = \left(\hat{C}_0 + \frac{\dot{m}\alpha}{i\omega D}\right) e^{\kappa_1 x} - \frac{\dot{m}\alpha}{i\omega_n D}$$

Example 14.3: Finite 1D domain with zero-order production

A fully saturated 10 m long porous medium domain has a bulk density 1500 kg/m^3 and porosity 0.3. Its groundwater moves at an average velocity of 1.0e–4 m/s. The domain initially contains a chemical species with 5 kg/m concentration, with distribution coefficient 0.0001 m^3/kg and hydrodynamic dispersion coefficient 0.001 m^2/s. At $x = 0$, the domain is exposed to a rectangular pulse concentration of 10 kg/m per unit area for nearly 20 h, while at $x = L$ the domain is impervious. Under some chemical or physical processes, the species undergoes either generation or decaying with a constant rate of $\mp$0.0001 kg/ms. The production or decay starts slowly upon the entry of the species from the boundary, and last for nearly 100 h. Figure 14.6 depicts the time distribution of the prescribed concentration at the boundary and the production/decay rate of the species. Compute the concentration of the species at 0.0, 1.0, 2.0, 5.0 and 9.0 m from the source due to both generation and decaying.

Solution

As outlined in the previous examples, we solve this problem by first transforming the input pulses in 0 from the time domain

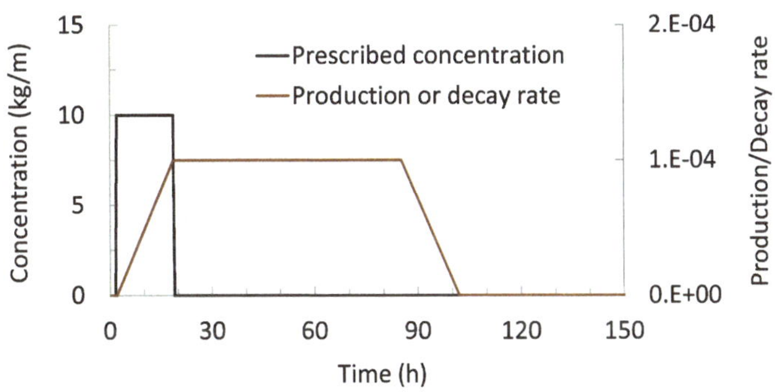

Fig. 14.6 Input species concentration and generation/decaying rate pulses

to the frequency domain using the FFT algorithm. Note that for generation, $\dot{m} < 0$, and for decay, $\dot{m} > 0$. The input concentration pulse is effectively Δc, which indicates that it is an addition to the initially existing concentration. However, since the Fourier transform does not recognize initial conditions, we first solve the problem by assuming zero initial concentration.

We solve the problem for every frequency using (14.40). Then, upon employing the IFFT algorithm, we obtain the solution in the time domain. The initial condition is then added algebraically to the computed results to get the final concentration of the species.

Figure 14.7 shows the computed species concentrations for the two scenarios: the chemical species is transported through the medium while being generated (left figure), or decayed (right figure). The spatial distribution of the species along the domain with the first scenario is shown in Fig. 14.8.

Should the domain be semi-infinite, the behavior of the species while being generated will have the time distribution shown in Fig. 14.9.

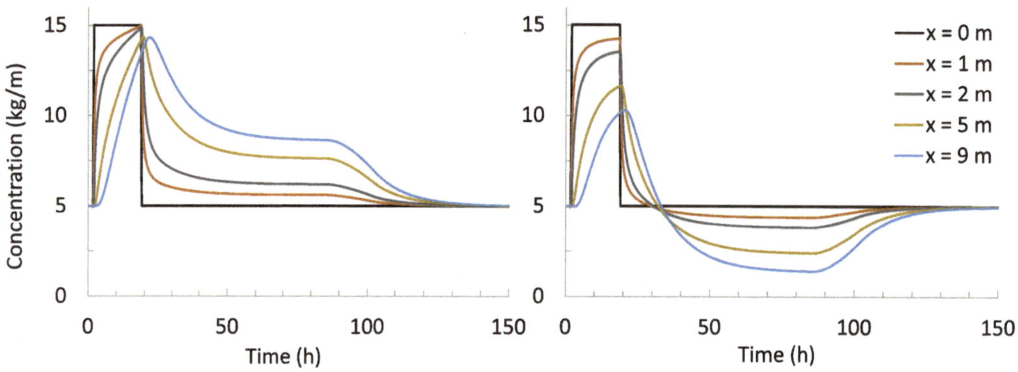

Fig. 14.7 Left: generated; Right: decaying

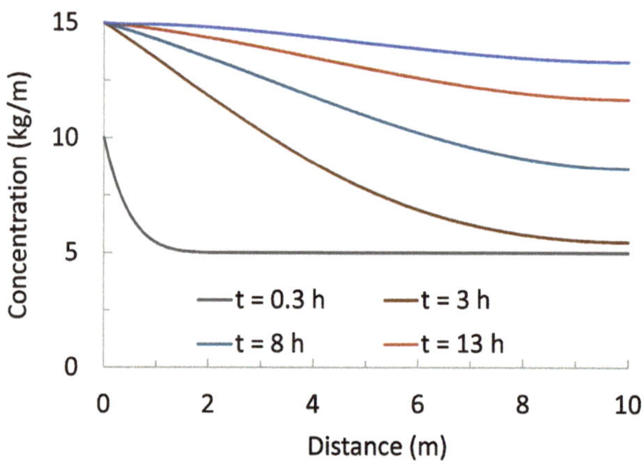

Fig. 14.8 Species concentration along the domain at different times

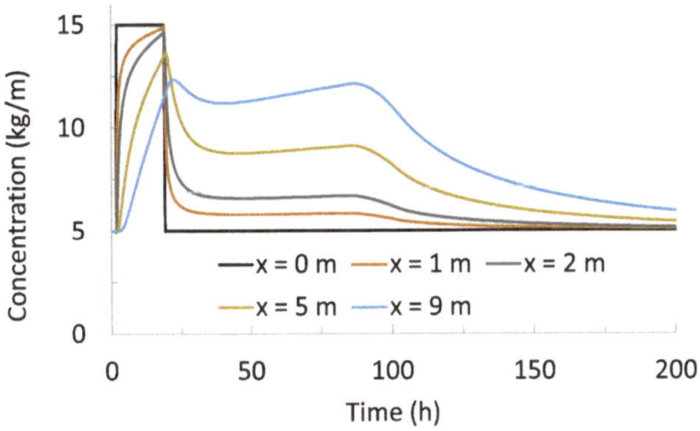

Fig. 14.9 Infinite domain with generation of the species

14.3.3 Finite 1D Domain: Transient Robin-Neumann Problem

The boundary value problem of a contaminant transport with zero-order production in a finite 1D domain, subjected to a Robin boundary condition on one side and a Neumann boundary condition, representing a non-pervious domain, on the other side is given by

$$\frac{1}{\alpha}\frac{\partial c}{\partial t} - \frac{\partial^2 c}{\partial x^2} + \frac{v}{D}\frac{\partial c}{\partial x} + \frac{\dot{m}}{D} = 0, \qquad 0 \le x \le L$$

$$\begin{aligned} \text{IC}: &\quad c(x,0) = 0 \\ \text{BC}: &\quad -D\left.\frac{\partial c}{\partial x}\right|_{x=0} = v\left[\,C_0 - c(0,t)\,\right] \\ &\quad \left.\frac{\partial c}{\partial x}\right|_{x=L} = 0 \end{aligned} \tag{14.41}$$

in which C_0 (kg m^{-3}) is a contamination source at $x = 0$.

General solution

Applying the Fourier transform to the BVP (14.41) gives

$$\frac{i\omega}{\alpha}\hat{c} - \frac{d^2\hat{c}}{dx^2} + \frac{v}{D}\frac{d\hat{c}}{dx} + \frac{\dot{m}}{D} = 0$$

$$\begin{aligned} \text{BC}: &\quad -D\left.\frac{d\hat{c}}{dx}\right|_{x=0} = v[\hat{C}_0 - \hat{c}(0,\omega)] \\ &\quad \left.\frac{d\hat{c}}{dx}\right|_{x=L} = 0 \end{aligned} \tag{14.42}$$

As discussed in the previous case, the general solution of this problem is given by

$$\hat{c} = a_1 e^{\kappa_1 x} + a_2 e^{\kappa_2 x} - \frac{\dot{m}\alpha}{i\omega D}$$

with

$$\kappa_1 = \frac{\alpha v - \sqrt{\alpha^2 v^2 + 4i\alpha\omega D^2}}{2\alpha D}, \quad \kappa_2 = \frac{\alpha v + \sqrt{\alpha^2 v^2 + 4i\alpha\omega D^2}}{2\alpha D} \tag{14.43}$$

Boundary conditions

Taking the derivative of (14.43) with respect to x, and solving for $x = 0$, the first boundary condition in (14.42) implies

$$a_1(v - D\kappa_1) + a_2(v - D\kappa_2) = v\left(\hat{C}_0 + \frac{\dot{m}\alpha}{i\omega D}\right)$$

At $x = L$, the second boundary condition in (14.42) implies

$$a_1\kappa_1 e^{\kappa_1 L} + a_2\kappa_2 e^{\kappa_2 L} = 0$$

Putting these two equations in matrix form and solving for the constants yields

$$a_1 = \frac{v\left(\hat{C}_0 + \frac{\dot{m}\alpha}{i\omega D}\right)}{(v - D\kappa_1) - (v - D\kappa_2)\frac{\kappa_1}{\kappa_2}e^{(\kappa_1-\kappa_2)L}}$$
$$a_2 = \frac{-v\left(\hat{C}_0 + \frac{\dot{m}\alpha}{i\omega D}\right)}{(v - D\kappa_1)\frac{\kappa_2}{\kappa_1}e^{(\kappa_2-\kappa_1)L} - (v - D\kappa_2)} \tag{14.44}$$

BVP Solution

Substituting (14.44) into (14.43) gives the complete solution of the boundary value problem (14.42) in the frequency domain, in the form

$$\hat{c}(x, \omega_n) = \frac{v\left(\hat{C}_0 + \frac{\dot{m}\alpha}{i\omega D}\right)}{(v - D\kappa_1) - (v - D\kappa_2)\frac{\kappa_1}{\kappa_2}e^{(\kappa_1-\kappa_2)L}}e^{\kappa_1 x} + \frac{-v\left(\hat{C}_0 + \frac{\dot{m}\alpha}{i\omega D}\right)}{(v - D\kappa_1)\frac{\kappa_2}{\kappa_1}e^{(\kappa_2-\kappa_1)L} - (v - D\kappa_2)}e^{\kappa_2 x} - \frac{\dot{m}\alpha}{i\omega_n D} \tag{14.45}$$

in which

$$\alpha = \frac{D}{R}$$

$$\kappa_1 = \frac{\alpha v - \sqrt{\alpha^2 v^2 + 4i\alpha\omega D^2}}{2\alpha D}, \quad \kappa_2 = \frac{\alpha v + \sqrt{\alpha^2 v^2 + 4i\alpha\omega D^2}}{2\alpha D}$$

Having solved the problem in the frequency domain, the contaminant distribution in the time domain can then be obtained using the inverse fast Fourier transform (IFFT) algorithm, such that

$$c(x,t) = \sum_n \hat{c}(x,\omega_n)e^{i\omega_n t}$$

14.4 Contaminant Transport with Zero-Order Production and First Order Decay

Contaminant transport with zero-order production and first-order decay encompasses a transport scenario where a chemical species is simultaneously subjected to production and decay processes. Production occurs at a constant rate, independent of the concentration, while the decay follows a first-order kinetics, meaning the rate of decay is proportional to the concentration.

This process is often expressed by what is called the Advection–Dispersion-Reaction (ADR) equation. This equation takes into account advection, dispersion, and reaction processes, allowing studying the combined effects of production and decaying on the distribution and fate of contaminants in various environmental systems. In this section, we proceed to solve this equation for various geometries and boundary conditions. First, we start with addressing the steady state condition because of its importance in practice.

14.4.1 Finite 1D Domain: Steady State Dirichlet–Neumann Problem

The boundary value problem of a steady-state ADR contaminant transport in a finite 1D domain, subjected to a Dirichlet boundary condition on one side and a Neumann boundary condition, representing a non-pervious domain, on the other side is given by

$$\frac{d^2c}{dx^2} - \frac{v}{D}\frac{dc}{dx} - \frac{\mu}{D}c + \frac{\dot{m}}{D} = 0, \qquad 0 \le x \le L$$

$$\text{BC:} \quad c(0) = C_0$$
$$\left.\frac{dc}{dx}\right|_{x=L} = 0 \tag{14.46}$$

where μ (s^{-1}) is some first-order decay coefficient.

General solution

The differential equation in (14.46) is nonhomogeneous due to the presence of the reaction term $\dot{m}$. Hence, the solution can be approached by solving the corresponding homogeneous and particular equations, such that

$$c = c_h + c_p$$

The corresponding homogenous differential equation of (14.46) is

$$\frac{d^2 c_h}{dx^2} - \frac{v}{D}\frac{dc_h}{dx} - \frac{\mu}{D}c_h = 0 \tag{14.47}$$

As the coefficient of this ODE is constant, the trial solution can be expressed as

$$c_h = ae^{\kappa x}$$

Substituting this trial solution into (14.47) gives

$$\kappa^2 - \frac{v}{D}\kappa - \frac{\mu}{D} = 0$$

This entails

$$\kappa_1 = \frac{v}{2D} + \sqrt{\frac{\mu}{D} + \left(\frac{v}{2D}\right)^2}, \quad \kappa_2 = \frac{v}{2D} - \sqrt{\frac{\mu}{D} + \left(\frac{v}{2D}\right)^2}$$

Following this, the general solution for the homogeneous equations can then be given by

$$c_h = a_1 e^{\kappa_1 x} + a_2 e^{\kappa_2 x}$$

The relevant particular differential equation of (14.46) is

$$\frac{d^2 c_p}{dx^2} - \frac{v}{D}\frac{dc_p}{dx} - \frac{\mu}{D}c_p + \frac{\dot{m}}{D} = 0 \tag{14.48}$$

As the particular term $\dot{m}/D$ is constant (not function of x), the method of undetermined coefficients assumes a trial function that can be any constant, such that

$$c_p = A$$

Substituting this trial solution into (14.48) gives

$$A = \frac{\dot{m}}{\mu}$$

Thus, the general solution of the particular equation is

$$c_p = \frac{\dot{m}}{\mu}$$

Combining the homogeneous and the particular solutions together, we obtain

$$c = a_1 e^{\kappa_1 x} + a_2 e^{\kappa_2 x} + \frac{\dot{m}}{\mu} \tag{14.49}$$

Boundary conditions

At $x = 0$, the first boundary condition in (14.46) implies

$$a_2 = C_0 - \frac{\dot{m}}{\mu} - a_1$$

Substituting a_2 into (14.49) yields

$$c = a_1(e^{\kappa_1 x} - e^{\kappa_2 x}) + C_0 e^{\kappa_2 x} - \frac{\dot{m}}{\mu} e^{\kappa_2 x} + \frac{\dot{m}}{\mu}$$

Taking the derivation of c with respect to x gives

$$\frac{dc}{dx} = a_1(\kappa_1 e^{\kappa_1 x} - \kappa_2 e^{\kappa_2 x}) + C_0 \kappa_2 e^{\kappa_2 x} - \frac{\dot{m}}{\mu} \kappa_2 e^{\kappa_2 x}$$

At $x = L$, the second boundary condition in (14.46) implies

$$a_1(\kappa_1 e^{\kappa_1 L} - \kappa_2 e^{\kappa_2 L}) + C_0 \kappa_2 e^{\kappa_2 L} - \frac{\dot{m}}{\mu} \kappa_2 e^{\kappa_2 L} = 0$$

This gives

$$a_1 = \frac{-C_0 \mu \kappa_2 e^{\kappa_2 L} + \dot{m} \kappa_2 e^{\kappa_2 L}}{\mu(\kappa_1 e^{\kappa_1 L} - \kappa_2 e^{\kappa_2 L})}$$

BVP Solution

Following this, the solution of the steady state boundary value problem (14.46) is therefore

$$c(x) = \frac{e^{\kappa_2 x} - e^{\kappa_1 x}}{\kappa_1 e^{\kappa_1 L} - \kappa_2 e^{\kappa_2 L}} (C_0 - \frac{\dot{m}}{\mu}) \kappa_2 e^{\kappa_2 L} + (C_0 - \frac{\dot{m}}{\mu}) e^{\kappa_2 x} + \frac{\dot{m}}{\mu}$$

with

$$\kappa_1 = \frac{v}{2D} + \sqrt{\frac{\mu}{D} + (\frac{v}{2D})^2}, \quad \kappa_2 = \frac{v}{2D} - \sqrt{\frac{\mu}{D} + (\frac{v}{2D})^2} \tag{14.50}$$

Special case: Semi-infinite domain

As $L \to \infty$, the domain becomes semi-infinite. In this limit, the exponents involving κ_1 become unbounded, necessitating the elimination of the first term on the right-hand side of (14.50). Consequently, the solution of this limiting case simplifies to

$$c(x) = \left(C_0 - \frac{\dot{m}}{\mu}\right) e^{\kappa_2 x} + \frac{\dot{m}}{\mu}$$

with (14.51)

$$\kappa_2 = \frac{v}{2D} - \sqrt{\frac{\mu}{D} + \left(\frac{v}{2D}\right)^2}$$

14.4.2 Finite 1D Domain: Transient Robin-Neumann Problem

The relevant boundary value problem of an ADR contaminant transport in a finite 1D domain, subjected to a Robin boundary condition on one side and a Neumann boundary condition, representing a non-pervious domain, on the other side is given by

$$\frac{1}{\alpha}\frac{\partial c}{\partial t} - \frac{\partial^2 c}{\partial x^2} + \frac{v}{D}\frac{\partial c}{\partial x} + \frac{\mu}{D}c - \frac{\dot{m}}{D} = 0, \qquad 0 \le x \le L$$

$$\begin{aligned} &\text{IC:} \quad c(x,0) = 0 \\ &\text{BC:} \quad -D\left.\frac{\partial c}{\partial x}\right|_{x=0} = v\,[\,C_0 - c\,(0,t)\,] \\ &\qquad\quad \left.\frac{\partial c}{\partial x}\right|_{x=L} = 0 \end{aligned} \tag{14.52}$$

with $\alpha = D/R$ and C_0 a contamination source.

General solution

Applying the Fourier transform to (14.52) gives

$$\begin{aligned} &\frac{i\omega}{\alpha}\hat{c} - \frac{d^2\hat{c}}{dx^2} + \frac{v}{D}\frac{d\hat{c}}{dx} + \frac{\mu}{D}\hat{c} - \frac{\dot{m}}{D} = 0 \\ &\text{BC:} \quad -D\left.\frac{d\hat{c}}{dx}\right|_{x=0} = v\,[\,\hat{C}_0 - \hat{c}\,(0,\omega)\,] \\ &\qquad\quad \left.\frac{d\hat{c}}{dx}\right|_{x=L} = 0 \end{aligned} \tag{14.53}$$

This is a nonhomogeneous problem, and the solution requires solving both the homogeneous equation and the particular equation, such that

$$\hat{c} = \hat{c}_h + \hat{c}_p \tag{14.54}$$

The corresponding homogenous differential equation of (14.53) is

$$\frac{i\omega}{\alpha}\hat{c}_h - \frac{d^2\hat{c}_h}{dx^2} + \frac{v}{D}\frac{d\hat{c}_h}{dx} + \frac{\mu}{D}\hat{c}_h = 0 \tag{14.55}$$

As the coefficients are constant, the trial solution can be expressed as

$$\hat{c}_h = a e^{\kappa x}$$

Substituting this trial solution into (14.55) gives

$$\kappa^2 - \frac{v}{D}\kappa - \left(\frac{i\omega}{\alpha} + \frac{\mu}{D}\right) = 0$$

14

Using the quadratic formula, the solution of this algebraic equation leads to

$$\kappa_1 = \frac{v}{2D} + \sqrt{\frac{\mu}{D} + \left(\frac{v}{2D}\right)^2 + \frac{i\omega}{\alpha}}, \quad \kappa_2 = \frac{v}{2D} - \sqrt{\frac{\mu}{D} + \left(\frac{v}{2D}\right)^2 + \frac{i\omega}{\alpha}}$$

Accordingly, the general solution of the homogeneous Eq. (14.55) is

$$\hat{c}_h = a_1 e^{\kappa_1 x} + a_2 e^{\kappa_2 x} \tag{14.56}$$

The relevant particular differential equation of (14.53) is

$$\frac{i\omega}{\alpha}\hat{c}_p - \frac{d^2\hat{c}_p}{dx^2} + \frac{v}{D}\frac{d\hat{c}_p}{dx} + \frac{\mu}{D}\hat{c}_p - \frac{\dot{m}}{D} = 0 \tag{14.57}$$

As the reaction term $\dot{m}/D$ is constant (not function of x, though can be a function of time), the trial solution can thus be

$$\hat{c}_p = A$$

Substituting this trial solution into (14.57) gives

$$A = \frac{\dot{m}}{D\left(\frac{i\omega}{\alpha} + \frac{\mu}{D}\right)}$$

Thus, the general solution of the particular differential Eq. (14.57) is

$$\hat{c}_p = \frac{\dot{m}}{D\left(\frac{i\omega}{\alpha} + \frac{\mu}{D}\right)}$$

Putting the homogeneous and particular solutions together gives

$$\hat{c} = a_1 e^{\kappa_1 x} + a_2 e^{\kappa_2 x} + \frac{\dot{m}}{D\left(\frac{i\omega}{\alpha} + \frac{\mu}{D}\right)} \tag{14.58}$$

Boundary conditions

Taking the derivative of (14.58) with respect to x, the first boundary condition in (14.53) implies, at $x = 0$,

$$a_1(v - D\kappa_1) + a_2(v - D\kappa_2) = v\left[\hat{C}_0 - \frac{\dot{m}}{D\left(\frac{i\omega}{\alpha} + \frac{\mu}{D}\right)}\right]$$

At $x = L$, the second boundary condition in (14.53) implies

$$a_1\kappa_1 e^{\kappa_1 L} + a_2\kappa_2 e^{\kappa_2 L} = 0 \tag{14.59}$$

Putting these two equations in a matrix form gives

$$\begin{pmatrix} v - D\kappa_1 & v - D\kappa_2 \\ \kappa_1 e^{\kappa_1 L} & \kappa_2 e^{\kappa_2 L} \end{pmatrix} \begin{pmatrix} a_1 \\ a_2 \end{pmatrix} = \begin{pmatrix} v \left[\hat{C}_0 - \frac{\dot{m}}{D\left(\frac{i\omega}{\alpha} + \frac{\mu}{D}\right)} \right] \\ 0 \end{pmatrix}$$

Solving for the constants vector yields

$$\begin{pmatrix} a_1 \\ a_2 \end{pmatrix} = \frac{1}{\Delta} \begin{pmatrix} \kappa_2 e^{\kappa_2 L} & -v + D\kappa_2 \\ -\kappa_1 e^{\kappa_1 L} & v - D\kappa_1 \end{pmatrix} \begin{pmatrix} v \left[\hat{C}_0 - \frac{\dot{m}}{D\left(\frac{i\omega}{\alpha} + \frac{\mu}{D}\right)} \right] \\ 0 \end{pmatrix}$$

$$\Delta = (v - D\kappa_1)\, \kappa_2 e^{\kappa_2 L} - (v - D\kappa_2)\, \kappa_1 e^{\kappa_1 L}$$

This entails

$$\begin{aligned} a_1 &= \frac{\kappa_2 v}{\Delta} \left[\hat{C}_0 - \frac{\dot{m}}{D\left(\frac{i\omega}{\alpha} + \frac{\mu}{D}\right)} \right] e^{\kappa_2 L} \\ a_2 &= -\frac{\kappa_1 v}{\Delta} \left[\hat{C}_0 - \frac{\dot{m}}{D\left(\frac{i\omega}{\alpha} + \frac{\mu}{D}\right)} \right] e^{\kappa_1 L} \end{aligned} \tag{14.60}$$

BVP Solution

Substituting (14.60) into the general solution (14.58) gives

$$\begin{aligned} \hat{c}(x, \omega_n) = &\frac{\kappa_2 v}{\Delta} \left[\hat{C}_0 - \frac{\dot{m}}{D\left(\frac{i\omega_n}{\alpha} + \frac{\mu}{D}\right)} \right] e^{\kappa_2 L} e^{\kappa_1 x} \\ &- \frac{\kappa_1 v}{\Delta} \left[\hat{C}_0 - \frac{\dot{m}}{D\left(\frac{i\omega_n}{\alpha} + \frac{\mu}{D}\right)} \right] e^{\kappa_1 L} e^{\kappa_2 x} + \frac{\dot{m}}{D\left(\frac{i\omega_n}{\alpha} + \frac{\mu}{D}\right)} \end{aligned} \tag{14.61}$$

where

$$\Delta = (v - D\kappa_1)\, \kappa_2 e^{\kappa_2 L} - (v - D\kappa_2)\, \kappa_1 e^{\kappa_1 L}$$

$$\alpha = \frac{D}{R}$$

$$\kappa_1 = \frac{v}{2D} + \sqrt{\frac{\mu}{D} + \left(\frac{v}{2D}\right)^2 + \frac{i\omega_n}{\alpha}}, \quad \kappa_2 = \frac{v}{2D} - \sqrt{\frac{\mu}{D} + \left(\frac{v}{2D}\right)^2 + \frac{i\omega_n}{\alpha}}$$

Having solved the problem in the frequency domain, the contaminant distribution in the time domain can then be calculated using the inverse fast Fourier transform (IFFT) algorithm, such that

$$c(x,t) = \sum_{n} \hat{c}(x,\omega_n)\, e^{i\omega_n t}$$

Special Case 1: Semi-infinite domain

At the limiting case where $L \to \infty$, the domain becomes semi-infinite, and the solution in the frequency domain results in

$$\hat{c}(x,\omega_n) = \frac{v}{v - D\kappa_2}\left[\hat{C}_0 - \frac{\dot{m}}{D\left(\dfrac{i\omega_n}{\alpha} + \dfrac{\mu}{D}\right)}\right] e^{\kappa_2 x} + \frac{\dot{m}}{D\left(\dfrac{i\omega_n}{\alpha} + \dfrac{\mu}{D}\right)}$$

Special Case 2: Steady state

As $\omega \to 0$, i.e. $t \to \infty$, the system reaches the steady state, and the solution becomes

$$c(x) = \frac{v}{v - D\kappa_2}\left(\hat{C}_0 - \frac{\dot{m}}{\mu}\right) e^{\kappa_2 x} + \frac{\dot{m}}{\mu}$$

with

$$\kappa_2 = \frac{v}{2D} - \sqrt{\frac{\mu}{D} + \left(\frac{v}{2D}\right)^2}$$

Example 14.4: Finite 1D with zero-order production and first order decay

A fully saturated 10 m long porous medium domain, having a bulk density 1500 kg/m^3 and porosity 0.3, is exposed at one of its boundaries to a chemical species with distribution coefficient 0.8 m^3/kg and hydrodynamic dispersion coefficient 1.5 m^2/s. The groundwater in the porous domain flows at an average velocity of 2.0e−1 m/s. No initial concentration of the species exists in the domain. At $x = 0$, the domain is exposed to a Robin-type boundary condition, as given in (14.53), while the boundary at $x = L$ it is bounded by an impervious layer. A rectangular pulse concentration of 10 kg/m enters the domain for nearly 20 h. Under some chemical or physical processes, the species simultaneously undergoes a zero-order production with a constant rate of 0.5 kg/ms and a first-order decay of 0.1 s^{-1}. Figure 14.10 depicts the time distribution of the prescribed concentration at the boundary, together with the production rate in the domain. The decay rate is continuous.

Compute the concentration of the species at 0.0, 1.0, 2.0, 5.0 and 9.0 m from the source.

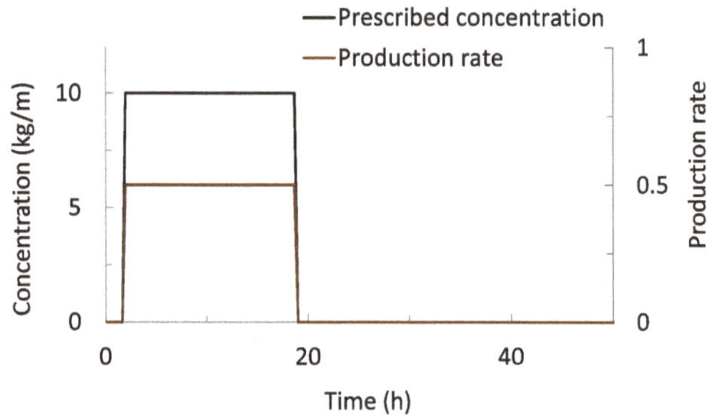

Fig. 14.10 Prescribed species concentration and production rate pulses

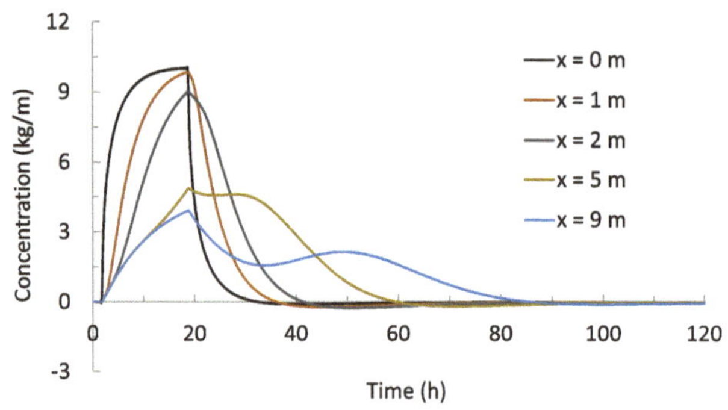

Fig. 14.11 Species concentrations for the ADR case

- **Solution**

Using the solution procedure described in the previous examples, the species concentration along the domain has been calculated and is presented in Fig. 14.11. The figure highlights the dynamic interplay between the production rate and the effect of decay over time and space.

14.4.3 Semi-infinite Domain with Interaction with Other Species

Contaminant transport with chemical interaction refers to the process where the transport of a contaminant in a medium is influenced by chemical reactions or physical interactions with other species present in the system.

The relevant boundary value problem for a semi-infinite domain with a Dirichlet boundary condition is

$$
D\frac{\partial^2 c}{\partial x^2} - v\frac{\partial c}{\partial x} - R\frac{\partial c}{\partial t} - \mu(c - c_s) + \dot{m} = 0
$$

$$
\begin{aligned}
&\text{IC:}\ \ c(x,t) = 0 \\
&\text{BC:}\ \ c(0,t) = C_0(t) \\
&\qquad\ \ c(\infty,t) = 0
\end{aligned}
\tag{14.62}
$$

where c_s (kg m^{-3}) is some species presence in the medium that has the affinity to interact with the contaminant species.

General solution

Applying the Fourier transform to (14.62) and rearranging leads to

$$
\frac{d^2\hat{c}}{dx^2} - \frac{v}{D}\frac{d\hat{c}}{dx} - \frac{i\omega}{\alpha}\hat{c} - \frac{\mu}{D}(\hat{c} - c_s) + \frac{\dot{m}}{D} = 0
$$

$$
\begin{aligned}
&\text{BC}:\ \hat{c}(0,\omega) = \hat{C}_0(\omega) \\
&\qquad\ \hat{c}(\infty,\omega) = 0
\end{aligned}
\tag{14.63}
$$

where $\alpha = D/R$.

The differential equation in (14.63) is nonhomogeneous, that can be solved via

$$
\hat{c} = \hat{c}_h + \hat{c}_p \tag{14.64}
$$

The corresponding homogeneous part of the differential equation in (14.63) is

$$
\frac{d^2\hat{c}_h}{dx^2} - \frac{v}{D}\frac{d\hat{c}_h}{dx} - \left(\frac{i\omega}{\alpha} + \frac{\mu}{D}\right)\hat{c}_h = 0 \tag{14.65}
$$

As usual, as the coefficients are constant, the trial solution can be

$$
\hat{c}_h = ae^{-\kappa x}
$$

Substituting this trial function into (14.65) leads to

$$
\kappa^2 - \frac{v}{D}\kappa + P = 0, \quad P = \frac{i\omega}{\alpha} + \frac{\mu}{D}
$$

Using the quadratic formula, the solution of this algebraic equation gives

$$
\kappa_1 = \frac{1}{2D}\left(-v + \sqrt{4D^2P + v^2}\right), \quad \kappa_2 = \frac{1}{2D}\left(-v - \sqrt{4D^2P + v^2}\right)
$$

Thus, the general solution of the homogeneous equation (14.65) is

$$
\hat{c}_h = a_1 e^{-i\kappa_1 x} + a_2 e^{-\kappa_2 x}
$$

The relevant particular equation of (14.63) is

$$\frac{d^2\hat{c}_p}{dx^2} - \frac{v}{D}\frac{d\hat{c}_p}{dx} - \left(\frac{i\omega}{\alpha} + \frac{\mu}{D}\right)\hat{c}_p = -\left(\frac{\mu}{D}c_s + \frac{\dot{m}}{D}\right) \tag{14.66}$$

As the nonhomogeneous term is constant, the trial solution might be

$$\hat{c}_p = A$$

Substituting this trial solution into (14.66) lead to

$$A = \frac{\mu c_s + \dot{m}}{\dfrac{i\omega D}{\alpha} + \mu}$$

Thus, the general solution of the particular Eq. (14.66) is

$$\hat{c}_p = \frac{\mu c_s + \dot{m}}{\dfrac{i\omega D}{\alpha} + \mu}$$

Putting the homogeneous and particular solutions together, the general solution of the problems is then

$$\hat{c} = a_1 e^{-\kappa_1 x} + a_2 e^{-\kappa_2 x} + \frac{\mu c_s + \dot{m}}{\dfrac{i\omega D}{\alpha} + \mu} \tag{14.67}$$

Boundary conditions

The second boundary condition in (14.63) entails $a_2 = 0$, otherwise the solution becomes unbounded at far distances, reducing the solution general (14.67) to

14

$$\hat{c} = a_1 e^{-\kappa_1 x} + \frac{\mu c_s + \dot{m}}{\dfrac{i\omega D}{\alpha} + \mu} \tag{14.68}$$

At $x = 0$, the first boundary condition in (14.63) implies

$$a_1 = \hat{C}_0 - \frac{\mu c_s + \dot{m}}{\dfrac{i\omega D}{\alpha} + \mu} \tag{14.69}$$

BVP Solution

Substituting (14.69) into (14.68) gives

$$\hat{c}\,(x, \omega_n) = \hat{C}_0 e^{-\kappa_1 x} + \frac{\mu c_s + \dot{m}}{\dfrac{i\omega_n D}{\alpha} + \mu}\left(1 - e^{-\kappa_1 x}\right) \tag{14.70}$$

Having obtained the solution in the frequency domain, the contaminant distribution in the time domain can then be calculated using the inverse fast Fourier transform (IFFT) algorithm, such that

$$c\,(x,t) = \sum_n \hat{c}\,(x,\omega_n)\,e^{i\omega_n t}$$

14.4.4 Retarded Solute Due to Adsorption

Adsorption refers to the process by which solutes adhere to solid surfaces, such as soil particles or sediment, resulting in reduced mobility compared to their movement in the water phase. This retardation effect plays a significant role in shaping the transport behavior of solute.

In a scenario where a dissolved solute in groundwater experiences retardation due to adsorption, its transport can be effectively described using a *two-domain approach*. This approach distinguishes between a mobile aqueous phase and an immobile solid phase. The governing transport equations for this case are expressed by an equation for the dissolved solute in the groundwater and another for the adsorbed solute in the solid matrix, such that

$$\begin{aligned} &\varphi\frac{\partial c}{\partial t} + q_w\frac{\partial c}{\partial x} - \varphi D\frac{\partial^2 c}{\partial x^2} = \varphi\mu_w c - \varphi\dot{m}_w \\ &\rho\frac{\partial s}{\partial t} = \mu_s\rho s - \rho\dot{m}_s \end{aligned} \tag{14.71}$$

where s (Kg m^{-3}) is the adsorbed concertation in the solid phase; φ is the volumetric moisture content; $q_w = \varphi v$, is the volumetric flux of the groundwater (ms^{-1}), with v the pore water velocity; ρ is the bulk density of the aquifer solid phase (adsorbent); μ_w and μ_s are rate constants for the first-order decay of the liquid and solid phases (s^{-1}); and $\dot{m}_w$ and $\dot{m}_s$ are rate constants for the zero-order production of these two phase (Kg m^{-3}s^{-1}).

Summing the two equations gives

$$\varphi\frac{\partial c}{\partial t} + \rho\frac{\partial s}{\partial t} + q_w\frac{\partial c}{\partial x} - \varphi D\frac{\partial^2 c}{\partial x^2} = \varphi\mu_w c + \mu_s\rho s - \varphi\dot{m}_w - \rho\dot{m}_s \tag{14.72}$$

If we assume that the adsorbed concentration is a linear function of the dissolved solute, expressed as

$$s = K_d c$$

where K_d is usually described as the *adsorption* or *distribution coefficient* (m^{3}kg^{-1}), the equation in (14.72) can be expressed as

$$\left(1 + \frac{\rho_s K_d}{\varphi}\right)\frac{\partial c}{\partial t} + v\frac{\partial c}{\partial x} - D\frac{\partial^2 c}{\partial x^2} = \left(\mu_w + \frac{\mu_s\rho K_d}{\varphi}\right)c - \varphi\dot{m}_w - \rho\dot{m}_s \tag{14.73}$$

where $v = q_w/\varphi$ is the pore water steady state velocity.

If we express

$$R = 1 + \frac{\rho_s K_d}{\varphi}$$
$$\mu = \mu_w + \frac{\mu_s \rho K_d}{\varphi} \tag{14.74}$$
$$\dot{m} = \dot{m}_w + \frac{\rho \dot{m}_s}{\varphi}$$

Equation (14.73) can be rewritten as

$$R\frac{\partial c}{\partial t} + v\frac{\partial c}{\partial x} - D\frac{\partial^2 c}{\partial x^2} = \mu c - \dot{m} \tag{14.75}$$

which is the standard ADR equation discussed in detail in ▶ Sect. 14.4. With the definitions provided in (14.74), the solution method outlined in that section can similarly be applied to describe contaminant transport affected by adsorption.

14.4.5 Multispecies Diffusion in a Finite Homogenous Domain

Diffusion of species in porous medium domains refers to the movement of the species driven primarily by their concentration gradients. In a subsurface environment, consider the presence of two interacting species whose transport is predominantly governed by diffusion. The two species may represent different types of contaminants, dissolved substances, or reactive agents influencing each other's behavior through concentration-dependent interactions. The interactions can occur through various mechanisms such as chemical reactions, sorption/desorption, or competitive adsorption. This mode of transport is often referred to as the *uniform transport* or *equilibrium transport*.

The boundary value problem of a diffusive transport of two interacting species in a finite 1D domain can be expressed as

$$\left.\begin{aligned} D_1\frac{\partial^2 c_1}{\partial x^2} - R_1\frac{\partial c_1}{\partial t} &= \lambda_1 c_1 - \lambda_2 c_2 \\ D_2\frac{\partial^2 c_2}{\partial x^2} - R_2\frac{\partial c_2}{\partial t} &= -\lambda_1 c_1 + \lambda_2 c_2 \end{aligned}\right\} \quad 0 \le x \le L$$

$$\begin{aligned} \text{IC:}\quad & c_i(x,0) = 0 \\ \text{BC:}\quad & c_1(0,t) = C_{01}(t) \\ & c_2(0,t) = C_{02}(t) \\ & c_i(L,t) = 0, \quad i \equiv 1,2 \end{aligned} \tag{14.76}$$

in which c_1 and c_2 (kg m^{-3}) are co-existing contaminant species, λ_1 and λ_2 (s^{-1}) are some given interaction coefficients, D (m^{2}s^{-1}) is the hydrodynamic dispersion coefficient, and R is defined as

$$R = 1 + \rho K_d / \varphi$$

with ρ (kg m^{-3}) the bulk density of the porous medium, K_d (m^{3}kg^{-1}) the distribution coefficient and φ the porosity.

General solution

Applying the Fourier transform to (14.76) yields

$$\begin{aligned} D_1 \frac{d^2 \hat{c}_1}{dx^2} - R_1 i\omega \hat{c}_1 &= \lambda_1 \hat{c}_1 - \lambda_2 \hat{c}_2 \\ D_2 \frac{d^2 \hat{c}_2}{dx^2} - R_2 i\omega \hat{c}_2 &= -\lambda_1 \hat{c}_1 + \lambda_2 \hat{c}_2 \end{aligned} \tag{14.77}$$

$$\begin{aligned} \text{BC: } \hat{c}_1(0, \omega) &= \hat{C}_{01}(\omega) \\ \hat{c}_2(0, \omega) &= \hat{C}_{02}(\omega) \\ \hat{c}_i(L, \omega) &= 0 \end{aligned}$$

As the coefficients of this coupled differential equations system are constant, the trial solution can be expressed as

$$\hat{c}_1 = a e^{-\kappa x}, \qquad \hat{c}_2 = b e^{-\kappa x}$$

where a, b and κ are constants need to be determined. Substituting these trial solutions into (14.77) and simplifying gives

$$\begin{aligned} D_1 a \kappa^2 - i\omega R_1 a - \lambda_1 a + \lambda_2 b &= 0 \\ D_2 b \kappa^2 - i\omega R_2 b + \lambda_1 a - \lambda_2 b &= 0 \end{aligned}$$

Expressing these equations in a matrix form yields

$$\begin{pmatrix} D_1 \kappa^2 - i\omega R_1 - \lambda_1 & \lambda_2 \\ \lambda_1 & D_2 \kappa^2 - i\omega R_2 - \lambda_2 \end{pmatrix} \begin{pmatrix} a \\ b \end{pmatrix} = 0 \tag{14.78}$$

As a and b cannot be zero, the solution of this coupled system of equations can only be obtained by letting the determinate of the matrix equals to zero. Obviously, the determinant of the matrix in (14.78) would yield a complex four degree polynomial of the form

$$m_2 \kappa^4 + m_1 \kappa^2 + m_0 = 0 \tag{14.79}$$

which denote the four eigenvalues of the system, for which there are four eigenfunction.

Following this, the general solution of the coupled differential equations in (14.77) can then be expressed as

$$\begin{aligned}\hat{c}_1 &= a_1 e^{-\kappa_1 x} + a_2 e^{-\kappa_2 x} + a_3 e^{-\kappa_3 x} + a_4 e^{-\kappa_4 x} \\ \hat{c}_2 &= b_1 e^{-\kappa_1 x} + b_2 e^{-\kappa_2 x} + b_3 e^{-\kappa_3 x} + b_4 e^{-\kappa_4 x}\end{aligned} \tag{14.80}$$

Since the two species are coupled, their constants must be related, expressed via the eigenvectors of the system. The constants a and b in (14.78) are related, such that

$$b_j = Y_i a_j$$

where

$$Y_j = \frac{-D_1 \kappa_j^2 + R_1 i\omega + \lambda_1}{\lambda_2}$$

with j denoting that for each eigenvalue κ_j there is a corresponding eigenvector. Accordingly, (14.80) can be expressed as

$$\begin{aligned}\hat{c}_1 &= a_1 e^{-\kappa_1 x} + a_2 e^{-\kappa_2 x} + a_3 e^{-\kappa_3 x} + a_4 e^{-\kappa_4 x} \\ \hat{c}_2 &= Y_1 a_1 e^{-\kappa_1 x} + Y_2 a_2 e^{-\kappa_2 x} + Y_3 a_3 e^{-\kappa_3 x} + Y_4 a_4 e^{-\kappa_4 x}\end{aligned} \tag{14.81}$$

Boundary conditions

At $x = 0$, the first and second boundary conditions in (14.77) imply

$$\begin{aligned}a_1 + a_2 + a_3 + a_4 &= \hat{C}_{01} \\ Y_1 a_1 + Y_2 a_2 + Y_3 a_3 + Y_4 a_4 &= \hat{C}_{02}\end{aligned}$$

At $x = L$, the third boundary condition in (14.77) implies

$$\begin{aligned}a_1 e^{-\kappa_1 L} + a_2 e^{-\kappa_2 L} + a_3 e^{-\kappa_3 L} + a_4 e^{-\kappa_4 L} &= 0 \\ Y_1 a_1 e^{-\kappa_1 L} + Y_2 a_2 e^{-\kappa_2 L} + Y_3 a_3 e^{-\kappa_3 L} + Y_4 a_4 e^{-\kappa_4 L} &= 0\end{aligned}$$

Including these four equations in a matrix form gives

$$\begin{pmatrix} 1 & 1 & 1 & 1 \\ Y_1 & Y_2 & Y_3 & Y_4 \\ e^{-\kappa_1 L} & e^{-\kappa_2 L} & e^{-\kappa_3 L} & e^{-\kappa_4 L} \\ Y_1 e^{-\kappa_1 L} & Y_2 e^{-\kappa_2 L} & Y_3 e^{-\kappa_3 L} & Y_4 e^{-\kappa_4 L} \end{pmatrix} \begin{pmatrix} a_1 \\ a_2 \\ a_3 \\ a_4 \end{pmatrix} = \begin{pmatrix} \hat{C}_{01} \\ \hat{C}_{02} \\ 0 \\ 0 \end{pmatrix} \tag{14.82}$$

from which the coefficients can be determined in terms of the prescribed boundary values, expressed in a compact form as

$$\mathbf{a} = \mathbf{HC}$$

where **H** is the inverse of the matrix in (14.82) and **C** is the right hand side vector.

14

BVP solution

Having determined the coefficients, together with the roots of (14.79), the solution of the boundary value problem in the frequency domain can be obtained via (14.81).

Then, using the inverse FFT, the solution in the time domain can be restored, such that

$$c_1(x,t) = \sum_n \hat{c}_1(x,\omega_n)\, e^{i\omega_n t}$$

$$c_2(x,t) = \sum_n \hat{c}_2(x,\omega_n)\, e^{i\omega_n t}$$

Example 14.5: Multispecie

A fully saturated 10 m long porous medium domain is exposed at one of its boundaries to two interacting chemical species with hydrodynamic dispersion coefficients $D_1 = 1.\mathrm{e}{-3}$ m^2/s and $D_2 = 1.\mathrm{e}{-2}$ m^2/s. The interaction coefficients are $\lambda_1 = 1.\mathrm{e}{-4}\ s^{-1}$ and $\lambda_2 = 1.\mathrm{e}{-5}\ s^{-1}$, with $R_1 = 5.0$ and $R_2 = 10.0$. No initial concentrations of the chemical species exist in the domain. At $x = 0$, a rectangular concentration pulse of 10 kg/m of each species enters the domain for nearly 18 h as depicted in Fig. 14.12. On the other side, at $x = L$, the concentrations are set to zero. Compute the concentration of both species at 0.0, 1.0 and 5.0 m from the boundary.

Solution

Using the solution method described in the previous examples, the species concentration along the domain has been calculated and is presented in Fig. 14.13. Apparently, the concentration of the species with higher diffusion rate spread more in the domain.

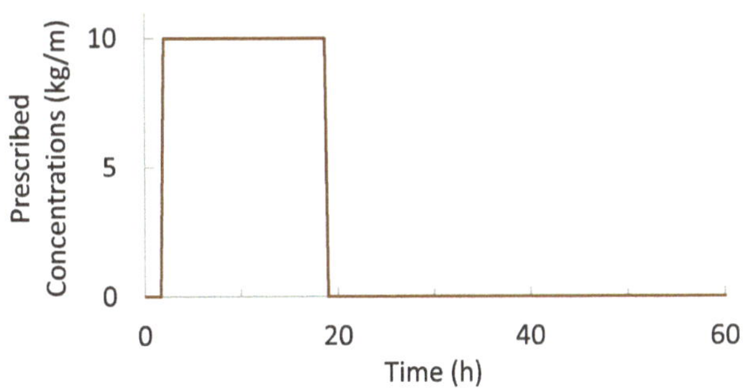

Fig. 14.12 Prescribed concentration for both species

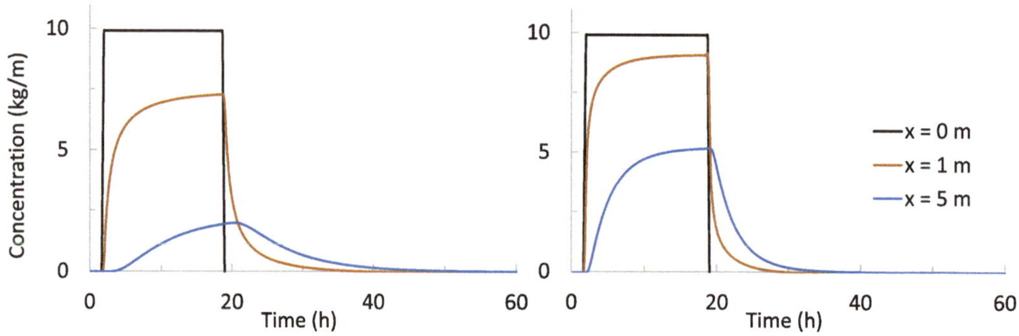

Fig. 14.13 Computed concentrations for c_1 (left) and c_2 (right)

14.4.6 Multispecies Diffusion in a Finite Homogenous Domain: Alternative Solution

Here we solve the boundary value problem which we have addressed in the previous section using the *diagonalization method*. This method is a procedural mathematical technique used for solving coupled differential equations by means of decoupling them. Solution by diagonalization is valid provided that an $n \times n$ matrix having n linearly independent eigenvectors does exist. That is, systems with matrices having repeated eigenvalues cannot be solved by this method.

To facilitate the algebra, we set $D_1 = D_2 = D$ and $R_1 = R_2 = R$. Following this, the boundary value problem (14.76) can be represented as

$$\begin{aligned} D\frac{\partial^2 c_1}{\partial x^2} - R\frac{\partial c_1}{\partial t} &= \lambda_1 c_1 - \lambda_2 c_2 \\ D\frac{\partial^2 c_2}{\partial x^2} - R\frac{\partial c_2}{\partial t} &= -\lambda_1 c_1 + \lambda_2 c_2 \end{aligned} \tag{14.83}$$

$$\begin{aligned} \text{IC}: &\ c_i(x,0) = 0 \\ \text{BC}: &\ c_1(0,t) = C_{01}(t) \\ &\ c_2(0,t) = C_{02}(t) \\ &\ c_i(L,t) = 0 \end{aligned}$$

General solution

Applying the Fourier transform to (14.83) yields

$$D\frac{d^2\hat{c}_1}{dx^2} + (-Ri\omega - \lambda_1)\,\hat{c}_1 + \lambda_2\hat{c}_2 = 0$$
$$D\frac{d^2\hat{c}_2}{dx^2} + (-Ri\omega - \lambda_2)\,\hat{c}_2 + \lambda_1\hat{c}_1 = 0 \tag{14.84}$$

$$\text{BC: } \hat{c}_1(0,\omega) = \hat{C}_{01}(\omega)$$
$$\hat{c}_2(0,\omega) = \hat{C}_{02}(\omega)$$
$$\hat{c}_i(L,\omega) = 0$$

The system of differential equation in (14.84) can be expressed concisely as

$$\mathbf{A}\mathbf{c}'' + \mathbf{B}\mathbf{c} = 0 \tag{14.85}$$

where

$$\mathbf{c} = \begin{pmatrix} \hat{c}_1 \\ \hat{c}_2 \end{pmatrix}, \quad \mathbf{A} = \begin{pmatrix} D & 0 \\ 0 & D \end{pmatrix}, \quad \mathbf{B} = \begin{pmatrix} -Ri\omega - \lambda_1 & \lambda_2 \\ \lambda_1 & -Ri\omega - \lambda_2 \end{pmatrix}$$

Since A can be inverted, (14.85) may be written as

$$\mathbf{c}'' + \mathbf{A}^{-1}\mathbf{B}\mathbf{c} = 0, \quad \text{or} \quad \mathbf{c}'' + \mathbf{M}\mathbf{c} = 0 \tag{14.86}$$

where

$$\mathbf{M} = \mathbf{A}^{-1}\mathbf{B} = \begin{pmatrix} \frac{1}{D} & 0 \\ 0 & \frac{1}{D} \end{pmatrix}\begin{pmatrix} -Ri\omega - \lambda_1 & \lambda_2 \\ \lambda_1 & -Ri\omega - \lambda_2 \end{pmatrix} = \begin{pmatrix} \frac{-Ri\omega - \lambda_1}{D} & \frac{\lambda_2}{D} \\ \frac{\lambda_1}{D} & \frac{-Ri\omega - \lambda_2}{D} \end{pmatrix}$$

The eigenvalues of this matrix, commonly referred to as the *coefficient matrix*, are

$$\mathbf{M}_{\text{eigen}} = \begin{pmatrix} \frac{-Ri\omega - \lambda_1 - \lambda_2}{D} \\ \frac{-iR\omega}{D} \end{pmatrix}$$

And the corresponding two eigenvectors are

$$\mathbf{P} = \begin{pmatrix} -1 & \frac{\lambda_2}{\lambda_1} \\ 1 & 1 \end{pmatrix}$$

Diagonalization

The diagonalization method first implies diagonalizing the coefficient matrix **M**. This is done by first letting $\mathbf{c} = \mathbf{PC}$ and then substituting it into the system of differential Eqs. (14.86), leading to

$$\mathbf{PC}'' + \mathbf{MPC} = 0, \quad \text{or} \quad \mathbf{C}'' + \mathbf{P}^{-1}\mathbf{MPC} = 0, \quad \text{or} \quad \mathbf{C}'' + \mathbf{DC} = 0 \tag{14.87}$$

where

$$\mathbf{D} = \mathbf{P}^{-1}\mathbf{M}\mathbf{P} = \begin{pmatrix} \dfrac{-iR\omega - \lambda_1 - \lambda_2}{D} & 0 \\ 0 & \dfrac{-iR\omega}{D} \end{pmatrix}$$

which is indeed the matrix that its entries on the diagonal are the eigenvalues, and the rest are zero.

Substituting $\mathbf{D}$ into the third equation in (14.87), the system of equations can now be expressed as

$$\mathbf{C}'' + \begin{pmatrix} \dfrac{-iR\omega - \lambda_1 - \lambda_2}{D} & 0 \\ 0 & \dfrac{-iR\omega}{D} \end{pmatrix} \mathbf{C} = 0, \quad \text{or}$$

$$\begin{pmatrix} C_1'' \\ C_2'' \end{pmatrix} + \begin{pmatrix} \dfrac{-iR\omega - \lambda_1 - \lambda_2}{D} & 0 \\ 0 & \dfrac{-iR\omega}{D} \end{pmatrix} \begin{pmatrix} C_1 \\ C_2 \end{pmatrix} = 0$$

Obviously, this system of equations is uncoupled and can be written as

$$C_1'' + \kappa_1^2 C_1 = 0, \qquad \kappa_1^2 = \frac{-iR\omega - \lambda_1 - \lambda_2}{D}$$

$$C_2'' + \kappa_2^2 C_2 = 0, \qquad \kappa_2^2 = \frac{-iR\omega}{D}$$

The solutions of these ODEs are

$$\begin{aligned} C_1 &= a_1 \sin(\kappa_1 x) + a_2 \cos(\kappa_1 x) \\ C_2 &= a_3 \sin(\kappa_2 x) + a_4 \cos(\kappa_2 x) \end{aligned} \tag{14.88}$$

Going back to the original variables, we write

14

$$\mathbf{c} = \mathbf{P}\mathbf{C} = \begin{pmatrix} -1 & \frac{\lambda_2}{\lambda_1} \\ 1 & 1 \end{pmatrix} \begin{pmatrix} C_1 \\ C_2 \end{pmatrix} = \begin{pmatrix} -C_1 + \frac{\lambda_2}{\lambda_1} C_2 \\ C_1 + C_2 \end{pmatrix} \tag{14.89}$$

By substituting C_1 and C_2 from (14.88) into the right-hand side of (14.89) we obtain the general solutions of $\widehat{c}_1$ and $\widehat{c}_2$, such that

$$\begin{aligned} \hat{c}_1(x) &= -[a_1 \sin(\kappa_1 x) + a_2 \cos(\kappa_1 x)] + \frac{\lambda_2}{\lambda_1}[a_3 \sin(\kappa_2 x) + a_4 \cos(\kappa_2 x)] \\ \hat{c}_2(x) &= a_1 \sin(\kappa_1 x) + a_2 \cos(\kappa_1 x) + a_3 \sin(\kappa_2 x) + a_4 \cos(\kappa_2 x) \end{aligned} \tag{14.90}$$

Boundary conditions

Applying the boundary conditions in (14.84) into (14.90) gives

$$-a_2 + \frac{\lambda_2}{\lambda_1} a_4 = \hat{C}_{01}$$

$$a_2 + a_4 = \hat{C}_{02}$$

$$-[a_1 \sin(\kappa_1 L) + a_2 \cos(\kappa_1 L)] + \frac{\lambda_2}{\lambda_1}[a_3 \sin(\kappa_2 L) + a_4 \cos(\kappa_2 L)] = 0$$

$$a_1 \sin(\kappa_1 L) + a_2 \cos(\kappa_1 L) + a_3 \sin(\kappa_2 L) + a_4 \cos(\kappa_2 L) = 0$$

Solving these equations simultaneously yields

$$a_1 = -\frac{\cos(\kappa_1 L)}{\sin(\kappa_1 L)} a_2, \quad a_2 = \frac{-\lambda_1 \hat{C}_{01} + \lambda_2 \hat{C}_{02}}{\lambda_1 + \lambda_2}$$

$$a_3 = -\frac{\cos(\kappa_2 L)}{\sin(\kappa_2 L)} a_4, \quad a_4 = \frac{\lambda_1(\hat{C}_{01} + \hat{C}_{02})}{\lambda_1 + \lambda_2} \tag{14.91}$$

with

$$\kappa_1 = \sqrt{\frac{-iR\omega - \lambda_1 - \lambda_2}{D}}, \quad \kappa_2 = \sqrt{\frac{-iR\omega}{D}}$$

Note that since κ_1 and κ_2 are complex, the trigonometric functions are evaluated by splitting them into their real and imaginary parts, such that

$$\sin(\kappa_i L) = \sin[\mathrm{Re}(\kappa_i)L]\cosh[\mathrm{Im}(\kappa_i)L] + i\cos[\mathrm{Re}(\kappa_i)L]\sinh[\mathrm{Im}(\kappa_i)L]$$

$$\cos(\kappa_i L) = \cos[\mathrm{Re}(\kappa_i)L]\cosh[\mathrm{Im}(\kappa_i)L] - i\sin[\mathrm{Re}(\kappa_i)L]\sinh[\mathrm{Im}(\kappa_i)L]$$

BVP solution

The solution of the BVP (14.83) in the frequency domain is therefore

$$\hat{c}_1(x) = -[a_1 \sin(\kappa_1 x) + a_2 \cos(\kappa_1 x)] + \frac{\lambda_2}{\lambda_1}[a_3 \sin(\kappa_2 x) + a_4 \cos(\kappa_2 x)]$$

$$\hat{c}_2(x) = a_1 \sin(\kappa_1 x) + a_2 \cos(\kappa_1 x) + a_3 \sin(\kappa_2 x) + a_4 \cos(\kappa_2 x) \tag{14.92}$$

where the coefficients $a_1 \cdots a_4$ are given in (14.91).

Using the inverse FFT, the solution in the time domain can be restored, such that

$$c_1(x,t) = \sum_n \hat{c}_1(x, \omega_n)\, e^{i\omega_n t}$$

$$c_2(x,t) = \sum_n \hat{c}_2(x, \omega_n)\, e^{i\omega_n t}$$

Appendix

In this appendix, we highlight the benefits of employing spectral analysis for solving linear boundary value problems, showcasing its advantage over alternative analytical solution techniques.

Specifically, we present an analytical solution proposed by van Genuchten and Alves (1982), addressing the transient Robin-Neumann boundary value problem given in ▶ Sect. 13.4.2. Upon closer examination of both solutions, the reader can readily recognize the simplicity and directness inherent in the spectral analysis method, compared to the one given here.

Van Genuchten and Alves published in 1982 a compendium of analytical solutions of one-dimensional convective–dispersive solute transport problems with varying complexities. The solutions are mostly conducted using the Laplace transforms. For the problem that we are examining here, they presented two solutions: one exact and another approximate. Here, we list both solutions as formulated by the authors.

Governing equation

$$R\frac{\partial c}{\partial t} = D\frac{\partial^2 c}{\partial x^2} - v\frac{\partial c}{\partial x} - \mu c + \gamma$$

Initial and boundary conditions

$$c(x,0) = c_i$$

$$\left(-D\frac{\partial c}{\partial x} + vc\right)\bigg|_{x=0} = \begin{cases} vC_0 & 0 < t < t_0 \\ 0 & t > t_0 \end{cases}$$

$$\frac{\partial c}{\partial x}\bigg|_{x=L} = 0$$

Analytical solution

$$c(x,t) = \begin{cases} \frac{\gamma}{\mu} + (c_i - \frac{\gamma}{\mu})A(x,t) + (C_0 - \frac{\gamma}{\mu})B(x,t) & 0 < t < t_0 \\ \frac{\gamma}{\mu} + (c_i - \frac{\gamma}{\mu})A(x,t) + (C_0 - \frac{\gamma}{\mu})B(x,t) - C_0 B(x, t - t_0) & t > t_0 \end{cases}$$

14

where

$$A(x,t) = \sum_{m=1}^{\infty} E(\beta_m, x) \exp\left[\frac{vx}{2D} - \frac{\mu t}{R} - \frac{v^2 t}{4DR} - \frac{\beta_m^2 D t}{L^2 R}\right]$$

$$B(x,t) = B_1(x) - B_2(x,t)$$

with

$$B_1(x) = 1 - \frac{\mu L^2}{D}\sum_{m=1}^{\infty} \frac{E(\beta_m, x)\exp\left(\frac{vx}{2D}\right)}{\left[\beta_m^2 + \left(\frac{vL}{2D}\right)^2 + \frac{\mu L^2}{D}\right]}$$

$$B_2(x,t) = \sum_{m=1}^{\infty} \frac{E(\beta_m, x)\left[\beta_m^2 + \left(\frac{vL}{2D}\right)^2\right]\exp\left[\frac{vx}{2D} - \frac{\mu t}{R} - \frac{v^2 t}{4DR} - \frac{\beta_m^2 D t}{L^2 R}\right]}{\beta_m^2 + \left(\frac{vL}{2D}\right)^2 + \frac{\mu L^2}{D}}$$

and

$$E(\beta_m, x) = \frac{\frac{2vL}{D}\beta_m\left[\beta_m \cos\left(\frac{\beta_m x}{L}\right) + \frac{vL}{2D}\sin\left(\frac{\beta_m x}{L}\right)\right]}{\left[\beta_m^2 + \left(\frac{vL}{2D}\right)^2 + \frac{vL}{D}\right]\left[\beta_m^2 + \left(\frac{vL}{2D}\right)^2\right]}$$

The eigenvalues β_m are the positive roots of

$$\beta_m \cot(\beta_m) - \frac{\beta_m^2 D}{vL} + \frac{vL}{4D} = 0$$

Van Genuchten and Alves stated that the term $B_1(x)$ converges much slower than the other terms, and provided an alternative form that is easier to evaluate, given by

$$B_1(x) = \frac{\exp\left[\frac{(v-u)x}{2D}\right] + \left(\frac{u-v}{u+v}\right)\exp\left[\frac{(v+u)x-2uL}{2D}\right]}{\frac{u+v}{2v} - \frac{(u-v)^2}{2v(u+v)}\exp\left(-\frac{uL}{D}\right)}$$

with

$$u = v\sqrt{1 + \frac{4\mu D}{v^2}}$$

Furthermore, according to van Genuchten and Alves, solutions involving infinite series, such as the one we are dealing with in this Appendix, converges slowly for relatively large values of the Peclet number, $P = vL/D$. For relatively large Peclet number problems, they provided an approximate solution, given hereafter.

Approximate solution

The approximate solution of this BVP is given by

$$\begin{aligned} A(x,t) = \exp(-\mu t/R)\Bigg\{ & 1 - \frac{1}{2}\operatorname{erfc}\left[\frac{Rx - vt}{2(DRt)^{1/2}}\right] \\ & - \left(\frac{v^2 t}{\pi DR}\right)^{1/2} \exp\left[-\frac{(Rx - vt)^2}{4DRt}\right] \\ & + \frac{1}{2}\left[1 + \frac{vx}{D} + \frac{v^2 t}{DR}\right] \exp(vx/D)\operatorname{erfc}\left[\frac{Rx + vt}{2(DRt)^{1/2}}\right] \\ & - \left(\frac{4v^2 t}{\pi DR}\right)^{1/2}\left[1 + \frac{v}{4D}\left(2L - x + \frac{vt}{R}\right)\right] \\ & \exp\left[\frac{vL}{D} - \frac{R}{4Dt}\left(2L - x + \frac{vt}{R}\right)^2\right] \\ & + \frac{v}{D}\left[2L - x + \frac{3vt}{2R} + \frac{v}{4D}\left(2L - x + \frac{vt}{R}\right)^2\right] \\ & \exp(vL/D)\operatorname{erfc}\left[\frac{R(2L - x) + vt}{2(DRt)^{1/2}}\right]\Bigg\} \end{aligned}$$

$$B(x,t) = B_3(x,t)/B_4(x)$$

where

$$
\begin{aligned}
B_3(x,t) = {} & \frac{v}{(v+u)} \exp\left[\frac{(v-u)\,x}{2D}\right] \operatorname{erfc}\left[\frac{Rx-ut}{2(DRt)^{1/2}}\right] \\
& + \frac{v}{(v-u)} \exp\left[\frac{(v+u)\,x}{2D}\right] \operatorname{erfc}\left[\frac{Rx+ut}{2(DRt)^{1/2}}\right] \\
& + \frac{v^2}{2\mu D} \exp\left(\frac{vx}{D}-\frac{\mu t}{R}\right) \operatorname{erfc}\left[\frac{Rx+vt}{2(DRt)^{1/2}}\right] \\
& + \frac{v^2}{2\mu D}\left[\frac{v(2L-x)}{D}+\frac{v^2 t}{DR}+3+\frac{v^2}{\mu D}\right] \\
& \exp\left(\frac{vL}{D}-\frac{\mu t}{R}\right) \operatorname{erfc}\left[\frac{R(2L-x)+vt}{2(DRt)^{1/2}}\right] \\
& - \frac{v^3}{\mu D}\left(\frac{t}{\pi DR}\right)^{1/2} \exp\left[\frac{vL}{D}-\frac{\mu t}{R}-\frac{R}{4Dt}\left(2L-x+\frac{vt}{R}\right)^2\right] \\
& + \frac{v\,(u-v)}{(u+v)^2} \exp\left[\frac{(v+u)x-2uL}{2D}\right] \operatorname{erfc}\left[\frac{R(2L-x)-ut}{2(DRt)^{1/2}}\right] \\
& - \frac{v\,(u+v)}{(u-v)^2} \exp\left[\frac{(v-u)x+2uL}{2D}\right] \operatorname{erfc}\left[\frac{R(2L-x)+ut}{2(DRt)^{1/2}}\right]
\end{aligned}
$$

$$
B_4(x) = 1 - \frac{(u-v)^2}{(u+v)^2} \exp\left(-uL/D\right)
$$

According to van Genuchten and Alves, this approximate solution is accurate if it meets any of these two conditions:

$$
\frac{vL}{D} > 5 + 40\frac{vt}{RL}, \quad \text{or} \quad \frac{vL}{D} > 100
$$

▪ Problems

1. Solve the boundary value problem in ▶ Sect. 14.2.3 by dividing the problem into two sub-problems, such that $c(x,t) = c_1(x,t) + c_2(x,t)$, with each having one of its boundaries being homogeneous.
2. Solve the steady state case in ▶ Sect. 14.3.1 with the following boundary conditions:

$$
\left.\frac{dc}{dx}\right|_{x=0} = q, \qquad c\,(L) = 0
$$

3. Solve the boundary value problem in ▶ Sect. 14.3.3 for a semi-infinite domain.
   Hint: Use the limit technique.
4. Verify the computed results of Example 14.4 (▶ Sect. 14.4.2) with the analytical solution given in the Appendix.
5. Solve the boundary value problem of ▶ Sect. 14.4.5 for a Neumann boundary condition at $x = 0$, while the domain is insulated at $x = L$.

14

6. Compare the computed concentrations obtained from the eigenfunction expansion in ▶ Sect. 14.4.5 and the diagonalization method of ▶ Sect. 14.4.6. Which method is more appealing, and why?
7. Go through the compendium of van Genuchten and Alves (available on the internet) and compare their solutions with the corresponding solutions given in this chapter. Explain the following:
   (a) Why solutions derived via the Fourier method are simpler than the given solutions in the compendium, which were derived using Laplace transform?
   (b) What is the main advantage of spectral analysis compared to Laplace transform?
   (c) What is the main shortcoming of spectral analysis compared to Laplace transform?

Reference

Van Genuchten MTh (1982) Analytical solutions of one dimensional convective dispersive solute transport equations. ResearchGate

Further Reading

Goltz M, Huang, J (2017) Analytical modeling of solute transport in groundwater. Wiley.

Van Genuchten MT (1981) Analytical solutions for chemical transport with simultaneous adsorption, zero-order production and first-order Decay. J Hydrol 49:213-233

White RE, Subramanian VR(2010) Computational methods in chemical engineering with maple. Springer

Two-Dimensional Contaminant Transport

R. Al Khoury and C. Kasbergen, *Analytical Subsurface Environment*,
https://doi.org/10.1007/978-3-031-93910-5_15

Contaminant transport occurs in many situations by pollutant spills which are accidentally released onto the ground surface. Factors such as the type of contaminant and soil characteristics influence how the contaminants infiltrate into the ground and spread and migrate into the groundwater, potentially affecting ecosystems and water resources. Modeling contaminant transport from a spill on the ground surface is crucial for understanding and predicting pollutants infiltrating and spreading in the subsurface environment. In this chapter, we model the infiltration of contaminants into the ground and their spreading with the groundwater flow. Contaminant transports with different attributes of production and decay are considered.

15.1 Mass Transport with No Production or Decay

Mass transport with no production or decay refers to a scenario where the concentration of a substance remains constant over time due to a balance between input and output mass of the substance. In this situation, there are no reactions or transformations affecting the substance concentration, making it a steady-state condition. This concept is commonly used in environmental and engineering contexts to describe systems where no net change in substance concentration occurs.

15.1.1 Circular Contaminant Spill on Ground Surface: Diffusion via Dirichlet

A circular contaminant spill on the ground surface describes an accidental release of pollutants or hazardous substances in an essentially circular pattern. The infiltration of the pollutant into the ground occurs effectively in a uniform axial-symmetric pattern in the radial direction. Using the polar coordinate system, this contamination process leads to infiltration, which is a function of r and z.

Here, we model diffusion caused by a contaminant spill on the ground surface with radius a (m), as schematically shown in Fig. 15.1. The Dirichlet boundary value problem of this case can be expressed as

$$D_r \frac{\partial^2 c}{\partial r^2} + \frac{D_r}{r}\frac{\partial c}{\partial r} + D_z \frac{\partial^2 c}{\partial z^2} - R\frac{\partial c}{\partial t} = 0$$

$$\begin{aligned} \text{IC}&: c(r, z, 0) = 0 \\ \text{BC}&: c(r, 0, t) = C_0(r, t), \quad 0 \le r \le a \\ & \quad c(\ell, z, t) = 0, \qquad \ell \gg a \end{aligned}$$

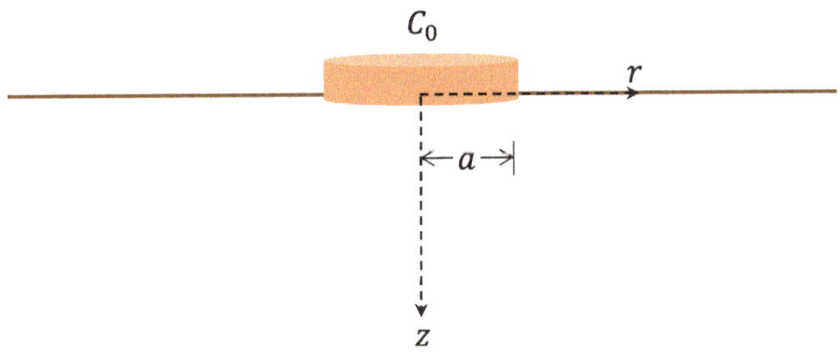

Fig. 15.1 A sketch of a circular spill on ground surface: Dirichlet problem

where D_r and D_z ($\mathrm{m^2 s^{-1}}$) are the hydrodynamic dispersion coefficients in the radial and vertical directions, respectively, R is a coefficient describing adsorption, and ℓ (m) is some far distance where the contaminant vanishes. C_0 ($\mathrm{kg\,m^{-3}}$) is a prescribed pollutant concentration on the ground surface, covering a circle with radius a.
Assuming $D_r = D_z = D$, and having $\alpha = D/R$, this BVP can be conveniently expressed as

$$\frac{\partial^2 c}{\partial r^2} + \frac{1}{r}\frac{\partial c}{\partial r} + \frac{\partial^2 c}{\partial z^2} - \frac{1}{\alpha}\frac{\partial c}{\partial t} = 0$$

$$\begin{aligned} &\text{IC}: \ c(r,z,0) = 0 \\ &\text{BC}: \ c(r,0,t) = C_0(r,t), \ 0 \le r \le a \\ &\qquad\ c(\ell,z,t) = 0, \qquad \ell \gg a \end{aligned} \tag{15.1}$$

General Solution

Applying the Fourier transform to (15.1), the boundary value problem in the frequency domain becomes

$$\frac{\partial^2 \hat{c}}{\partial r^2} + \frac{1}{r}\frac{\partial \hat{c}}{\partial r} + \frac{\partial^2 \hat{c}}{\partial z^2} - \frac{i\omega}{\alpha}\hat{c} = 0$$

$$\begin{aligned} &\text{BC}: \ \hat{c}(r,0,\omega) = \hat{C}_0(r,\omega), \ 0 \le r \le a \\ &\qquad\ \hat{c}(\ell,z,\omega) = 0, \qquad \ell >> a \end{aligned} \tag{15.2}$$

Using the separation of variables, we undertake

$$\hat{c}(r,z) = R(r)\,Z(z)$$

Taking the first and second derivatives of this function with respect to r and z gives

$$\frac{\partial \hat{c}}{\partial r} = Z(z)\frac{\partial R(r)}{\partial r}$$

$$\frac{\partial^2 \hat{c}}{\partial r^2} = Z(z)\frac{\partial^2 R(r)}{\partial r^2}$$

$$\frac{\partial^2 \hat{c}}{\partial z^2} = R(r)\frac{\partial^2 Z(z)}{\partial z^2}$$

Substituting these identities into (15.2) and dividing by $R(r)Z(z)$ gives

$$\frac{1}{R(r)}\frac{\partial^2 R(r)}{\partial r^2} + \frac{1}{R(r)}\frac{1}{r}\frac{\partial R(r)}{\partial r} + \frac{1}{Z(z)}\frac{\partial^2 Z(z)}{\partial z^2} - \frac{i\omega}{\alpha} = 0$$

This equation can be divided into two separate equations by equating them to some constant, say λ, yielding

$$\begin{aligned} &\frac{d^2R(r)}{dr^2} + \frac{1}{r}\frac{dR(r)}{dr} - \lambda R(r) = 0 \\ &\frac{d^2Z(z)}{dz^2} + (\lambda - \frac{i\omega}{\alpha})\, Z(z) = 0 \end{aligned} \tag{15.3}$$

The solution of these equations depends on λ:
If $\lambda = 0$, the general solution of $R(r)$ is

$$R(r) = c_1 + c_2 \ln r \tag{15.4}$$

If $\lambda = \kappa^2 > 0$, the solution becomes

$$R(r) = c_3 I_0(\kappa r) + c_4 K_0(\kappa r) \tag{15.5}$$

where I_0 and K_0 are the modified Bessel functions of the first and second order.
If $\lambda = -\kappa^2 < 0$, the general solution results in

$$R(r) = c_5 J_0(\kappa r) + c_6 Y_0(\kappa r) \tag{15.6}$$

where J_0 and Y_0 are the Bessel functions of the first and second order.

To solve the problem using the separation of variables together with the eigenfunction expansion, we seek a solution that is periodic. Accordingly, from these possible general solutions, we can dismiss (15.5) as inherently non-periodic, unless $c_3 = c_4 = 0$. For $\lambda = 0$, we must have $c_2 = 0$, because, otherwise, the solution in (15.4) becomes unbounded at $r = 0$. Similarly for (15.6), $c_6 = 0$, since $Y_0(\kappa r)$ is unbounded at $r = 0$, leading the general solution to become

15

$$\begin{aligned} R(r) &= c_1, && \lambda = 0 \\ R(r) &= c_5 J_0(\kappa r), && \lambda = -\kappa^2 \end{aligned}$$

Yet, since none of the roots of the Bessel function J_0 is zero, $\lambda = 0$ is not of use, giving

$$R(r) = c_5 J_0(\kappa r), \quad \lambda = -\kappa^2 \tag{15.7}$$

Having determined that only $\lambda = -\kappa^2$ eigenvalues are relevant, the differential equation of $Z(z)$ in (15.3) then becomes

$$\frac{d^2Z(z)}{dz^2} - \left(\kappa^2 + \frac{i\omega}{\alpha}\right) Z(z) = 0 \tag{15.8}$$

Considering that the domain goes to infinity in the z-direction, the solution of (15.8) is simply given by

$$Z(z) = c_7 e^{-\kappa_z z}, \quad \kappa_z = \sqrt{\kappa^2 + i\omega \big/ \alpha}$$

Collecting the two solutions of $R(r)$ and $Z(z)$, the general solution of the differential equation in (15.2) would then lead to

$$\hat{c} = A e^{-\kappa_z z} J_0(\kappa r), \quad \kappa_z = \sqrt{\kappa^2 + i\omega \big/ \alpha} \tag{15.9}$$

in which $A = c_5 c_7$.

Boundary Conditions

The second boundary condition in (15.2) implies that at some distance far from the spill, ℓ, the contaminant vanishes. This applies to all z, leading (15.9) to be expressed as

$$J_0(\kappa \ell) = 0 \tag{15.10}$$

This condition can be satisfied at the infinitely many positive roots μ_m of J_0 that implies $\kappa\ell = \mu_m$. Thus, $\kappa = \kappa_m = \mu_m/\ell$ represent the eigenvalues of the system, with

$$J_0(\kappa_m r) \equiv J_0\left(\frac{\mu_m}{\ell} r\right) \tag{15.11}$$

are their corresponding eigenfunctions.

Since the problem is linear, the superposition principle then entails

$$\hat{c}\,(r, z, \omega) = \sum\nolimits_m A_m e^{-\kappa_z z} J_0(\kappa_m r) \tag{15.12}$$

At $z = 0$, the first boundary condition in (15.2) implies

$$\sum\nolimits_m A_m J_0(\kappa_m r) = \hat{C}_0(r), \quad 0 \le r \le a$$

This is a Bessel-Fourier series with its coefficient determined by (see ▶ Chap. 2 and 3)

$$A_m = \frac{2}{\ell^2 J_1^2(\mu_m)} \int_0^a r\,\hat{C}_0(r)\, J_0(\kappa_m r)\, dr \tag{15.13}$$

BVP Solution

The solution of the BVP (15.2) for a given frequency ω_n is thus

$$\hat{c}\,(r, z, \omega_n) = \sum\nolimits_m A_m e^{-\kappa_z z} J_0(\kappa_m r)$$

with

$$\begin{aligned} A_m &= \frac{2}{\ell^2 J_1^2(\mu_m)} \int_0^a r\,\hat{C}_0(r)\,J_0(\kappa_m r)\,dr \\ \kappa_m &= \frac{\mu_m}{\ell} \\ \kappa_z &= \sqrt{\kappa_m^2 + \frac{i\omega_n}{\alpha}} \end{aligned} \tag{15.14}$$

In the special case where $\hat{C}_0$ is constant, the Fourier-Bessel coefficient becomes

$$A_m = \frac{2a\hat{C}_0}{\ell\mu_m J_1^2(\mu_m)} J_1(\kappa_m a) \tag{15.15}$$

Having done that, the solution in the time domain can readily be obtained using the inverse fast Fourier transform (IFFT) algorithm, as

$$c\,(r, z, t) = \sum\nolimits_n \hat{c}\,(r, z, \omega_n)\, e^{i\omega_n t}$$

■ **Example 15.1: Circular spill**

A circular contaminant, 0.5 m in radius, has been accidentally spilled on a soil surface that released a pollutant with a time varying concentration $C_0(t)$ shown in ◘ Fig. 15.2. The soil mass has a bulk density 1500 kg/m^3 and porosity 0.35. The pollutant is characterized by a distribution coefficient 0.1 m^3/kg and a hydrodynamic dispersion coefficient 1.5 m^2/d. Compute the concentration of the pollutant at the following (r, z) coordinates: (0.0, 0.25), (0.0, 0.5), (0.0, 1.0), (0.0, 1.5) m; as well as at (0.75, 0.25), (0.75, 0.5), (0.75, 1.0), (0.75, 1.5) m. Also show the contour plots for the contaminant growth after 30 days and 100 days.

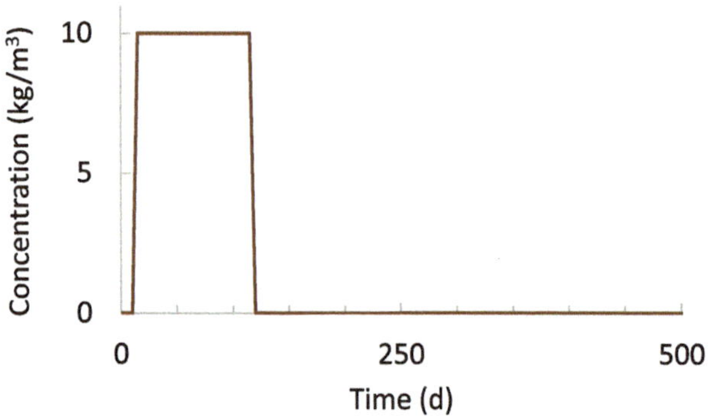

◘ **Fig. 15.2** Input spill concentration in the time domain

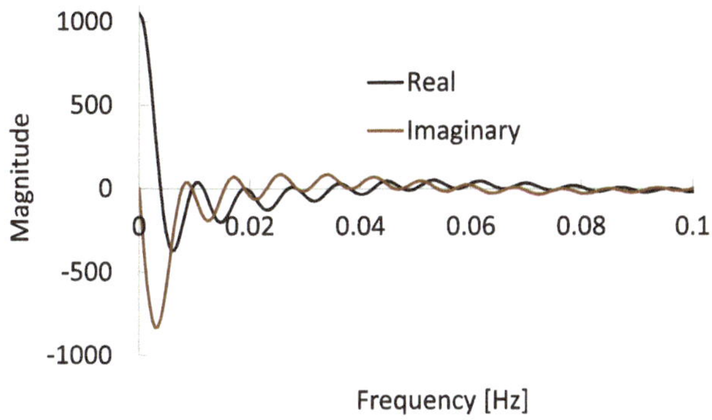

Fig. 15.3 Input spill signal in the frequency domain

Solution

We start the solution by transforming the time signal of the input concentration, given in Fig. 15.2, to the frequency domain using the forward fast Fourier transform (FFT) algorithm. The FFT is executed in this case using 2048 samples, with a 1 day sampling rate. The resulting frequency spectrum of the input signal is shown in Fig. 15.3. In light of this transformation, the magnitude of the input on each frequency $\hat{C}_0(\omega_n)$ is then used in (15.14), where the Fourier coefficient is determined via (15.15). Subsequently, upon calculating the pollutant concentration for all frequencies, the transformation back to the time domain is carried out using the inverse fast Fourier transform (IFFT).

The computational results for the first set of coordinates are given in Fig. 15.4, and for the second set of coordinates are given in Fig. 15.5. The two figures clearly show the time lag and concentration decay with further distances. Figure 15.6 displays the contour plots after 30 days and after 100 days. This figure gives a clear indication of the spill transport in the ground with time.

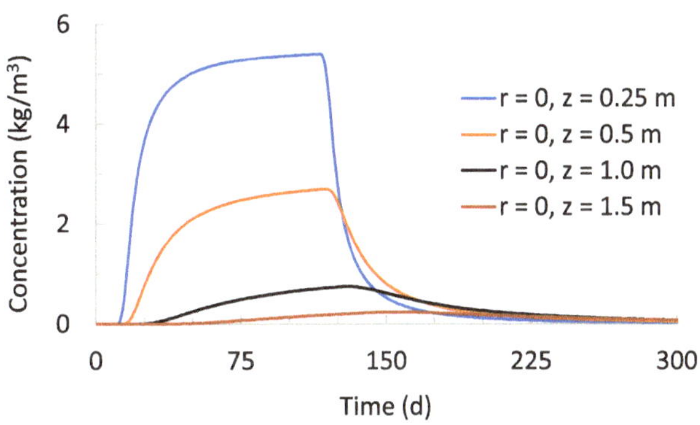

Fig. 15.4 Computed concentration at $r = 0$ m with different depths

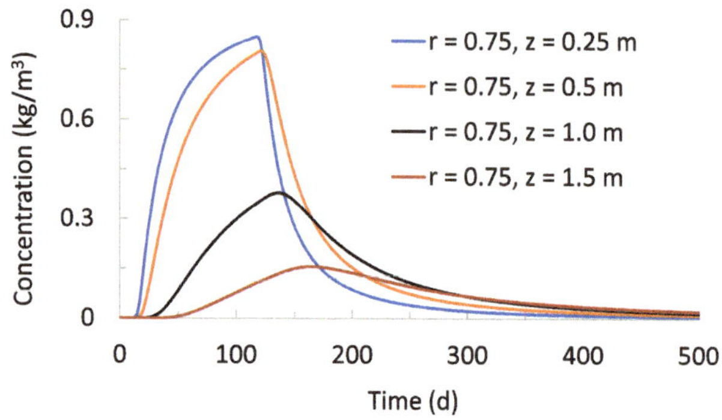

Fig. 15.5 Computed concentration at $r = 0.75$ m with different depths

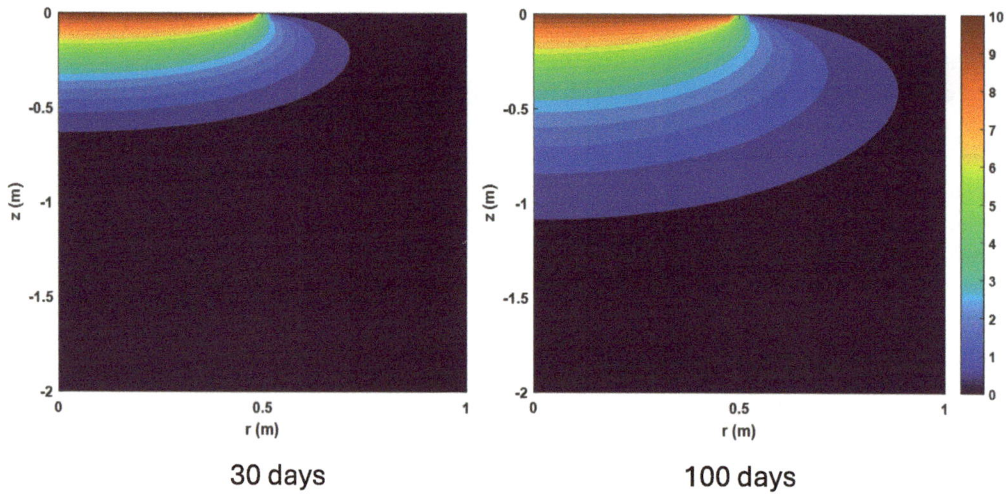

Fig. 15.6 Contour plots of the species concentration after 30 days (Left) and after 100 days (Right)

15

15.1.2 Circular Contaminant Spill on Ground Surface: Diffusion via Neumann

Here, we simulate the diffusion of the contaminant spill scenario which we addressed in the previous section, but with a Neumann boundary condition. Figure 15.7 shows schematically the spill and the geometry.

The boundary value problem of this scenario, assuming $D_r = D_z = D$ and $\alpha = D/R$, can be expressed as

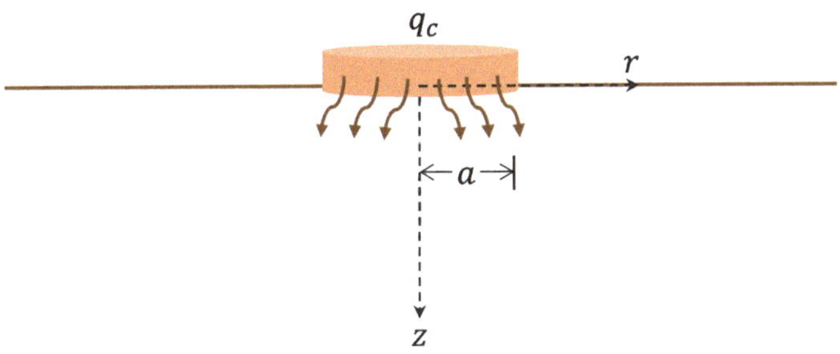

Fig. 15.7 A sketch of a circular spill on ground surface: Neumann problem

$$\frac{\partial^2 c}{\partial r^2} + \frac{1}{r}\frac{\partial c}{\partial r} + \frac{\partial^2 c}{\partial z^2} - \frac{1}{\alpha}\frac{\partial c}{\partial t} = 0$$

$$\begin{aligned} &\text{IC:} \quad c(r,z,0) = 0 \\ &\text{BC:} \quad -D\pi a^2 \left.\frac{\partial c}{\partial z}\right|_{z=0} = q_c(r,t), \; 0 \le r \le a \\ &\qquad c(\ell, z, t) = 0, \qquad \ell >> a \end{aligned} \tag{15.16}$$

where a (m) is the radius of the contaminant circle, and ℓ (m) is some far distance, where the contaminant vanishes. q_c (kgs^{-1}) is an applied pollutant flux on the ground surface, covering a circle with radius a.

BVP Solution

Applying the Fourier transform to (15.16) results in

$$\begin{aligned} &\frac{\partial^2 \hat{c}}{\partial r^2} + \frac{1}{r}\frac{\partial \hat{c}}{\partial r} + \frac{\partial^2 \hat{c}}{\partial z^2} - \frac{i\omega}{\alpha}\hat{c} = 0 \\ &\text{BC:} \quad -D\pi a^2 \left.\frac{\partial \hat{c}}{\partial z}\right|_{z=0} = \hat{q}_c(r,\omega), \; 0 \le r \le a \\ &\qquad \hat{c}(\ell, z, \omega) = 0, \qquad \ell \gg a \end{aligned} \tag{15.17}$$

Using the separation of variables, and following the procedure given in the previous case, the solution of this PDE, considering the second boundary condition, is

$$\begin{aligned} &\hat{c}(r,z,\omega) = \sum\nolimits_m A_m e^{-\kappa_z z} J_0(\kappa_m r) \\ &\kappa_m = \frac{\mu_m}{\ell} \\ &\kappa_z = \sqrt{\kappa_m^2 + \frac{i\omega}{\alpha}} \end{aligned} \tag{15.18}$$

where μ_m are the positive roots of the Bessel function J_0. Taking the derivative of (15.18) with respect to z gives

$$\frac{\partial \hat{c}}{\partial z} = \sum\nolimits_m -\kappa_z A_m e^{-i\kappa_z z} J_0(\kappa_m r)$$

At $z = 0$, the first boundary condition in (15.17) implies

$$\sum_m D\pi a^2 \kappa_z A_m J_0(\kappa_m r) = \hat{q}_c(r,\omega), \quad 0 \le r \le a$$

This is a Bessel-Fourier series with its coefficients determined by

$$A_m = \frac{2}{\kappa_z D\pi a^2 \ell^2 J_1^2(\mu_m)} \int_0^a r\,\hat{q}_c(r,\omega)\,J_0(\kappa_m r)\,dr$$

Having determined the coefficients, the solution of the boundary value problem (15.17) for a given frequency ω_n is thus

$$\hat{c}\,(r,z,\omega_n) = \sum_m A_m e^{-\kappa_z z} J_0(\kappa_m r)$$

with

$$A_m = \frac{2}{\kappa_z D\pi a^2 \ell^2 J_1^2(\mu_m)} \int_0^a r\,\hat{q}_c(r,\omega)\,J_0(\kappa_m r)dr \tag{15.19}$$

$$\kappa_m = \frac{\mu_m}{l}, \quad \kappa_z = \sqrt{\kappa_m^2 + \frac{i\omega_n}{\alpha}}$$

In the special case where $\hat{q}_c$ is spatially constant, the Fourier-Bessel coefficients become

$$A_m = \frac{2\hat{q}_c(\omega)}{\kappa_z D\pi a\, l\, \mu_m J_1^2(\mu_m)} J_1(\kappa_m a)$$

Having the solution in the frequency domain, the solution in the time domain can readily be reconstructed using the inverse fast Fourier transform (IFFT) algorithm, as

$$c\,(r,z,t) = \sum_n \hat{c}\,(r,z,\omega_n) e^{i\omega_n t}$$

15

15.1.3 Diffusion-Advection Contaminant Transport in 2D Planar Domain

In this section, we model a contaminant transport in a two-dimensional planar domain subjected to a circular contaminant source with radius a. A groundwater moving with velocity $-v$ (ms^{-1}) along the x-axis passes through the contaminant source and dispersing it via diffusion and advection, as schematically shown in Fig. 15.8.

The boundary value problem of this case with a Dirichlet boundary condition can be expressed as

$$D_x \frac{\partial^2 c}{\partial x^2} + D_y \frac{\partial^2 c}{\partial y^2} + v\frac{\partial c}{\partial x} - R\frac{\partial c}{\partial t} = 0$$

$$\begin{aligned} &\text{IC}: \ c\,(x,y,t) = 0 \\ &\text{BC}: \ c\,(x,y,t)|_{(x^2+y^2=a^2)} = C_0(t) \\ &\qquad c\,(x,y,t)|_{(a,\theta)} = c\,(x,y,t)|_{(a,\theta+2\pi)}, \ \text{Periodicity condition} \end{aligned} \tag{15.20}$$

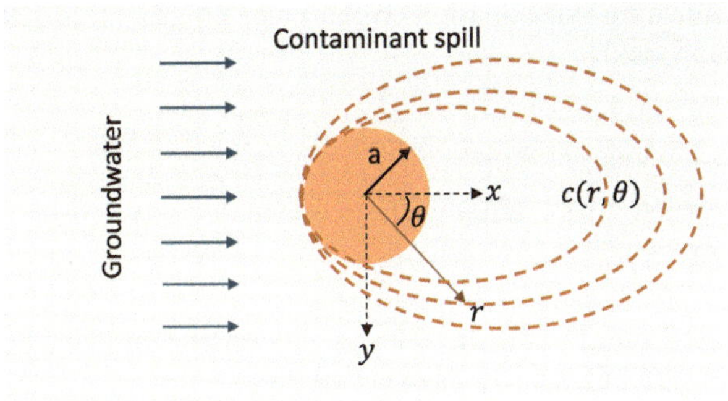

Fig. 15.8 A schematic presentation of contaminant transport in a planar domain

where D_x and D_y (m^2s^{-1}) are the hydrodynamic dispersion coefficients in the x- and y-directions, respectively, R is a coefficient describing adsorption, and C_0 (kg m^{-3}) is a prescribed pollutant concentration in the ground surface, covering a circle with radius $a = \sqrt{x^2 + y^2}$.

If we assume $D_x = D_y = D$, the PDE in (15.20) can be written as

$$\frac{\partial^2 c}{\partial x^2} + \frac{\partial^2 c}{\partial y^2} + \frac{v}{D}\frac{\partial c}{\partial x} - \frac{1}{\alpha}\frac{\partial c}{\partial t} = 0, \qquad \alpha = \frac{D}{R} \tag{15.21}$$

Since the contaminant source is circular, it is convenient to express (15.21) in the polar coordinate system. However, the presence of the third component in this equation complicates the conversion from the Cartesian to the polar coordinate system. To facilitate this conversion, following the approach of Carslaw and Jaeger (1959), we propose a trial solution for (15.21) as

$$c\,(x, y, t) = u\,(x, y, t)\, e^{\eta x} \tag{15.22}$$

Taking the first and second derivatives of (15.22) with respect to x and y, and then substituting them into (15.21) gives

$$\frac{\partial^2 u}{\partial x^2} + \frac{\partial^2 u}{\partial y^2} + \left(2\eta + \frac{v}{D}\right)\frac{\partial u}{\partial x} + \left(\eta^2 + \frac{v}{D}\eta\right) u - \frac{1}{\alpha}\frac{\partial u}{\partial t} = 0 \tag{15.23}$$

The spatial derivative on the third term of this equation can be eliminated if we let $\eta = -v/2D$, leading the trial solution (15.22) to become

$$c\,(x, y, t) = u\,(x, y, t)\, e^{-vx/2D} \tag{15.24}$$

and the PDE (15.23) to reduce to

$$\frac{\partial^2 u}{\partial x^2} + \frac{\partial^2 u}{\partial y^2} - \frac{v^2}{4D^2}u - \frac{1}{\alpha}\frac{\partial u}{\partial t} = 0 \tag{15.25}$$

Transforming this equation to the polar coordinate system in xy-plane yields

$$\frac{\partial^2 u}{\partial r^2}+\frac{1}{r}\frac{\partial u}{\partial r}+\frac{1}{r^2}\frac{\partial^2 u}{\partial \theta^2}-\frac{v^2}{4D^2}u-\frac{1}{\alpha}\frac{\partial u}{\partial t}=0 \tag{15.26}$$

for which we employed $x=r\cos\theta$, $y=r\sin\theta$ and $r=\sqrt{x^2+y^2}$.

Based on (15.24), the relevant Dirichlet boundary condition for (15.26) at $x=a\cos\theta$ is therefore

$$u(a,\theta,t)\,e^{-va\cos(\theta)/2D}=C_0(t) \tag{15.27}$$

Similarly, the periodic boundary condition in (15.20) implies

$$u(a,\theta,t)=u(a,\theta+2\pi,t) \tag{15.28}$$

General Solution

Applying the Fourier transform to (15.26) yields

$$\frac{\partial^2 \hat{u}}{\partial r^2}+\frac{1}{r}\frac{\partial \hat{u}}{\partial r}+\frac{1}{r^2}\frac{\partial^2 \hat{u}}{\partial \theta^2}-\kappa^2\hat{u}=0,\quad \kappa^2=\frac{v^2}{4\alpha^2}+\frac{i\omega}{\alpha} \tag{15.29}$$

where $\hat{u}\equiv\hat{u}(r,\theta,\omega)$, with ω the angular frequency. The solution of this differential equation can be obtained using the separation of variables, which can be expressed as

$$\hat{u}(r,\theta)=R(r)\,F(\theta)\equiv RF \tag{15.30}$$

Substituting this formula into (15.29), dividing by $R(r)F(\theta)$ and equating both terms to a constant, say λ, gives

$$\frac{r^2}{R}\frac{d^2R}{dr^2}+\frac{r}{R}\frac{dR}{dr}-r^2\kappa^2=-\frac{1}{F}\frac{d^2F}{d\theta^2}=\lambda \tag{15.31}$$

The relevant F-dependent boundary value problem in (15.31) is

$$F''+\lambda F=0,\quad F(\theta)=F(\theta+2\pi) \tag{15.32}$$

which is evidently 2π-periodic. This problem is not a regular Sturm-Liouville, yet, due to the periodicity, it generates eigenvalues and eigenfunctions. The three possible general solutions of (15.32) are

$$\begin{aligned}
&F(\theta)=c_1+c_2\theta, && \lambda=0\\
&F(\theta)=c_1\cosh(m\theta)+c_2\sinh(m\theta), && \lambda=-m^2<0\\
&F(\theta)=c_1\cos(m\theta)+c_2\sin(m\theta), && \lambda=m^2>0
\end{aligned} \tag{15.33}$$

Apparently the second solution in (15.33) is not periodic and thus can be dismissed. The first solution is periodic if $c_2=0$, with the remaining solution $F(\theta)=c_1, c_1\neq 0$. Hence, $\lambda=0$ is an eigenvalue. Finally, the third solution in (15.33) can be 2π -periodic if we take $=1,2,\ldots$. Hence, considering $\lambda=0$, the eigenvalues of this BVP are $\lambda_m = m^2, m = 0, 1, 2, \ldots$, leading the solution to be expressed as

$$F(\theta) = \sum_m a_m \cos(m\theta) + b_m \sin(m\theta), \qquad m = 0,\ 1,\ 2, \cdots \tag{15.34}$$

Letting $\lambda = m^2$, the R-dependent differential equation in (15.31) is given by

$$r^2 \frac{d^2 R}{dr^2} + r \frac{dR}{dr} - (r^2 \kappa^2 + m^2) R = 0$$

The solution of this ordinary differential equation is expressed in terms of the modified Bessel functions of first kind $I_m(\kappa r)$, and second kind $K_m(\kappa r)$ (Abramowitz and Stegun 1972). However, since $I_m(\kappa r)$ is unbounded at far distances, the solution reduces to

$$R(r, \omega) = K_m(\kappa r) \tag{15.35}$$

Substituting (15.34) and (15.35) into (15.30) leads to

$$\hat{u}(r, \theta, \omega) = \sum_m [a_m \cos(m\theta) + b_m \sin(m\theta)]\, K_m(\kappa r), \qquad m = 0,\ 1,\ 2, \cdots \tag{15.36}$$

Boundary Condition

Applying the Fourier transform to the boundary condition in (15.27) gives

$$\hat{u}(a, \theta, \omega) = \hat{C}_0(\omega)\, e^{v\, a \cos(\theta)/\, 2D} \tag{15.37}$$

At $r = a$, (15.36) becomes

$$\hat{u}(a, \theta, \omega) = \sum_m [a_m \cos(m\theta) + b_m \sin(m\theta)]\, K_m(\kappa a) \tag{15.38}$$

Equating (15.37) to (15.38) yields

$$\hat{C}_0(\omega)\, e^{v\, a \cos(\theta)/\, 2D} = \sum_m [a_m \cos(m\theta) + b_m \sin(m\theta)]\, K_m(\kappa a) \tag{15.39}$$

This equation is a trigonometric Fourier series with its coefficients expressed as

$$\begin{aligned}
a_0 &= \frac{\hat{C}_0(\omega)}{K_0(\kappa a)} \frac{1}{2\pi} \int_0^{2\pi} e^{v\, a \cos(\theta)/\, 2D} d\theta, && m = 0 \\
a_m &= \frac{\hat{C}_0(\omega)}{K_m(\kappa a)} \frac{1}{\pi} \int_0^{2\pi} e^{v\, a \cos(\theta)/\, 2D} \cos(m\theta)\, d\theta, && m \ge 1 \\
b_m &= \frac{\hat{C}_0(\omega)}{K_m(\kappa a)} \frac{1}{\pi} \int_0^{2\pi} e^{v\, a \cos(\theta)/\, 2D} \sin(m\theta)\, d\theta, && m \ge 1 \\
&= 0
\end{aligned} \tag{15.40}$$

Since $\hat{C}_0$ is constant (dependent only on frequency), the integral of the b_m coefficients is zero for all m. Conversely, the integrals for the a_m coefficients have closed-form expressions, represented in terms of the modified Bessel function of the first kind (Abramowitz and Stegun 1972, p. 376), such that

$$\begin{aligned} a_0 &= \frac{\hat{C}_0(\omega)}{K_0(\kappa a)} I_0\left(\frac{v\,a}{2D}\right), & m &= 0 \\ a_m &= \frac{2\hat{C}_0(\omega)}{K_m(\kappa a)} I_m\left(\frac{v\,a}{2D}\right), & m &\geq 1 \end{aligned} \tag{15.41}$$

in which I_m is the modified Bessel function of the first kind of order m.

BVP Solution

Following the above derivation, the solution of the boundary value problem for a given frequency ω_n can then be expressed as

$$\hat{u}\,(r,\theta,\omega_n) = \sum_m \varepsilon_m \frac{\hat{C}_0(\omega)}{K_m(ka)} I_m\left(\frac{va}{2D}\right) \cos\,(m\theta)\, K_m(\kappa r)$$

with

$$\varepsilon_m = \begin{cases} 1 & m = 0 \\ 2 & m \geq 1 \end{cases}, \qquad \kappa = \sqrt{\frac{v^2}{4D^2} + \frac{i\omega_n}{\alpha}} \tag{15.42}$$

Applying the inverse Fourier transform, the solution in the time domain can be retrieved via

$$u\,(r,\theta,t) = \sum_n \hat{u}\,(r,\theta,\omega_n)\, e^{i\omega_n t}$$

Having solved the problem for $u(r,\theta,t)$, the contaminant concentration in the domain can then be determined via

15

$$c\,(r,\theta,t) = u\,(r,\theta,t)\, e^{-v\,r\cos\theta/\,2D} \tag{15.43}$$

15.2 Mass Transport with Zero-Order Production in a 2D Planar Domain

Mass transport with zero-order production refers to a chemical reaction where the rate of product formation remains constant regardless of the reactant concentration, resulting in a linear increase in product over time.

The boundary value problem of a contaminant transport with zero-order production in an infinite domain, moving with velocity $-v$ along the x-axis, assuming $D_x = D_y = D$, can be expressed as

$$\frac{\partial^2 c}{\partial x^2}+\frac{\partial^2 c}{\partial y^2}+\frac{v}{D}\frac{\partial c}{\partial x}-\frac{1}{\alpha}\frac{\partial c}{\partial t}+\frac{\dot{m}}{D}=0, \qquad \alpha=\frac{D}{R}$$

$$\begin{aligned} \text{IC}: &\ c(x,y,t)=0 \\ \text{BC}: &\ c(x,y,t)|_{(x^2+y^2=a^2)}=C_0(t) \\ &\ c(x,y,t)|_{(a,\theta)}=c(x,y,t)|_{(a,\theta+2\pi)} \end{aligned} \tag{15.44}$$

where $\dot{m}$ is the production rate of the species.

General Solution

As for the previous case, expressing the trial solution as $c(x,y,t)=u(x,y,t)e^{\eta x}$, and carrying out the derivation, we obtain

$$c(x,y,t)=u(x,y,t)\,e^{-vx/2D} \tag{15.45}$$

and

$$\frac{\partial^2 u}{\partial x^2}+\frac{\partial^2 u}{\partial y^2}-\frac{v^2}{4D^2}u-\frac{1}{\alpha}\frac{\partial u}{\partial t}+\frac{\dot{m}}{D}e^{vx/2D}=0$$

Transforming this equation to the polar coordinate system in the xy-plane yields

$$\frac{\partial^2 u}{\partial r^2}+\frac{1}{r}\frac{\partial u}{\partial r}+\frac{1}{r^2}\frac{\partial^2 u}{\partial \theta^2}-\frac{v^2}{4D^2}u-\frac{1}{\alpha}\frac{\partial u}{\partial t}+\frac{\dot{m}}{D}e^{vr\cos(\theta)/2D}=0 \tag{15.46}$$

This is a nonhomogeneous equation due to the presence of $\dot{m}$. As discussed before, the solution of nonhomogeneous differential equations can be attained by solving the associated homogeneous and particular equations, such that

$$u=u_h+u_p \tag{15.47}$$

The corresponding homogeneous equation of (15.46) is

$$\frac{\partial^2 u_h}{\partial r^2}+\frac{1}{r}\frac{\partial u_h}{\partial r}+\frac{1}{r^2}\frac{\partial^2 u_h}{\partial \theta^2}-\frac{v^2}{4D^2}u_h-\frac{1}{\alpha}\frac{\partial u_h}{\partial t}=0 \tag{15.48}$$

Applying the Fourier transform to this partial differential equation leads to

$$\frac{\partial^2 \hat{u}_h}{\partial r^2}+\frac{1}{r}\frac{\partial \hat{u}_h}{\partial r}+\frac{1}{r^2}\frac{\partial^2 \hat{u}_h}{\partial \theta^2}-\kappa^2\hat{u}_h=0, \qquad \kappa^2=\frac{v^2}{4D^2}+\frac{i\omega}{\alpha}$$

The general solution of this equation, considering the periodicity condition in (15.44), has been undertaken in detail in the previous section. It results in

$$\hat{u}_h(r,\theta,\omega)=\sum_m\left[a_m\cos(m\theta)+b_m\sin(m\theta)\right]K_m(\kappa r), \qquad m=0,\ 1,\ 2,\cdots \tag{15.49}$$

The particular part of (15.46) is

$$\frac{\partial^2 u_p}{\partial r^2}+\frac{1}{r}\frac{\partial u_p}{\partial r}+\frac{1}{r^2}\frac{\partial^2 u_p}{\partial \theta^2}-\frac{v^2}{4D^2}u_p-\frac{1}{\alpha}\frac{\partial u_p}{\partial t}+\frac{\dot{m}}{D}e^{v\,r\cos(\theta)/2D}=0 \tag{15.50}$$

Applying the Fourier transform to this equation yields

$$\frac{\partial^2 \hat{u}_p}{\partial r^2}+\frac{1}{r}\frac{\partial \hat{u}_p}{\partial r}+\frac{1}{r^2}\frac{\partial^2 \hat{u}_p}{\partial \theta^2}-\kappa^2\hat{u}_p+\frac{\dot{m}}{D}e^{v\,r\cos(\theta)/2D}=0, \qquad \kappa^2=\frac{v^2}{4D^2}+\frac{i\omega}{\alpha} \tag{15.51}$$

Since the nonhomogeneous term is exponential, the trial solution may have the form

$$\hat{u}_p=Ae^{v\,r\cos(\theta)/2D} \tag{15.52}$$

Substituting this trial solution into (15.51) and processing the algebra, it leads to

$$A=\frac{4\dot{m}D}{4\kappa^2D^2-v^2}$$

Substituting this constant into (15.52) yields

$$\hat{u}_p=\frac{4\dot{m}D}{4\kappa^2D^2-v^2}e^{v\,r\cos(\theta)/2D} \tag{15.53}$$

Substituting (15.49) and (15.53) into (15.47), the general solution in the frequency domain can thus be expressed as

$$\hat{u}(r,\theta,\omega)=\sum\nolimits_m\left[a_m\cos(m\theta)+b_m\sin(m\theta)\right]K_m(\kappa r)+\frac{4\dot{m}D}{4\kappa^2D^2-v^2}e^{v\,r\cos(\theta)/2D} \tag{15.54}$$

Boundary Conditions

By applying the Fourier transform to the boundary condition in (15.44) and utilizing (15.45), we can express the boundary condition in the $\hat{u}$-space, as

$$\hat{u}(r,\theta,\omega)\,e^{-v\,r\cos(\theta)/2D}\Big|_{r=a}=\hat{C}_0(\omega) \tag{15.55}$$

Substituting (15.54) into (15.55) and solving for $r=a$, we obtain

$$\sum\nolimits_m\left[a_m\cos(m\theta)+b_m\sin(m\theta)\right]K_m(\kappa a)+\frac{4\dot{m}D}{4\kappa^2D^2-v^2}e^{v\,a\cos(\theta)/2D}=\hat{C}_0(\omega)e^{v\,a\cos(\theta)/2D}$$

This can be written as

$$\sum\nolimits_m a_m\cos(m\theta)+b_m\sin(m\theta)=\frac{f(\theta)}{K_m(\kappa a)}$$

with

$$f(\theta)=\hat{G}(\omega)\,e^{v\,a\cos(\theta)/2D},\quad \hat{G}(\omega)=\hat{C}_0(\omega)-\frac{4\dot{m}D}{4\kappa^2D^2-v^2} \tag{15.56}$$

15

which is a typical trigonometric Fourier series with its coefficients expressed as

$$
\begin{aligned}
a_0 &= \frac{\hat{G}(\omega)}{2\pi K_0(\kappa a)} \int_0^{2\pi} e^{\nu a \cos(\theta)/2D} d\theta, && m = 0 \\
a_m &= \frac{\hat{G}(\omega)}{\pi K_m(\kappa a)} \int_0^{2\pi} e^{\nu a \cos(\theta)/2D} \cos(m\theta)\, d\theta, && m \geq 1 \\
b_m &= \frac{\hat{G}(\omega)}{\pi K_m(\kappa a)} \int_0^{2\pi} e^{\nu a \cos(\theta)/2D} \sin(m\theta)\, d\theta, && m \geq 1
\end{aligned}
\tag{15.57}
$$

where for a constant $\hat{C}_0$ (not being a function of r), the integral of the b_m coefficients vanish for all m. However, the integrals in a_m coefficients have a closed-form solution, expressed in terms of the modified Bessel function of the first kind of 0 order (I_0) and m order (I_m), such that

$$
\begin{aligned}
a_0 &= \frac{\hat{G}(\omega)}{K_0(\kappa a)} I_0\left(\frac{\nu a}{2D}\right), && m = 0 \\
a_m &= \frac{2\hat{G}(\omega)}{K_m(\kappa a)} I_m\left(\frac{\nu a}{2D}\right), && m \geq 1
\end{aligned}
\tag{15.58}
$$

BVP Solution

Inserting (15.58) into (15.54) yields the solution of this boundary value problem for any given frequency ω_n, as

$$
\begin{aligned}
&\hat{u}(r, \theta, \omega_n) = \sum\nolimits_m \varepsilon_m \frac{\hat{G}(\omega_n)}{K_m(\kappa a)} I_m\left(\frac{\nu a}{2D}\right) \cos(m\theta)\, K_m(\kappa r) + \frac{4\dot{m}D}{4\kappa^2 D^2 - \nu^2} e^{\nu r \cos(\theta)/2D} \\
&\text{with} \\
&\varepsilon_m = \begin{cases} 1 & m = 0 \\ 2 & m \geq 1 \end{cases}, \\
&\hat{G}(\omega_n) = \hat{C}_0(\omega_n) - \frac{4\dot{m}D}{4\kappa^2 D^2 - \nu^2} \\
&\kappa^2 = \frac{\nu^2}{4D^2} + \frac{i\omega_n}{\alpha}
\end{aligned}
\tag{15.59}
$$

Applying the inverse Fourier transform produces the solution in the time domain, such that

$$
u(r, \theta, t) = \sum\nolimits_n \hat{u}(r, \theta, \omega_n)\, e^{i\omega_n t}
$$

Then, having reconstructed $u(r, \theta, t)$ to the time domain, the contaminant concentration in the time domain can readily be determined via

$$
c(r, \theta, t) = u(r, \theta, t)\, e^{-\nu r \cos(\theta)/2D} \tag{15.60}
$$

15.3 Mass Transport with Zero-Order Production and First-Order Decay in a 2D Planar Domain

Mass transport with zero-order production and first-order decay describes a process where the pollutant experiences a constant rate of production $\dot{m}$, and, simultaneously, a decay proportional to its concentration μc.

The boundary value problem of this type of contaminant transport in a 2D planar domain, moving with velocity $-v$ along the x-axis, can be expressed as

$$D_x \frac{\partial^2 c}{\partial x^2} + D_y \frac{\partial^2 c}{\partial y^2} + v \frac{\partial c}{\partial x} - R \frac{\partial c}{\partial t} - \mu c + \dot{m} = 0$$

$$\begin{aligned} \text{IC}: \ & c\,(x, y, t) = 0 \\ \text{BC}: \ & c\,(a, \theta, t) = C_0(t) \\ & c\,(a, \theta, t) = c\,(a, \theta + 2\pi, t) \end{aligned} \tag{15.61}$$

If we assume that $D_x = D_y = D$, the PDE in (15.61) can be reduced to

$$\frac{\partial^2 c}{\partial x^2} + \frac{\partial^2 c}{\partial y^2} + \frac{v}{D}\frac{\partial c}{\partial x} - \frac{1}{\alpha}\frac{\partial c}{\partial t} - \frac{\mu}{D} c + \frac{\dot{m}}{D} = 0, \qquad \alpha = \frac{D}{R}$$

General Solution

As for the previous case, expressing the involved state variable as $c(x, y, t) = u(x, y, t)e^{\eta x}$, carrying out the derivation, and then transforming the resulted equation to the polar coordinate system, we obtain

$$c\,(r, \theta, t) = u\,(r, \theta, t) e^{-v r \cos(\theta)/\,2D} \tag{15.62}$$

and

$$\frac{\partial^2 u}{\partial r^2} + \frac{1}{r}\frac{\partial u}{\partial r} + \frac{1}{r^2}\frac{\partial^2 u}{\partial \theta^2} - \left(\frac{\mu}{D} + \frac{v^2}{4D^2}\right) u - \frac{1}{\alpha}\frac{\partial u}{\partial t} + \frac{\dot{m}}{D} e^{v r \cos(\theta)/\,2D} = 0 \tag{15.63}$$

which is a nonhomogeneous equation, with its solution is given by

$$u = u_h + u_p \tag{15.64}$$

where u_h represents the homogeneous solution and u_p represents the particular solution.

The homogeneous part of (15.63) is

$$\frac{\partial^2 u_h}{\partial r^2} + \frac{1}{r}\frac{\partial u_h}{\partial r} + \frac{1}{r^2}\frac{\partial^2 u_h}{\partial \theta^2} - \left(\frac{\mu}{D} + \frac{v^2}{4D^2}\right) u_h - \frac{1}{\alpha}\frac{\partial u_h}{\partial t} = 0$$

Applying the Fourier transform to this PDE leads to

$$\frac{\partial^2 \hat{u}_h}{\partial r^2} + \frac{1}{r}\frac{\partial \hat{u}_h}{\partial r} + \frac{1}{r^2}\frac{\partial^2 \hat{u}_h}{\partial \theta^2} - \kappa^2 \hat{u}_h = 0, \qquad \kappa^2 = \frac{\mu}{D} + \frac{v^2}{4D^2} + \frac{i\omega}{\alpha} \tag{15.65}$$

The general solution of (15.65), considering the periodicity condition in (15.61), is given in detail in ▶ Sect. 15.1. It leads to

$$\hat{u}_h(r,\theta,\omega) = \sum\nolimits_m \left[a_m \cos(m\theta) + b_m \sin(m\theta)\right] K_m(\kappa r), \qquad m = 0,\ 1,\ 2, \cdots \tag{15.66}$$

The particular part of (15.63) is

$$\frac{\partial^2 u_p}{\partial r^2} + \frac{1}{r}\frac{\partial u_p}{\partial r} + \frac{1}{r^2}\frac{\partial^2 u_p}{\partial \theta^2} - \left(\frac{\mu}{D} + \frac{v^2}{4D^2}\right) u_p - \frac{1}{\alpha}\frac{\partial u_p}{\partial t} + \frac{\dot{m}}{D} e^{v\, r\cos(\theta)/\,2D} = 0 \tag{15.67}$$

Carrying out the Fourier transform and following the derivation procedure given in the previous case, we arrive at

$$\hat{u}_p = \frac{4\dot{m}D}{4\kappa^2 D^2 - v^2} e^{v\, r\cos(\theta)/2D} \tag{15.68}$$

Substituting (15.66) and (15.68) into (15.64), the general solution in the frequency domain becomes

$$\hat{u}\,(r,\theta,\omega) = \sum\nolimits_m \left[a_m \cos(m\theta) + b_m \sin(m\theta)\right] K_m(\kappa r) + \frac{4\dot{m}D}{4\kappa^2 D^2 - v^2} e^{v\, r\cos(\theta)/2D} \tag{15.69}$$

which, as can readily be noticed, resembles the case with zero-order production (15.59). The difference, however, lies in the definition of κ, see (15.65).

BVP Solution

As the boundary conditions of this case is similar to the boundary conditions of the previous case, and since the general solution (15.69) is also similar (except for κ), then the solution of this BVP in the frequency domain would lead to

$$\hat{u}\,(r,\theta,\omega_n) = \sum\nolimits_m \varepsilon_m \frac{\hat{G}(\omega_n)}{K_m(ka)} I_m(\frac{va}{2D}) \cos(m\theta)\, K_m(\kappa r) + \frac{4\dot{m}D}{4\kappa^2 D^2 - v^2} e^{v\, r\cos(\theta)/2D}$$

$$\varepsilon_m = \begin{cases} 1 & m = 0 \\ 2 & m \geq 1 \end{cases}$$

$$\hat{G}(\omega_n) = \hat{C}_0(\omega_n) - \frac{4\dot{m}D}{4\kappa^2 D^2 - v^2}, \quad \kappa^2 = \frac{\mu}{D} + \frac{v^2}{4D^2} + \frac{i\omega_n}{\alpha} \tag{15.70}$$

Applying the inverse Fourier transform to this equation gives

$$u\,(r,\theta,t) = \sum\nolimits_n \hat{u}\,(r,\theta,\omega_n)\, e^{i\omega_n t}$$

Having the solution of $u(r,\theta,t)$, the contaminant concentration in the time domain can be readily determined via

$$c\,(r,\theta,t) = u\,(r,\theta,t)\,e^{-v\,r\cos(\theta)/\,2D} \tag{15.71}$$

15.4 Mass Transport in 2D Planar Domain Due to an Infinite Line Contaminant Source

Although simulating the contamination source as a cylinder is more physically appropriate, some engineering literature models the contamination source as a line. This can be achieved by taking the limit of the cylindrical source as its radius approaches zero, i.e. $a \to 0$.

In Sects. 15.2 and 15.3, we determined that the solution to the contaminant transport problem due to a moving cylindrical source is given by

$$\hat{u}\,(r,\theta,\omega) = \begin{cases} \dfrac{\hat{G}(\omega)}{K_0(\kappa a)} I_0(\dfrac{va}{2D})\, K_0(\kappa r) + \dfrac{4\dot{m}D}{4\kappa^2 D^2 - v^2} e^{v\,r\cos(\theta)/2D}, & m = 0 \\ \sum_m 2\dfrac{\hat{G}(\omega)}{K_m(\kappa a)} I_m(\dfrac{va}{2D}) \cos(m\theta)\, K_m(\kappa r) + \dfrac{4\dot{m}D}{4\kappa^2 D^2 - v^2} e^{v\,r\cos(\theta)/2D}, & m = 1,\ 2, \cdots \end{cases} \tag{15.72}$$

As the radius of the contamination source approaches zero, the infinite cylindrical contaminant source would represent an infinite line source, which, in a plane, corresponds to a point source. To analyze this, we first define the limits of the terms in (15.72). As $a \to 0$, the following limits are relevant:

$$\begin{aligned} &\lim_{z\to 0} I_0(z) = 1 \\ &\lim_{z\to 0^+} K_0(z) \simeq -\ln(z) - \gamma + \ln(2) + O(z^2) \\ &\lim_{z\to 0^+} \frac{I_m(z)}{K_m(z)} = 0 \end{aligned} \tag{15.73}$$

where γ is the *Euler-Mascheroni constant*, approximately equal to 0.57721 (see Wolfram Alpha Webpage for the second limit approximation). The 0^+ in the second and third limiting cases indicates that z is a small argument.

Applying these limits to (15.72) reveals that the terms with $m > 0$ are all zero, and the solution to the moving infinite line contamination source for a given frequency ω_n leads to

$$\hat{u}\,(r,\theta,\omega_n) = -\frac{\hat{G}(\omega_n)}{\ln(\kappa a) + \gamma - \ln(2)} K_0(\kappa r) + \frac{4\dot{m}D}{4\kappa^2 D^2 - v^2} e^{v\,r\cos(\theta)/2D} \tag{15.74}$$

with κa, by definition, a small argument, depending on the accuracy of the computer. For a standard PC, a source radius

a as small as the limit of double–precision arithmetic (on the order of 1e-300) can be used. It is important to note that employing a small source radius a in the cylindrical formulation (15.72) yields results nearly identical to those obtained from (15.74). However, if the goal is to study the effect of the line source, the advantage of using the line source formulation lies in avoiding the summation over the Fourier series terms.

Having solved the problem in the frequency domain, applying the inverse fast Fourier transform would reconstruct the solution in the time domain, such that

$$u(r,\theta,t) = \sum_n \hat{u}(r,\theta,\omega_n)\, e^{i\omega_n t}$$

Having the solution of $u(r,\theta,t)$, the contaminant concentration in the time domain can be readily determined via

$$c(r,\theta,t) = u(r,\theta,t)e^{-v\, r \cos(\theta)/\,2D} \qquad (15.75)$$

Note that depending on the definition of κ, the solution of a line source for either the zero-order production (▶ Sect. 15.2) or the zero-order production and first-order decay (▶ Sect. 15.3) can be obtained, such that

$$\kappa^2 = \frac{v^2}{4D^2} + \frac{i\omega_n}{\alpha}, \qquad \text{zero - order production}$$

$$\kappa^2 = \frac{\mu}{D} + \frac{v^2}{4D^2} + \frac{i\omega_n}{\alpha}, \qquad \text{zero - order production and first - order decay}$$

15.5 Mass Transport with Zero-Order Production and First-Order Decay in a 2D Planar Domain: Robin Boundary Condition

In the context of mass transport with zero-order production and first-order decay in a 2D planar domain, the Robin boundary condition defines the exchange of the transported substance at the boundary of the contaminant source. This condition accounts for both diffusive flux and prescribed conditions at the boundaries, allowing for a comprehensive modeling of the system behavior including production, decay, and boundary interactions.

The boundary value problem of a contaminant transport in an infinite, homogeneous, isotropic domain, moving with velocity $-v$ along the x-axis, undergoing Robin boundary condition can be expressed as

$$\frac{\partial^2 c}{\partial x^2} + \frac{\partial^2 c}{\partial y^2} + \frac{v}{D}\frac{\partial c}{\partial x} - \frac{1}{\alpha}\frac{\partial c}{\partial t} - \frac{\mu}{D} c + \frac{\dot{m}}{D} = 0, \qquad \alpha = \frac{D}{R}$$

$$\begin{aligned} \text{IC}: &\ c\,(x, y, t) = 0 \\ \text{BC}: &\ \left.(-2\pi a\, D\frac{\partial c}{\partial r} + h\, c)\right|_{r=a} = h\, C_0(t) \\ &\ c\,(a, \theta, t) = c\,(a, \theta + 2\pi, t) \end{aligned} \qquad (15.76)$$

where $h(m^2/s)$ is some mass exchange coefficient.

General Solution

In the previous sections we found that upon adopting a trial function of the form

$$c\,(x, y, t) = u\,(x, y, t)\, e^{-v\, r \cos(\theta)/\,2D} \qquad (15.77)$$

the general solution of this problem in the frequency domain would lead to

$$\hat{u}\,(r, \theta, \omega) = \sum\nolimits_m [a_m \cos(m\theta) + b_m \sin(m\theta)]\, K_m(\kappa r) + \frac{4\dot{m}D}{4\kappa^2 D^2 - v^2} e^{v\, r\cos(\theta)/2D} \qquad (15.78)$$

Boundary Condition

Using the chain rule, we can write the Robin boundary condition in (15.76) in the u–space, which, in the frequency domain, yields

$$\left.\left\{-2\pi aD\left(\frac{\partial \hat{u}}{\partial r} + \hat{u}\frac{r\cos(\theta)}{2D}\right) e^{-vr\cos(\theta)/2D} + h\hat{u}e^{vr\cos(\theta)/2D}\right\}\right|_{r=a} = h\hat{C}_0(\omega) \qquad (15.79)$$

Transforming (15.77) to the frequency domain and solving at $r = a$, the boundary condition in (15.79) implies

$$\sum\nolimits_m a_m \cos(m\theta) + b_m \sin(m\theta) = \frac{1}{G} f(\theta)$$

with

$$f(\theta) = 4h\left\{\pi aD\dot{m}\cos(\theta) + \left(D^2\kappa^2 - \frac{v^2}{4}\right)\hat{C}_0 - \dot{m}D\right\} e^{va\cos(\theta)/2D}$$

$$G = \left[2\pi aD\kappa K_{m+1}(\kappa a) - (2\pi mD + \pi av\cos(\theta) + h)K_m(\kappa a)\right]\left[4D^2\kappa^2 - v^2\right] \qquad (15.80)$$

This function is a typical trigonometric Fourier series with its coefficients given by

$$a_0 = \frac{1}{2\pi G}\int_0^{2\pi} f(\theta)\, d\theta, \qquad m = 0$$

$$a_m = \frac{1}{\pi G}\int_0^{2\pi} f(\theta)\, \cos(m\theta)\, d\theta, \qquad m \geq 1$$

$$b_m = \frac{1}{\pi G}\int_0^{2\pi} f(\theta)\, \sin(m\theta)\, d\theta, \qquad m \geq 1 \tag{15.81}$$

BVP Solution

The solution of this boundary value problem for a given frequency ω_n is thus

$$\hat{u}(r,\theta,\omega_n) = \sum\nolimits_m \left[a_m \cos(m\theta) + b_m \sin(m\theta)\right] K_m(\kappa r) + \frac{4\dot{m}D}{4\kappa^2 D^2 - v^2} e^{v\,r\cos(\theta)/2D}, \qquad m = 0, 1, \ldots \tag{15.82}$$

with

$$\kappa = \sqrt{\frac{\mu}{D} + \frac{v^2}{4D^2} + \frac{i\omega_n}{\alpha}}$$

and the Fourier coefficients are given in (15.81).

Applying the inverse Fourier transform to (15.82) gives the solution in the time domain, such that

$$u(r,\theta,t) = \sum\nolimits_n \hat{u}(r,\theta,\omega_n)\, e^{i\omega_n t}$$

Having the solution of $u(r,\theta,t)$, then upon utilizing (15.77) the contaminant distribution in the domain can be determined via

$$c(r,\theta,t) = u(r,\theta,t)\, e^{-v\,r\cos(\theta)/2D} \tag{15.83}$$

■ **Example 15.2 Contaminant spill with Robin-type boundary condition**

A circular contaminant source, 2 m in radius, releases a pollutant with a time varying concentration $C_0(t)$ shown in Fig. 15.9. The soil mass has a bulk density 1800 kg/m^3 and porosity 0.35. It encompasses groundwater moving at an average speed

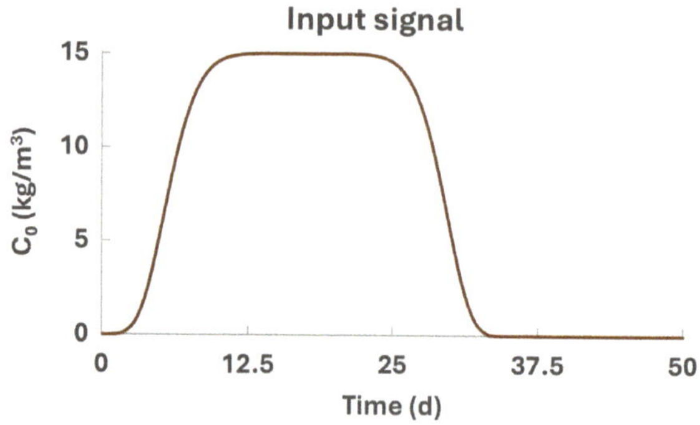

Fig. 15.9 Time distribution of the input concentration at the source

of 1.5 m/d. The pollutant is characterized by no adsorption with a hydrodynamic dispersion coefficient of 2 m^2/d. It disperses in the domain as a result of a Robin-type boundary condition, with a mass exchange coefficient h of 3 m^2/d. Compute the concentration of the pollutant for these four scenarios:

1. The pollutant exhibits no generation, neither decay.
2. The pollutant exhibits generation rate of 1 $kg/(m^3 \cdot d)$ with a time distribution similar to that given in Fig. 15.9, but shows no decay.
3. The pollutant exhibits no generation but exhibits a constant decay rate of 0.1 per day.
4. The pollutant exhibits a production rate of 1 $kg/(m^3 \cdot d)$ with a time distribution similar to that presented in Fig. 15.9, and at the same time a constant decay rate of 1 per day.

Solution

15

As outlined in the previous example, the solution of this case is conducted by first transforming the input signals of the pollutant $C_0(t)$ and its generation $\dot{m}$ from the time domain to the frequency domain. Then upon employing (15.82), the solution of the pollutant concentration in the frequency domain can be obtained for every frequency. After that, by employing the inverse FFT, the solution in the time domain is obtained, from which the pollutant concentration anywhere in the domain can be calculated via (15.83).

Figure 15.10 show the pollutant concentrations with time for the four scenarios after, and Fig. 15.11 demonstrates the spatial distribution for the scenario with production and decay.

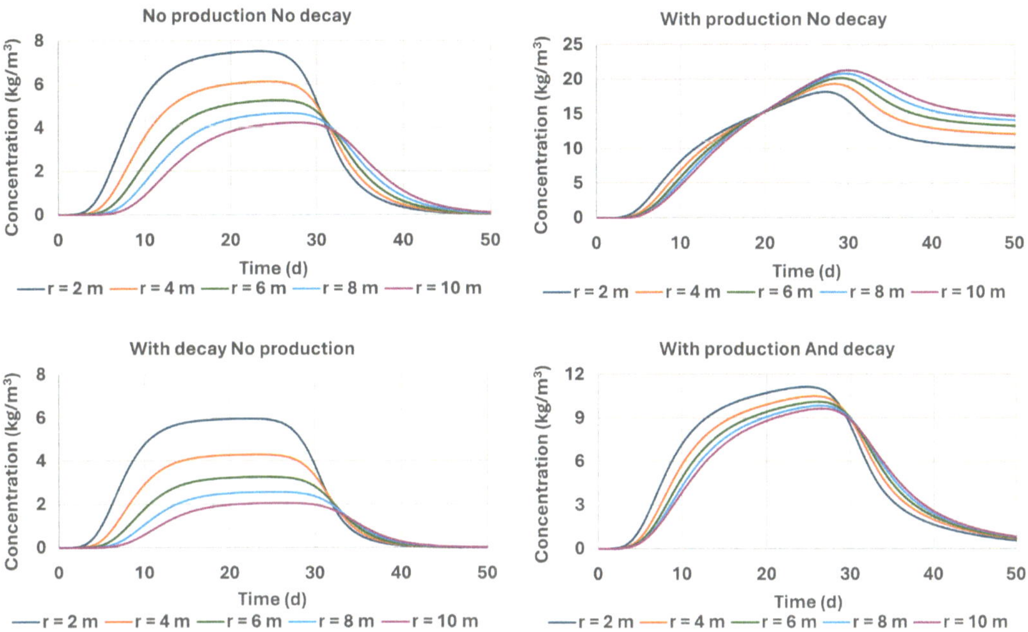

Fig. 15.10 Temporal concentration distribution for the four scenarios

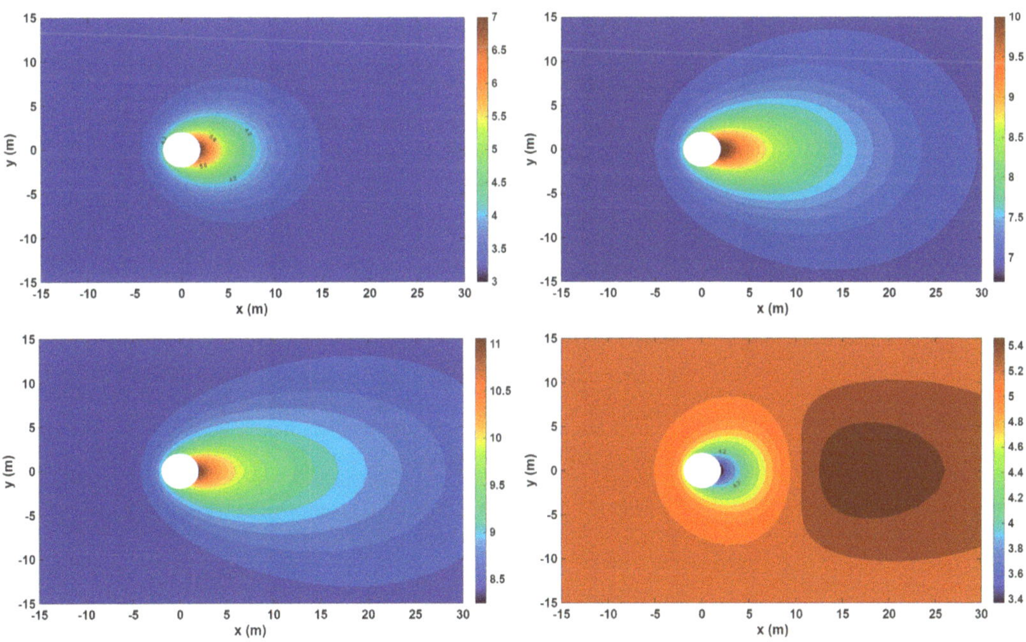

Fig. 15.11 Temporal concentration distribution for scenario with generation and decay after 10, 18, 25 and 35 days

Problems

1. Consider a circular contaminant spill with concentration $c(r, z, t)$ that undergoes neither production nor decay. The spill has a radius a and is located on the surface of a soil layer. It exerts a Robin-type boundary condition of the soil layer surface. Formulate the corresponding boundary value problem of this scenario.
   Hint: A relevant Robin boundary condition can be

   $$-\pi a^2 D \frac{\partial c(r, 0, t)}{\partial r} = b\left[C_0(t) - c(r, 0, t)\right]$$

   where $C_0(t)$ represents a concentration of a pollutant already existing in the domain, and b is a coefficient describing the exchange of mass between the environment and the domain.
2. Solve the governing equations of the scenario on Problem 1 using the Fourier method.
3. In addressing the diffusion-advection problems in two-dimensional planar domains, we employed a negative groundwater flow $(-v)$. Try to derive one of the 2D planar cases by considering a positive groundwater velocity.
4. Solve the mass transport with zero-order production and first-order decay in a 2D planar subjected to a Neumann boundary condition.
5. Implement the cylindrical contaminant scenario provided in ▶ Sect. 15.3 and its limiting case in ▶ Sect. 15.4 in a computer code and compare the results if you set the radius of the cylindrical source to 1e-10 m.

Reference

Abramowitz M, Stegun IA (1972) Handbook of mathematical functions. Dover Publications, Inc.

Carslaw HS, Jaeger JC (1959) Conduction of heat in solids, 2nd edn. Oxford University Press

WolframAlpha computational intelligence. ▶ https://www.wolframalpha.com/input?i=BesselK%5B0%2Cx%5D+series+at+x+-%3E+0

Further Reading

Goltz M, Huang J (2017). Analytical modeling of solute transport in groundwater. Wiley

White RE, Subramanian VR (2010) Computational methods in chemical engineering with maple. Springer

16 Contaminant Transport in Fractured Porous Medium Domains

R. Al Khoury and C. Kasbergen, *Analytical Subsurface Environment*,
https://doi.org/10.1007/978-3-031-93910-5_16

Contaminant transport in fractured porous medium domains refers to the study of how pollutants move through geological formations that consist of both a porous matrix, such as soils and sediments, and fractures, like cracks and fissures in rock. These environments are complex since the fractures often provide pathways for rapid movement of contaminants, while the porous matrix can store and slowly release pollutants over time. Understanding this transport is crucial for managing groundwater quality, predicting the spread of contamination, and designing remediation strategies.

Researchers use mathematical models and simulations to analyse the transport mechanisms, considering factors like fluid flow, chemical reactions, and the physical properties of the media. The multi-domain approach is widely used in hydrogeology, poromechanics, and environmental engineering to model fluid flow and contaminant transport in fractured rock formations, and other heterogeneous porous media. Various multi-domain approaches with different levels of complexity have been introduced, with two essential methods commonly in use: the dual porosity approach and the dual permeability approach.

In this chapter, we simulate various scenarios encompassing diverse complexities and boundary conditions, primarily focusing on the dual porosity model. Yet, in the last section, we examine a contaminant transport scenario using the dual permeability model. It is made general such that other scenarios can be derived from it. The models consider a range of factors, including advection, dispersion, diffusion, sorption, and biochemical reactions.

16.1 Dual Porosity Approach

16

The dual porosity approach simulates the porous medium domain as consisting of two overlapping continua: the matrix and the fractures. The matrix features lower permeability but higher storage capacity, making it responsible for storing the majority of the contaminant. Conversely, fractures are highly permeable pathways that facilitate the primary routes for fluid flow and rapid contaminant transport.

In many rock formations, the permeability ratio between the fractures and the matrix is so high that it is reasonable to simulate the fluid in the fractures as *mobile*, and the fluid in the matrix as *immobile*. Contaminants in the mobile domain can move freely, largely experiencing advection. On the other hand, contaminants within the immobile domain are effectively trapped and cannot move as readily as those in the mobile domain. The immobile domain acts as a storage region for contaminants, and their release from this region occurs through slow processes such as sorption and diffusion.

Mathematically, the dual porosity transport model, also known as the *double porosity model* or the *two-region transport model*, assumes that the liquid phase is partitioned into a mobile component and immobile component. The transfer of the solute between the two regions occurs via a first-order concentration exchange process. Even though the dual porosity model is designed to describe physical diffusion between mobile and immobile components, its governing equations are also applicable to *adsorption-desorption* processes in reactive transport.

The governing equations of a nonequilibrium transport in a multi-dimensional domain is (refer to Toride et al. 1993)

$$\begin{aligned}
&\left(1+\frac{f\rho K_d}{\varphi_m}\right)\frac{\partial c_m}{\partial t}+v\cdot\nabla c_m-\nabla\cdot(\nabla D c_m)=-\frac{b}{\varphi_m}(c_m-c_{im})- \\
&\qquad\left(\mu_{w,m}+\frac{f\rho K_d}{\varphi_m}\mu_{s,m}\right)c_m+\dot{m}_{w,m}+\frac{f\rho}{\varphi_m}\dot{m}_{s,m} \\
&\left(1+\frac{(1-f)\rho K_d}{\varphi_{im}}\right)\frac{\partial c_{im}}{\partial t}=\frac{b}{\varphi_{im}}(c_m-c_{im})- \\
&\qquad\left(\mu_{w,im}+\frac{(1-f)\rho K_d}{\varphi_{im}}\mu_{s,im}\right)c_{im}+\dot{m}_{w,im}+\frac{(1-f)\rho}{\varphi_{im}}\dot{m}_{s,im}
\end{aligned} \tag{16.1}$$

with

$$\varphi=\varphi_m+\varphi_{im}$$

where the subscripts m and im denote the mobile and immobile liquid components, and w and s denote the liquid and sorbed concentrations, respectively; f is the sorption fraction; φ is the porosity; b (s^{-1}) is a mass transfer coefficient governing the solute rate of exchange between the mobile and immobile liquid components; μ (s^{-1}) is a first-order decay rate; $\dot{m}$ ($\mathrm{kg\,m^{-3}\,s^{-1}}$) is a zero-order production rate; D ($\mathrm{m^2s^{-1}}$) is the hydrodynamic dispersion coefficient; v ($\mathrm{ms^{-1}}$) is the fluid velocity, and ρ ($\mathrm{kg\,m^{-3}}$) is the associated mass density.

In a more compact form, the governing equations can be written as

$$\begin{aligned}
R_m\frac{\partial c_m}{\partial t}+v\cdot\nabla c_m-\nabla\cdot(\nabla D c_m)&=-b_m(c_m-c_{im})-\mu_m c_m+\dot{m}_m \\
R_{im}\frac{\partial c_{im}}{\partial t}&=b_{im}(c_m-c_{im})-\mu_{im}c_{im}+\dot{m}_{im}
\end{aligned} \tag{16.2}$$

where

$$\begin{aligned}
R_m&=1+\frac{f\rho K_d}{\varphi_m}, & R_{im}&=1+\frac{(1-f)\rho K_d}{\varphi_{im}} \\
b_m&=\frac{b}{\varphi_m}, & b_{im}&=\frac{b}{\varphi_{im}} \\
\mu_m&=\mu_{w,m}+\frac{f\rho K_d}{\varphi_m}\mu_{s,m}, & \mu_{im}&=\mu_{w,im}+\frac{(1-f)\rho K_d}{\varphi_{im}}\mu_{s,im} \\
\dot{m}_m&=\dot{m}_{w,m}+\frac{f\rho}{\varphi_m}\dot{m}_{s,m}, & \dot{m}_{im}&=\dot{m}_{w,im}+\frac{(1-f)\rho}{\varphi_{im}}\dot{m}_{s,im}
\end{aligned}$$

with K_d ($\mathrm{m^3\,kg^{-1}}$) the *distribution coefficient*, also known as the *partition coefficient*. For a nonadsorption case, $R = 1$.

In the following, we will solve this system of equations for one-dimensional and multi-dimensional domains, considering various transport mechanisms and boundary conditions.

16.1.1 1D Domain: First-Order Decay with No Production: Robin Problem

The boundary value problem of a mobile-immobile contaminant transport in a semi-infinite 1D fractured domain, exhibiting first-order decay with no production on both components, and subjected to a Robin boundary condition at $x = 0$, is

$$\begin{aligned}
&R_m \frac{\partial c_m}{\partial t} + v\frac{\partial c_m}{\partial x} - D\frac{\partial^2 c_m}{\partial x^2} = -b_m(c_m - c_{im}) - \mu_m c_m \\
&R_{im}\frac{\partial c_{im}}{\partial t} = b_{im}(c_m - c_{im}) - \mu_{im} c_{im} \\
&\text{IC:} \quad c_m(x,0) = c_{im}(x,0) = 0 \\
&\text{BC}: \left(-D\frac{\partial c_m}{\partial x} + v c_m\right)\bigg|_{x=0} = vC_0(t) \\
&\qquad c_m(\infty, t) = 0
\end{aligned} \tag{16.3}$$

where C_0 (kg m^{-3} s^{-1}) is the prescribed contaminant quantity at the boundary.

General Solution

Applying the Fourier transform to (16.3) yields

$$\begin{aligned}
&R_m i\omega \hat{c}_m + v\frac{d\hat{c}_m}{dx} - D\frac{d^2\hat{c}_m}{dx^2} = -b_m\left(\hat{c}_m - \hat{c}_{im}\right) - \mu_m \hat{c}_m \\
&R_{im} i\omega \hat{c}_{im} = b_{im}\left(\hat{c}_m - \hat{c}_{im}\right) - \mu_{im}\hat{c}_{im} \\
&\text{BC}: \left(-D\frac{\partial \hat{c}_m}{\partial x} + v\hat{c}_m\right)\bigg|_{x=0} = v\hat{C}_0(\omega) \\
&\qquad \hat{c}_m(\infty, 0) = 0
\end{aligned} \tag{16.4}$$

where it can be noticed that the first-order differential equation for c_{im} in (16.3) has been converted to an algebraic equation, as given in the second equation of (16.4). Solving this equation algebraically gives

$$\begin{aligned}
&\hat{c}_{im} = \alpha_{im}\hat{c}_m \\
&\text{with} \\
&\alpha_{im} = \frac{b_{im}}{R_{im} i\omega + b_{im} + \mu_{im}}
\end{aligned} \tag{16.5}$$

Substituting this relationship into the first differential equation in (16.4) yields

$$D\frac{d^2\hat{c}_m}{dx^2} - v\frac{d\hat{c}_m}{dx} + (-i\omega R_m - b_m + b_m\alpha_{im} - \mu_m)\hat{c}_m = 0 \tag{16.6}$$

16

As the coefficients are constant, the trial solution of this differential equation can be

$$\hat{c}_m = ae^{-\kappa x}$$

Substituting this trial solution into (16.6), and simplifying, gives

$$D\kappa^2 + v\kappa + g_{im} = 0, \quad g_{im} = -i\omega R_m - b_m(1 - \alpha_{im}) - \mu_m$$

Solving this equation for κ using the quadratic formula, results to

$$\kappa_1 = \frac{-v + \sqrt{v^2 - 4Dg_{im}}}{2D}, \quad \kappa_2 = \frac{-v - \sqrt{v^2 - 4Dg_{im}}}{2D}$$

where κ_1 and κ_2 represent the two eigenvalues of the system. Following this, the general solution of (16.6) is given by

$$\hat{c}_m = a_1 e^{-\kappa_1 x} + a_2 e^{-\kappa_2 x} \tag{16.7}$$

Boundary Conditions

At $x \to \infty$, the second boundary condition in (16.4) implies $a_2 = 0$; otherwise the solution becomes unbounded. This leads the general solution (16.7) to become

$$\hat{c}_m = a_1 e^{-\kappa_1 x} \tag{16.8}$$

Taking the derivative of (16.8) with respect to x, the first boundary condition in (16.4) implies at $x = 0$

$$a_1 = \frac{v\hat{C}_0(\omega)}{\kappa_1 D + v} \tag{16.9}$$

BVP Solution

Substituting (16.9) into (16.8) yields the solution of this BVP for a given frequency ω_n as

$$\hat{c}_m = \frac{v\hat{C}_0(\omega_n)}{\kappa_1 D + v} e^{-\kappa_1 x}, \quad \kappa_1 = \frac{-v + \sqrt{v^2 - 4Dg_{im}}}{2D} \tag{16.10}$$

Employing (16.5), the concentration of the immobile contaminant can then be expressed as

$$\hat{c}_{im} = \frac{b_{im}}{R_{im} i\omega_n + b_{im} + \mu_{im}} \hat{c}_m \tag{16.11}$$

Having solved the problem in the frequency domain, the solution in the time domain can then be obtained via the inverse FFT algorithm, such that

$$c_m(x,t) = \sum_n \hat{c}_m(x,\omega_n)e^{i\omega_n t}$$
$$c_{im}(x,t) = \sum_n \hat{c}_{im}(x,\omega_n)e^{i\omega_n t}$$

16.1.2 1D Domain: First-Order Decay with Zero-Order Production: Dirichlet Problem

The boundary value problem of a coupled mobile-immobile contaminant transport, exhibiting zero-order production and first-order decay, in a semi-infinite 1D domain bounded by a Dirichlet boundary condition is expressed as

$$\begin{aligned}
&R_m\frac{\partial c_m}{\partial t} + v\frac{\partial c_m}{\partial x} - D\frac{\partial^2 c_m}{\partial x^2} = -b_m(c_m - c_{im}) - \mu_m c_m + \dot{m}_m\\
&R_{im}\frac{\partial c_{im}}{\partial t} = b_{im}(c_m - c_{im}) - \mu_{im}c_{im} + \dot{m}_{im}\\
&\text{IC:} \quad c_m(x,0) = c_{im}(x,0) = 0\\
&\text{BC:} \quad c_m(0,t) = C_0(t)\\
&\qquad\;\; c_m(\infty,t) = 0
\end{aligned} \tag{16.12}$$

General Solution

Applying the Fourier transform to (16.12) results in

$$\begin{aligned}
&R_m i\omega\hat{c}_m + v\frac{d\hat{c}_m}{dx} - D\frac{d^2\hat{c}_m}{dx^2} = -b_m\left(\hat{c}_m - \hat{c}_{im}\right) - \mu_m\hat{c}_m + \hat{\dot{m}}_m\\
&R_{im} i\omega\hat{c}_{im} = b_{im}\left(\hat{c}_m - \hat{c}_{im}\right) - \mu_{im}\hat{c}_{im} + \hat{\dot{m}}_{im}\\
&\text{BC:} \quad \hat{c}_m(0,\omega) = \hat{C}_0(\omega)\\
&\qquad\;\; \hat{c}_m(\infty,\omega) = 0
\end{aligned} \tag{16.13}$$

Apparently this BVP is nonhomogeneous due to the presence of $\dot{m}_m$ and $\dot{m}_{im}$ terms in the governing differential equations. Thus, the solution must be conducted by finding the complementary solutions $\hat{c}_{mh}$ and $\hat{c}_{imh}$ of the corresponding homogeneous equations and the particular solutions $\hat{c}_{mp}$ and $\hat{c}_{imp}$ of the nonhomogeneous equation, such that

$$\begin{aligned}
\hat{c}_m &= \hat{c}_{mh} + \hat{c}_{mp}\\
\hat{c}_{im} &= \hat{c}_{imh} + \hat{c}_{imp}
\end{aligned} \tag{16.14}$$

Considering the boundary condition at far distance, the solution of the homogeneous equations in the frequency domain was given in detail in the preceding section, that led to

$$
\begin{aligned}
\hat{c}_{mh} &= a_1 e^{-\kappa_1 x}, \quad \kappa_1 = \frac{-v + \sqrt{v^2 - 4Dg_{im}}}{2D} \\
\hat{c}_{imh} &= \frac{b_{im}}{R_{im} i\omega + b_{im} + \mu_{im}} a_1 e^{-\kappa_1 x}
\end{aligned} \tag{16.15}
$$

where a_1 is a constant needs to be determined from the boundary conditions.

The relevant nonhomogeneous system of equations of (16.13) is

$$
\begin{aligned}
R_m i\omega \hat{c}_{mp} + v\frac{d\hat{c}_{mp}}{dx} - D\frac{d^2\hat{c}_{mp}}{dx^2} &= -b_m\left(\hat{c}_{mp} - \hat{c}_{imp}\right) - \mu_m \hat{c}_{mp} + \hat{\dot{m}}_m \\
R_{im} i\omega \hat{c}_{imp} &= b_{im}\left(\hat{c}_{mp} - \hat{c}_{imp}\right) - \mu_{im}\hat{c}_{imp} + \hat{\dot{m}}_{im}
\end{aligned} \tag{16.16}
$$

As the nonhomogeneous terms are constant, the particular solution of these equations can also be constants, of the form

$$
\begin{aligned}
\hat{c}_{mp} &= A \\
\hat{c}_{imp} &= B
\end{aligned} \tag{16.17}
$$

Substituting these trial solutions into (16.16) gives

$$
\begin{aligned}
R_m i\omega A &= -b_m(A - B) - \mu_m A + \hat{\dot{m}}_m \\
R_{im} i\omega B &= b_{im}(A - B) - \mu_{im} B + \hat{\dot{m}}_{im}
\end{aligned}
$$

Rearranging, and putting these equations in matrix format, leads to

$$
\begin{pmatrix} R_m i\omega + b_m + \mu_m & -b_m \\ -b_{im} & R_{im} i\omega + b_{im} + \mu_{im} \end{pmatrix} \begin{pmatrix} A \\ B \end{pmatrix} = \begin{pmatrix} \hat{\dot{m}}_m \\ \hat{\dot{m}}_{im} \end{pmatrix}
$$

Solving for the constants A and B yields

$$
\begin{aligned}
\begin{pmatrix} A \\ B \end{pmatrix} &= \frac{1}{\Delta} \begin{pmatrix} R_{im} i\omega + b_{im} + \mu_{im} & b_m \\ b_{im} & R_m i\omega + b_m + \mu_m \end{pmatrix} \begin{pmatrix} \hat{\dot{m}}_m \\ \hat{\dot{m}}_{im} \end{pmatrix} \\
\Delta &= (R_m i\omega + b_m + \mu_m)(R_{im} i\omega + b_{im} + \mu_{im}) - b_m b_{im}
\end{aligned}
$$

from which we obtain

$$
\begin{aligned}
A &= \frac{1}{\Delta}\left[(R_{im} i\omega + b_{im} + \mu_{im})\hat{\dot{m}}_m + b_m \hat{\dot{m}}_{im}\right] \\
B &= \frac{1}{\Delta}\left[b_{im}\hat{\dot{m}}_m + (R_m i\omega + b_m + \mu_m)\hat{\dot{m}}_{im}\right]
\end{aligned}
$$

Substituting these constants into (16.17) gives the particular solutions of the nonhomogeneous equations, such that

$$
\begin{aligned}
\hat{c}_{mp} &= \frac{1}{\Delta}\left[(R_{im} i\omega + b_{im} + \mu_{im})\hat{\dot{m}}_m + b_m \hat{\dot{m}}_{im}\right] \\
\hat{c}_{imp} &= \frac{1}{\Delta}\left[b_{im}\hat{\dot{m}}_m + (R_m i\omega + b_m + \mu_m)\hat{\dot{m}}_{im}\right]
\end{aligned} \tag{16.18}
$$

Substituting (16.15) and (16.18) into (16.14), the general solutions of (16.13) result in

$$\hat{c}_m = a_1 e^{-\kappa_1 x} + \frac{1}{\Delta}\left[(R_{im} i\omega + b_{im} + \mu_{im})\hat{\dot{m}}_m + b_m \hat{\dot{m}}_{im}\right]$$
$$\hat{c}_{im} = \frac{b_{im}}{R_{im} i\omega + b_{im} + \mu_{im}} a_1 e^{-\kappa_1 x} + \frac{1}{\Delta}\left[b_{im}\hat{\dot{m}}_m + (R_m i\omega + b_m + \mu_m)\hat{\dot{m}}_{im}\right] \tag{16.19}$$

Boundary Condition

At $x = 0$, the first boundary condition in (16.13) implies

$$a_1 = \hat{C}_0(\omega) - \frac{1}{\Delta}\left[(R_{im} i\omega + b_{im} + \mu_{im})\hat{\dot{m}}_m + b_m \hat{\dot{m}}_{im}\right]$$

BVP Solution

Collecting solutions and terms, the solution of this boundary value problem for a given frequency ω_n can therefore be expressed as

$$\hat{c}_m = a_1 e^{-\kappa_1 x} + \frac{1}{\Delta}\left[(R_{im} i\omega_n + b_{im} + \mu_{im})\hat{\dot{m}}_m + b_m \hat{\dot{m}}_{im}\right]$$
$$\hat{c}_{im} = \frac{b_{im}}{R_{im} i\omega_n + b_{im} + \mu_{im}} a_1 e^{-\kappa_1 x} + \frac{1}{\Delta}\left[b_{im}\hat{\dot{m}}_m + (R_m i\omega_n + b_m + \mu_m)\hat{\dot{m}}_{im}\right] \tag{16.20}$$

with

$$a_1 = \hat{C}_0(\omega_n) - \frac{1}{\Delta}\left[(R_{im} i\omega_n + b_{im} + \mu_{im})\hat{\dot{m}}_m + b_m \hat{\dot{m}}_{im}\right]$$
$$\Delta = (R_m i\omega_n + b_m + \mu_m)(R_{im} i\omega_n + b_{im} + \mu_{im}) - b_m b_{im}$$
$$\kappa_1 = \frac{-v + \sqrt{v^2 - 4Dg_{im}}}{2D}$$

Having solved the system in the frequency domain, the solution in the time domain can then be obtained via the inverse FFT algorithm, such that

$$c_m(x,t) = \sum_n \hat{c}_m(x,\omega_n) e^{i\omega_n t}$$
$$c_{im}(x,t) = \sum_n \hat{c}_{im}(x,\omega_n) e^{i\omega_n t}$$

16

- **Example 16.1: 1D first-order decay with zero-order production**

A fully saturated fractured porous medium domain, having a bulk density 1500 kg/m^3 and porosities $\varphi_m = \varphi_{im} = 0.2$, is exposed at one of its boundaries to a chemical species with distribution coefficient 0.1 m^3/kg and hydrodynamic dispersion coefficient 0.8 m^2/s. The groundwater in the porous domain flows at an average velocity of 1.0e-4 m/s. No initial concentration of a chemical species exists in the domain. At $x = 0$ the domain is exposed to a rectangular pulse concentration of $100\ \mathrm{kg/m^3}$ for nearly 18 hours. Under some chemical or physical processes, the species simultaneously undergoes first-order decay of $\mu_{w,m} = \mu_{w,im} = 0.001\ s^{-1}$ and $\mu_{s,m} = \mu_{s,im} = 0.0001\ s^{-1}$ and zero-order production rates with maximum values $\dot{m}_{w,m} = \dot{m}_{w,im} = 5\ \mathrm{kg/m^3s}$ and $\dot{m}_{s,m} = 0.01\ \mathrm{kg/m^3s}$, $\dot{m}_{s,im} = 0$. Figure 16.1 shows the time distribution of the prescribed concentration at the boundary, together with the production rates in the domain. Compute the concentration of the species at 0.0, 1.0, and 10.0 m for two different species mass exchange coefficients: $b = 0.008$ and $0.08\ s^{-1}$.

Solution

To solve this problem using the spectral analysis, we follow these steps:

(a) Transform the input concentration pulse and the generation/production rate pulses from the time domain to the frequency domain using the forward fast Fourier transform (FFT).
(b) Solve the problem for all involved frequencies using (16.20).
(c) Transform the computed concentrations back to the time domain using the inverse fast Fourier transform algorithm (IFFT).

Figure 16.2 shows the computed concentrations of the mobile and immobile species for the $b = 0.008 s^{-1}$ case. This figure clearly demonstrates that the chemical species is transported through the medium while being generated. Figure 16.3 shows the computed concentrations of the mobile and immobile species for the $b = 0.08 s^{-1}$ case. This figure emphasizes that the larger mass exchange coefficient leads to more mass of the chemical species to be trapped in the solid matrix.

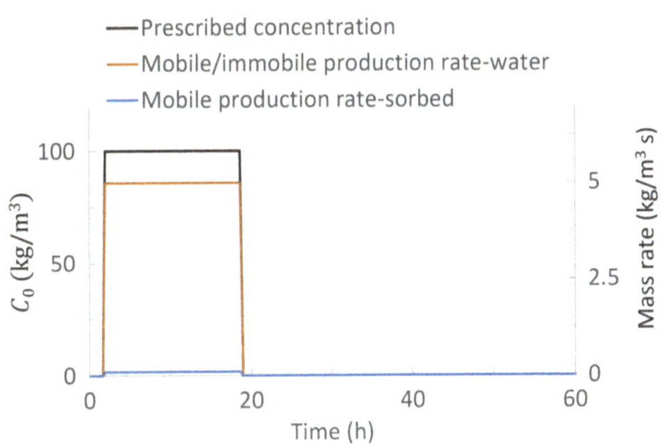

Fig. 16.1 Prescribed species concentration and its production rates

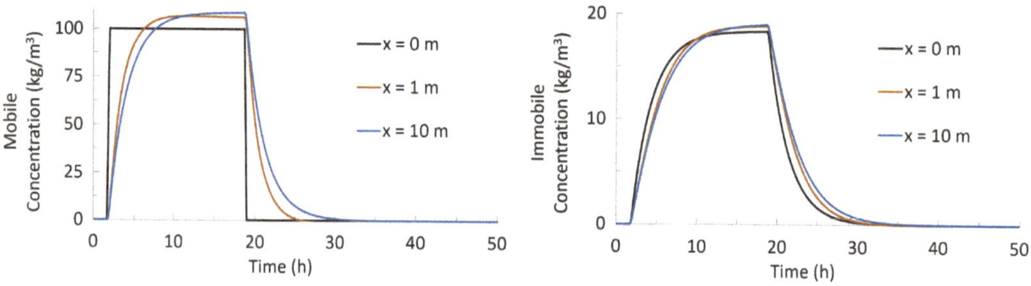

Fig. 16.2 Mobile and immobile concentrations for $b = 0.008\,s^{-1}$

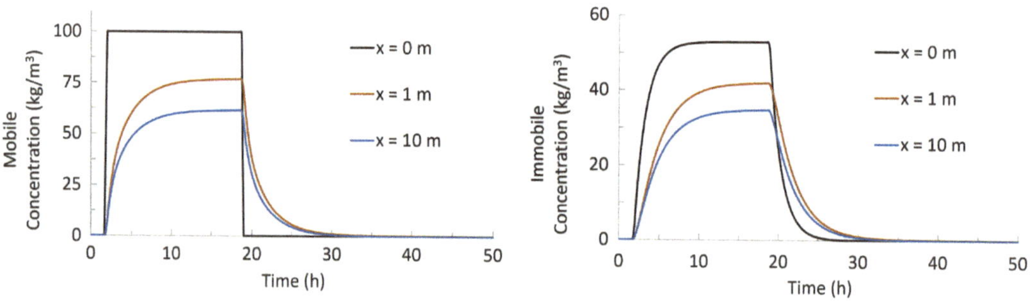

Fig. 16.3 Mobile and immobile concentrations for $b = 0.08\,s^{-1}$

16.1.3 2D Planar domain: Cylindrical contaminant source with first order decay and zero order production: Dirichlet Problem

The boundary value problem of a contaminant exhibiting first-order decay with zero-order production, moving in a dual porosity infinite domain embodying groundwater flow passing across a cylindrical contaminant source, can be expressed in the Cartesian coordinate system as

$$
\begin{aligned}
&R_m\frac{\partial c_m}{\partial t} - v\frac{\partial c_m}{\partial x} - D\frac{\partial^2 c_m}{\partial x^2} - D\frac{\partial^2 c_m}{\partial y^2} = -b_m(c_m - c_{im}) - \mu_m c_m + \dot{m}_m \\
&R_{im}\frac{\partial c_{im}}{\partial t} = b_{im}(c_m - c_{im}) - \mu_{im} c_{im} + \dot{m}_{im} \\
&\text{IC}: \quad c_m(x,y,0) = c_{im}(x,y,0) = 0 \\
&\text{BC}: \quad c_m(x,y,t)|_{(x^2+y^2=a^2)} = C_0(t) \\
&\qquad\quad c_m(x,y,t)|_{(a,\theta)} = c_m(x,y,t)|_{(a,\theta+2\pi)} \\
&\qquad\quad c_{im}(x,y,t)|_{(a,\theta)} = c_{im}(x,y,t)|_{(a,\theta+2\pi)}
\end{aligned}
\tag{16.21}
$$

where we assume $D_x = D_y = D$ and the groundwater is moving with an average velocity $v_x = -v$. R is a coefficient describing adsorption, and C_0 (kg m^{-3}) is a prescribed pollutant concentration in the ground surface, covering a circle with radius $a = \sqrt{x^2 + y^2}$.

General Solution

Since the contaminant source is circular, it is convenient to express (16.21) in the polar coordinate system. However, the presence of the third component in this equation complicates the conversion from the Cartesian to the polar coordinate system. To facilitate this conversion, following the approach of Carslaw and Jaeger (1959), we propose a trial solution of the form

$$c_m(x,y,t) = u(x,y,t)e^{\eta x} \tag{16.22}$$

Taking the derivative of (16.22) with respect to x and y, we obtain

$$\begin{aligned}
\frac{\partial c_m}{\partial x} &= u\eta e^{\eta x} + \frac{\partial u}{\partial x}e^{\eta x} \\
\frac{\partial^2 c_m}{\partial x^2} &= \frac{\partial^2 u}{\partial x^2}e^{\eta x} + 2\frac{\partial u}{\partial x}\eta e^{\eta x} + u\eta^2 e^{\eta x} \\
\frac{\partial^2 c_m}{\partial y^2} &= \frac{\partial^2 u}{\partial y^2}e^{\eta x}
\end{aligned}$$

Substituting these identities into the first PDE in (16.21), and rearranging, gives

$$\begin{aligned}
&\frac{1}{\alpha}\frac{\partial u}{\partial t} - \frac{\partial^2 u}{\partial x^2} - \frac{\partial^2 u}{\partial y^2} - \left(\frac{v}{D} + 2\eta\right)\frac{\partial u}{\partial x} + \left(-\frac{v}{D}\eta - \eta^2 + \frac{b_m}{D} + \frac{\mu_m}{D}\right)u \\
&\quad - \frac{b_m}{D}c_{im}e^{-\eta x} - \frac{\dot{m}_m}{D}e^{-\eta x} = 0
\end{aligned} \tag{16.23}$$

where $\alpha_m = D/R_m$.The spatial derivative on the fourth term of this equation can be eliminated if we let $\eta = -v/2D$. This leads to

$$c_m(x,y,t) = u(x,y,t)e^{-vx/2D} \tag{16.24}$$

and

$$\frac{\partial^2 u}{\partial x^2} + \frac{\partial^2 u}{\partial y^2} - \left(\frac{v^2}{4D^2} + \frac{\mu_m}{D} + \frac{b_m}{D}\right)u - \frac{1}{\alpha}\frac{\partial u}{\partial t} + \frac{b_m}{D}c_{im}e^{vx/2D} + \frac{\dot{m}_m}{D}e^{vx/2D} = 0 \tag{16.25}$$

Transforming this equation to the polar coordinate system in xy-plane, the governing equations can now be expressed as

$$\begin{aligned}
&\frac{\partial^2 u}{\partial r^2} + \frac{1}{r}\frac{\partial u}{\partial r} + \frac{1}{r^2}\frac{\partial^2 u}{\partial \theta^2} - \left(\frac{v^2}{4D^2} + \frac{\mu_m}{D} + \frac{b_m}{D}\right)u \\
&\quad - \frac{1}{\alpha}\frac{\partial u}{\partial t} + \frac{b_m}{D}c_{im}e^{vr\cos(\theta)/2D} + \frac{\dot{m}_m}{D}e^{vr\cos(\theta)/2D} = 0 \qquad (16.26) \\
&R_{im}\frac{\partial c_{im}}{\partial t} = b_{im}\left(ue^{-vr\cos(\theta)/2D} - c_{im}\right) - \mu_{im}c_{im} + \dot{m}_{im}
\end{aligned}$$

in which we considered

$$c_m(r,\theta,t) = u(r,\theta,t)e^{-vr\cos(\theta)/2D} \tag{16.27}$$

and for which we employed $x = r\cos\theta$, $y = r\sin\theta$ and $r = \sqrt{x^2 + y^2}$.

These equations are nonhomogeneous due to the presence of $\dot{m}_m$ and $\dot{m}_{im}$. Solutions for this kind of differential equations can be expressed in terms of their corresponding complementary functions u_h and c_{im_h}, and their particular solutions u_p and c_{im_p}, such that

$$\begin{aligned} u &= u_h + u_p \\ c_{im} &= c_{im_h} + c_{im_p} \end{aligned} \tag{16.28}$$

The relevant homogeneous system of equations of (16.26) is

$$\frac{\partial^2 u_h}{\partial r^2} + \frac{1}{r}\frac{\partial u_h}{\partial r} + \frac{1}{r^2}\frac{\partial^2 u_h}{\partial \theta^2} - \left(\frac{v^2}{4D^2} + \frac{\mu_m}{D} + \frac{b_m}{D}\right)u_h - \frac{1}{\alpha}\frac{\partial u_h}{\partial t} + \frac{b_m}{D}c_{im_h}e^{vr\cos(\theta)/2D} = 0$$

$$R_{im}\frac{\partial c_{im_h}}{\partial t} = b_{im}\left(u_h e^{-vr\cos(\theta)/2D} - c_{im_h}\right) - \mu_{im}c_{im_h}$$

Applying the Fourier transform to this system of equations gives

$$\begin{aligned} &\frac{\partial^2 \hat{u}_h}{\partial r^2} + \frac{1}{r}\frac{\partial \hat{u}_h}{\partial r} + \frac{1}{r^2}\frac{\partial^2 \hat{u}_h}{\partial \theta^2} - \left(\frac{v^2}{4D^2} + \frac{\mu_m}{D} + \frac{b_m}{D} + \frac{i\omega}{\alpha}\right)\hat{u}_h + \frac{b_m}{D}\hat{c}_{im_h}e^{vr\cos(\theta)/2D} = 0 \\ &i\omega R_{im}\hat{c}_{im_h} = b_{im}\left(\hat{u}_h e^{-vr\cos(\theta)/2D} - \hat{c}_{im_h}\right) - \mu_{im}\hat{c}_{im_h} \end{aligned} \tag{16.29}$$

Solving the second equation in (16.29) algebraically leads to

$$\hat{c}_{im_h} = \alpha_{im}\hat{u}_h e^{-vr\cos(\theta)/2D}$$

with

$$\alpha_{im} = \frac{b_{im}}{i\omega R_{im} + b_{im} + \mu_{im}} \tag{16.30}$$

Substituting (16.30) into the first equation in (16.29) yields

$$\frac{\partial^2 \hat{u}_h}{\partial r^2} + \frac{1}{r}\frac{\partial \hat{u}_h}{\partial r} + \frac{1}{r^2}\frac{\partial^2 \hat{u}_h}{\partial \theta^2} - \kappa^2\hat{u}_h = 0$$

with

$$\kappa^2 = \frac{v^2}{4D^2} + \frac{\mu_m}{D} + \frac{b_m}{D} + \frac{i\omega}{\alpha} - \frac{b_m}{D}\alpha_{im} \tag{16.31}$$

This partial differential equation can be solved using the separation of variables, which can be expressed as

$$\hat{u}_h(r,\theta) = R(r)F(\theta) \equiv RF \tag{16.32}$$

Substituting this formula into (16.31), dividing by $R(r)F(\theta)$ and equating both terms to a constant, say λ, gives

16

$$\frac{r^2}{R}\frac{d^2R}{dr^2}+\frac{r}{R}\frac{dR}{dr}-r^2\kappa^2=-\frac{1}{F}\frac{d^2F}{d\theta^2}=\lambda \tag{16.33}$$

The relevant F-dependent boundary value problem in (16.33) is

$$F''+\lambda F=0,\quad F(\theta)=F(\theta+2\pi) \tag{16.34}$$

which is evidently 2π-periodic. This problem is not a regular Sturm-Liouville, but due to its periodicity, it generates eigenvalues and eigenfunctions. The three possible general solutions of (16.34) are

$$\begin{aligned} F(\theta)&=c_1+c_2\theta, & \lambda&=0\\ F(\theta)&=c_1\cosh(s\theta)+c_2\sinh(s\theta), & \lambda&=-s^2<0\\ F(\theta)&=c_1\cos(s\theta)+c_2\sin(s\theta), & \lambda&=s^2>0 \end{aligned} \tag{16.35}$$

Apparently the second solution in (16.35) is not periodic and thus can be dismissed. The first solution is periodic if $c_2=0$, with the remaining solution $F(\theta)=c_1, c_1\neq 0$. Hence, $\lambda=0$ is an eigenvalue. Finally, the third solution in (16.35) can be 2π –periodic if we take $s=1,\ 2,\ \ldots$. Hence, considering $\lambda=0$, the eigenvalues of this BVP are $\lambda_s=s^2,\ s=0,\ 1,\ 2,\ \ldots$, leading the solution to be expressed as

$$F(\theta)=\sum_s a_s\cos(s\theta)+b_s\sin(s\theta),\quad s=0,1,2,\cdots \tag{16.36}$$

Letting $\lambda=s^2$, the R-dependent differential equation in (16.33) is given by

$$r^2\frac{d^2R}{dr^2}+r\frac{dR}{dr}-\left(r^2\kappa^2+s^2\right)R=0$$

The solution of this ordinary differential equation is expressed in terms of the modified Bessel functions of first kind $I_s(\kappa r)$, and second kind $K_s(\kappa r)$ (Abramowitz and Stegun 1972). However, since $I_s(\kappa r)$ is unbounded at far distances, the solution reduces to

$$R(r,\omega)=K_s(\kappa r) \tag{16.37}$$

Substituting (16.36) and (16.37) into (16.32) leads to

$$\hat{u}_h(r,\theta,\omega)=\sum_s[a_s\cos(s\theta)+b_s\sin(s\theta)]K_s(\kappa r),\quad s=0,1,2,\ldots \tag{16.38}$$

from which

$$\hat{c}_{im_h}=\alpha_{im}\sum_s[a_s\cos(s\theta)+b_s\sin(s\theta)]K_s(\kappa r)e^{-vr\cos(\theta)/2D},\quad s=0,1,2,\ldots \tag{16.39}$$

The relevant particular equations of (16.26) are

$$\frac{\partial^2 u_p}{\partial r^2}+\frac{1}{r}\frac{\partial u_p}{\partial r}+\frac{1}{r^2}\frac{\partial^2 u_p}{\partial \theta^2}-\left(\frac{v^2}{4D^2}+\frac{\mu_m}{D}+\frac{b_m}{D}\right)u_p$$
$$-\frac{1}{\alpha}\frac{\partial u_p}{\partial t}+\frac{b_m}{D}c_{imp}e^{vr\cos(\theta)/2D}+\frac{\dot{m}_m}{D}e^{vr\cos(\theta)/2D}=0$$
$$R_{im}\frac{\partial c_{im_p}}{\partial t}=b_{im}\left(u_p e^{-vr\cos(\theta)/2D}-c_{im_p}\right)-\mu_{im}c_{im_p}+\dot{m}_{im}$$

Applying the Fourier transform to these equations, and rearranging, gives

$$\frac{\partial^2 \hat{u}_p}{\partial r^2}+\frac{1}{r}\frac{\partial \hat{u}_p}{\partial r}+\frac{1}{r^2}\frac{\partial^2 \hat{u}_p}{\partial \theta^2}-\left(\frac{v^2}{4D^2}+\frac{\mu_m}{D}+\frac{b_m}{D}+\frac{i\omega}{\alpha}\right)u_p$$
$$+\frac{b_m}{D}\hat{c}_{im_p}e^{vr\cos(\theta)/2D}+\frac{\dot{m}_m}{D}e^{vr\cos(\theta)/2D}=0$$
$$\hat{c}_{im_p}=\alpha_{im}\hat{u}_p e^{-vr\cos(\theta)/2D}+g_{im} \tag{16.40}$$

with

$$g_{im}=\frac{\dot{m}_{im}}{R_{im}i\omega+b_{im}+\mu_{im}}$$

Substituting $\hat{c}_{imp}$ into the first equation of (16.40) leads to

$$\frac{\partial^2 \hat{u}_p}{\partial r^2}+\frac{1}{r}\frac{\partial \hat{u}_p}{\partial r}+\frac{1}{r^2}\frac{\partial^2 \hat{u}_p}{\partial \theta^2}-\kappa^2\hat{u}_p+\left(\frac{\dot{m}_m}{D}+\frac{b_m}{D}g_{im}\right)e^{vr\cos(\theta)/2D}=0$$

with

$$\kappa^2=\frac{v^2}{4D^2}+\frac{\mu_m}{D}+\frac{b_m}{D}+\frac{i\omega}{\alpha}-\frac{b_m}{D}\alpha_{im} \tag{16.41}$$
$$g_{im}=\frac{\dot{m}_{im}}{R_{im}i\omega+b_{im}+\mu_{im}}$$

As the nonhomogeneous term is exponential, the trial solution can be

$$\hat{u}_p=Ae^{vr\cos(\theta)/2D} \tag{16.42}$$

Substituting this relationship into (16.41) and rearranging gives

$$A=\frac{4D(\dot{m}_m+b_m g_{im})}{4\kappa^2D^2-v^2} \tag{16.43}$$

Substituting (16.43) into (16.42) leads the particular solutions to become

$$\hat{u}_p=\frac{4D(\dot{m}_m+b_m g_{im})}{4\kappa^2D^2-v^2}e^{vr\cos(\theta)/2D} \tag{16.44}$$

and

$$\hat{c}_{im_p}=\alpha_{im}\left[\frac{4D(\dot{m}_m+b_m g_{im})}{4\kappa^2D^2-v^2}\right]+g_{im} \tag{16.45}$$

Collecting the homogeneous and the particular parts, the general solutions would then be expressed as

$$\hat{u}(r,\theta,\omega) = \sum_s [a_s \cos(s\theta) + b_s \sin(s\theta)]K_s(\kappa r) + \frac{4D(\dot{m}_m + b_m g_{im})}{4\kappa^2 D^2 - \nu^2} e^{\nu r \cos(\theta)/2D}$$

and

$$\hat{c}_{im} = \alpha_{im} \sum_s [a_s \cos(s\theta) + b_s \sin(s\theta)]K_s(\kappa r)e^{-\nu r \cos(\theta)/2D} + \alpha_{im}\left[\frac{4D(\dot{m}_m + b_m g_{im})}{4\kappa^2 D^2 - \nu^2}\right] + g_{im} \tag{16.46}$$

Boundary Condition

Recalling the first boundary condition in (16.21) and transforming it to the frequency domain yields

$$\hat{c}_m(a,\theta,\omega) = \hat{C}_0(\omega)$$

or

$$\hat{u}(a,\theta,\omega) = \hat{C}_0(\omega)e^{\nu a \cos(\theta)/2D}$$

Solving the first equation in (16.46) at $r = a$, this boundary condition leads to

$$\sum_s [a_s \cos(s\theta) + b_s \sin(s\theta)]K_s(\kappa a) + \frac{4D(\dot{m}_m + b_m g_{im})}{4\kappa^2 D^2 - \nu^2} e^{\nu a \cos(\theta)/2D} = \hat{C}_0(\omega)e^{\nu a \cos(\theta)/2D}$$

Rearranging this equation results in

$$\sum_s a_s \cos(s\theta) + b_s \sin(s\theta) = Fe^{\nu a \cos(\theta)/2D}$$

with

$$F = \frac{1}{K_s(\kappa a)}\left[\hat{C}_0(\omega) - \frac{4D(\dot{m}_m + b_m g_{im})}{4\kappa^2 D^2 - \nu^2}\right] \tag{16.47}$$

which is a typical trigonometric Fourier series with its coefficients given by

$$a_0 = \frac{1}{2\pi}\int_0^{2\pi} Fe^{\nu a \cos(\theta)/2D} d\theta, \quad s = 0$$
$$a_s = \frac{1}{\pi}\int_0^{2\pi} Fe^{\nu a \cos(\theta)/2D} \cos(s\theta)d\theta, \quad s \geq 1$$
$$b_s = \frac{1}{\pi}\int_0^{2\pi} Fe^{\nu a \cos(\theta)/2D} \sin(s\theta)d\theta, \quad s \geq 1$$

BVP Solution

The solution of this boundary value problem for a given frequency ω_n is expressed as

$$\hat{u}(r,\theta,\omega_n) = \sum_s [a_s \cos(s\theta) + b_s \sin(s\theta)] K_s(\kappa r) + \frac{4D(\dot{m}_m + b_m g_{im})}{4\kappa^2 D^2 - v^2} e^{vr\cos(\theta)/2D}$$

$$\hat{c}_{im}(r,\theta,\omega_n) = \alpha_{im} \sum_s [a_s \cos(s\theta) + b_s \sin(s\theta)] K_s(\kappa r) e^{-vr\cos(\theta)/2D} + \alpha_{im} \left[\frac{4D(\dot{m}_m + b_m g_{im})}{4\kappa^2 D^2 - v^2} \right] + g_{im}$$

with

$$a_0 = \frac{1}{2\pi} \int_0^{2\pi} F e^{va\cos(\theta)/2D} d\theta, \quad s = 0$$

$$a_s = \frac{1}{\pi} \int_0^{2\pi} F e^{va\cos(\theta)/2D} \cos(s\theta) d\theta, \quad s \geq 1$$

$$b_s = \frac{1}{\pi} \int_0^{2\pi} F e^{va\cos(\theta)/2D} \sin(s\theta) d\theta, \quad s \geq 1$$

$$F = \frac{1}{K_s(\kappa a)} \left[\hat{C}_0(\omega_n) - \frac{4D(\dot{m}_m + b_m g_{im})}{4\kappa^2 D^2 - v^2} \right]$$

$$g_{im} = \frac{\dot{m}_{im}}{i\omega_n R_{im} + b_{im} + \mu_{im}}$$

$$\kappa^2 = \frac{v^2}{4D^2} + \frac{\mu_m}{D} + \frac{b_m}{D} + \frac{i\omega_n}{\alpha} - \frac{b_m}{D}\alpha_{im}$$

$$\alpha_{im} = \frac{b_{im}}{i\omega_n R_{im} + b_{im} + \mu_{im}}$$

Having solved the problem in the frequency domain, the solution in the time domain can then be obtained via the inverse FFT algorithm, such that

$$u(r,\theta,t) = \sum_n \hat{u}(r,\theta,\omega_n) e^{i\omega_n t}$$

$$c_{im}(r,\theta,t) = \sum_n \hat{c}_{im}(r,\theta,\omega_n) e^{i\omega_n t}$$

Furthermore, having solved for $u(r,\theta,t)$, the concentration of the mobile contaminant can readily be determined from (16.27), as

$$c_m(r,\theta,t) = u(r,\theta,t) e^{-vr\cos(\theta)/2D}$$

16.1.4 2D planar Domain: Cylindrical Contaminant Source with First-Order Decay and Zero-Order Production: Neumann Problem

We here consider a contaminant transport in a dual porosity infinite domain induced by a cylindrical contaminant source with a Neumann boundary condition effect. The contaminant exhibits first-order decay and zero-order produc-

16

tion, and moving with velocity $-v$ along the x-axis. Assuming $D_x = D_y = D$, the boundary value problem of this scenario can be expressed in the Cartesian coordinate system as

$$
\begin{aligned}
&R_m \frac{\partial c_m}{\partial t} - v\frac{\partial c_m}{\partial x} - D\frac{\partial^2 c_m}{\partial x^2} - D\frac{\partial^2 c_m}{\partial y^2} = -b_m(c_m - c_{im}) - \mu_m c_m + \dot{m}_m \\
&R_{im}\frac{\partial c_{im}}{\partial t} = b_{im}(c_m - c_{im}) - \mu_{im}c_{im} + \dot{m}_{im} \\
&\text{IC}: \quad c_m(x,y,0) = c_{im}(x,y,0) = 0 \\
&\text{BC}: \quad -2\pi a D\frac{\partial c_m}{\partial r}\bigg|_{\sqrt{x^2+y^2}=a} = q(t) \\
&\qquad c_m(x,y,t)|_{(a,\theta)} = c_m(x,y,t)|_{(a,\theta+2\pi)} \\
&\qquad c_{im}(x,y,t)|_{(a,\theta)} = c_{im}(x,y,t)|_{(a,\theta+2\pi)}
\end{aligned} \tag{16.48}
$$

General Solution

As demonstrated in the previous section, the general solution to the partial differential equations in (16.48), in the frequency domain, is given by

$$
\begin{aligned}
\hat{u}(r,\theta,\omega) &= \sum_s [a_s\cos(s\theta) + b_s\sin(s\theta)]K_s(\kappa r) \\
&\quad + \frac{4D(\dot{m}_m + b_m g_{im})}{4\kappa^2 D^2 - v^2} e^{vr\cos(\theta)/2D}, \qquad s = 0,1,2,\cdots \\
\hat{c}_{im} &= \alpha_{im}\sum_s [a_s\cos(s\theta) + b_s\sin(s\theta)]K_s(\kappa r)e^{-vr\cos(\theta)/2D} \\
&\quad + \alpha_{im}\left[\frac{4D(\dot{m}_m + b_m g_{im})}{4\kappa^2 D^2 - v^2}\right] + g_{im}, \qquad s = 0,1,2,\cdots
\end{aligned} \tag{16.49}
$$

with

$$
\hat{c}_m(r,\theta,\omega) = \hat{u}(r,\theta,\omega)e^{-vr\cos(\theta)/2D} \tag{16.50}
$$

Boundary Condition

Transforming the first boundary condition in (16.48) to the frequency domain yields

$$
-2\pi a D\frac{\partial \hat{c}_m}{\partial r}\bigg|_{r=a} = \hat{q}(\omega)
$$

Using the product and chain rules on (16.50) results in

$$
\frac{\partial \hat{c}_m}{\partial r} = \left(\frac{\partial \hat{u}}{\partial r} - \frac{v\cos(\theta)}{2D}\hat{u}\right)e^{-vr\cos(\theta)/2D}
$$

Thus, at $r = a$, upon utilizing (16.49), this boundary condition implies

$$\pi\left[a_s\cos(s\theta) + b_s\sin(s\theta)\right]\left[\begin{array}{l}2\mathrm{K}_{s+1}(\kappa a)D\kappa a + av\mathrm{K}_s(\kappa a)\cos(\theta)\\ -2sD\mathrm{K}_s(\kappa a)\end{array}\right]e^{-va\cos(\theta)/2D} = \hat{q}(\omega) \tag{16.51}$$

This can be written as

$$\sum_s [a_s\cos(s\theta) + b_s\sin(s\theta)] = f(\theta)\, e^{v\, a\cos(\theta)/2D}$$

with

$$f(\theta) = \frac{\hat{q}(\omega)}{\pi\left[2D\kappa a\mathrm{K}_{s+1}(\kappa a) + \{av\cos(\theta) - 2sD\}\mathrm{K}_s(\kappa a)\right]}$$

which is a typical trigonometric Fourier series with its coefficients given by

$$\begin{aligned}
a_0 &= \frac{1}{2\pi}\int_0^{2\pi} f(\theta)e^{va\cos(\theta)/2D}d\theta, && s = 0\\
a_s &= \frac{1}{\pi}\int_0^{2\pi} f(\theta)e^{va\cos(\theta)/2D}\cos(s\theta)d\theta, && s \geq 1\\
b_s &= \frac{1}{\pi}\int_0^{2\pi} f(\theta)e^{va\cos(\theta)/2D}\sin(s\theta)d\theta, && s \geq 1
\end{aligned}$$

BVP Solution

The solution of this boundary value problem for a given frequency ω_n is thus given by

$$\begin{aligned}
\hat{u}(r,\theta,\omega_n) &= \sum_s [a_s\cos(s\theta) + b_s\sin(s\theta)]K_s(\kappa r)\\
&\quad + \frac{4D(\dot{m}_m + b_m g_{im})}{4\kappa^2D^2 - v^2}e^{vr\cos(\theta)/2D}\\
\hat{c}_{im}(r,\theta,\omega_n) &= \alpha_{im}\sum_s [a_s\cos(s\theta) + b_s\sin(s\theta)]K_s(\kappa r)e^{-vr\cos(\theta)/2D}\\
&\quad + \alpha_{im}\left[\frac{4D(\dot{m}_m + b_m g_{im})}{4\kappa^2D^2 - v^2}\right] + g_{im}
\end{aligned} \tag{16.52}$$

16

with

$$
\begin{aligned}
a_0 &= \frac{1}{2\pi}\int_0^{2\pi} f(\theta)e^{va\cos(\theta)/2D}d\theta, && s=0\\
a_s &= \frac{1}{\pi}\int_0^{2\pi} f(\theta)e^{va\cos(\theta)/2D}\cos(s\theta)d\theta, && s\geq 1\\
b_s &= \frac{1}{\pi}\int_0^{2\pi} f(\theta)e^{va\cos(\theta)/2D}\sin(s\theta)d\theta, && s\geq 1\\
f(\theta) &= \frac{\hat{q}(\omega)}{\pi\left[2D\kappa a\,\mathrm{K}_{s+1}(\kappa a)+\{av\cos(\theta)-2sD\}K_s(\kappa a)\right]}\\
g_{im} &= \frac{\dot{m}_{im}}{R_{im}i\omega_n+b_{im}+\mu_{im}}\\
\kappa^2 &= \frac{v^2}{4D^2}+\frac{\mu_m}{D}+\frac{b_m}{D}+\frac{i\omega_n}{\alpha}-\frac{b_m}{D}\alpha_{im}\\
\alpha_{im} &= \frac{b_{im}}{i\omega_n R_{im}+b_{im}+\mu_{im}}
\end{aligned}
$$

Having solved for the frequency domain, the solution in the time domain can then be obtained via the inverse FFT algorithm, such that

$$
\begin{aligned}
u(r,\theta,t) &= \sum_n \hat{u}(r,\theta,\omega_n)e^{i\omega_n t}\\
c_{im}(r,\theta,t) &= \sum_n \hat{c}_{im}(r,\theta,\omega_n)e^{i\omega_n t}
\end{aligned}
$$

Furthermore, after obtaining the solution for $u(r,\theta,t)$, the concentration of the mobile contaminant can be determined using (16.50), which, in the time domain, is expressed as

$$
c_m(r,\theta,t) = u(r,\theta,t)e^{-vr\cos(\theta)/2D} \tag{16.53}
$$

- **Example 16.2 Planar 2D first-order decay and zero-order production**

A fully saturated porous medium domain, having a bulk density 1800 kg/m^3 and porosities $\varphi_m = 0.35$ and $\varphi_{im} = 0.15$, is characterized as a dual porosity domain. It is exposed to a 1.5 m in radius cylindrical contaminant with a hydrodynamic dispersion coefficient 1.25 m^2/d. The groundwater in the porous domain flows at an average velocity of 0.5 m/d. No initial concentration of the contaminant species exists in the domain. At the circumference of the contaminant source, the domain is exposed to a pulse with a concentration flux of 20 kg/m · s for nearly 365 days, see Fig. 16.4. The mass exchange coefficients are $b_m = b_{im} = 0.001\,d^{-1}$. We study four possible scenarios:

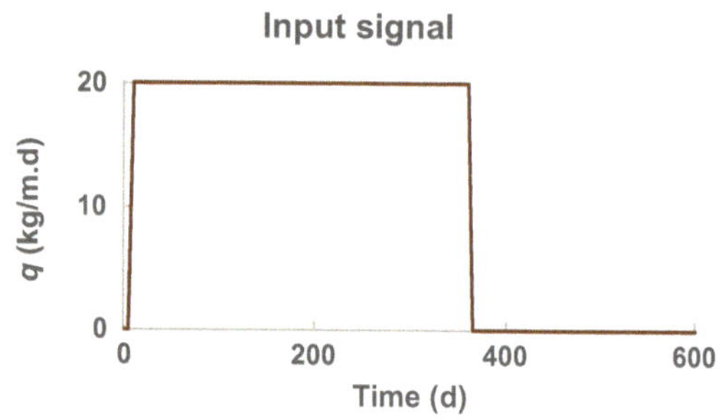

Fig. 16.4 Input concentration flux

1. The species exhibits no production or decay.
2. The species undergoes first-order decay of $\mu_{w,m} = \mu_{w,im} = 0.01\, d^{-1}$ and $\mu_{s,m} = \mu_{s,im} = 0$; and no production rates.
3. The species exhibits production with rates of $\dot{m}_{w,m} = \dot{m}_{w,im} = 0.01\,\text{kg/m}^3\text{s}$ and $\dot{m}_{s,m} = \dot{m}_{s,im} = 0$.
4. The species exhibits production with rates $\dot{m}_{w,m} = \dot{m}_{w,im} = 0.01\,\text{kg/m}^3\text{s}$ and $\dot{m}_{s,m} = \dot{m}_{s,im} = 0.0$; and simultaneously undergoes first-order decay of $\mu_{w,m} = \mu_{w,im} = 0.01\, d^{-1}$ and $\mu_{s,m} = \mu_{s,im} = 0$.

Solution

As for Example 16.1, we address this problem by:

(a) Transforming the input concentration pulse to the frequency domain using the forward fast Fourier transform (FFT). Here we used 4096 FFT samples with a time step of 0.5 d.

(b) Solving the problem in the frequency domain using (16.52).

(c) Transforming the computed concentrations back to the time domain using the inverse fast Fourier transform (IFFT) algorithm, followed by (16.53).

16

Figure 16.5 shows the temporal concentration of the mobile and immobile species for the four possible scenarios. Figures 16.6 and 16.7 shows the distribution of the show the spatial distribution of the mobile and immobile species, respectively, for the scenario involving bot production and decay.

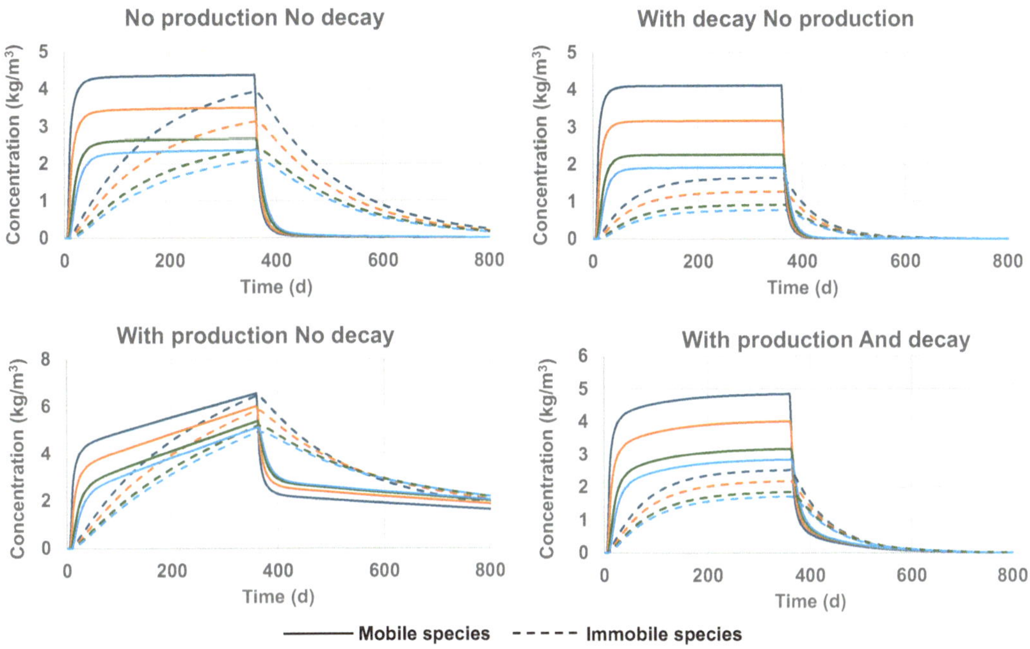

Fig. 16.5 Four scenarios of the temporal distribution for mobile and immobile species

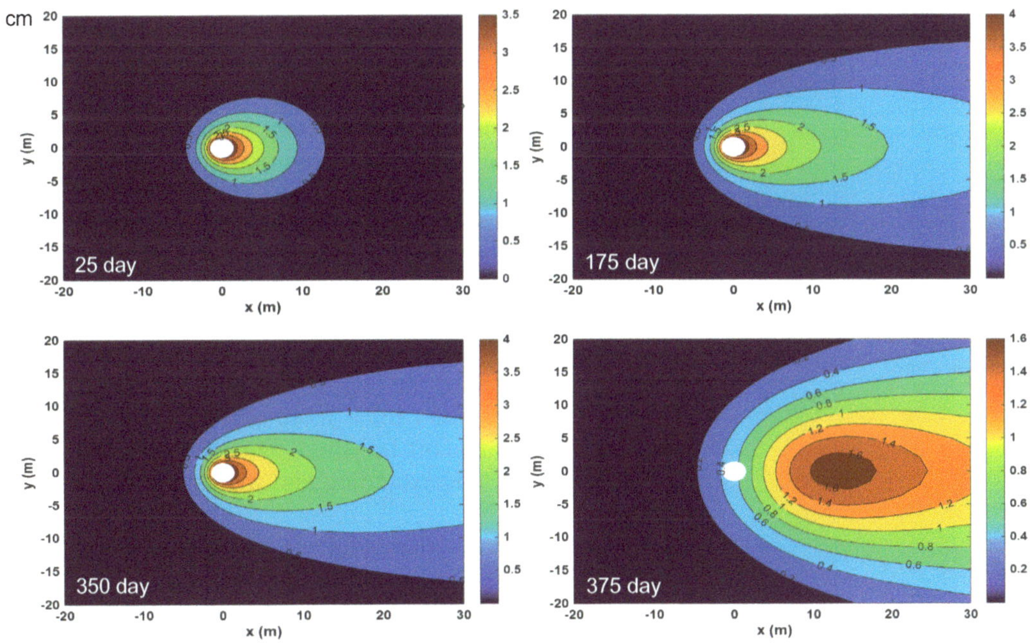

Fig. 16.6 Mobile concentration at different times

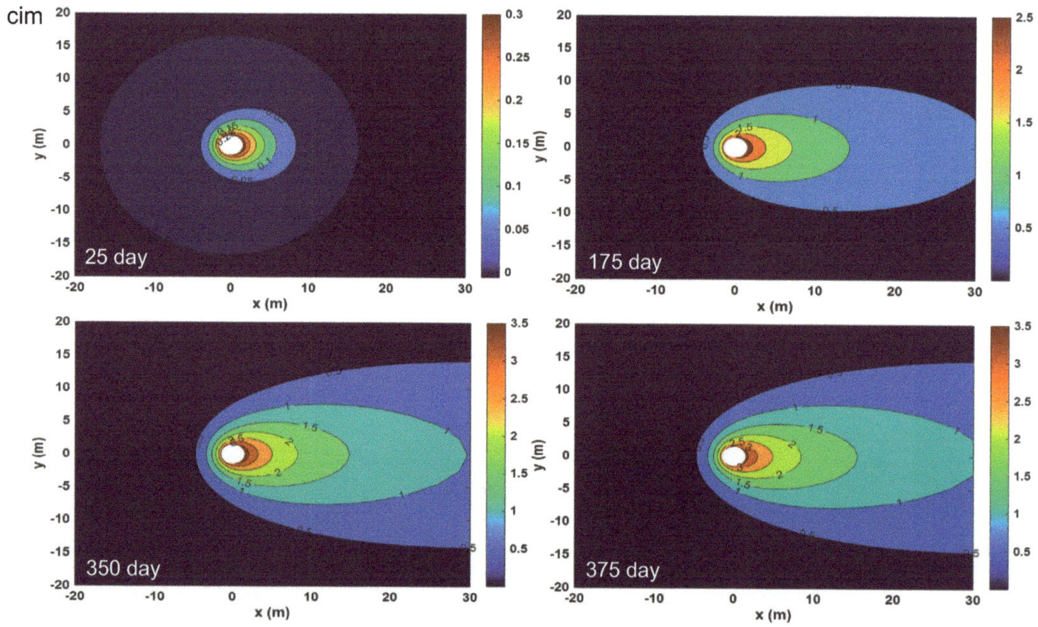

Fig. 16.7 Immobile concentration at different times

16.2 Dual Permeability Approach

The dual permeability method, also known as the *double permeability* method, is a conceptual approach used in environmental engineering and hydrogeology to describe fluid flow and contaminant transport in heterogeneous porous medium domains. Unlike the traditional *equivalent porous medium* approach, also termed *single permeability* or *single domain* approach, this technique considers two distinct flow domains within the porous medium: the solid matrix and the fracture network. The solid matrix represents part of the domain with a finer, low-permeability material, while the fracture network represents the remaining part of the domain having interconnected pathways with higher permeability. This dual nature allows for a more accurate representation of complex contaminant transport phenomena, making it a useful concept in understanding and managing groundwater pollution.

The governing equations of a non-equilibrium transport in a heterogeneous 1D domain is

$$\begin{aligned} R_m \frac{\partial c_m}{\partial t} + v_m \frac{\partial c_m}{\partial x} - D_m \frac{\partial^2 c_m}{\partial x^2} &= -b_m \left(c_m - c_f\right) - \mu_m c_m \\ R_f \frac{\partial c_f}{\partial t} + v_f \frac{\partial c_f}{\partial x} - D_f \frac{\partial^2 c_f}{\partial x^2} &= b_f \left(c_m - c_f\right) - \mu_f c_f \end{aligned} \tag{16.54}$$

where the subscripts m and f refer to the species concentration in the matrix and fracture components, respectively. The physical parameters in (16.54) are given in ▶ Sect. 16.1, with the subscript im replaced by f to denote the fractured network.

16.2.1 1D Domain: First-Order Decay with No Production: Dirichlet Problem

The boundary value problem of a contaminant transport in a double permeability semi-infinite domain with first-order decay on both components, subjected to a Dirichlet boundary condition, is

$$
\begin{aligned}
& R_m \frac{\partial c_m}{\partial t} + v_m \frac{\partial c_m}{\partial x} - D_m \frac{\partial^2 c_m}{\partial x^2} = -\alpha\left(c_m - c_f\right) - \mu_m c_m \\
& R_f \frac{\partial c_f}{\partial t} + v_f \frac{\partial c_f}{\partial x} - D_f \frac{\partial^2 c_f}{\partial x^2} = \alpha\left(c_m - c_f\right) - \mu_f c_f \\
& \text{IC}: \quad c_m(x,0) = c_f(x,0) = 0 \\
& \text{BC}: \quad c_m(0,t) = C_{m0}(t) \\
& \qquad\;\; c_f(0,t) = C_{f0}(t) \\
& \qquad\;\; c_m(\infty,t) = c_f(\infty,t) = 0
\end{aligned}
\tag{16.55}
$$

where $\alpha = b_m = b_f$ representing a mass exchange coefficient.

General Solution

Transforming (16.55) to the frequency domain yields

$$
\begin{aligned}
& D_m \frac{d^2 \hat{c}_m}{dx^2} - v_m \frac{d\hat{c}_m}{dx} - (i\omega R_m + \alpha + \mu_m)\hat{c}_m + \alpha \hat{c}_f = 0 \\
& D_f \frac{d^2 \hat{c}_f}{dx^2} - v_f \frac{d\hat{c}_f}{dx} - \left(i\omega R_f + \alpha + \mu_f\right)\hat{c}_f + \alpha \hat{c}_m = 0 \\
& \text{BC}: \hat{c}_m(0,\omega) = \hat{C}_{m0}(\omega) \\
& \qquad \hat{c}_f(0,\omega) = \hat{C}_{f0}(\omega) \\
& \qquad \hat{c}_m(\infty,\omega) = \hat{c}_f(\infty,\omega) = 0
\end{aligned}
\tag{16.56}
$$

As the coefficients of the differential equations in (16.56) are constants, the trial solutions can be

$$
\begin{aligned}
\hat{c}_m &= a e^{-\kappa x} \\
\hat{c}_f &= b e^{-\kappa x}
\end{aligned}
$$

Substituting these solutions into the DEs in (16.56), and rearranging, yields

$$
\begin{aligned}
& D_m \kappa^2 a + v_m \kappa a - (i\omega R_m + \alpha + \mu_m) a + \alpha b = 0 \\
& D_f \kappa^2 b + v_f \kappa b - \left(i\omega R_f + \alpha + \mu_f\right) b + \alpha a = 0
\end{aligned}
$$

Putting these two equations in a matrix format, we get

$$\begin{pmatrix} G_m & \alpha \\ \alpha & G_f \end{pmatrix} \begin{pmatrix} a \\ b \end{pmatrix} = \begin{pmatrix} 0 \\ 0 \end{pmatrix} \tag{16.57}$$

with

$$G_m = D_m \kappa^2 + v_m \kappa - (i\omega R_m + \alpha + \mu_m)$$
$$G_f = D_f \kappa^2 + v_f \kappa - \left(i\omega R_f + \alpha + \mu_f\right)$$

The solution of (16.57) is valid only if the determinate of the matrix is set to zero, resulting in a fourth order polynomial equation of the form

$$m_4 \kappa^4 + m_3 \kappa^3 + m_2 \kappa^2 + + m_1 \kappa + m_0 = 0$$

where κ represents the four eigenvalues of the system. Following this, the general solutions of both mediums become

$$\begin{aligned} \hat{c}_m &= a_1 e^{-\kappa_1 x} + a_2 e^{-\kappa_2 x} + a_3 e^{-\kappa_3 x} + a_4 e^{-\kappa_4 x} \\ \hat{c}_f &= b_1 e^{-\kappa_1 x} + b_2 e^{-\kappa_2 x} + b_3 e^{-\kappa_3 x} + b_4 e^{-\kappa_4 x} \end{aligned} \tag{16.58}$$

Boundary Conditions

The third boundary conditions in (16.56) imply that the coefficients associated with the positive exponential terms in (16.58) must be discarded, as they result in an unbounded solution at far distances. This entails

$$\begin{aligned} \hat{c}_m &= a_1 e^{-\kappa_1 x} + a_2 e^{-\kappa_2 x} \\ \hat{c}_f &= b_1 e^{-\kappa_1 x} + b_2 e^{-\kappa_2 x} \end{aligned} \tag{16.59}$$

Since the two species are coupled, their constants must be related, expressed via the eigenvectors of the system. The constants a and b in (16.57) are related via

16

$$b_j = Y_i a_j$$

where

$$Y_j = -\frac{G_m}{\alpha}$$

with j denoting that for each eigenvalue κ_j there is a corresponding eigenvector. Following this, (16.59) can be expressed as

$$\begin{aligned} \hat{c}_m &= a_1 e^{-\kappa_1 x} + a_2 e^{-\kappa_2 x} \\ \hat{c}_f &= Y_1 a_1 e^{-\kappa_1 x} + Y_2 a_2 e^{-\kappa_2 x} \end{aligned} \tag{16.60}$$

At $x = 0$, the first and second boundary conditions in (16.56) imply

$$a_1 + a_2 = \hat{C}_{0m}$$
$$Y_1 a_1 + Y_2 a_2 = \hat{C}_{0f}$$

Solving these two equations simultaneously leads to

$$a_1 = \frac{\hat{C}_{0f} - Y_2 \hat{C}_{0m}}{Y_1 - Y_2} \qquad (16.61)$$
$$a_2 = \hat{C}_{0m} - a_1$$

BVP Solution

Substituting (16.61) into (16.60), the solution of the BVP in (16.56) for a given frequency ω_n can be expressed as

$$\hat{c}_m(x, \omega_n) = \frac{\hat{C}_{0f} - Y_2 \hat{C}_{0m}}{Y_1 - Y_2} \left(e^{-\kappa_1 x} - e^{-\kappa_2 x}\right) + \hat{C}_{0m} e^{-\kappa_2 x}$$
$$\hat{c}_f(x, \omega_n) = \frac{\hat{C}_{0f} - Y_2 \hat{C}_{0m}}{Y_1 - Y_2} \left(Y_1 e^{-\kappa_1 x} - Y_2 e^{-\kappa_2 x}\right) + Y_2 \hat{C}_{0m} e^{-\kappa_2 x} \qquad (16.62)$$

with

$$Y_j = -\frac{1}{\alpha}\left[D_m \kappa_j^2 + v_m \kappa_j - (i\omega R_m + \alpha + \mu_m)\right]$$

Following this, the solution in the time domain can then be obtained via the inverse FFT algorithm, such that

$$c_m(x, t) = \sum_n \hat{c}_m(x, \omega_n) e^{i\omega_n t}$$
$$c_f(x, t) = \sum_n \hat{c}_f(x, \omega_n) e^{i\omega_n t}$$

- **Example 16.3: 1D dual permeability**

A fully saturated fractured porous medium domain, having a bulk density 1500 kg/m^3 and porosities $\varphi_m = 0.15$, $\varphi_f = 0.5$, is exposed at one of its boundaries to a chemical species with $D_m = D_f = 0.01\,\text{m}^2/\text{s}$, $K_d = 1\,\text{m}^3/\text{kg}$ and fraction $f = 0.3$. The groundwater in the porous matrix moves at an average velocity of 1.0e-5 m/s, and in the fractured domain it flows at 0.1 m/s. No initial concentration of the chemical exists in the domain. At $x = 0$, the domain is exposed to a rectangular pulse concentration of 10 kg/m^3 for nearly 18 h. Under some chemical or physical processes, the species simultaneously undergoes first-order decay of $\mu_{w,m} = \mu_{w,f} = 0.0001\, s^{-1}$ and $\mu_{s,m} = \mu_{s,f} = 0.00001\, s^{-1}$ with no production. Compute the concentration of the species at 0.0, 0.5, 1.0, and 3.0 m for $b = 0.008\, s^{-1}$.

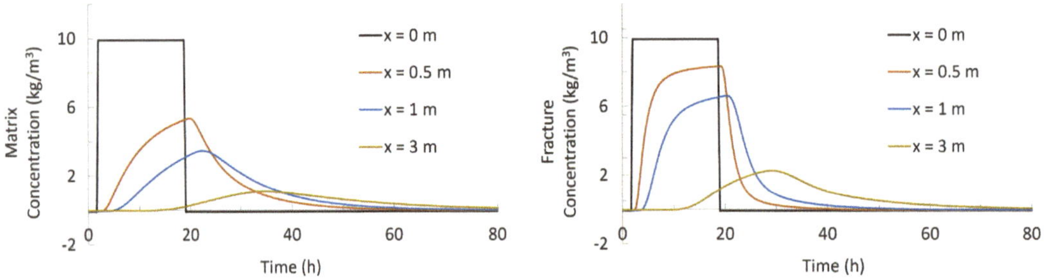

Fig. 16.8 Species concentration at the porous matrix (left) and fractured domain (right)

Solution

As for Example 16.1 we address this problem by:

(a) Transforming the input concentration pulse to the frequency domain using the forward fast Fourier transform (FFT).

(b) Solving the problem in the frequency domain using (16.62).

(c) Transforming the computed concentrations back to the time domain using the inverse fast Fourier transform algorithm (IFFT).

Figure 16.8 shows the computed concentrations at the porous matrix and the fracture domain, where it can be seen that the fractures carry more species, especially at short distances.

Problems

1. For what kind of aquifers the multi-domain models are necessary?
2. What is the main difference between the dual porosity model and the dual permeability model? In what case the two models give close results?
3. Solve the boundary value problem in ▶ Sect. 16.1.1 with a Neumann boundary condition.
4. Solve the boundary value problem in ▶ Sect. 16.1.2 with a Robin boundary condition. Then, solve the problem in Example 16.1 with a Robin boundary condition. Make assumptions for the surrounding concentration and the interaction coefficient.
5. Solve the boundary value problem in ▶ Sect. 16.1.4 with a Robin boundary condition.
6. Solve the boundary value problem in ▶ Sect. 16.2.1 with a Robin boundary condition.
7. Solve the boundary value problem in ▶ Sect. 16.2.1 in a finite domain $0 < x < L$.

8. Solve Example 16.2 for a case where the Dirichlet boundary condition is prescribed only on the porous matrix.

   Hint: Set $C_{f0}(t) = 0$ in (16.55).

References

Abramowitz and Stegun (1972) Handbook of mathematical functions. Dover Publications, Inc.

Carslaw HS, Jaeger JC (1959) Conduction of heat in solids, 2nd edn. Oxford University Press

Toride N, Leij FJ, van Genuchten MT (1993) A comprehensive set of analytical solutions for nonequilibrium solute transport with first-order decay and zero-order production. Water Resour Res 29(7):2167–2182

Supplementary Information

Appendix – 2

R. Al Khoury and C. Kasbergen, *Analytical Subsurface Environment*,
https://doi.org/10.1007/978-3-031-93910-5

Appendix

Computer Coding

In this appendix, we outline the basic procedure required to implement the mathematical models presented throughout this book in computer codes. The focus is placed on utilizing the FORTRAN programming language to develop these codes. Each model discussed in the book can be written as a separate FORTRAN subroutine, which can then be called from a main program. We here introduce a program algorithm consisting of six building blocks that are applicable to all cases discussed in the book. To illustrate the construction of a subroutine, we provide a detailed example that includes the complete coding for one of the mathematical models featured in ► Chap. 11.

We have designed the coding structure to be adaptable across various programming languages, ensuring the algorithm validity beyond FORTRAN. The provided code example can be easily modified for use in several other programming languages, including Python, MATLAB, and C++.

A.1 Algorithm

The algorithm presented in this appendix comprises six building blocks, each designed to address a specific aspect of the mathematical model under study. Box A.1 lists an algorithm applicable to all subroutines. It includes the associated FORTRAN commands necessary for building the computer codes. The building blocks are defined as follows:

1. **Initialization**: This block handles the initial setup, including variable declarations, parameter definitions, and the opening of files.
2. **Forward FFT**: This block manages the transformation of the input data from the time domain to the frequency domain using the forward fast Fourier transform (FFT).
3. **Core Computation**: This block implements the mathematical operations necessary to compute the state variables in the frequency domain. It loops over all frequencies up to the Nyquist frequency for the requested output points in the domain.
4. **Inverse FFT**: This block manages the transformation of the computed results from the frequency domain to the time domain using the inverse fast Fourier transform (IFFT). This includes calculating the mirrored complex conjugates of the computed state variable for frequencies beyond the Nyquist frequency. The computed state vari-

ables with time and space are then written to output files that can be used by Excel or other software to visualize the model responses.

5. **Output Generation**: This block deals with the creation and formatting of the output results for the two-dimensional cases, making them suitable for contour plotting.
6. **Closing**: This block ensures the deallocation of variables and closing of files.

Together, these building blocks form a unified and adaptable framework for implementing the mathematical models in various programming environments.

Box A.1 Program Algorithm for Subroutines

```
SUBROUTINE BOOK_EXAMPLE()
! Building Block 1: Initialization. (variable declarations, opening files and reading data)
DECLARE necessary variables
OPEN input/output files
READ model parameters and time input signals from file
ALLOCATE data variables
! Building Block 2: Forward FFT
CALL FFT (input time signal, frequency domain output)
! Building Block 3: Core computation. (computing state variable in frequency domain)
DO n = 1, N/2 ( N : number of frequencies, and N/2 is the Nyquist frequency )
   ! Loop over output locations
   DO ioutp =1, NOUTP ( NOUTP = number of output locations )
        Compute the state variable (example, temperature response û(x, ω_n) ) in
        the frequency domain using the BVP solution
   END DO
END DO
! Building Block 4: Inverse FFT
DO ioutp=1, NOUTP ( NOUTP = number of output locations )
  Calculate the mirrored complex conjugates of the computed state variable for all
  frequencies
      CALL IFFT (input frequency responses, time domain output)
      WRITE results (write response to tim-file)
END DO
! Building Block 5: Output generation (Optional )
CREATE output for contour plots
! Building Block 6: Closing
DEALLOCATE data variables
CLOSE input/output files
END SUBROUTINE
```

A.2 Code Example

Consider a two-dimensional plane strain consolidation of a soil mass subjected to pressure U_0 (Pa), exerted by an earth dam with half width a (m), see Fig. A.1. The excess pore pressure generated by this load has been derived in detail in Sect. 11.5. It gave

$$\hat{u}_e(x, \omega_n) = \sum_m A_m \cos \kappa_x x \; e^{-\kappa_z z / \sqrt{c_z}}, \quad m = 0, 1, 2, \cdots$$

with

$$A_0 = \frac{1}{l} \int_0^a \hat{U}_0(x)\, dx, \quad m = 0$$

$$A_m = \frac{2}{l} \int_0^a \hat{U}_0(x) \cos(\kappa_x x)\, dx, \quad m = 1, 2, \cdots$$

$$\kappa_x = \frac{1}{l}\left(\frac{\pi}{2} + m\pi\right)$$

$$\kappa_z = \sqrt{c_x \kappa_x^2 + i\omega_n}$$

where c_x and c_z ($m^2 s^{-1}$) are the coefficients of consolidation in the horizontal and vertical directions, respectively, and l (m) is some distance far from the load where consolidation is negligible.

The solution in the time domain is obtained using the inverse fast Fourier transform (IFFT) algorithm, as

$$u_e(x, z, t) = \sum_n \hat{u}_e(x, z, \omega_n)\, e^{i\omega_n t}$$

Construct a FORTRAN code for this model.

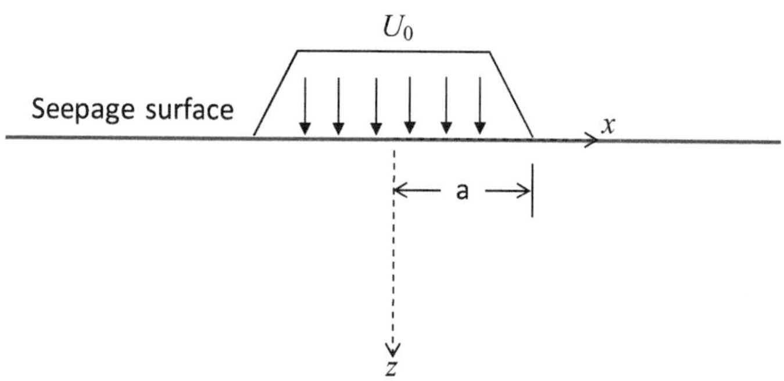

Figure A.1 Coding example of a consolidation model

Solution

Below, we provide a block-by-block code to implement the theory of the 2D plane strain consolidation discussed above.

Building Block 1: Initialization

In this building block we declare variables, open files, read input data and reserve memory for allocatable variables. Although FORTRAN permits implicit declaration of variables, it is more appropriate to declare them explicitly using the IMPLICIT NONE command. We use double precision floating-point numbers (REAL8) and double precision complex numbers (COMPLEX16).

```
      SUBROUTINE BOOK_EXAMPLE()
C  2D Consolidation: a plane strain condition
C  DECLARATION OF VARIABLES                                    ! LINE 1
      IMPLICIT NONE
      integer     :: ISTEP,NT,IOUTP,NOUTP,IOUTPUT,IERR
      INTEGER     :: I,J,M,N,MM,NN,NP2,FDAT,FIN,FOUT,FFRQ1,FFRQ2,FTMP
      INTEGER     :: NX,NZ,NTIME,START_INDEX,NDIM
      INTEGER     :: ISRADIAL  ! ( 0 = CARTESIAN; 1 = POLAR )
      real*8      :: A,L,OMEGAN,FREQN,DT,UREAL,UIMAG
      real*8      :: CX,CZ,PI,ZERO,FT,UJUNK1,UJUNK2,KAPPAM
      complex*16 :: ICI,KAPPAZ,U0_HAT,AM
      integer, allocatable :: TIMESTEPS(:)
      logical, external    :: INTIMESTEPLIST
      real*8, allocatable :: X(:),Z(:),TIM(:),UETIM(:,:)
      complex*16, allocatable :: UE_HAT(:,:),ASUM(:)
```

Next in this building block is the opening of the necessary input and output files. A user-created **<Name>.dat** file listing the model input data is opened, and a **<Name>.tim** file for storing the computed results in the time domain is generated and opened. Other files are also generated and opened for intermediate data storage.

```
C  OPEN FILES                                                  ! LINE 2
      fdat=70; fin=71; fout=72; ffrq1=73; ffrq2=74; ftmp=75
      OPEN(fdat,FILE='example_12.5.dat',FORM='FORMATTED',
     &                                        STATUS='UNKNOWN')
      OPEN(fin,FILE='example_12.5.in',FORM='FORMATTED',
     &                                        STATUS='UNKNOWN')
      OPEN(ffrq1,FILE='example_12.5.frq1',FORM='FORMATTED',
     &                                        STATUS='UNKNOWN')
      OPEN(ftmp,FILE='example_12.5.ftmp',FORM='FORMATTED',
     &                                        STATUS='UNKNOWN')
      OPEN(ffrq2,FILE='example_12.5.frq2',FORM='FORMATTED',
     &                                        STATUS='UNKNOWN')
      OPEN(fout,FILE='example_12.5.tim',FORM='FORMATTED',
     &                                        STATUS='UNKNOWN')
```

Following this, the code reads the input parameters and the time signals from the **<Name>.dat** file. An example of this file is provided in the Box A.2 below. The **<Name>.dat** file is created by the user and includes involved parameters, output points and the time steps (TIMESTEPS) when the response of the output points are needed to be written in the output file **<Name>.tim**. The remaining parameters should be self-explanatory when compared with the example theory.

```
C  READ INPUT PARAMETERS AND SIGNALS                               ! LINE 3
      READ (fdat,*) NP2           ! Number of frequencies
      READ (fdat,*) A,L,DT,MM     ! DT: sample time; MM: number of Fourier components
      READ (fdat,*) CX,CZ         ! Consolidation coefficients

      READ (FDAT,*) NTIME          !  NUMBER of OUTPUT TIME STEPS
      ALLOCATE(TIMESTEPS(NTIME),STAT=IERR)
      IF (IERR.NE.0) THEN
        WRITE (*,*) 'CANNOT ALLOCATE ARRAY TIMESTEPS'
        STOP
      ENDIF
      READ (FDAT,*) TIMESTEPS(1:NTIME)
      READ (FDAT,*) NX,NZ     ! NUMBER of OUTPUT LOCATIONS ALONG X and Z
      NOUTP=(NX+1)*(NZ+1)
      ALLOCATE(X(NOUTP),Z(NOUTP),STAT=IERR)
      IF (IERR.NE.0) THEN
        WRITE (*,*) 'CANNOT ALLOCATE ARRAYS X AND Z'
        STOP
      ENDIF
      CALL GENERATE_GRIDPOINTS(fdat,2,NX,X,NZ,Z) ! CREATE (X,Z)-GRID

      READ (fdat,*) NT         ! NUMBER OF TIME POINTS FOR U0(t)
      DO N=1,NT
        READ (fdat,*) ISTEP,FT
        WRITE (fin,*) ISTEP*DT,FT    ! WRITE fin-file
      ENDDO
```

Box A.2 Example of <Name>.dat file

```
4096                          ! NR OF FREQUENCIES
5.0 200.0 3600.0 1000         ! A,L,DT,MM (NR OF COS FOURIER)
1.0e-4 1.0e-5                 ! CX,CZ
4                             ! NR OF OUTPUT TIMESTEPS
32 112 122 162                ! OUTPUT TIMESTEPS
150 100                       ! NR OF X, NR OF Z (GRID POINTS)
0.0 15.                       ! MIN X, MAX X (WINDOW SIZE in X)
0.0 10.                       ! MIN Z, MAX Z (WINDOW SIZE in Z)
6                             ! NR OF POINTS FOR U0(t)
0     0.0                     ! TIMESTEP, U0 VALUE
10    0.0
12    10.0
112   10.0
114   0.0
4096 0.0
```

Next, we define the involved constants and, if needed, the model parameters. In the meanwhile, we allocate all allocatable variables.

```
C  CONSTANTS AND RELATED/DERIVED MODEL PARAMETERS
      ZERO=0.0d0
      ICI=(0.0d0,1.0d0); PI=4.0d0*DATAN(1.0d0)
C  ALLOCATE VARIABLES                                               ! LINE 4
      ALLOCATE(UE_HAT(NP2/2,NOUTP),TIM(NP2),
     &         UETIM(NP2,NOUTP),ASUM(NOUTP),STAT=IERR)
      IF (IERR.NE.0) THEN
        WRITE (*,*) 'NO ENOUGH MEMORY TO ALLOCATE UE_HAT'
        STOP
      ENDIF
C
```

Building Block 2: Forward FFT

In this building block we transform the input data from the time domain to the frequency domain using the forward fast Fourier transform (FFT).

```
C  TRANSFORM INPUT SIGNALS TO THE FREQUENCY DOMAIN                ! LINE 5
      rewind(fin); rewind(ffrql);
      CALL FourierTransform(dt,np2,fin,ffrql)
C
```

Building Block 3: Core Computation

In this building block we conduct the mathematical operations to compute the response of the state variable for every frequency and output locations specified in the **<Name>.dat** file.

```
C  CORE COMPUTATIION
      ASUM(1:NOUTP)=(0.0d0,0.0d0)
      rewind(ffrql);
      READ (ffrql,*) NP2
      READ (ffrql,*) FREQN,UJUNK1,UJUNK2
      DO N=1,NP2/2         ! NP2/2 is the Nyquist frequency         ! LINE 6
        READ (ffrql,*) FREQN,UREAL,UIMAG
        U0_HAT=DCMPLX(UREAL,UIMAG)
        OMEGAN=2.0d0*PI*FREQN
        DO IOUTP=1,NOUTP                                           ! LINE 7
          UE_HAT(N,IOUTP)=(0.0d0,0.0d0)
          DO M=0,MM
            KAPPAM=(PI/2.0d0+M*PI)/L
            KAPPAZ=CDSQRT(ICI*OMEGAN+CX*KAPPAM*KAPPAM)
            AM=2.0d0*U0_HAT*DSIN(KAPPAM*A)/(KAPPAM*L)
            UE_HAT(N,IOUTP)=UE_HAT(N,IOUTP)+
     &                      AM*DCOS(KAPPAM*X(IOUTP))*
     &                      CDEXP(-KAPPAZ*Z(IOUTP)/DSQRT(CZ))
          ENDDO
          ASUM(IOUTP)=ASUM(IOUTP)+UE_HAT(N,IOUTP)
        ENDDO                                                     ! LINE 11
      ENDDO                                                       ! LINE 12
      ASUM(1:NOUTP)=-2.0d0*ASUM(1:NOUTP)
C
```

Building Block 4: Inverse FFT

In this building block, we transform the computed state variable back to the time domain using the inverse fast Fourier transform (IFFT). As shown in the above code, the state variable is computed for all frequencies up to the Nyquist frequency (NP2/2). To account for the remaining frequencies, we generate the mirrored conjugates of the computed state variable before transforming the data back to the time domain.

```
C  WRITE FFRQ2 FILE AND TRANSFORM U_HAT TO TIME DOMAIN
      ioutput=1
      DO IOUTP=1,NOUTP                                        ! LINE 11
        rewind(ffrq1); rewind(ftmp); rewind(ffrq2);
        READ(ffrq1,*) NP2
        WRITE (ffrq2,*) NP2
        READ (ffrq1,*) FREQN,UJUNK1,UJUNK2
        WRITE (ffrq2,*) FREQN,DREAL(ASUM(IOUTP)),ZERO
        DO N=1,NP2/2
          READ (ffrq1,*) FREQN,UJUNK1,UJUNK2
          WRITE (ffrq2,*) FREQN,DREAL(UE_HAT(N,IOUTP)),
     &                                   DIMAG(UE_HAT(N,IOUTP))
        ENDDO
        DO N=1,NP2/2-1                                        ! LINE 12
          READ (ffrq1,*) FREQN,UJUNK1,UJUNK2
          WRITE (ffrq2,*) FREQN,
     &       DREAL(UE_HAT(NP2/2-N,IOUTP)),-DIMAG(UE_HAT(NP2/2-N,IOUTP))
        ENDDO
        rewind(ffrq2)
        CALL InvFastFourier(ioutput,ffrq2,ftmp)               ! LINE 13

        rewind(ftmp)
        DO N=1,NP2
          READ(ftmp,*) TIM(N),UETIM(N,IOUTP)
        ENDDO
      ENDDO
C
```

Building Block 5: Output Generation

In this building block we create output file containing the computed state variable response data in time for all the grid points specified in the **<Name>.dat** file to be read by MATLAB. A MATLAB script is made to generate contour plots for all specified times in the TIMESTEPS of the **<Name>.dat** file. Refer to Sect. A.3 below.

```
C CREATE CONTOURPLOT DATA TO BE USED IN MATLAB                ! LINE 16
      rewind(fout)
      NDIM=2; ISRADIAL=0
      WRITE (fout,*) NDIM,ISRADIAL
      WRITE (fout,*) 'x ','(m) ','z ','(m) ',NTIME,NX,NZ
      START_INDEX=1
      DO N=1,NP2
        IF (INTIMESTEPLIST(N,START_INDEX,NTIME,TIMESTEPS)) THEN
           WRITE (fout,*) TIM(N)
           IOUTP=0
           DO I=1,NX+1
             DO J=1,NZ+1
               IOUTP=IOUTP+1
               WRITE (fout,*) X(IOUTP),-Z(IOUTP),UETIM(N,IOUTP),' 0'
             ENDDO
           ENDDO
        ENDIF
      ENDDO
C
```

Building Block 6: Closing Program

In this building block we start closing the subroutine by first deallocating the allocated variables and closing all opened files.

```
C DEALLOCATE VARIABLES                                                  ! LINE 17
      DEALLOCATE(TIMESTEPS)
      DEALLOCATE(UE_HAT,TIM,UETIM,X,Z,ASUM)

C CLOSE FILES                                                           ! LINE 18
      CLOSE(fdat); CLOSE(fin); CLOSE(fout);
      CLOSE(ffrq1); CLOSE(ffrq2); CLOSE(ftmp)
C
      END SUBROUTINE
C
```

MATLAB Code

Here, we outline a procedure for presenting computed results generated by FORTRAN using MATLAB plotting features. The goal is to guide the reader through the script in a step-by-step manner, highlighting the key components and their functions in creating visual representations of the data.

Step-by-Step MATLAB script

Cleaning the Workspace and Command Window

```
clear; clc;
```

Reading <Name>.tim file generated by FORTRAN

This file contains the computed results. In here, we define in the file ***chooseML.dat*** which is assigned to the variable '**fname**'.

```
file0=fopen('chooseML.dat','r');
fname=fscanf(file0,"%s",[1]);
fclose(file0);

file=fopen(fname,'r');
```

Reading plot dimensions and type of coordinates from file

In **<Name>.tim** file we specify the type of plots that have to be generated.

'***ndim***': represents the dimension of the plot (1 is 1D, 2 is 2D).

'***isradial***': represents the type of the plot. If Cartesian *isradial*=0, and if Polar *isradial*=1.

```
ndim=fscanf(file,"%d",[1]);
isradial=fscanf(file,"%d",[1]);

if (ndim==1)
  # coding for 1D plots
elseif (ndim==2)
  # coding for 2D contour plots
end
```

1D Plot (*ndim* = 1)

The plot label ***'xlab'*** is assembled from *'xvar'* and *'xunit'*.

'xvar': the name of variable on the x-axis.

'xunit': the unit of the x-axis.

'ntime': the number of timesteps to be plotted.

'na': the number of grid points in x-direction. The total number of grid points is defined by ***'noutp'***.

```
xvar=fscanf(file,"%s",[1]);          xunit=fscanf(-
file,"%s",[1]);
xlab=strcat(xvar," ",xunit);
ntime=fscanf(file,"%d",[1]);
na=fscanf(file,"%d",[1]);

noutp=na+1;
```

Initializing time and spatial coordinates arrays

```
ttime=[1:ntime];

X1=[1:(Na+1)]; Y1=[1:(Na+1)];
```

Looping through time steps to read and plot data

```
for nt = 1:ntime
    # for every time step, read and plot the data
end

ttime(nt)=fscanf(file,"%f",[1]);
for i = 1:(na+1)
    X1(i)=fscanf(file,"%e",[1]);
    Y1(i)=fscanf(file,"%e",[1]);
    fscanf(file,'\n');
end
```

Plotting data for current time step

```
figure(nt); plot(X1,Y1);

xlabel(xlab);
```

2D Plot (*Ndim* = 2)

The plot labels ***'xlab'*** and ***'ylab'*** are, respectively, assembled from *'xvar'*, *'xunit'* and *'yvar'*,*'yunit'*.

'xvar': the name of variable on the x-axis.

'xunit': the unit of the x-axis.

'*yvar*': the name of variable on the y-axis.
'*yunit*': the unit of the y-axis.
'*ntime*': the number of timesteps to be plotted.
'*na*': the number of grid points in x-direction.
'*nb*': the number of grid points in y-direction. The total number of grid points is defined by **'*noutp*'**.

```
xvar=fscanf(file,"%s",[1]);           xunit=fscanf(-
file,"%s",[1]);
yvar=fscanf(file,"%s",[1]);           yunit=fscanf(-
file,"%s",[1]);
xlab=strcat(xvar," ",xunit); ylab=strcat(yvar,"
",yunit);
ntime=fscanf(file,"%d",[1]);
na=fscanf(file,"%d",[1]);                 nb=fscanf(-
file,"%d",[1]);

noutp=(na+1)*(nb+1);
```

Initializing time and spatial coordinates arrays

```
ttime=[1:ntime];

A=zeros(na+1,nb+1);            B=zeros(na+1,nb+1);
V=zeros(na+1,nb+1);
```

Looping through time steps to read and plot data

```
for nt = 1:ntime
    # for every time step, read and plot the data
end
```

The output data is read in for every grid point (x, y). The x- and y-coordinates are stored in **A** and **B** arrays for all grid indexes (i, j). The computed state variables are stored in **V** array. The variable **'*exclude*'** is used to overrule the computed value at a grid point to an undefined value (NaN: Not a Number), if applicable.

```
ttime(nt)=fscanf(file,"%f",1);
for i = 1:(na+1)
   for j = 1:(nb+1)
      A(i,j)=fscanf(file,"%e",1);
      B(i,j)=fscanf(file,"%e",1);
      V(i,j)=fscanf(file,"%e",1);
      exclude=fscanf(file,"%d",1);
      if (exclude~=0)
        V(i,j)=NaN;
      end
      fscanf(file,'\n');
   end
end
```

Plotting data for current time step

```
figure(nt);
```

Contour type

Two contour plots can be produced: Cartesian (*isradial*=0) or Polar (*isradial*=1). If a polar coordinate plot is needed, the polar coordinates are first converted to the Cartesian coordinates using the MATLAB function ***'pol2cart'***. The preparation of the filled contour plot is done via the MATLAB function ***'contourf'***.

```
if (isradial==1)
   [xAxis, yAxis] = pol2cart(B,A);
   [C,h] = contourf(squeeze(xAxis), squeeze(yAxis), squeeze(V(:,:)));
else
    [C,h] = contourf(squeeze(A), squeeze(B), squeeze(V(:,:)));
end
```

Adding Labels, Contour Labels, and Color Mapping to the Plot

```
xlabel(xlab); ylabel(ylab);
clabel(C,h,'FontSize',8);

colormap('turbo'); colorbar;
```

Closing tim-file

```
fclose(file);
```

Box A.3 below lists the full script for MATLAB plotting.

Box A.3 MATLAB plotting script

Full MATLAB Plotting Script

```
clear; clc;
file0=fopen('chooseML.dat','r');
fname=fscanf(file0,"%s",[1]);
fclose(file0);
file=fopen(fname,'r');
ndim=fscanf(file,"%d",[1]);
isradial=fscanf(file,"%d",[1]);
if (ndim==1)
  xvar=fscanf(file,"%s",[1]); xunit=fscanf(file,"%s",[1]);
  xlab=strcat(xvar," ",xunit);
  ntime=fscanf(file,"%d",[1]);
  na=fscanf(file,"%d",[1]);
  noutp=na+1;
  ttime=[1:ntime];
  X1=[1:(na+1)]; Y1=[1:(na+1)];
  for nt = 1:ntime
    ttime(nt)=fscanf(file,"%f",[1]);
    for i = 1:(na+1)
      X1(i)=fscanf(file,"%e",[1]);
      Y1(i)=fscanf(file,"%e",[1]);
      fscanf(file,'\n');
    end
    figure(nt); plot(X1,Y1);
    xlabel(xlab);
  end
elseif (ndim==2)
  xvar=fscanf(file,"%s",[1]); xunit=fscanf(file,"%s",[1]);
  yvar=fscanf(file,"%s",[1]); yunit=fscanf(file,"%s",[1]);
  xlab=strcat(xvar," ",xunit); ylab=strcat(yvar," ",yunit);
  ntime=fscanf(file,"%d",[1]);
  na=fscanf(file,"%d",[1]); nb=fscanf(file,"%d",[1]);
  noutp=(na+1)*(nb+1);
  ttime=[1:ntime];
  A=zeros(na+1,nb+1); B=zeros(na+1,nb+1); V=zeros(na+1,nb+1);
  for nt = 1:ntime
    ttime(nt)=fscanf(file,"%f",1);
    for i = 1:(na+1)
      for j = 1:(nb+1)
        A(i,j)=fscanf(file,"%e",1);
        B(i,j)=fscanf(file,"%e",1);
        V(i,j)=fscanf(file,"%e",1);
        exclude=fscanf(file,"%d",1);
        if (exclude~=0)
          V(i,j)=NaN;
        end
        fscanf(file,'\n');
      end
    end
    figure(nt);
    if (isradial==1)
      [xAxis, yAxis] = pol2cart(B,A);
      [C,h] = contourf(squeeze(xAxis), squeeze(yAxis), squeeze(V(:,:)));
    else
      [C,h] = contourf(squeeze(A), squeeze(B), squeeze(V(:,:)));
    end
    xlabel(xlab); ylabel(ylab);
    clabel(C,h,'FontSize',8);
    colormap('turbo'); colorbar;
  end
end
fclose(file);
```

Index

A

B

C

D

E

F

G

H

I

K

L

S

T

U

V

W

Z

Kolibri 360 GmbH
Ferdinand-Jühlke-Straße 7
99095 Erfurt, Deutschland
produktsicherheit@kolibri360.de